BIOTECHNOLOGY IN ANIMAL HUSBANDRY
VOLUME 5

FOCUS ON BIOTECHNOLOGY

Volume 5

Series Editors

MARCEL HOFMAN

Centre for Veterinary and Agrochemical Research, Tervuren, Belgium

JOZEF ANNÉ

Rega Institute, University of Leuven, Belgium

Volume Editors

R. RENAVILLE and A. BURNY

Gembloux Agricultural University, Gembloux, Belgium

COLOPHON

Focus on Biotechnology is an open-ended series of reference volumes produced by Kluwer Academic Publishers BV in co-operation with the Branche Belge de la Société de Chimie Industrielle a.s.b.l.

The initiative has been taken in conjunction with the Ninth European Congress on Biotechnology. ECB9 has been supported by the Commission of the European Communities, the General Directorate for Technology, Research and Energy of the Wallonia Region, Belgium and J. Chabert, Minister for Economy of the Brussels Capital Region.

Biotechnology in Animal Husbandry
Volume 5

Edited by

R. RENAVILLE and A. BURNY
Gembloux Agricultural University,
Gembloux, Belgium

KLUWER ACADEMIC PUBLISHERS
DORDRECHT / BOSTON / LONDON

A C.I.P. Catalogue record for this book is available from the Library of Congress.

ISBN 0-7923-6851-7

Published by Kluwer Academic Publishers,
P.O. Box 17, 3300 AA Dordrecht, The Netherlands.

Sold and distributed in North, Central and South America
by Kluwer Academic Publishers,
101 Philip Drive, Norwell, MA 02061, U.S.A.

In all other countries, sold and distributed
by Kluwer Academic Publishers,
P.O. Box 322, 3300 AH Dordrecht, The Netherlands.

Printed on acid-free paper

Printed in the Netherlands.

PREFACE

RENAVILLE R. AND BURNY A.

In the past decade, many impressive advances were made in a number of scientific disciplines that have led to the discovery and development of exciting new biotechnologies that offer the potential to improve the production efficiency of animal agriculture. Nevertheless, while recent progress seems extremely rapid, it is impressive to recall the first attempts to culture (1880) and transfer embryos (1891) took place in the late 1800s.

The application of biotechnologies to farm animals has been the subject of numerous, and often heated, debates. Most of this controversy and bad public perception result from animal cloning, nuclear transfer (Dolly) and other genetic manipulations which could be applied to humans. But, reducing animal biotechnology to these sole applications is extremely restrictive. Indeed, biotechnology, of fundamental importance to the food industry, could be defined as any technology that exploits the biochemical activities of living organisms or their products to create industrial products and processes.

Biotechnology is perceived by many as a collection of very recent procedures but they forget that biotechnology has been around almost throughout human history. Animal breeding, applied to farm livestock such as dairy or beef cattle, or to companion animals such as the many different breeds of dogs, is a type of biotechnology that has been going on for many centuries. Breeders have selected animals that show particular characteristics or traits and used them in breeding programmes to ensure that these traits are retained.

Identification of restriction enzymes that cut DNA chains at specific sites (1970), methods for replicating genes (1973) and nucleic acid sequencing (1977) provided tools for a "new" biotechnology involved with gene identification, gene manipulation and transfer between species. With these techniques, it is now possible to identify how different genes control different characteristics. Scientists are "mapping" the genes on the chromosomes, so that they can see where genes are located and the extent to which they tend to be co-inherited. This will help identify which combinations give the traits that are desired. We now know that this type of conventional breeding of animals involves swapping hundreds of genes, most of which are unidentified. Some traits are determined by a single gene (associated with disease when mutations occur) and these are relatively easy to study. But others, including commercially important traits (ETL/QTL) such as lactation, growth rate and feed conversion efficiency, are controlled by many genes working together. Also, accelerated progress in animal breeding should be possible once the proper genes will be identified, appropriate markers will be found. However, identification of associations between QTL/ETL or candidate gene markers with animal productions or diseases require development of new quantitative genetic strategies. Moreover, embryo transfer, semen sexing, gene transfer, cloning, ... could affect the selection scheme in the near future and require adaptive strategy.

The successful transfer of the growth hormone gene from rats to mice (Palmiter *et al.*, 1982, Nature, Lond., 300:612-615) stimulated considerable interest and substantial research into development of new applications involving manipulation of subcellular components rather than complete organisms. To date, although an impressive body of new knowledge has been acquired and laboratory successes are numerous, biotechnology is still somewhat long on promises but short on performances that contribute to substantial improvements in commercial livestock farming. As the methodology for molecular genetics is refined, expression of introduced genes can be regulated in recipients. Another approach involves identifying and isolating the genes coding for specific proteins that are deficient in certain human diseases. The ideal techniques allow regulating expression of the "foreign" gene or genes so they are only active in mammary tissue; for example, ewes, does or cows that successfully express the introduced trait should secrete the human protein in their milk, providing opportunity for isolation of the compound for therapeutic use. Similarly, immunoglobulin genes or even genes from other species can be introduced into chickens. The antibodies or pharmaceutical proteins resulting from such treatments are concentrated in the egg yolk and can be harvested for use in humans or other animals.

Introduction of foreign genes into bacteria or hybridoma lines so that these altered cells will produce new compounds for harvest is another aspect of biotechnology. Many companies initiated programs to examine application of biotechnology to the somatotropic axis. One of the most extensively investigated technology in animal husbandry is recombinant porcine (pST) and bovine (bST) somatotropin. It is clear that the approach of elevating somatotropin concentration in blood results in a remarkable change in growth and lactation. However, application of these first biotechnology products in agriculture generated a heated debate between USA where bST is accepted and Europe where consumer associations are opposed to use of bST.

One possible 'friendly consumer' alternative method to promote growth or lactation is not to administer but rather to modulate the producing cells. The advent of technology allowing the generation of specifically targeted and engineered antibodies may make such regulation possible and can be considered as an alternative strategy with the potential to make important impact in animal science. Immunomodulatory approaches are attractive for at least three reasons. Firstly, the concept of vaccination is still perceived as 'acceptable' by the consumer when compared with the use of hormones or transgenic animals. With vaccine technology, it is easy to certificate the origin of the products (detection of induced antibodies). Secondly, the approach is economically cheaper than hormonal treatment (the technique involves an amplified response in vivo to small amounts of immunogen) and one of a long duration (weeks to months)). And thirdly, with epitope mapping and peptide synthesis, stimulation of the immune system can now be obtained by using very small molecules without any 'direct hormonal activities'.

Another biotechnological approach consists in manipulating the genome of micro-organisms to reduce pathogenicity (the ability to cause disease) while enhancing antigenicity (the ability to stimulate immunity), providing a methodology to create more effective vaccines. Perhaps even more futuristic and exciting is the prospect for genetic engineering of rumen micro-organisms to enhance their ability to degrade cellulose. Other strains could possibly be developed to break down highly ligninized materials *in vitro* to liberate cellulose or even convert the basic material in to starch. Cereal grain and sugar cane production generate several billion tons of crop residue every year which, if treated to improve digestibility, would alleviate much of the livestock malnutrition commonly encountered in lesser developed regions. Recombinant DNA technology might also be used to produce relatively inexpensive phytases which could be incorporated into animal diets to make phosphorus, which is usually tightly bound into the plant tissue, more readily available for digestion.

Finally, biotechnology must be cost-effective. This should translate into fewer mature breeding animals generating lower quantities of manure with less potential for pollution. The use of biotechnological processes, particularly genetic modification, is extremely important in devising new ways to increase food production, improve nutrient content, and provide better processing or storage characteristics. It follows that when new foods or food components are developed using biotechnology, there are both national legal requirements and consumer expectations for effective systems and procedures to assess the safety of food components for consumption.

As editors, we wish to thank those authors who contributed to this book and helped us to bring this endeavour to fruition.

TABLE OF CONTENTS

IMPLICATION OF A STRUCTURAL MOTIF IN THE INSTABILITY OF A TOXIC PROTEIN : THE PRION

LINS, L. [1]**, CHARLOTEAUX, B.**[2]**, THOMAS, A.** [1] **AND BRASSEUR, R.** [2]
[1]INSERM U410, Faculté X. Bichat, 75870 Paris Cedex 18, France
[2]Centre de Biophysique Moléculaire Numérique, Faculté des Sciences Agronomiques de Gembloux, 5030 Gembloux, Belgium

Abstract

Transmissible spongiform encephalopathies (TSE's) are neurodegenerative diseases found in animals as well as in men. They are due to infection by non conventional transmission agents called « prions ». The protein only hypothesis stipulates that the agent of the prion diseases is the PrPsc protein found in the inoculum. The sequence of this protein is strictly identical to a normal protein called PrPc. The only difference between PrPsc and PrPc is in the conformation : while the latter is mainly helical, the β-sheet content is significantly increased in PrPsc, with a concomitant loss of helical structure. It is postulated that the disease is due to a structural conversion PrPc→PrPsc induced by PrPsc present in the infectious fractions. By molecular modelling, we detected a so-called "tilted peptide" in the PrP sequence. Those short fragments have destabilizing properties, due to an asymmetric distribution of their hydrophobicity when helical. They were initially found in viral fusion proteins and later in series of proteins with various functions. Tilted peptides were shown to be structurally labile, their conformation depending upon the environment. We postulate that the PrP tilted fragment could be involved in the PrP neurotoxicity, and more generally, that tilted peptides could activate misfolding processes.

1. Introduction

Prion diseases have been described in man and animals since more than 20 years, with a renewed interest due to the « mad cow disease » epidemic (Table 1). The economic and agricultural impacts pointed out the urgent need to better understand the molecular determinants of those devastating neurodegenerative diseases.

R. Renaville and A. Burny (eds.), Biotechnology in Animal Husbandry, 15–32.

In this chapter, we will decipher how molecular modelling approaches can help to investigate the mechanisms inferred in the development of such diseases and more specifically in the structural lability of proteins.

Table 1. *The prion diseases.*

Diseases	Transmission
Human diseases	
Kuru	Infection through cannibalism
Iatrogenic Creuztfeldt-Jakob disease (CJD)	Infection from prion-contaminated HGH (human growth hormone)
familial CJD	Germ-line mutation in PrP gene
Variant CJD	Infection from BSE ?
Sporadic CJD	Somatic mutation or spontaneous conversion PrPc→PrPsc
Gerstmann-Straüssler-Scheinker (GSS) syndrome	Germ-line mutation in PrP gene
Fatal familial insomnia	Germ-line mutation in PrP gene
Animal diseases	
Scrapie (sheep and goats)	Infection in genetically susceptible sheep
Bovine spongiform encephalopathy (BSE)	Infection through prion-contaminated feeding
transmissible Mink encephalopathy (TME)	Infection through prion-contaminated feeding
Chronic wasting disease (CWD) (mule deer, and elk)	Unknown
Feline spongiform encephalopahty (FSE)	Infection through prion-contaminated feeding
Exotic ungulate encephalopathy (oryx, greater kudu, nyala)	Infection through prion-contaminated feeding

2. Introduction to Prions

Transmissible spongiform encephalopathies (TSE's) are neurodegenerative diseases found in animals as well as in men. They can be familial (inherited), sporadic or transmissible (Table 1). They are due to infection by non conventional transmission agents called « prions ».

Those diseases were first considered as being induced by non conventional or slow-acting viruses. However, no detectable viral agent has been evidenced up to now (Kellings *et al.*, 1992, 1994).

The various prion diseases have common features: clinically, they have asymptomatic long incubation times, the neuropathology is characterised by a neuronal vacuolisation and proteins aggregate in the brain with no detectable and/or specific immune response, no mark of any micro-organism in the infected brains despite high infectious titres.

3. Nature of the Non-Conventional Transmission Agents

The exact nature of the transmission agent still remains controversial. Every attempt to extract an infectious particle have led to fractions enriched in a protein called PrPsc. Any process altering nucleic acids have no effect on the infectivity (Alper *et al.*, 1967; Latarjet *et al.*, 1970). On the contrary, all protein-denaturing processes significantly decrease the infectious titre (Prusiner, 1981; Prusiner, 1982). These observations firmly suggest an exclusive proteinaceous composition for the TSE infectious agent (Prusiner, 1982). The term « prion » (*pro*teinaceous *in*fection particles) was introduced to clearly distinguish TSE's infectious particles from viruses and viroids.

The « protein only » hypothesis first suggested by Griffith in 1967 (Griffith, 1967) and further developed by Prusiner and colleagues (Prusiner, 1982, 1999) for more than 15 years, stipulates that the prion diseases are caused by the PrPsc proteins present in the purified infectious fractions.

Surprisingly, the primary sequence of this protein is strictly identical to that of a normal cell protein called PrPc.

4. The PrP Protein

PrP is a protein of about 250 residues (253 aa for human PrP) and is currently present in mammals. It is encoded by a gene on chromosome 20 containing two exons. The second exon codes for the whole sequence (Basler *et al.*, 1986).

Human PrP has a 22 amino acid signal peptide that is cleaved in the endoplasmic reticulum . The N-terminal domain (residues' 51-91) also contains octarepeats (Stahl *et al.*, 1993). The native protein has a disulphide bridge between Cys 179 and Cys 214 (Figure 1).

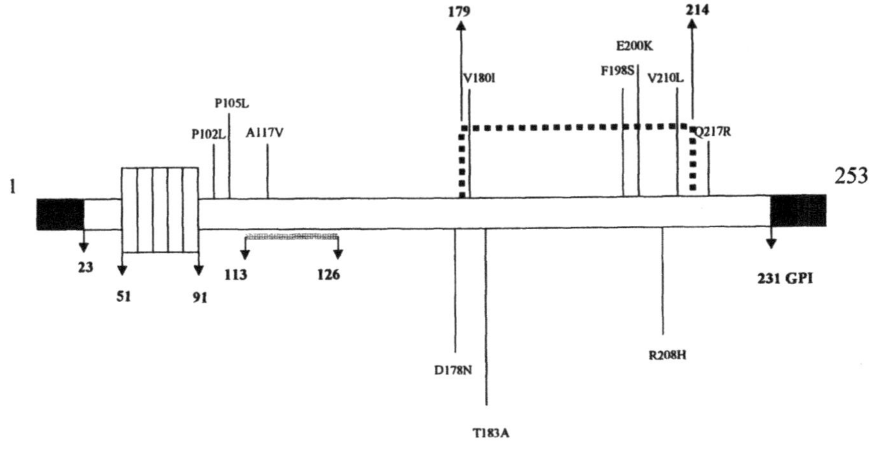

Figure 1. *Schematic representation of the human PrP sequence. Grey blocks represent residues that are cut to generate the mature form; dotted line represents the disulfide bridge (C179-C214). Mutations involved in prion diseases are indicated. (51-91 domain : 5 octapeptide repeats; 113-126 domain : conserved hydrophobic domain).*

In the mature form of PrP, the 23 last residues are replaced by a glycosyl phosphatidylinositol moiety (GPI) that anchors the protein to the exterior cellular membrane (Stahl *et al.*, 1987) (Figure 1). PrP is thus exposed at the cell surface and then re-internalized and degraded. Some authors suggest that the protein could be recycled after re-internalization (Liemann and Glockshuber, 1998).

The sequence of PrP is well conserved in mammals, suggesting the conservation of some important functions through evolution. Two domains of PrP are remarkably conserved : the N-terminal octarepeat region and the 110-130 central domain, rich in Gly and Ala residues (Figure 1). While insertions of extra repeats have been detected in patients with familial disease, it does not seem that the naturally-occurring deletion of single repeats causes the disease (Prusiner and Scott, 1997). Another interesting feature is that all mutations leading to familial forms of prion diseases are located in the 100-200 region of the protein (Figure 1), suggesting a crucial role of that domain in the development of the disease. A point mutation in the central domain (A117V) is linked to the GSS (Gerstmann-Scheinker-Straussler) syndrome.

5. What is the Difference between PrPc and PrPsc ?

No chemical modification was evidenced, even after exhaustive studies (Stahl *et al.* , 1993; Prusiner, 1996 , 1998). The only difference between PrPc and PrPsc is in the conformation.

It is generally accepted that one amino acid sequence specifies a unique biologically active conformation of protein, the native structure, and that this structure corresponds to the minimal energy (Anfinsen, 1973). In the case of prions, one primary sequence, that of PrP, can adopt at least two different conformations with different properties (Prusiner, 1982 ; 1998, 1999).

It was found by spectroscopic studies that, while PrPc is predominantly α-helical, the content of PrPsc significantly increases in β-sheet and decreases in α-helix (Caughey *et al.*, 1991 ; Safar *et al.*, 1993 ; Pan *et al.*, 1993). These observations suggest a α to β conversion during PrPsc formation.

A first model of PrPc was proposed by Huang *et al.* (1995), based on molecular modelling and structural data of PrP fragments (Figure 2A). In this model, PrPc was predicted to shape as a helix bundle (Figure 2A). Recently, the tertiary structure of two recombinant C-terminal fragments, PrPc 121-231 (Riek *et al.*, 1996) (Figure 2B) and PrPc 90-231(James *et al.*, 1997) (Figure 2C), were resolved using 2D NMR. In these structures, the existence of two C-terminal helices, in the 170-230 region, is confirmed, but the presence of a short two-stranded antiparallel β sheet is observed. This β-structure could have a role in the α to β conversion occurring during PrPsc formation.

The hydrophobic core of PrPc should be involved in the α to β conversion : deletion of the 113-126 residues, adjacent to the PrPc 128-131 β strand inhibits the PrPsc formation (Prusiner, 1998). In the pathogenic form, this region was shown to form a β-strand (Huang *et al.*, 1995) (Figure 2D).

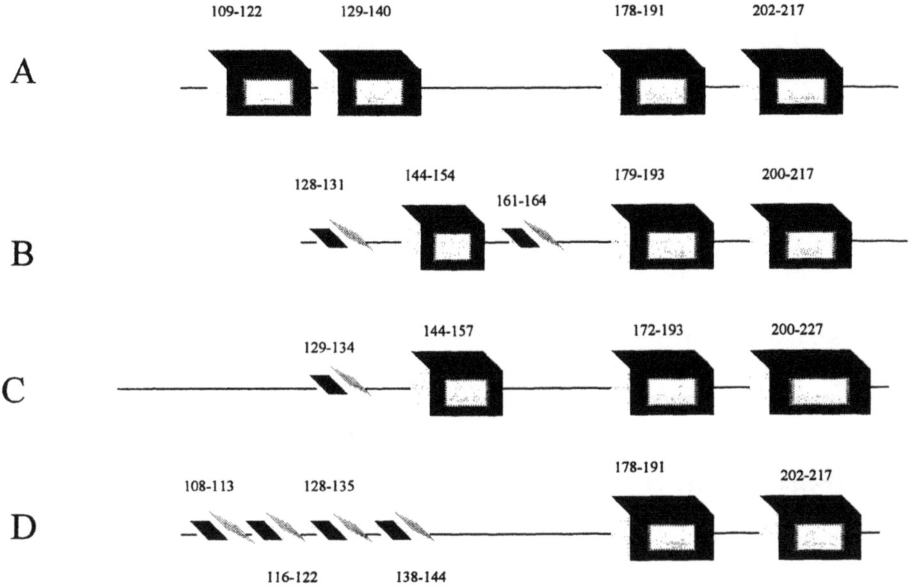

Figure 2. *Proposed structures of PrPc and PrPsc. Blocks represent helices and arrows, β-strands. A = PrPc model of Huang* et al. *(1995); B = NMR structure of mouse 121-231 recombinant PrPc (Riek* et al., *1996); C = NMR structure of hamster 90-231 recombinant PrPc (James* et al., *1997); D = PrPsc model of Huang* et al. *(1995).*

The α to β transition induces severe modifications of the PrP physicochemical properties. While PrPc is soluble in non-denaturing detergents, PrPsc is not ; PrPc is readily digested by proteases whereas PrPsc is partially resistant (Prusiner, 1999). The resulting PrPsc protease-resistant fragment is called PrP27-30 since its molecular weight is about 27-30 kD (Prusiner *et al.*, 1984 ; Basler *et al.*, 1986). It corresponds to the hydrolysis of approximately 65 residues from the amino terminus of the protein. This fragment is able to polymerise into amyloid plaques (Pan *et al.*, 1993).

6. Role of PrP

The normal role of PrPc remains unknown. Ablation of PrP gene in mice (PrP0/0) does not affect the development of these animals (Bueler *et al.*, 1993). However, a perturbation of the waking/sleeping cycles together with a premature ageing of Purkinje cells was reported in one study (Sakaguchi *et al.*, 1996). PrP0/0 mice are resistant to prions and no trace of infection was detected in their brains. This event strongly supports that PrPc plays a receptor role for the infectious agent and that PrPsc in the inoculum should directly interact with the surface-exposed PrPc of the host. The pathogenic transconformation should thus result from dimer formations suggesting that the protein conformation should be considered as a coding system for the pathological information. Since the absence of PrPc in mice inhibits the development of the disease , the latter should be the consequence of PrPsc accumulation rather than the impairment of PrPc function (Bueler *et al.*, 1992). However, the way PrPsc induces neuronal death and vacuolisation is still unknown.

7. Localisation of PrPc and PrPsc

In mammals, PrP is mainly expressed in neurons. The cellular localisation of PrPsc differs from that of PrPc. PrPsc is mainly cytoplasmic, while PrPc is exposed at the cell surface. Authors agree with the fact that PrPsc derives from PrPc and that the conversion occurs during the re-internalization of PrPc. Acquisition of the partial protease resistance during this process could lead to the accumulation of the protein in the lysosomes. The latter would then fuse into vacuoles.

8. Inter-Species Barrier and Prion Propagation

The sequence of the prion protein constitutes the inter-species barrier. The closest the pathogenic prion protein and the host PrP protein are, the highest the susceptibility to the disease is.

The transmission of prions between species is usually a stochastic process characterised by prolonged incubation times during the first passage to the new host (Pattison, 1965). Subsequent passages in homologous hosts result in reduced incubation times. The prions aggregated during infection are the host PrP and not that of the PrPsc molecules in the inoculum (Bockman *et al.*, 1987). Almost all our current knowledge arises from transgenic studies. For example, the insertion of the syrian hamster PrP gene in mice induces abrogation of the barrier normally observed between those two species (Scott *et al.*, 1989). Directed mutagenesis experiments have shown that the residues important to the inter-species determinism are located in the central domain which is also involved in the transconformation process (Prusiner, 1997).

From transgenic mice studies, prion propagation should involve the formation of a complex between PrPsc and the host PrPc (Prusiner *et al.*, 1990). It was recently observed that dimerisation is not sufficient to induce a structural conversion (Telling *et al.*, 1995). Propagation might require the presence of a complementary factor called protein X, that is not identified yet and that could be a molecular chaperone catalysing conformational changes (Debburman *et al.*, 1997).

9. Tilted Peptides : Structural Motif Potentially Involved in Destabilisation of Organised Systems

Prion diseases are characterised by the occurrence of the conformational change of a non pathogenic protein, PrPc, to a pathogenic isoform, PrPsc. This event highlights to some extent a structural instability of the PrP protein.

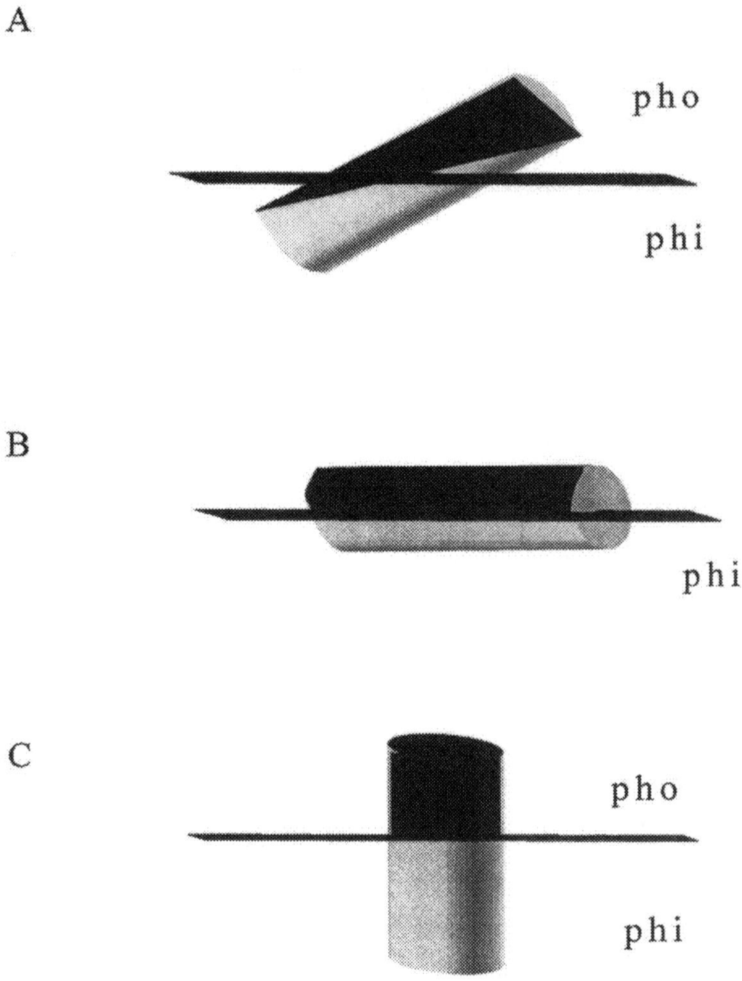

Figure 3. *A = schematic representation of a tilted peptide oriented at the hydrophobic (pho)/hydrophilic (phi) interface (black plane). The hydrophobicity gradient is schematically represented in dark grey. B and C = orientation at the interface of mutants in which the hydrophobicity gradient is lost. (B = parallel mutant; C = perpendicular mutant).*

Few years ago, we detected by molecular modelling short fragments of sequence (10 to 20 residues) having destabilising properties. Such a fragment was detected in the conserved central region of PrP, between residues 118 and 135 (Pillot *et al.*, 1997).

Those short fragments are characterised by an asymmetric distribution of hydrophobicity when helical (Brasseur, 1991 ; 2000 ; Brasseur *et al.*, 1997) (Figure 3A). They adopt a tilted orientation when they interact with a hydrophobic/ hydrophilic interface such as the lipid/water interface (Figure 3A). From there, they were named « tilted peptides ».

Their destabilising capacities were suggested when we discovered the first tilted peptides in viral glycoproteins such as Gp32 of SIV (Simian Immunodeficiency Virus) and Gp30 of BLV (Bovine Leukaemia Virus) involved in virus fusion to its host (Brasseur, 1991 ; Horth *et al.*,1991).

The SIV tilted peptide, which is the N-terminal 12 residues of Gp32, was experimentally shown to induce the fusion of liposomes *in vitro* and to be involved in the formation of polynucleated cells (syncytia), a feature of viral fusion (Martin *et al.*, 1991 ; Horth *et al.*, 1991). To involve the hydrophobicity gradient in the mechanism, mutations were predicted that should abolish this gradient (Figure 3BC). Those mutants do not induce liposome fusion nor syncytia formation and the fusion is recovered with new mutants in which the hydrophobicity gradient is restored (Martin *et al.*, 1991 ; Horth *et al.*, 1991). Therefore, it was concluded that the tilted orientation of those peculiar fragments rather than their amino acid sequence is important for fusion. To explain their fusogenic activity, we suggested that the oblique orientation into membranes could destabilise the parallelism of lipid chains (Martin *et al.*, 1991 ; Horth *et al.*, 1991).

Other tilted peptides were then discovered in various proteins, such as lipolytic enzymes (Brasseur *et al.*, 1997), membrane proteins (Brasseur *et al.*, 1997; Rahman *et al.*,1997), signal sequences (Talmud *et al.*, 1996) and in neurotoxic proteins, notably in the Aβ protein involved in Alzheimer's disease (Pillot *et al.*, 1996) and in the PrP protein (Pillot *et al.*, 1997) (Table 2).

10. Molecular Modelling Strategy for the Detection of Tilted Peptides

We developed a strategy for the prediction of oblique peptides from protein sequences (Brasseur, 2000) (Figure 4). The hydrophobicity gradient is first detected using the Jahnig (Jahnig,1990) and HCA (Hydrophobic cluster Analysis) (Gaboriaud *et al.*, 1987) methods. In the former, a 17-residue window is moved along the sequence. The mean hydrophobicity (Hα) of the window central residue is calculated taking into account its neighbours, close in terms of secondary structure (here, a α-helix). In the plot of hydrophobicity versus sequence, an increasing (or decreasing) oscillating curve is indicative of a tilted hydrophobicity gradient (Figure 4).

The HCA method is based on a 2D helical representation of the sequence. Hydrophobic clusters are detected and their shapes and lengths are predictive of secondary structures. Tilted peptides are detected as triangular clusters (Figure 4). Hydrophilic fragments are rejected because we postulate that tilted peptide must be sufficiently hydrophobic to insert in a membrane.

A classical α-helix structure is then imposed and conformations of the side chains are optimised by energy minimisation taking into account the presence of a hydrophobic/hydrophilic interface (Brasseur, 1990). The structure of minimal energy is then oriented at the interface separating the hydrophobic phase from the hydrophilic one. The tilt (δ) is defined as the angle between the helix axis and the interface plane, and is comprised between 30° and 60° (Brasseur, 1990). Calculation of the repartition of the hydrophobic and hydrophilic isopotential outer layers around the peptide allows visualising the hydrophobicity gradient (Brasseur, 1991).

The last step computes the interaction of the tilted peptide with phospholipids (Brasseur, 1990). In this procedure, the position of the peptide is frozen and the lipid molecule is moved along and around the peptide (rotations and translations-see arrows in Figure 4). For each position, the energy of interaction (Van der Waals, electrostatic, torsional and hydrophobic interactions) is calculated and the energies of all the positions are stored in a hypermatrix. The position of the first lipid is the lowest energy complex ; a second molecule is inserted as the next energetically favourable position in the hypermatrix, taking into account the presence of the first lipid. For the next lipids, the same process is repeated until the peptide is completely surrounded with lipids.

The assembly of tilted peptide with lipids shows that the peptide is able to disturb the parallelism of lipid acyl chains (Figure 5B). This perturbation could induce different mechanisms such as the fusion of lipidic domains, the perturbation of the function of membrane protein and could induce apoptosis,...

11. The 118-135 Tilted Peptide of PrP

Figure 5A shows the orientation of the 118-135 fragment at the interface. As with tilted peptides of fusion proteins, the assembly of this peptide with lipids induces a perturbation of the lipid organisation (Figure 5B). Parallel and perpendicular mutants were calculated in order to abolish the hydrophobicity gradient (Pillot *et al.*, 1997).

Experimentally, the 118-135 fragment is able to induce the fusion of lipid phases and the mixing of the liposome aqueous compartments (Pillot *et al.*, 1997). The parallel and perpendicular mutants are obtained by residue permutation. They are unable to induce the same processes.

It is worth noting that this 118-135 fragment encompasses the putative transmembrane segment of PrP (approximately residues 110 to 135). Actually, a marginal transmembrane form of PrP (ctmPrP) is normally present in brain (Hedge *et al.*, 1998). The level of ctmPrP increases with some mutations in the 110-135 domain such as the A117V mutant associated with the GSS syndrome. In infectious prion diseases, the time course of PrPsc accumulation is closely followed by the increased generation of ctmPrP, suggesting that the production of ctmPrP could be a common pathway of neurodegeneration in transmissible and inherited prion diseases (Hedge *et al.*, 1999). Transgenic mice expressing high levels of PrP transmembrane forms exhibit neuronal vacuolisation and neuronal death very similar to those found in experimental prion diseases.

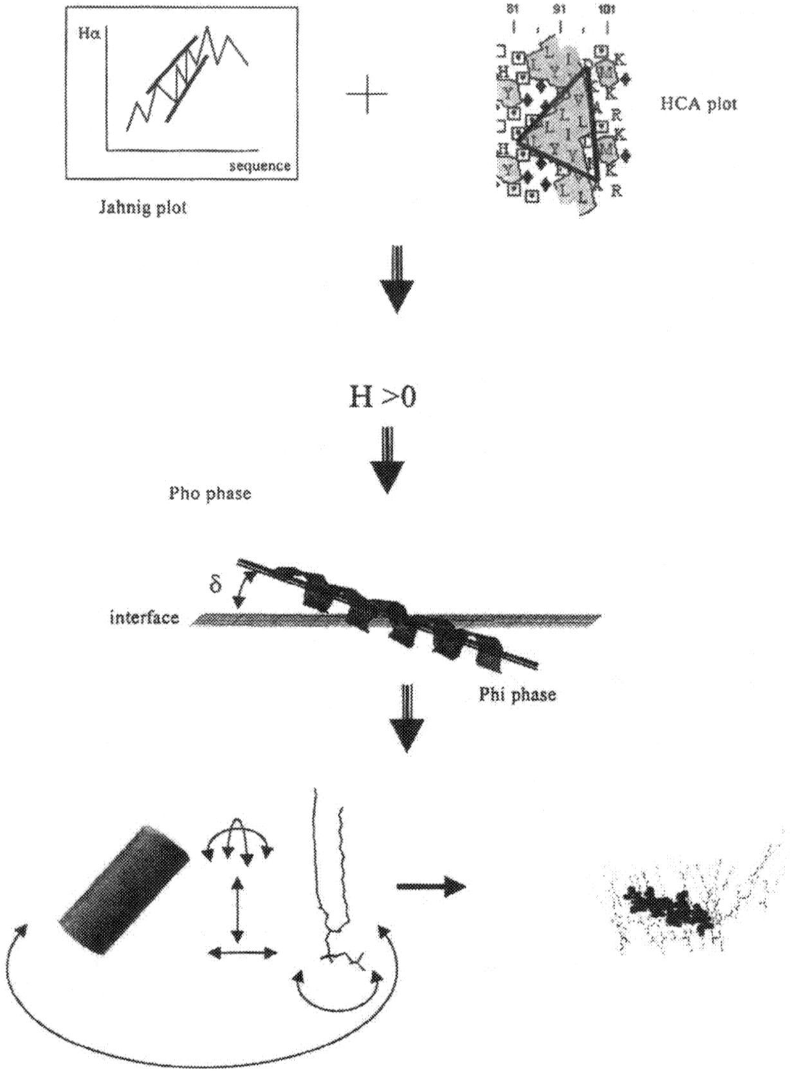

Figure 4. *Detection of tilted peptides (see text for more details) : the sequence is analysed by the Jahnig and the HCA methods. The selected segments with positive hydrophobicity are then constructed as α-helices and minimised at the lipid/water interface. The peptides are then assembled with lipids. The arrows represent all movements tested.*

Table 2. *General characteristics of several tilted peptides from the literature.*

Protein	Nb a.a.	Mapping in the sequence	Sequence	pho/phi ratio	Angle	Experimental data					Biological process
						1	2	3	4	5	
SIV	12	528-539	GVFVLGFLGFLA	0.93	30-53	X	X	X	X	X	virus fusion
HIV	12	478-489	AVGIGALFLGFL	0.91	45	X	X	X	X		virus fusion
measles virus	12		FAGVVLAGAALG	0.79	77						virus fusion
NDV	18	104-121	FIGAIIGSVALGVATAAG	0.75	73						virus fusion
RSV	17		FLGFLLGVGSAIASGVA	0.75	63						virus fusion
Sendai virus	17		FFGAVIGTIALGVATSA	0.71	72						virus fusion
BLV	12	269-280	SPVAALTLGLAL	0.64	56-60				X	X	virus fusion
MLV	17		GPVSLTLALLLGGLTMG	0.63	63						virus fusion
Ebola	17	524-540	GAAIGLAWIPYFGPAAE	0.60	45						virus fusion
influenza HA-2	20	1-20	GLFGAIAGFIENGWEGMIDG	0.51	47	X		X			virus fusion
B hepatitis S prot	16	1-16	MENITSGFLGPLLVLQ	0.45	57						virus fusion
PrP	18	118-135	AGAVVGGLGGYMLGSAMS	0.57	25		X	X			neurotoxic
α amyloid	14	29-42	GAIIGLMVGGVVIA	0.89	50		X	X	X		neurotoxic
meltrine α	14	591-603	VIGTNAVSIETNIE	0.30	50						myoblast fusion
yeast invertase PS	19	1-19	MLLQAFLFLLAGFAAKISA	0.64	55					X	signal peptide
ApoB 100 PS	12		RPALLALLALPA	0.476	45					X	signal peptide
human Apo A-II	13	58-70	TELVNFLSYFVEL	0.44	30-35	X	X	X	X		lipid metabolism
CETP	16	461-476	FGFPEHLLVDFLQSLS	0.40	30						lipid transport
LCAT	13	56-68	DFFTIWLDLNMFL	0.55	40		X	X	X		lypolytic enzyme
HLP	13	234-246	FLELYRHIAQHGF	0.19	47						lypolytic enzyme
LPL	13	218-230	IGEAIRVIAERGL	0.16	55						lypolytic enzyme
1bct	18	195-212	GAGIVPLNIETLLFMVLD	0.58	31						membrane insertion
1bct	16	177-192	VTVVLWSAYPVVWLIG	0.75	65						membrane insertion
Fertilin	17	83-99	DSTKCGKLICTGISSIP	0.14	58				X		Spermato-zoid fusion

(Column 1 = the proteins in which tilted peptides were detected; column 2 = the length of the tilted pepti-de; column 3 = mapping in the sequence; column 4 = the sequence; column 5 = the mean hydrophobicity; column 6 = the calculated tilts (in degrees); column 7 = experimental data when they exist (1 = FTIR-ATR: a suitable technique to obtain an experimental value of the tilt angle; 2 = mixing of lipid phases (liposome agregation or fusion); 3 = mixing of aqueous phases; 4 = liposome sizing: both 3 and 4 are measures of liposome fusion; 5 = mutations were introduced, the protein was expressed and a functional activity was measured); column 8 = the biological processes in which the tilted peptides are or could be involved

A

B

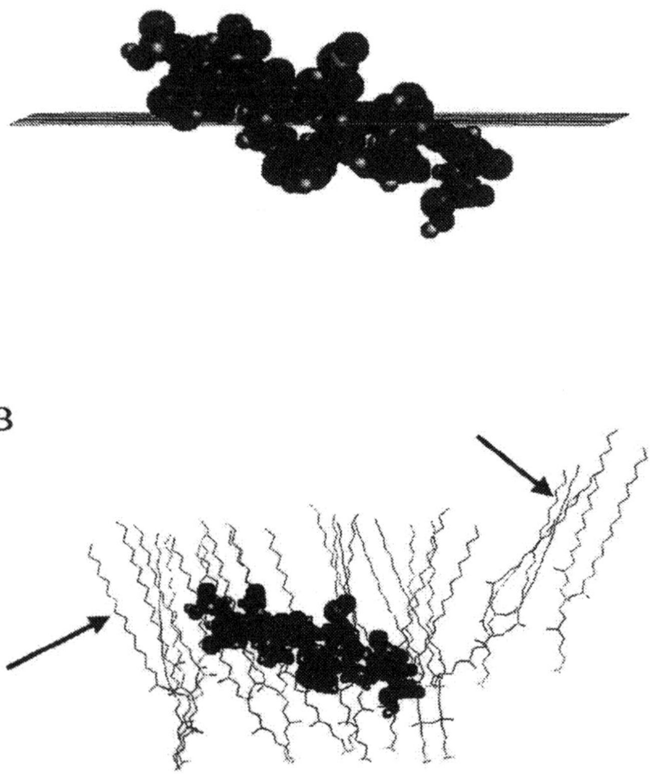

Figure 5. *A = The 118-135 PrP fragment oriented at the interface (black plane). The hydrophobic phase is upper and the hydrophilic one is below. B = Assembling of the 118-135 PrP tilted peptide with DPPE (dipalmitoylphosphatidylethanolamine). Arrows show the perturbation of the lipid acyl chain organisation.*

In a very recent work, Dormont and his colleagues have shown that the 118-135 peptide exhibits major neurotoxic effects, inducing neuronal death by apoptosis via direct membrane perturbation (Haik *et al.*, submitted). These data suggest that the fusogenic properties of the 118-135 PrP tilted peptide might contribute to the neurotoxicity by destabilising cell membranes.

On the other hand, the location of such a destabilising fragment within the PrP domain involved in transconformation also suggests its participation in the conformational transition PrPc → PrPsc by facilitating the destabilisation of the prion protein core.

Like other tilted peptides, such as the SIV or Aβ oblique fragments (Lins *et al.*, 1999), the 118-135 peptide appears structurally labile. Either in the NMR PrPc structures or in the PrPc theoretical model, the fragment is a mixture: a αhelix-turn-αhelix in the model and a β-strand-turn-α-helix motif in the NMR structures. This suggests a high structural mobility. A conformational change from α to β could occur after insertion in lipids or interaction with another PrP domain. The conformational lability of the PrP tilted peptide could thus also be important in the PrPc to PrPsc conversion.

12. Generalisation of the Structural Instability Notion

Many proteins possess a tilted peptide (Table 2). We have suggested that they are involved in different processes implying the disruption of a hydrophobic/hydrophilic interface. The generic role of tilted peptides would be to decrease the free energy of the process leading to the transformation of one structural state to another (Figure 6) (Brasseur, 2000). These could be physiological as well as pathogenic processes (Peuvot *et al.*, 1999).

One physiological process that potentially involves an oblique peptide is the spermatozoid-egg fusion during fecondation. Actually, fertilin α, a membrane protein located on the spermatozoid head shares common properties with viral fusion peptides. We have determined by computer modelling the presence of a tilted peptide that induces the destabilisation of liposome membranes, as shown by 31P NMR (Schanck *et al.*, 1998).

Secretion or membrane insertion of proteins could also be assisted by tilted peptides. The fragments were evidenced in signal sequences of different secreted proteins (Talmud *et al.*, 1996) or of members of G-coupled receptors.

Enzymes are important physiological elements for the cell. Lipolytic enzymes hydrolyse lipids that are located in the core of the substrate. So, the first step of the enzyme reaction requires the destabilisation of this substrate. For different enzymes (LPL, LCAT..), this is achieved by an oblique-orientated peptide (Brasseur *et al.*, 1997).

Tilted peptides are also involved in pathogenic events. As already mentioned, they were identified in many enveloped viruses (SIV, HIV, BLV, hemagglutinin, ebola, rubella virus, hepatitis B virus,...-Table 2). They appear to facilitate the entry of the virus into host cells by fusion of the viral outer layer with the host plasma membrane.

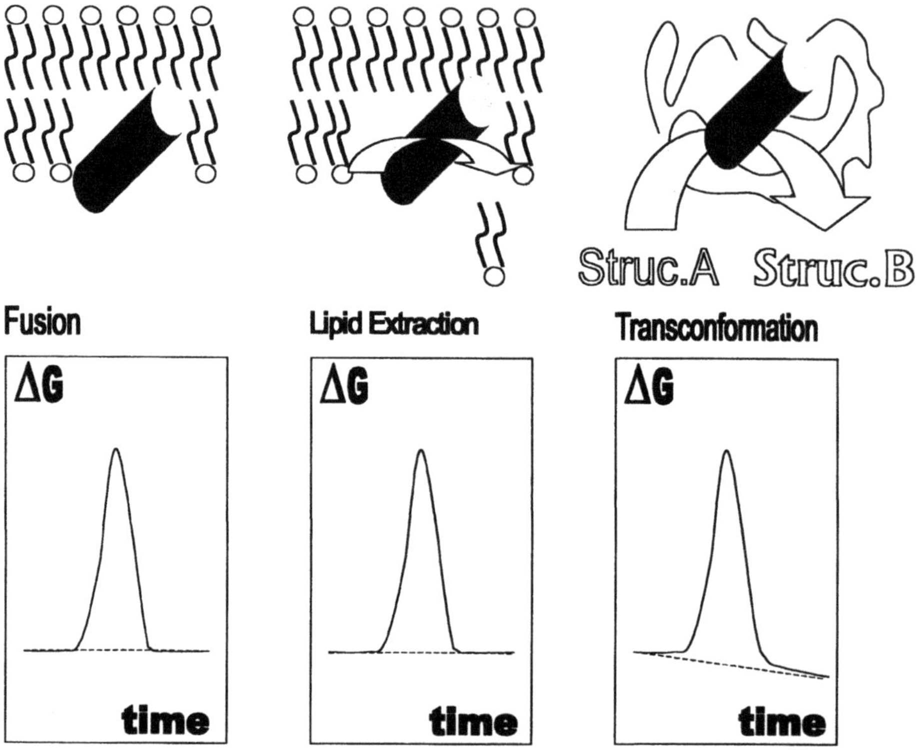

Figure 6. *Scheme of the free energy decrease induced by tilted peptides in various processes.*

Neurotoxicity could also be mediated by tilted peptides, as suggested for prion diseases. Alzheimer's disease is characterised by the presence of amyloid plaques in the brain of patients. These plaques are formed by the aggregation of a protein called Aβ, a 39-43 residues' protein. This protein is present in the brain of healthy individuals. Like in prion diseases, the pathological aggregation is due to structural changes. By molecular modelling, we identified a tilted peptide in the carboxy-terminal domain of Aβ (Brasseur *et al.*, 1997). This fragment has fusogenic activities that could account for Aβ neurotoxicity, either by the direct perturbation of the neuronal cell membrane or by the mediation of pathogenic transconformational events as suggested for the PrP tilted peptide.

Alzheimer's and prion diseases belong to an intriguing group of disorders that has been under the light of medical and structural biology studies since few years : the conformational diseases (Kelly, 1996 ; Carrell and Gooptu, 1998 ; Dobson, 1999).
Each of these various diseases seems to primarily arise from a conformational instability of the native form of the protein, leading to ordered aggregation and tissue deposition. Despite the fact that all proteins involved in these kinds of diseases, including lysozyme, transthyretin, serpin, insulin etc., and of course prion and Aβ, have

different native folds the amyloid fibrils in which they aggregate are very similar in their general appearance. They are organised beta structures.

One common feature of the native proteins is that they can be in vitro converted to the misfolded conformation by exposing them to mildy denaturing conditions, i.e. by changing the protein environment. In vivo, the rate of this conversion would be enhanced by the naturally-occuring mutations. Since they are unstable fragments, the tilted peptides could be the catalyst of such phenomena.

13. Conclusions

In this chapter, we have underlined the importance of peculiar fragments of proteins, the tilted peptides, in a more general context of protein structure instability. The destabilising behaviour of tilted peptides highly depends on the protein environment that is to say the medium. Therefore, changing the natural medium of a protein by introducing it in a new environment might have unexpected drastic effects, one effect could be illustrated in the BSE epidemy.

Acknowledgments

R.Brasseur is Research Director at the National Funds for Scientific Research of Belgium (FNRS). This work was supported by the "Interuniversity Poles of Attraction Programme-Belgian State, Prime Minister's Office-Federal Office for Scientific, Technical and Cultural Affairs" contract N° P4/03, the Loterie Nationale, the Funds for Industrial and Agricultural Research (FRIA) and the National Funds for Scientific Research of Belgium (FNRS).

References

Alper, T., Cramp, W.A., Haig, D.A. and Clarke, M.C. (1967) Does the agent of scrapie replicate without nucleic acid ? *Nature 214*, 764-766.

Anfinsen, C.B. (1973) Principles that govern the folding of protein chains, *Science* 181 :223-230

Basler, K., Oesch, B., Scott, M., Westaway, D., Walchli, M., Groth D.F., McKinley, M.P., Prusiner, S.B. and Weissmann, C. (1986) Scrapie and cellular PrP isoforms are encoded by the same chromosomal gene, *Cell 46*, 417-428.

Bockman, J.M., Prusiner, S.B., Tateishi, J. and Kingsbury, D.T. (1987) Immunoblotting of Creutzfeldt-Jakob disease prion proteins : host species-specific epitopes, *Ann. Neurol. 21,* 589-595.

Borchelt, D.R., Scott, M., Taraboulos, A., Stahl, N. and Prusiner, S.B. (1990) Scrapie and cellular prion proteins differ in their kinetics of synthesis and topology in cultured cells, *J. cell. Biol. 110,* 743-752.

Brasseur, R. (2000) Tilted peptides : a motif for membrane destabilization (hypothesis), *Mol. Membr. Biol. 17*, 31-40.

Brasseur, R. (1990) *Molecular description of biological membrane components by computer-aided conformational analysis* pp 203-220 CRC Press, Boca Raton, FL.

Brasseur, R.(1991). Differentiation of lipid-associating helices by use of three-dimensional molecular hydrophobicity potential calculations, *J.Biol.Chem. 266*, 16120-16127.

Brasseur, R., Pillot, T., Lins, L., Vandekerckhove, J. and Rosseneu, M. (1997) Peptides in membranes: tipping the balance of membrane stability, *TIBS. 22*, 167-171.

Bueler, H., Aguzzi, A., Sailer, A., Greiner, R.A., Autenried, P., Aguet, M. and Weissmann, C. (1993) Mice devoid of PrP are resistant to scrapie, *Cell* 73 :1339-1347.

Bueler, H., Fischer, M., Lang, Y., Bluethmann, H., Lipp, H.P., DeArmond, S.J., Prusiner, S.B., Aguet, M. and Weissmann, C. (1992) Normal development and behaviour of mice lacking the neuronal cell-surface PrP protein, *Nature 356*, 577-582.

Carrell, R.W. and Gooptu, B. (1998) Conformational changes and disease – serpins, prions and Alzheimer's, *Curr. Opin. Struct. Biol. 8,* 799-809.

Caughey, B.W., Dong, A., Bhat, K.S., Ernst, D., Hayes, S.F. and Caughey, W.S. (1991) Secondary structure analysis of the scrapie-associated protein PrP 27-30 in water by infrared spectroscopy, *Biochemistry 30,* 7672-7680.

DebBurman, S.K., Raymond, G.J., Caughey, B. and Lindquist, S. (1997) Chaperone-supervised conversion of prion protein to its protease-resistant form, *Proc. Natl. Acad. Sci. USA 94*, 13938-13943.

Dobson, C.M. (1999) Protein misfolding, evolution and disease, *TIBS 24,* 329-332.

Gaboriaud, C., Bissery, V., Benchetrit, T. and Mornon, J.P. (1987) Hydrophobic cluster analysis : an efficient way to compare and analyse amino acid sequences, *FEBS Lett. 224*, 149-155.

Griffith, J.S. (1967) Self-replication and scrapie, *Nature 215,* 1043-1044.

Haïk, S., Peyrin, J.M., Lins, L., Rosseneu, M., Brasseur, R., Langeveld, J.P., Tagliavini, F., Deslys, J.P., Lasmézas, C. and Dormont, D. Neurotoxicity of the transmembrane domain of the prion protein, submitted

Hegde, R., Mastrianni, J.A., Scott, R., DeFea, K.A., Tremblay, P., Torchia, M., DeArmond, S.J., Prusiner, S.B. and Lingappa, V.R. (1998) A transmembrane form of the prion protein in neurodegenerative disease, *Science 279*, 827-834.

Hegde, R., Tremblay, P., Groth, D., DeArmond, S.J., Prusiner, S.B. and Lingappa, V.R. (1999) Transmissible and genetic prion disease share a common pathway of neurodegeneration, *Nature 402,* 822-826.

Horth, M., Lambrecht, B., Khim, M.C., Bex, F., Thiriart, C., Ruysschaert, J.M., Burny, A., and Brasseur, R. (1991) Theoretical and functional analysis of the SIV fusion peptide, *EMBO J. 10*, 2747-2755.

Huang, Z., Prusiner, S.B. and Cohen, F.E. (1995) Scrapie prions : a 3D model of an infectious fragment, *Fold. Design 1,* 13-19.

Jahnig, F. (1990) Structure prediction of membrane proteins are not that bad, *TIBS 15,* 93-95.

James, T.L. , Liu, H., Ulyanov, N.B., Farr-Stones, S., Zhang, H., Donne, D.G., Kaneko, K., Groth, D., Mehlhorn, I., Prusiner, S.B. and Cohen, F.E. (1997) Solution structure of a 142-residue recombinanrt protein corresponding to the infectious fragment of the scrapie isoform, *Proc. Natl. Acad. Sci. USA 94,* 10086-10091.

Kaneko, K., Vey, M., Scott, M., Pilkuhn, S., Cohen, F.E. and Prusiner, S.B. (1997) COOH-terminal sequence of the cellular prion protein directs subcellular trafficking and controls conversion into the scrapie isoform, *Proc. Natl. Acad. Sci. USA 94*, 2333-2338.

Kellings, K., Meyer, N., Mirenda, C., Prusiner, S.B. and Riesner, D. (1992) Further analysis of nucleic acids in purified scrapie prion preparations by improved return refocussing gel electrophoresis (RRGE), *J. Gen. Virol. 73*, 1025-1029.

Kellings, K., Prusiner, S.B. and Riesner, D. (1994) Nucleic acids in prion preparation :unspecific background or essential component ? *Philos. R. Soc. Lond. B. Biol. Sci. 343*, 425-430.

Kelly, J.W. (1996) Alternative conformations of amyloidogenic proteins govern their behavior, *Curr. Opin. Struct. Biol. 6*, 11-17.

Latarjet, R., Muel, B., Haig, D.A., Clarke, M.C. and Alper, T. (1970) Inactivation of the scrapie agent by near monochromatic ultraviolet light, *Nature 227*, 1341-1343.

Liemman, S. and Glockshuber, R. (1998) Transmissible spongiform encephalopathies, *Biochem. Biophys. Res. Commun. 250*, 187-193.

Lins, L., Thomas, A., Pillot, T., Vanderkerckhove, J., Rosseneu, M. and Brasseur, R. (1999) Molecular determinants of the interaction between the C-terminal domain of Alzheimer's β-amyloid peptide and apolipoprotein E α-helices, *J. Neurochem. 73*, 758-769.

Martin, I., Defrise-Quertain, F., Mandiau, V., Nielsen, N.M., Saermark, T., Burny, A., Brasseur, R., Ruysschaert, J.M. and Vandenbranden, M. (1991) Fusogenic activity of SIV peptides located in the GP32 NH2 terminal domain, *Biochem. Biophys. Res. Comm. 175*, 872-879.

Pan, K.M., Baldwin, M., Nguyen, J., Gasset, M., Serban, A., Groth, D., Mehlhorn, I., Huang, Z., Fletterick, R.J., Cohen, F.E. and Prusiner, S.B. (1993) Conversion of α helices into β-sheets features in the formation of the scrapie prion proteins, *Proc. Natl. Acad. Sci. USA 90,* 10962-10966.

Pattison, I.H. (1965) Experiments with scrapie with special reference to the nature of the agent and the pathology of the disease, in D.C. Gajdusek et al (eds.), *Slow, latent and temperate virus infections,* NINDB Monogr.2, US Government Printing Office, Washington D.C., pp 249-257.

Peuvot, J., Schanck, A., Lins, L. and Brasseur, R. (1999) Are the fusion processes involved in birth, life and death of the cell depending on tilted insertion of peptides into membranes ? *J. Theoretical Biol. 198,* 173-181.

Pillot, T., Goethals, M., Vanloo, B., Talussot, C., Brasseur, R., Vandekerckhove, J., Rosseneu, M. and Lins, L. (1996) Fusogenic properties of the C-terminal domain of the Alzheimer beta- amyloid peptide, *J.Biol.Chem. 271,* 28757-28765.

Pillot, T., Lins, L., Goethals, M., Vanloo, B., Vandekerckhove, J., Rosseneu, M. and Brasseur, R (1997) The 118-135 peptide of the human prion protein forms amyloid fibrils and induces liposome fusion, *J. Mol. Biol., 274,* 381-393.

Prusiner, S.B. (1982) Novel proteinaceous infectious particles cause scrapie, *Science 216,* 136-144.

Prusiner, S.B. (1996) Molecular biology and pathogenesis of prion diseases, *TIBS 21,* 482-487.

Prusiner, S.B. (1997) Prion diseases and BSE crisis, *Science 278,* 245-251.

Prusiner, S.B. (1998) Prions, *Proc. Natl. Acad. Sci . USA 95,* 13363-13383.

Prusiner, S.B. (1999) in *Prion biology and Diseases,* Cold Spring Harbor, New York

Prusiner, S.B., Groth, D.F., Bolton, D.C., Kent, S.B. and Hood, L.E. (1984) Purification and structural studies of a major scrapie prion protein, *Cell 38,* 127-134.

Prusiner, S.B., Mc Kinley, M.P., Groth, D.F., Bowman, K.A., Mock, N.I., Cochran, S.P. and Masiarz F.R. (1981) Scrapie agent contains a hydrophobic proteins, *Proc. Natl. Acad. Sci. USA, 78,* 6675-6679.

Prusiner, S.B., McKinley, M.P., Bowman, K.A., Bolton, D.C., Bendheim, P.E., Groth, D.F. and Glenner, G.G. (1983) Scrapie prions aggregate to form amyloid-like birefringent rods, *Cell 35,* 349-358.

Prusiner, S.B., Scott, M., Foster, D., Pan, K.M., Groth, D., Mirenda, C., Torchia, M., Yang, S.L., Serban, D., Carlson, G.A., Hoppe, P.C., Westaway, D. and DeArmond, S.J. (1990) Transgenic studies implicate interactions between homologous PrP isoforms in scrapie prion replication, *Cell 63,* 673-686.

Rahman, M., Lins, L., Thomas-Soumarmon, A. and Brasseur, R. (1997) Are amphipathic assymetric peptides ubiquitous structures for membrane destabilisation ? *J. Mol. Model. 3,* 203-215.

Riek, R., Hornemann, S., Wider, G., Billeter, M., Glockshuber, R. and Wüthrich, K. (1996) NMR structure of the mouse prion protein domain PrP (121-231), *Nature 382,* 180-182.

Rogers, m, Yehiely, F, Scott, M and Prusiner S.B. (1993) Conversion of truncated and elongated prion proteins into the scrapuie isoform in cultured cells, *Proc. Natl. Acad. Sci. USA 90,* 3182-3186.

Safar, J., Roller, P.P., Gajdusek, D.C. and Gibbs, C.J. (1993) Conformational transitions, dissociation and unfolding of scrapie amyloid (prion) protein, *J. Biol. Chem. 268,* 20276-20284.

Sakagushi, S., Katamine, S., Nishida, N., Moriuchi, R., Shigematsu, K., Sugimoto, T., Nakatani, A., Kataoka, Y., Houtani, T., Shirabe, S., Okada, H., Hasegawa, A., Smiyamoto, T. and Noda, T. (1996) Loss of cerebral Purkinje cells in aged mice homozygous for a disrupted PrP gene, *Nature 380,* 528-531.

Schanck, A., Brasseur, R. and Peuvot, J. (1998) Destabilization of a model membrane by a predicted fusion peptide of fertilin α ,*J. Chim. Phys. 95,* 467-473.

Scott, M., Foster, D., Mirenda, C., Serban, D., Coufal, F., Walchli, M., Torchia, M., Groth, D., Carlson, G., DeArmond S.J., Westaway, D. and Prusiner, S.B. (1989) Transgenic mice expressing hamster prion protein produce species-specific scrapie infectivity and amyloid plaques, *Cell 59,* 847-857.

Stahl, N., Baldwin, M.A., Teplow, D.B., Hood, L., Gibson, B.W., Burlingame, A.L. and Prusiner, S.B. (1993) Structural analysis of the scrapie prion protein using mass spectrometry and amino acid sequencing, *Biochemistry 32,* 1991-2002.

Talmud, P., Lins, L. and Brasseur, R. (1996) Prediction of signal peptide functional properties : a study of the orientation and angle of insertion of yeast invertase mutants and human apo B signal peptide variants, *Prot. Eng. 9,* 317-321.

Taraboulos, A., Scott, M., Semenov, A., Avrahami, D., Laszlo, L. and Prusiner, S.B. (1995) Cholesterol depletion and modification of COOH-terminal targetting sequence of the prion protein inhibits formation of the scrapie isoform, *J. Cell. Biol. 129,* 121-132.

Telling, G.C., Scott, M., Mastrianni, J., Gabizon, R., Torchia, M., Cohen, F.E., DeArmond, S.J. and Prusiner, S.B. (1995) Prion propagation in mice expressing human and chimeric transgenes implicates the interaction of cellular PrP with another protein, *Cell 83*, 79-90.

Vey, M., Pilkuhn, S., Wille, H., Nixon, R., DeArmond, S.J., Smart, E.J., Anderson, R.G., Taraboulos, A. and Prusiner, S.B. (1996) Subcellular colocalization of the cellular and scrapie prion proteins in caveolae-like membranous domains, *Proc. Natl. Acad. Sci. USA 93*, 14945-14949.

IMPACT OF BIOTECHNOLOGY ON ANIMAL BREEDING AND GENETIC PROGRESS

GENGLER N. AND DRUET T.
National Fund for Scientific Research, B-1000 Brussels, Belgium
Animal Science Unit, Gembloux Agricultural University,
B-5030 Gembloux, Belgium.

Abstract

Animal breeding is a field related to a whole range of biotechnologies. The impact of a biotechnology can be measured by the influence it has on genetic progress. According to the type of biotechnology considered, different component of genetic progress may be affected: accuracy of prediction, generation interval, intensity of selection and genetic variance. The first type of biotechnologies affects the efficiency of male and female reproduction: artificial insemination, multiple ovulation, in-vitro-fertilization, ova pick-up, embryo-transfer, twining, sexing of semen and embryos cloning and selfing. The impact of these technologies is mainly in the enhanced distribution of superior germplasm and the selection intensity, but also in the accuracy obtained when testing animals. In the past, artificial insemination has been a very successful biotechnology, enhancing greatly the genetic progress. A secondary, negative, impact is that these biotechnologies affect indirectly genetic diversity and therefore reduce genetic variance. A second group of biotechnologies can improve determination of the genetic merit of animals. These are all the techniques relate to quantitative or economical trait loci (QTL/ETL), their detection and use. Their main feature is the early availability in life, therefore allowing an earlier and more accurate selection. Two direction of research exists: detection of markers for the unknown QTL and direct use of a potential candidate genes as QTL/ETL. QTL/ETL will have a major impact on animal breeding especially if their use in future breeding programs can be optimized. A last type of biotechnologies with a large potential to affect animal breeding in the future are those with the ability to transform artificially DNA. The impact of these technologies is however still no very clear especially as gene expression and other issues remain unsolved. Biotechnology had, has and will have a major impact on animal breeding and genetic progress. To a certain extend animal breeding is a very promising field to use biotechnology as the past has already proven.

R. Renaville and A. Burny (eds.), Biotechnology in Animal Husbandry, 33–45.

1. Introduction

Precise definition of biotechnology is an arduous and difficult task. Therefore not only one definition, but a whole range exists from very general to very restricted ones. One broad definition could be to define biotechnologies as a group of technologies that are based on the use and mostly on the transformation of living organisms to provide services and products. Animal breeding on its own is therefore by this definition a biotechnology because its ultimate goal is the transformation of animal germplasm in order to create for animal production new generations of animals that are superior to current ones.

The definition of superior is here rather large as indeed focus of animal breeding has been shifting in recent years rapidly from only short-term production goals to functionality of animals, reduction of production costs, consumer perception and quality of products and therefore to overall sustainability and long-term economic return of animal production.

Classical animal breeding has two major characteristics. First, it uses the existing genetic variability, it does not create it artificially. Introduction of new alleles or characteristics is done by cross-breeding of different populations and/or selection of animals in existing populations. Second, detection of superior animals is done using phenotype and advanced statistical methods (BLUP) allowing the separation of genetic and environmental effects.

2. Biotechnologies and Animal Breeding

Widespread use of some biotechnologies in animal husbandry resulted (Van Vleck, 1981) and will result in a major impact on genetic progress. Different types of biotechnologies may have influence on both animal breeding and the resulting genetic progress through three main ways:

- Biotechnologies can affect efficiency of reproduction and therefore also selection programs: artificial insemination, embryo transfer, sexing, cloning and other related techniques (e.g., Ruane and Thompson, 1991; van Vleck, 1981).
- Biotechnologies can improve determination of genetic values of animal: genetic markers, candidate genes and other related techniques (e.g., Georges *et al.*, 1995; Renaville *et al.*, 1997).
- Biotechnologies can transform artificially the genome at the DNA level: genetic engineering, gene transfer and related techniques (e.g., Solter, 1981).

The objective of this study was to review the importance of the most common current and likely future biotechnologies on animal breeding and genetic progress.

3. Some Basics of Animal Breeding and Related Issues

3.1. GENETIC GAIN EXPRESSED AS SUPERIORITY OF SELECTED GROUPS

In order to be able to consider what effect biotechnologies may have, it is important to formalize equations that describe genetic gain ΔG (Falconer, 1989):

$$\Delta G = r_{G\hat{G}} i \sigma_G \tag{1}$$

where $r_{G\hat{G}}$ is the correlation between the actual and predicted additive genetic value, also called accuracy of evaluation, i is the selection intensity (expressed as the standard normal selection differential) and σ_G is the genetic standard deviation. Biotechnologies will act mainly by allowing an increase in selection intensity (by enhancing reproductive efficiency of animals) and by increasing the accuracy of evaluation (by additional information or increase in number of progenies). However, intensive use of some biotechnologies could also affect genetic variance (e.g., increasing the inbreeding trend or reducing genetic diversity).

The equation (1) can be modified to reflect genetic progress by time period (year) $\Delta g = \Delta G / yr$:

$$\Delta g = \Delta G / yr = \frac{\Delta G}{l} = \frac{r_{G\hat{G}} i \sigma_G}{l} \tag{2}$$

where l is the generation interval in years. In reducing generation interval, some biotechnologies could then increase genetic progress by time period.

Formula (2) is only theoretical because based on the hypothesis of equal accuracies, selection intensities and generation intervals in different selection paths. In reality this hypothesis is obviously not true, especially if one considers males and females. There are four different selection pathways (van Tassel and van Vleck, 1991): sire of sire (SS), sire of dam (SD), dam of sire (DS), dam of dam (DD) and formula (2) can be developed (Rendel and Robertson, 1950):

$$\Delta g = \frac{\Delta G_{SS} + \Delta G_{SD} + \Delta G_{DS} + \Delta G_{DD}}{l_{SS} + l_{SD} + l_{DS} + l_{DD}} \tag{3}$$

3.2. PROGENY DIFFERENCES

The equations given above reflect long-term and population wide genetic gain. Breeders are also interested in genetic gain in a short-term perspective. To express short-term gain, progeny superiority formula (1) can be modified to be (van Vleck, 1981):

$$\Delta G_P = \frac{\Delta G_S + \Delta G_D}{2} = \frac{r_{G_S\hat{G}_S} i_S \sigma_G + r_{G_D\hat{G}_D} i_D \sigma_G}{2} \quad (4)$$

Equation (4) reflects population wide expected progeny difference. For a given animal i the expected additive genetic value would be (Mrode, 1996):

$$\hat{a}_i = \frac{\hat{a}_{i_S} + \hat{a}_{i_D}}{2} \quad (5)$$

In most situations equation (5) will also be close to the expected phenotypic progeny difference. Often studies on impact of biotechnologies are limited to genetic additive aspects. However, the expected progeny difference will be affected by inbreeding and non-additive genetic effects (e.g., dominance effects). Both these parameters can heavily be influenced by biotechnologies (van Raden and Hoeschele, 1991). A more general formula (6) assessing the total superiority of a given animal i (Δp_i) without own records should include both these effects:

$$\Delta p_i = \frac{\hat{a}_{i_S} + \hat{a}_{i_D}}{2} + \hat{d}_{i_S,i_D} + \hat{\alpha}_F F_i \quad (6)$$

where $\hat{d}_{i_S,i_D}$ is the parental dominance effect [9], $\hat{\alpha}_F$ is the inbreeding depression per percent inbreeding and F_i is the inbreeding coefficient in percent. This formula could be extended especially towards molecular information as we will see later. Other environmental, genetic, and/or genetic x environmental effects could be introduced (van Raden and Hoeschele, 1991). In most polygenic cases the three terms in (6) will be enough, other genetic terms as additive x additive or additive x dominance effects are considered to be small (Misztal *et al.*, 1998) or can not be correctly estimated and environmental or genetic x environmental effects are considered to be negligible in most current animal breeding situations.

The concepts of parental dominance and inbreeding need some explanations. Parental dominance (Hoeschele and van Raden, 1991) effects reflect the interaction between alleles coming from sires and dams at a given loci. Intensive use of biotechnologies can create larger sire x dam families and therefore increase eventual necessity to take these effects into account. Inbreeding of an animal i reflects percentage of homozygotic pairs of alleles and is computed as the probability that the two alleles at a given loci are descendent from one allele of a common ancestor of sire and dam (Weigel and Lin, 2000). Inbreeding is also an important issue because a lot of biotechnologies will result in fewer active breeding animals or even new types of animals that have higher or even extreme inbreeding coefficient.

4. Biotechnologies Affecting Mainly Efficiency of Reproduction

Genetic improvement is highly dependent on efficiency of reproduction in order to disseminate superior germplasm. Early biotechnologies focused very strongly on the improvement of this biological function. Indeed, a lot of economically important animal species had very limited reproduction capacity.

4.1. MALES: ARTIFICIAL INSEMINATION

Artificial insemination (AI) can be considered as the first large scale reproduction biotechnology. Since its earliest commercial days in the 1930's AI has become an extremely common way to breed females at least in bovine, porcine and similar species (Foote, 1981). Most of its development occurred in dairy cattle and dairy production, therefore we focus on the impact of AI in this animal production. If its first use was often more to avoid diseases, its real impact, as appears nowadays, was on genetic progress (Foote, 1981).

AI acts on genetic progress in several ways. First AI allows to heavily increase the selection intensity (van Vleck, 1981). The number of required bulls to breed the available cows is heavily reduced. At the same time, the widespread use of tested older and the progeny testing of younger bulls allow to achieve very accurate estimations of breeding values. In dairy cattle, frozen semen of AI bulls can also be easily shipped, therefore it disseminates the favorable alleles in a wider population across countries but also world-wide. A surprising side effect of this seems to be that despite heavy selection genetic variances are not yet decreasing in most species undergoing heavy selection due to AI. A direct consequence of all these facts was a strong increase of the genetic progress as described by formula (1) and in dairy cattle (van Vleck, 1981).

However, some undesired side effects can appear. The heavy use of the best males resulted in a strong increase in inbreeding and a lost of genetic diversity. For the moment the annual increase in the inbreeding trend in the USA in Holsteins dairy cows is estimated to be near to 0.5 % (Wiggans *et al.,* 2000)! Also examples, like the Holstein bull Skalsumer Sunny Boy being used at least for over 1,000,000 first inseminations, show the risks of rapidly shrinking breeding populations. Recent research in France (Maigel *et al.*, 1996) showed that the local Holstein population with over 5 million animals was in reality behaving as if this population consisted of less than 100 unrelated animals!

4.2. FEMALES: MULTIPLE OVULATION, IN-VITRO FERTILIZATION, OVA PICK-UP, EMBRYO-TRANSFER AND TWINING

Multiple ovulation (MO) and embryo transfer (ET, often called together MOET) is a procedure that is similar to AI, but affects the reproduction ability of females (Seidel and Seidel, 1981). MOET is therefore a biotechnology that makes species like cattle multiparous, allowing the best cows to have more than the natural number of descendants. Its most important feature is the increase of selection intensity in the female selection paths (Ruane and Thompson, 1991), especially the selection of bull dams less but better dams therefore improving the bull dam pathway as shown in formula (2). Unfortunately good females, especially in dairy cattle, are often subject to preferential treatment to let them appear even better and it was and is always quite difficult to identify a truly superior female animal (van Vleck, 1999).

By application of MOET techniques, the number of full-sibs families increases (van Raden *et al.*, 1992). Therefore, it is more important to include dominance effects in order to get more precise breeding values (van Raden *et al.*, 1992). The gain of accuracy of breeding values through inclusion of dominance effects is quite important because animals produced by MOET techniques are often used as top-reproducers.

The splitting of embryos is a way to artificially produce twins. It is a rather old technique that was used as early as during the late 1980's to create dairy AI bulls (Jackbuilt Great Divide-ETS and Duplicate-ETS) (ABSglobal). This technique which is nowadays quiet often used for superior females increases the number of potential descendants per cow. Unfortunately survivability of embryos is affected.

Two other modern enhancements to MOET are in-vitro fertilization (IVF) and ova pick-up (OPU). IVF does not play in cattle for example the role it has in human reproductive biotechnologies. But it can help produce embryos from some females that do not react properly to MO, or that are in general poor health. Its implications for genetic improvement are low. OPU however can greatly reduce generation interval as it provides a method to use very young animals in selection schemes.

An interesting way to use MOET and related techniques is in nucleus breeding schemes. In such schemes parts of the whole population are subject to better performance recording eliminating preferential treatment and more intense (e.g., MOET) and earlier (e.g., OPU) selection.

Despite a lot of very positive simulations, closed and therefore totally disconnected from the whole population, closed nucleus herds were not successful in cattle breeding. In swine and poultry most breeding companies run apparently closed nucleus schemes. But one might suspect that these schemes are in reality often not totally closed, but subject to the introduction of external superior animals. Totally open nuclei which are basically combinations of future bull dam station testing and MOET technologies seem to be a very promising way to enhance genetic progress in cattle because they combine optimal or at least better bull dam selection with classical progeny testing of the produced sires.

Finally, a very special type of biotechnology affecting female reproduction is twining. Especially in beef cattle were obtaining more than one calf per calving can greatly enhance production efficiency this technology can be useful. At the same time twining and ovulation rate, a rather closely related trait, can be both measured and are partially genetic. Therefore the Meat Animal Research Center in Clay Center, Nebraska has initiated a selection experience in this field (van Vleck and Gregory, 1996). They were able to increase the rate of twins from around 3% to over 20% in less than 20 years (van Vleck and Gregory, 1996).

4.3. SEXING

Sexing of semen and of embryos provides in species and production circumstances where one sex is preferred, a way to produce the wanted type of animal (Betteridge *et al.*, 1981; Johnson *et al.*, 1998). Sexing of embryos is a technique that is rather reliable nowadays and currently used on a rather large extent. Some progress was recently made in the field of sexed semen. One problem is the price of sexing and the eventually reduced fertility of this semen. The other issue is that sexing has to be economic what means that the extra-price to be paid for sexed semen has to be in relation with the extra income.

4.4. CLONING AND SELFING

An apparently new technique that is widely covered in the media is cloning. In fact nucleus transfer is not really a new technique. The real new point is that it allows now to clone an adult mammal (Campbell *et al.*, 1996). This is indeed a major enhancement compared to the original techniques that required embryonic nucleus (McKinnell, 1981).

From the animal breeding point, cloning can have multiple aspects and the assessment of its economical value is still rather uncertain (Dematawewa and Berger, 1998). First it is a possibility to multiply phenotypically outstanding individuals. Unfortunately, even performances done by the same animal show a correlation, often called repeatability, of less than 1 (van Vleck, 1999). Therefore multiplying a phenotypically outstanding individual is by no means a guaranty that the resulting clones will be outstanding. Again phenotypically outstanding animals are also very sensitive to preferential treatment. Similarly, only animals with precisely estimated genetic merits could be interesting candidates for cloning, this means especially sires and here doubts exist why we should clone an outstanding sire that can reproduce easily through AI? Use of cloning should require precise evaluation of each clone family, therefore a lot of clone-testing with resulting costs. An advantage of cloning, is that we can integrate dominance effects into models (every clone of the same family has the same dominance effect) and reproduce them.

However, reproduction through clones would heavily increase selection intensity in the first generation, or better limit the population used for reproduction, but afterwards would result in no or severely reduced genetic progress as we duplicate always the same animals leading to the more fundamental following problem. Cloning is by definition conservation of existing germplasm. This means that, except for mutations and aging of DNA, at least nuclar DNA will stay the same. But genetic improvement is based on the creation of new, eventually favorable, combinations of parental germplasm at every mating. Dairy cattle selection shows that this is a highly efficient way to work. It is usually, in large populations such as Holsteins, easy to find sons that easily surpass their sires. Another more theoretical argument against cloning in general animal improvement is that extensive use of clones, as in plants, reduces genetic variation and therefore long-term progress that can be achieved. But cloning has truly high potential to multiply new genotypes obtained by natural and artificial means.

From all biotechnologies, cloning is certainly the one mostly reducing diversity. We already discussed above the impact of such a reduction on selection but reduction of diversity can also have other dangerous consequences. First, heavily selected animals produce optimally only in specific conditions. If environmental (including new techniques) conditions change, then those animals would show poor adaptability to new conditions and their would be no way back. This means that cloning forbids any change in selection path because animals are fixed, not adaptable to new needs, new economical conditions, they will not resist to new disease.

Selfing is based on the fusion of two X gametes from a bull and the creation of what could be called selfs (van Vleck, 1981). This would create animals that are potentially highly expressive of the parental genotype. Some technical issues are obviously not yet solved. But for testing of animals that are not expressing phenotypes, as dairy bulls, this can be a clear alternative. Unfortunately such animals are 50% inbred, therefore we have here a major problem because this means parents need to be totally free of any lethal or sublethal alleles. Current theory to account for inbreeding but also dominance effects under inbreeding in such extreme animals might not work.

5. Biotechnologies Improving Determination of Genetic Values of Animals

Estimation of breeding values is for the moment essentially done by analysis of the phenotype or the phenotypic performances of the animals. Advanced statistical methods (BLUP) are used to separate genetic and environmental effects, relating also animals to each other according to their genetic relationships (Henderson, 1984). Unfortunately the phenotype can not always be recorded due to physiological (e.g., bulls give no milk, boars give no piglets, dairy cows calf only around 2 years of age) or other reasons (e.g., recording expensive as for beef quality traits), or the recording of the phenotype is inaccurate or imprecise (e.g., calving ease). Since several years a large progress has been made in the understanding and description of DNA and the actions of certain genes. Under the influence of this research animal breeders started to investigate if variations in DNA (polymorphism) of some animals can be linked to differences in production or other economically important traits (e.g., milk production). But there are several basic problems inherent to whole genome. First, one has to determine the functional structure of the chromatin. Then inside this structure functional, often called genes, and unfunctional regions can be spotted.

Main advantages of QTL/ETL in breeding programs are: 1) an increase in accuracy in selection through additional information directly related to the genotype; 2) a possibility to reduce generation interval by adding a new selection stage at earlier age because QTL/ETL allow to make observations that are not sex or age dependent. QTL/ETL could also increase the efficiency of introgression or be used in the prediction of heterosis by genetic distance.

The detection, the use of QTL/ETL and their potential role in animal genetics and breeding will be emphasized.

5.1. DETECTION OF QUANTITATIVE OR ECONOMICAL TRAIT LOCI

This step basically consists in the study of potential links between DNA polymorphisms and production or similar traits. The idea is that a specific gene called hereafter QTL/ETL (quantitative or economical trait loci) is responsible for part of the phenotype. Two different approaches are currently used to detect QTL/ETL. If a known gene is used as candidate to be the QTL/ETL the method is called candidate gene approach. If the DNA polymorphism is only considered being the marker for a DNA region containing the QTL/ETL the method is called genetic marker approach (e.g., Georges *et al.*, 1995).

The candidate gene method (e.g., Renaville *et al.*, 1997) is rather dependent on a priori knowledge on a genetic and physiological level. First, genes and their DNA sequences need to be known. Then selection of the right gene should be based on good knowledge of its position inside a metabolic pathway. Additive effects for the different alleles for these candidate genes are then compared to each other in order to establish allele substitution effects. This can be done using different techniques and on a population or family level.

Genetic markers are also based on known DNA polymorphism. But contrary to candidate genes these polymorphisms are mostly not functional DNA regions, but nonfunctional. The markers used are mostly microsafellites, regions that are highly polymorphic. Through the observation of transmission of markers over generations, transmission of marked DNA parts can be traced through pedigree and linked to phenotypic differences. This QTL/ETL detection method is basically family oriented.

5.2. USE OF QUANTITATIVE OR ECONOMICAL TRAIT LOCI

Detection of QTL/ETL is one point, but the main issue is their integration in current breeding programs. Larzul *et al.* (1997) gave some details on the potential impact of its use. Also, different technical problems are linked to QTL/ETL use.

First detection of QTL/ETL is done in families or sub-populations, therefore their use in the whole population has to be done carefully or restricted to known families where their segregation is known.

QTL/ETL detection is done on a single trait basis, therefore only for certain traits, seldom for groups of traits. Therefore pleiotropic effects on other traits can not be excluded and have to be studied carefully. A real-life example is the muscular hypertrophy gene. This gene produces not only double muscled animals but seems to have a pleiotropic effect on several malformations.

Also current research is focusing on additive QTL/ETL effects. But, the remaining polymorphism at QTL/ETL levels needs to be explained. One possible answer is that there might be favorable dominance effects that keep different alleles in the population despite heavy selection.

A last important issue is that in a lot of situations molecular information is only available for some animals.

QTL/ETL could be used in many ways. Different methods could differ by the use or not of a priori levels of QTL/ETL effects, or by the joint estimation of polygenic and

QTL/ETL effects. Obviously the second way to proceed would be ideal. Some obvious uses of QTL/ETL could be as follows:

First, a priori knowledge of the level of QTL/ETL effects in families/sub populations/populations could be used for animals without any other information than their parent average (formula (5)) extending formula (6). This is typically the use most people think about for the moment. This could for example be used to identify earlier superior AI test sires. A second use of QTL/ETL is typically inside nucleus breeding schemes where also joint evaluations might become possible. In the near future joint evaluations including additive and dominance QTL/ETL effects might become feasible even in rather large populations. Then the predicted phenotype would be:

$$\Delta p_i = \frac{\hat{a}_{i_S} + \hat{a}_{i_D}}{2} + \hat{d}_{i_S,i_D} + \hat{\alpha}_F F_i + \sum \hat{q}_{a_i} + \sum \hat{q}_{d_i} \quad (7)$$

All polygenic effects are reduced for the QTL effects of the animal i where $\sum \hat{q}_{a_i}$ represents the sum of the additive and $\sum \hat{q}_{d_i}$ the sum of the dominance QTL effects.

Such developments will then be one element in the development of new mating schemes emphasizing optimal use of all information, and improved data exchange through new information technologies as INTERNET (Misztal *et al.*, 1998). Figure 1 modified from Misztal *et al.* (1998) gives an overview how such a scheme could be structured.

6. Biotechnologies Transforming Artificially the Genome at the DNA Level

Until this point all presented biotechnologies were based on naturally existing DNA and tried to evaluate, use, disseminate and adapt it optimally, but without intervening artificially on it. Genetic engineering however exists and can be used in superior animals and plants. Here we enter a rather difficult field and the real impact of these technologies on animal breeding may be not yet known. In fact genetic engineering in domestic species is a way to suppress species frontiers that limited introgression of certain genes (alleles) into populations. For example until now only crossbreeding stress negative Large White sows with stress positive Pietrain boars and backcrossing towards Pietrain but retaining the stress-negative alleles in the population allowed to obtain stress negative Pietrain pigs (Leroy and Verleyen, 1998). In the future desired genes or alleles can be directly introduced. For example then some dairy cows may be starting to produce high value protein or have no lactose. Such special animals could then be multiplied through cloning. Direct introduction of QTL/ETL in population will be, but not quickly a widespread technique. Indeed the real major issues with genetic engineering is gene expression and simple gene transfer is not enough. In fact current experience with domestic or other higher animals seem to show that expression of introduced genes is a major problem.

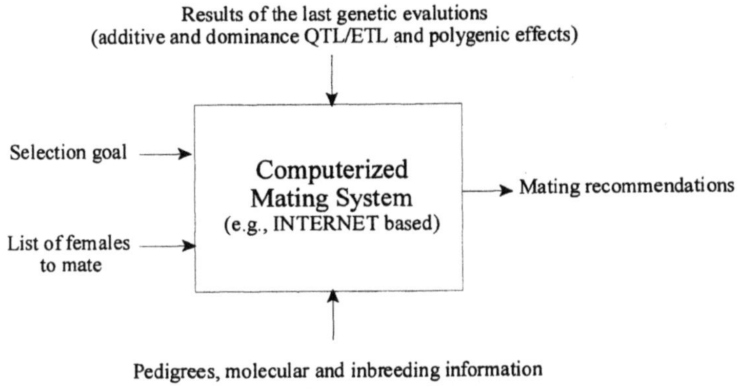

Figure 1. *Computerized mating system showing the potential use of QTL/ETL.*

7. Conclusions and Implications

There is a natural link between biotechnologies and animal improvement which is in itself a biotechnology. This link was therefore very important and animal breeding and genetic progress have been greatly enhanced by the use of reproductive biotechnologies as artificial insemination and embryo transfer. These old technologies are under continuous development leading to sexing of semen and embryos and cloning. Their influences are especially in the improved selection intensities and the more rapid dissemination of superior germplasm. But they have some negative side effects especially through increasing inbreeding and decreasing genetic diversity.

Currently, muchresearch is focusing on the detection of quantitative or economical trait loci. The ultimate goal of this is to improve accuracy of detection of genetic merit. Despite large research efforts these techniques will probably never explain all the genetic variation. Therefore they need to be integrated in polygenic evaluations. Also the real issue in the future will be the evaluation and introduction of molecular information into genetic evaluation and their use in selection schemes that will be taking advantage of advanced computerized mating schemes.

Direct and artificial modification of the germplasm of domestic animal is the current ultimate type of biotechnology. These genetic engineering techniques provide ways to adapt genetic material to specific needs but some technical, but also ethical issues are not yet solved.

Biotechnology had, has and will have a major impact on animal breeding and genetic progress. To a certain extent animal breeding is a very promising field to use biotechnology as the past has already proven.

References

ABSglobal, http://www.absglobal.com/about.htm.

Betteridge, K.J., Hare, W.C.D. and Singh, L.H. (1981) Approaches to sex selection in farm animals, in B.G. Brackett, G.E. Seidel and S.M. Seidel (eds.), *New technologies in Animal Breeding*, Academic Press, London, pp 109-125.

Campbell, K.H.S., McWhir, J., Ritchie, W.A. and Wilmut, I. (1996) Sheep cloned by nuclear transfer from a cultured cell line, *Nature 380*, 64-66.

Dematawewa, C.M.B. and Berger, P.J. (1998) Break-even cost of cloning in genetic improvement of dairy cattle, *J. Dairy Sci. 81*, 1136-1147.

Falconer, D.S. (1989) *Introduction to quantitative genetics, 3rd edition*, Longman, New York.

Foote, R.H. (1981) The artificial insemination industry, in B.G. Brackett, G.E. Seidel and S.M. Seidel (eds.), *New technologies in Animal Breeding*, Academic Press, London, pp 14-39.

Georges, M., Nielsen, D., Mackinnon, M., Mishra, A., Okimoto, R., Pasquino, A.T., Sargeant, L.S., Sorenson, A., Steele, Zhao, X., Womack, J.E. and Hoeschele, I. (1995) Mapping quantitative trait loci controlling milk production in dairy cattle by exploiting progeny testing, *Genetics 139*, 907.

Henderson, C.R. (1984) *Applications of linear models in animal breeding*, University of Guelph, Guelph, Canada.

Hoeschele, I. and VanRaden, P.M. (1991) Rapid inversion of dominance relationship matrices for noninbred populations by including sire by dam subclass effects, *J Dairy Sci 74*, 557-569.

Johnson, L.A., Welch, G.R. and Rens, W. (1998) The Beltsville sperm sexing technology: high-speed sperm sorting gives improved sperm output for in vitro fertilization and AI, *J. Anim. Sci. 77 Suppl 2*, 213-220.

Larzul, C., Manfredi, E. and Elsen, J.-M. (1997) Potential gain from including major gene information in breeding value estimation, *Genet. Sel. Evol. 29*, 161-184.

Leroy, P. and Verleyen, V. (1998) The new stress negative Piétrain line developed at the Faculty of Veterinary Medecine of the University of Liège. http://www.ulg.ac.be/fmv/rehal/ReHal9801.html

Maignel, L., Boichard, D. and Verrier, E. (1996) Genetic variability of French dairy breeds estimated from pedigree information, *Int. Bull. Eval. Serv. Bull n° 14*, 49-54.

McKinnell, R.G. (1981) Amphibian nuclear transplantation: state of the art, in B.G. Brackett, G.E. Seidel and S.M. Seidel (eds.), *New technologies in Animal Breeding*, Academic Press, London, pp 163-180.

Misztal, I., Varona, L., Culberston, M., Bertrand, J.K., Mabry, J., Lawlor, T.J., Van Tassel, C.P. and Gengler, N. (1998) Studies on the value of incorporating the effect of dominance in genetic evaluations of dairy cattle, beef cattle and swine, *Biotechnol. Agron. Soc. Environ. 2*, 227-233.

Mrode, R.A. (1996) *Linear models for the prediction of animal breeding values*, CAB International, Wallingford, UK.

Renaville, R., Gengler, N., Vrech, E., Prandi, A., Massart, S., Corradini, C., Bertozzi, C., Mortiaux, F., Burny, A. and Portetelle, D. (1997) Pit-1 Gene Polymorphism, milk yield, and conformation traits for Italian Holstein-Friesian bulls, *J. Dairy Sci. 80*, 3431-3438.

Rendel, J.M. and Roberston, A. (1950) Estimation of genetic gain in milk yield by selection in a closed herd of dairy cattle, *J. Genetics 50*, 1-8.

Ruane, J. and Thompson, R. (1991) Comparison of simulated and theoretical results in adult MOET nucleus schemes for dairy cattle, *Livest. Prod. Sci. 28*, 1-20.

Seidel, G.E. and Seidel, S.M. (1981) The embryo transfer industry, in B.G. Brackett, G.E. Seidel and S.M. Seidel (eds.), *New technologies in Animal Breeding*, Academic Press, London, pp 41-80.

Solter, D. (1981) Gene transfer in mammalian cells, in B.G. Brackett, G.E. Seidel and S.M. Seidel (eds.), *New technologies in Animal Breeding*, Academic Press, London, pp 201-220.

VanRaden, P.M. and Hoeschele, I. (1991) Rapid inversion of additive by additive relationship matrices by including sire-dam combination effects, *J. Dairy Sci. 74*, 570.

VanRaden, P.M., Lawlor, T.J., Short, T.H. and Hoeschele, I. (1992) Use of reproductive technology to estimate variances and predict effects of gene interactions, *J. Dairy Sci. 75*, 2892-2901.

Van Tassell, C.P. and Van Vleck, L.D. (1991) Estimates of genetic selection differentials and generation intervals for four paths of selection, *J. Dairy Sci. 74*, 1078-1086.

Van Vleck, L.D. (1981) Potential genetic impact of artificial insemination, sex selection, embryo transfer, cloning, and selfing in dairy cattle, in B.G. Brackett, G.E. Seidel and S.M. Seidel (eds.), *New technologies in Animal Breeding*, Academic Press, London, pp 221-242.

Van Vleck, L.D. (1999) Implications of cloning for breed improvement strategies: are traditional methods of animal improvement obsolete? *J. Anim. Sci. 77, Suppl 2*, 111-121.
Van Vleck, L.D. and Gregory, K.E. (1996) Genetic trend and environmental effects in a population of cattle selected for twinning, *J. Anim. Sci. 74*, 522-528.
Weigel, K.A. and Lin, S.W. (2000) Use of computerized mate selection programs to control inbreeding of Holstein and Jersey cattle in the next generation, *J. Dairy Sci. 83*, 822-828.
Wiggans, G.R., VanRaden, P.M. and Smith, L.A. (2000) Expected inbreeding coefficients: their current and future use. http://aipl.arsusda.gov/memos/html/inbredcof.html.

MARKER GENES IN FARM ANIMALS

PARMENTIER I., PORTETELLE D., BERTOZZI C., HAEZEBROECK V., PIRARD M. AND RENAVILLE R.
Animal and Microbial Biology Unit, Gembloux Agricultural University, Belgium

Abstract

Identification of mutations in genes responsible for traits of interest or in a locus marker for such genes will allow geneticists to implement the Marker Assisted Selection (MAS) approach. The main advantages of using genetic criteria for selection is that such data can be obtained early in life of animals (e.g.: at birth) and in each sex. When a locus shows a preponderant action and explains the major part of variation of a production trait production, it is called a Quantitative Trait Locus (QTL). At the present time, the two major strategies developed to detect such QTL are the candidate gene approach and the positional genetic approach. The present review summarises the main results obtained in this area by both approaches and their possible application in term of selection of farm animals.

1. Introduction

Classical selection programmes are based on phenotypic observations. Unfortunately, in some cases, this approach is not efficient because phenotypes are expressed late in life or in one sex only or determined after the death of the animal. In these cases use of the genotype as a mean of selection will be a powerful tool that can be used early in life of the animal and in each sex. Marker-assisted selection (MAS) can be defined as the use of DNA sequences, that are associated with improved production traits, to supplement or replace phenotypic selection. The basis of this approach is the polymorphic nature of DNA. Due to advances in molecular biology, it is now possible to detect more easily DNA mutations. These can directly affect the functionality of the coded protein by changing splice sites, RNA stability, rate and regulation of gene transciption, or amino acid sequence of the gene product. Mutations can also affect, not the gene itself, but a region closely linked to it that co-segregates with it. In the latter, the linked gene is generally unknown. In both cases, if one of the alleles of the polymorphic gene is associated with a production trait, it is called a marker and can be used in a MAS

R. Renaville and A. Burny (eds.), Biotechnology in Animal Husbandry, 47–64.

program. Two approaches exist to determine whether a locus is a marker: the candidate gene approach and the positional cloning approach.

1.1. THE CANDIDATE GENE APPROACH

This strategy consists in studying different genes potentially involved in the physiological process (e.g.: milk proteins synthesis) and determining for each gene, the allele that results in the interesting phenotype (e.g.: rennet coagulation properties). It requires the knowledge of the genetic control of the studied physiological process (e.g.: role of the somatotropic axis in regulation of lactation). Briefly, the identification of polymorphic sites and the evaluation of the allelic differences of the candidate genes, among animals of different phenotypic merit, offer potential for easy identification of gene markers associated with phenotypic merit. Various studies have shown that a number of single genes associated with mammary growth, development, and functions are excellent candidates for linkage relationships with quantitative traits of economic importance.

1.2. THE QTL DETECTION WITH GENETIC MARKERS (THE POSITIONAL CLONING APPROACH)

The detection, with genetic markers, of chromosomal segments responsible for a fraction of genetic variability of a trait is a completely different approach and does not require any prior knowledge about the genes involved. Its principle is the following: during meiosis, each haploid gamete receives n (n=30 in cattle) chromosomes resulting from a random sampling of large segments within each parental chromosome pair. As a consequence, two DNA sequences (or locus) located close to each other on the same parental chromosome are likely to be transmitted together to the gametes and then to the progeny. Let us consider two linked polymorphic loci called M and Q, and a double heterozygote parent MQ/mq. This notation means that the alleles M and Q are on the same chromosome (the grand paternal one, for instance) and m and q are on the homologous chromosome (the grand maternal one). The number of parental progeny, ie progeny receiving a parental (MQ or mq) chromosomal segment, is larger than the number of recombinant progeny, ie progeny receiving a recombined (Mq or mQ) chromosomal segment resulting from a crossing-over. When recombinations are rare enough, ie when the loci are close enough, M and m are mostly associated to Q and q, respectively, in the progeny of this parent. If the Q locus affects the trait, the effect of substitution of allele q by allele Q could be detected by comparing the two groups of progeny having received M and m. Thus, in this case, although the marker M is not involved in the genetic determinism of the trait, it can detect the QTL through its genetic linkage.

The efficiency of this method primarily depends on the characteristics of the markers:

- As QTL could be located anywhere, markers are expected to be spread over the entire genome;

- In order to characterise and trace each parental chromosome, the marker should be polymorphic, ie it should have at least 2, and possibly more, distinctable alleles;
- The determination of the marker genotypes should be possible, easy and cheap.

These constraints explain why this approach described in the early 60's (Neiman-Soensen and Robertson, 1961) was applied only recently. In fact, two basic discoveries opened the way to QTL detection: firstly, many polymorphic sequences, and particularly microsatellites (Weber and May, 1989) were discovered over the complete genome; and secondly the polymerase chain reaction (PCR) method provided a means to exhibit this polymorphism. In the last decade, a tremendous international effort has been put to build a medium-density genetic map for most domestic animal species.

Pratically, the strategy using marker-QTL associations to localise a QTL can be divided in the following steps: 1) identification of chromosomal regions containing QTL of interest (10-20 cM), 2) specification of QTL location within these regions (5 cM), 3) identification of markers in tight linkage with the QTL (1-2 cM) and 4) identification of a potential candidate gene in this region. This method of mapping QTL to progressively narrower chromosomal regions, using a battery of microsatellite markers, until a suitable candidate gene is identified is called the positional cloning method.

2. Genetic Markers Identified in the Bovine Species

2.1. LACTATION

2.1.1. The Candidate Gene Approach

As reviewed by Bauman (1999) and Baldi (1999), administration of recombinant growth hormone to dairy animals dramatically increases milk production and demonstrates the critical role of the somatotropic axis in regulation of lactation. Accordingly, the results presented in the next section focus on the possible relationships between candidate genes, associated with the somatotropic axis, and quantitative and qualitative milk production traits.

2.1.1.1. Somatocrinin (GHRH) and GHRH Receptor Polymorphisms. In the cascade of events that control the pituitary GH secretion, the hypothalamic factor GHRH and its receptor directly control the GH synthesis by the somatotropic cells. In 1995, Moody *et al.* (1995a) described a restriction fragment length polymorphism (RFLP) in the GHRH gene using the *HaeIII* restriction enzyme. In a preliminary study on 89 artificial insemination (AI) Holstein-Friesian bulls, we noted that the rare genotype (AA, 7.7%) obtained after PCR-RFLP *HaeIII* restriction enzyme analysis is significantly favourable for fat percentage and fat yield. However, this observation must be confirmed on a higher number of animals. Recently, Connor *et al.* (1998) reported a polymorphism at the GHRH receptor gene (located on chromosome 4) using *Eco57I* enzyme that could modify GHRH stimulation of the somatotropic cells (Connor *et al.*, 1998).

2.1.1.2. Pituitary-Specific Transcription Factor (Pit-1) Polymorphism. Biochemical and ontogenetic studies have shown that Pit-1 is the critical cell-specific transcription factor for activating expression of the prolactin (PRL) and GH genes in the anterior pituitary gland (Review in Tuggle and Trenkle, 1996). Pit-1 has subsequently been shown to be capable of activate other pituitary genes, including the Pit-1 gene itself (Chen *et al.*, 1990, Rhodes *et al.*, 1993), the GHRH receptor gene (Lin *et al.*, 1992), the gene coding for the β subunit of thyroid-stimulating hormone (TSH) (Haugen *et al.*, 1993, Steinfelder *et al.*, 1991) and the pituitary β2 TSH receptor (Wood *et al.*, 1994).

The regulatory regions recognized by Pit-1 within the PRL, GH and Pit-1 gene promoters have been demonstrated to be required for appropriate pituitary-specific expression of these genes both in cultured cell lines and transgenic animals (Lira *et al.*, 1988, Lira *et al.*, 1993 and Crewshaw 1989). In humans, different mutations of the Pit-1 gene have been also reported in patients with hypopituitarism (Pfäffle *et al.*, 1992) or with sporadic combined pituitary hormone defiency (Radovick *et al.*, 1992).

Pit-1 is a 291-amino-acid protein with a DNA-binding POU domain which is characteristic of a family of proteins, including Pit-1, Oct-1, Oct-2, Unc-86 (Herr *et al.*, 1988). The POU domain is divided into two well-conserved regions. The C-terminal half, called the POU-homeodomain, encodes a low-affinity DNA-binding domain, whereas the N-terminal half, called the POU-specific domain, confers high affinity to the POU domain and participates in protein-protein interactions (Tuggle and Trenkle, 1996).

In cattle, Woolard *et al.* (1994) reported a *HinfI* RFLP in exon 6 of the Pit-1 gene which has been assigned to chromosome 1 (Moody *et al.*, 1995b). Digestion of the PCR products with the enzyme revealed two alleles: A (*HinfI* -) and B (*HinfI* +). The frequencies of the three genotypes generated by the two alleles in a sample of 89 Italian AI Holstein-Friesian sires were 2.2, 31.5 and 66.3 % for AA , AB and BB, respectively (Renaville *et al.*, 1997). In this study, fixed and mixed linear models fitted on daughter yield deviations for milk and on deregressed proofs for conformation traits showed a superior effect of allele A for milk and protein yields, inferior for fat percentage, and superior for body depth, angularity and rear leg set. Recently, an additional study has been realised in our laboratory on 1100 AI Holstein bulls commercialised by Semex Alliance (Canada) (unpublished data). In this sample, the frequency of the AA genotype was higher than in the Italian bulls (9 % *vs* 2.2). The frequencies of the two other patterns were respectively 48% for AB and 42% for BB patterns respectively. The statistical analysis revealed similar associations as in our first study, except for fat yield, with a significant superiority of the A allele on the B allele for milk yield, protein yield and for fat yield. These preliminary investigations indicate that Pit-1 is a promising new possibility to select for increased protein and milk yield through selection for the A allele. However, this hypothesis must be validated on a large number of animals and for other breeds before definitive conclusions can be drawn.

2.1.1.3. Growth Hormone Polymorphism. Considering its essential role in lactation process, the GH gene (associated to chromosome region 19q26-qter in bovine (Hediger *et al.*, 1990)) is a potential target for studies of molecular variation in association with genetic merit of dairy breeds. Using different restriction enzymes, numerous polymorphisms have been reported in the bovine species.

Due to an insertion of a T at position +837 and a C-G transition at position +838, a *MspI* RFLP is present in the third intron of the bovine GH gene (Zhang *et al.*, 1993). Høj *et al.* (1993) observed a difference in allele frequencies at the GH loci between two Danish and Norwegian lines, the *MspI* (-) haplotype being more frequent in the line selected for milk fat. Lee *et al.* (1993, 1994) found also a significant association of *MspI (-)* with higher fat yield in Holstein cows. In a recent study, Lagziel *et al.* (1999) found a significant increasing effect of *MspI (-)* allele on protein percentage and kg of protein per year. For these authors, the observed effects are due to a locus in linkage with GH but at some distance of it. In contrast, for Yao *et al.* (1996), who have detected this polymorphic site by the SSCP method, the *MspI(+)* allele has been found favourable for milk, fat and protein yield. In this study, the authors found an other polymorphism, due to an A-to-C transversion which changes the codon AGG to CGG in the fifth exon of the gene, in linkage disequilibrium with the *MspI* mutation. For Furu *et al.* (1988), the *MspI* (+/-) pattern, located in the third intron, is associated with a 0.9 kb insertion/deletion in the 3' flanking region, potentially carrying transcription regulator sites. For these authors, this polymorphism does not result in differences of indicators for genetic merit in 100 Holstein AI sires.

With the PCR-RFLP technique using the *AluI* restriction enzyme, Lucy *et al.* (1993) revealed two alleles responsible for alternative forms of bovine GH with a Leu or Val residue at position 127. Injection of recombinant bovine GH with a valine residue corresponding at the Leu/Val polymorphic locus to lactating cows has been shown to improve fat corrected milk production more than GH-Leu^{127} administration (Eppard *et al.*, 1992). This observation suggests that the Val allele genotype is better for milk production. On the other hand, Lucy *et al.* (1993) reported a significant higher predicted transmitting ability (PTA) for milk yield for Val/Val genotype Jersey cows, but no significant effect was found in their samples of Holstein, Guernsey, Ayrshire or Jersey bulls. However, Furu *et al.* (1998), Schlee *et al.* (1994a) and van der Werf *et al.* (1998) observed no significant effect of this Leu/Val polymorphism on milk trait breeding values of Simmental, Holstein or Jersey sires. On the other hand, Lee *et al.* (1996) and Lucy *et al.* (1993) reported a decreased milk yield associated with the valine variant of the GH gene in Holstein cows. Further studies by Schlee *et al.* (1994b) in German Black and White bulls revealed that animals homozygous for the leucine variant had higher plasma levels of GH than their heterozygous counterparts.

A third RFLP using the *TaqI* restriction enzyme was described in the bovine GH gene by Rocha *et al.* (1992). In milk production, Falaki *et al.* (1996, 1997) reported an association between this GH-*TaqI* polymorphism and milk traits for Simmental cows but not for Holstein-Friesian bulls. The GH-*TaqI* polymorphism is due to an insertion/deletion of about 1000 bp in the 3' end of the gene, between the *EcoRI* and the *TaqI* restriction site (Hoj *et al.*, 1993; Hallerman *et al.*, 1987).

In the bovine GH gene, there are also other variation sites that could be interesting for association studies with milk production traits. Hecht and Geldermann (1996) have shown 7 mutations in the GH gene. Six of them have been identified in the 5'-flanking region and one in intron I. Some of these variable sites are also potential binding sites for trans-acting factors (CAAT/enhancer binding protein, polyoma virus enhancer A binding protein 3, thyroid hormone response element) and therefore possibly involved in the expression of the growth hormone gene. Recently, Rodrigues *et al.* (1998), have identified a polymorphic site in the promoter region of the GH gene. This polymorphism consists in the absence of an AAG trinucleotide localized 9 nucleotides upstream from the TATAAA sequence. Unfortunately, in these two last studies, the association between production traits and the polymorphism was not examined.

Therefore, it appears that selection for milk production traits based upon conflicting GH genotype data is not a promising way. The major problem to definitely conclude on the opportunity to select animals on GH genotype is often due to the limited number of genotyped animals in the published studies.

2.1.1.4. Growth hormone Receptor Polymorphism. The cDNA of the bovine transmembrane GHR has been sequenced by Hauser *et al.* (1990). It consists of 9 exons and is associated with chromosome 20 (Moody *et al.*, 1995c). As deduced from the human nucleotide sequence, the GHR is a protein of 620 amino acids comprising an extracellular hormone binding domain of 246 amino acids, a single 24 amino acid transmembrane region and a long cytoplasmic domain. The extracellular domain contains seven cysteine residues and five potential N-linked glycosylation sites and two boxes important for the signal transduction after binding of the hormone to its receptor (Leung *et al.*, 1987, Wang *et al.*, 1995).

After hybridization with a homologous cDNA probe containing the coding sequence for the intracellular C-terminal part of the receptor, we have identified six restriction enzyme *TaqI* bands (7.1, 6.2, 5.7, 5.4, 4.2 and 3.3 kb) that give nine genotypes in Holstein-Friesian AI bulls (Falaki *et al.*, 1996). The effect of this polymorphism on values for milk protein percentage was highly significant ($P < 0.005$) and favorable for the rare (6.6%) 5.7- and 5.4-kb patterns.

Recently, Moisio *et al.* (1998) detected three variants in the 3' flanking region of the bGHR: GHR_{311}, GHR_{320}, GHR_{325}. The longer alleles GHR_{320} and GHR_{325} are more frequent in the selected Finnish dairy breeds, whereas the shortest allele GHR_{311} predominates in the native breeds.

2.1.1.5. IGF-I and IGF-I Receptor Polymorphisms. Mapped on chromosome 5 (Miller *et al.*, 1991), the IGF-I gene showed a *SnaBI* polymorphism in the 5'-flanking region (Ge *et al.*, 1997). However, in 152 Holstein bulls from select sires and AI organizations, PTAs for fat, protein, and net merit did not tend to be significantly related to a IGF-I/*SnaBI* polymorphic form (Hines *et al.*, 1998). Moreover, in the same study, it was reported that no *Sna*BI polymorphism at IGF-I gene and *AluI* polymorphism at GH codon 127 interaction effects were detected. Moreover, Ge *et al.* (1997), using the Single Strand Conformational Polymorphism (SSCP) method, noted a significant difference in allele frequencies in seventy-six genotyped Angus cattle from lines selected for high or low blood serum IGF-I concentration that could influence the lactation process. In 1996, Moody *et al.* (1996a) have located the IGF-I receptor on bovine chromosome 21. Digestion of IGF-I receptor PCR products with *TaqI* restriction enzyme revealed a polymorphism with two alleles (A and B). However, these authors concluded that usefulness of this IGF-I receptor polymorphism in studies designed to identify economically important quantitative trait loci in cattle may be limited because of the low B allele frequency and because this allele is only present in Bos indicus cattle. To our knowledge, no new information have been published to validate or not this hypothesis.

2.1.2. The positional Cloning Approach

In order to generate linkage disequilibrium between the alleles of the QTL and the genetic marker, most studies have crossed inbred lines to construct F1 individuals heterozygous for both loci. When this method is not a viable option (in dairy cattle for example), alternative options exist. The progeny of a single sire, heterozygous for one or more genetic markers may be analysed. If a specific sire is also heterozygous for a linked QTL, then the progeny of this sire that inherit different paternal alleles will have different means for the quantitative trait. This method, the 'daughter design', has been applied to dairy cattle data (reviewed by Hines, 1990) and extensively analysed (Soller and Genizi, 1978; Weller *et al.*, 1990). In 1990, Weller *et al.* proposed an alternative 'granddaughter design', in which sons of a sire heterozygous for the genetic marker are genotyped, while the quantitative traits are analysed using the granddaughter records. Although the magnitude of the QTL effect measured in the granddaughter design will be only half of the effect measured in the daughter design, the number of progeny analysed for quantitative traits can be much greater. For many population structures, it is possible to genotype less than half as many individuals and still obtain greater power with the granddaughter design than with the daughter design (Weller *et al.*, 1990). Using the granddaughter design, Andersson-Eklund and Renkel (1993) reported a linkage between the amylase-1 locus and a quantitative trait locus influencing fat content in milk. In an other report, the granddaughter design was used to study chromosome substitution effects associated with κ-casein and β-lactoglobulin in Holstein cattle (Cowan *et al.*, 1992). The most severe limitation on analyses of this type has been the lack of suitable genetic markers. Most previous studies on dairy cattle have used blood groups, blood proteins and milk proteins. The number of such markers is limited. Presently, two published linkage maps exist for cattle (Barendse *et al.*, 1994; Bishop *et al.*, 1994), that cover most of the genome and contain a large selection of microsatellite markers. With the availability of highly informative marker maps, it has recently become possible to map quantitative trait loci underlying the genetic variation of multifactorial traits (Paterson, 1995). Although microsatellite linkage maps with intervals of 5-10 cM are sufficient for identification of loci with large effects on phenotype, characterization of loci with smaller effects, such as QTL, require higher resolution (1-2 cM average interval) (Matise *et al.*, 1994).

The first report of detection of QTL effects in dairy cattle with DNA microsatellites was due to Ron *et al.* in 1994. In 1995, Georges *et al.* identified five chromosomes showing strong evidence for the presence of QTL affecting milk yield and composition : chromosome 1, 6, 9, 10 et 20. The QTL on chromosome 6 has been confirmed by independent studies perfomed in the Holstein-Friesian (Kühn *et al.*, 1996, Spelman *et al.*, 1996, Ashwell *et al.*, 1998) as well as other dairy cattle breeds (Gomez-Raya *et al.*, 1996; Velmala *et al.*, 1997, Grisart *et al.*, 1998). Independant evidence for effect on milk yield and composition on chromosome 9 and 10 have been reported, respectively by Villki *et al.* (1997) and Ron *et al.* (1998). Arranz *et al.* (1998) confirmed the genuine nature of the QTL, mapping to chromosome 20, affecting milk yield and composition and segregating in Holstein-Friesian elite dairy cattle. This study identifies at least two QTL alleles, one of which causes an increase in milk volume accompanied by a dilution in fat and protein constituents. The corresponding gene(s) could act by affecting the osmolarity of milk either via lactose or other milk osmolarly active components. Confirmation of this QTL warrants investment into its fine-mapping towards its positional cloning. Interestingly, the mapped interval is known to contain the genes coding for the receptors of growth hormone and prolactin. Recently, a QTL with major effect on milk yield and composition has been identified on the centromeric end of chromosome 14, independently by Coppieters *et al.* (1998) and Ron *et al.* (1998) and Ashwell *et al.*, 1998). Several other studies have been designed in an attempt to map QTL affecting production in dairy cattle. Heyen *et al.* (1998), indicated potential QTL for yield and percentage traits on chromosomes 1, 2, 3, 7 and 14. In their study, Kalm *et al.* (1998) found QTL affecting milk yield on chromosomes 2, 5, 18 and 23, QTL affecting fat yield on chromosomes 2, 5 and 16, QTL affecting protein yield on chromosome 10, 16 and 23. Finally, using multiple marker mapping of QTL for milk production traits on chromosome 1 in Canadian Holstein bulls, Nadesalingam *et al.* (1998) found a QTL affecting the milk and protein yield and suspected a possible linkage with the described Pit-1 polymorphism (Nadesalingam, personal communication).

The somatic cells score (SCS) is an other milk production trait under study at present because it accounts for the largest reduction in producer income (Weigler *et al.*, 1990). The SCS have been used by the dairy industry as a measure of intramammary health because SCS are positively correlated with mastitis and because somatic cells have an important role in mammary gland defense. The bovine major histocompatibility complex (MHC), or bovine leukocyte antigen (BoLA) class II genotyping system was first chosen as a marker for SSC by van Eijk *et al.* (1992). The reasons for this choice are that this marker is highly polymorphic and plays an important role in immunity. Several studies of association between alleles at the BoLA class II DRB 3.2 locus and SCS parameter show that certain alleles appear to be associated with susceptibility and other alleles appear to be associated with resistance to mastitis or its effects (e.g.: elevated SCS) (Aarestrup *et al.*, 1995, Ashwell *et al.*, 1996, Dietz *et al.*, 1997; Kelm *et al.*, 1997, Starkenburg *et al.*, 1997, Sharif *et al.*, 1998). As some results of these studies are conflicting, they must be confirmed.

2.2. MEAT PRODUCTION

Interest in the identification of loci influencing carcass composition (fat deposition sites, lean tissue yield, ...) and quality (muscle tenderness, ...) is linked with the importance of these characteristics for the determination of carcass value and consumer satisfaction. However, many of these characteristics cannot be measured in live animals inducing difficulties for conventional selection programmes. Integration of markers associated with these characteristics will thus greatly facilitated selection.

In the case of meat production, the GH gene was also considered as a possible candidate gene. Rocha *et al.* (1992) reported an association between a *TaqI* RFLP polymorphism and birth weight and shoulders width of calves. For Sneyers *et al.*, (1994) the *TaqI* RFLP was associated with growth performance at the 7th and 13th months of age for Belgian Blue White bulls. In a recent study, Taylor *et al.* (1998) presented an approach to evaluate the support for GH1 gene as QTL for growth and carcass composition within the context of genome-wide-map-based cloning strategies. To establish candidacy, a bacterial artificial chromosome (BAC) clone containing a putative candidate gene is physically assigned to an anchored linkage map in order to localise the gene relative to an identified QTL effect. Microsatellite loci derived from BAC clones containing an established candidate gene are integrated into linkage map facilitating the evaluation by interval analysis of the statistical support for QTL identity. In their results, the authors conclude that, with the exception of ether extractable fat and adjusted fat, there is no compelling evidence for association between breed-specific Angus and Brahman GH1 alleles and any of the growth, carcass composition and quality characters. Nevertheless, this strategy defines an efficient and cost effective approach for the positional candidate cloning of QTL in domesticated livestock species. An other important finding in the area of meat production is the positional cloning of the muscular hypertrophy (mh) gene responsible for the double-muscling nature of the Belgian Blue White (BBW) (Charlier *et al.*, 1995 and Grobet *et al.*, 1997). The doubled-muscling animals have leaner carcasses than those not doubled-muscling and exhibit greater muscle (about 20%) mass with less fat. Their hyperplasia is due to an increase in the number of muscle fibers rather than the individual diameter. In the case of BBW, the authors reported an 11-bp deletion coding sequence for the bioactive carboxy-terminal domain of the myostatin gene that inactivates the protein. Several other studies of positional cloning approach have identified a QTL for meat production traits on chromosome 5 (Davis *et al.*, 1998, Stone *et al.*, 1999 and Casas *et al.*, 2000). This QTL is located near the IGF-I gene. Moody *et al.* (1996b) found an association between IGF-I and growth in Hereford cattle, suggesting the possibility that this or a neighboring gene could be associated with growth. Other chromosomes were identified as sites of QTL : QTL for birth weight was reported on chromosome 6 (Davis *et al.*, 1998 et Casas *et al.*, 2000); QTL for fat deposition traits were detected on chromosome 14 (Ashwell *et al.*, 1998, Casas *et al.*, 2000); two QTL for marbling on chromosomes 17 and 27 (Casas *et al.*, 2000), QTL for meat tenderness was detected on the telomeric end of chromosome 29 (Casas *et al.*, 2000).

3. Genetic Markers Identified in Pigs

3.1. THE CANDIDATE GENE APPROACH

In 1995, Nielsen *et al.* suggested differences in the transcriptional activity between GH gene variants, which might eventually cause higher plasma GH concentrations and higher growth rates. After, two studies (Larsen *et al.*, 1995 and Knorr *et al.*, 1997) indicated that the GH locus plays a major role in defining the genetic differences between Meishan and Piétrain, but not between Wild Boar and Piétrain, or Wild Boar and Large White pigs.

Pit-1 was chosen as a candidate gene to investigate its association with growth and carcass traits in pigs. In a recent study, the *MspI* polymorphism was found to be associated with birth weight and carcass fat (Yu *et al.*; 1995). On the contrary, the *RsaI* polymorphism showed no significant differences for lean-to-fat ratio (Stancekova *et al.*, 1999).

The receptor of the prolactin hormone was also investigated as a candidate gene for reproduction traits in pigs and was found to be associated with increased litter size (Rotschild *et al.*, 1998).

3.2. THE POSITIONAL CLONING APPROACH

An economically important use of markers in pigs was for the prediction of the halothane (Hal) genotype. The halothane locus is the best studied region of the procine genome. This locus controls the so-called Porcine Stress Syndrome which includes susceptibility to malignant hyperthermia and poor meat quality. In 1991, Fujii *et al.* identified a C to T mutation at nucleotide position 1843 in the porcine RYR gene. To day, an accurate DNA-based test is used. A successful MAS was applied in Sweden to reduce the incidence of the Halothane gene prior to the identification of the causative mutation (Gahne *et al.*, 1985 and review in Andersson *et al.*, 1998). A second major gene of economical importance in pigs is the Rendement Napole (RN) gene that influences meat quality. This gene has been mapped using a set of markers, represented by microsatellites and blood proteins (Mariani *et al.*, 1996). Recently, the RN loci is more accurately assigned to the 15q24-q25 region by Millan *et al.* (1996) and confirmed by Törnsten *et al.* (1998). While geneticists are seeking the causal mutation, the linked DNA markers identified so far, might be already used in a Marker Assisted Selection program.

Presently, low-resolution genetic maps have been published for pigs (Ellegren *et al.*, 1994; Rohrer *et al.*, 1994; Archibald *et al.*, 1995). The average interval between markers for these maps range from 5,5cM (Rohrer *et al.*, 1994) to 13,3 cM (Barendse *et al.*, 1994). The ability to set up test mating, the generation interval and the number of progeny per litter make swine an attractive model to identify QTL of economic importance as well as those underlying several genetically based human diseases. Microsatellites anchored on low-resolution maps have identified regions that influence disease resistance (Edfors-Lilja *et al.*, 1995), growth rate and fat deposition (Andersson *et al.*, 1994), reproduction traits (Paszek *et al.*, 1998, Bidanel *et al.*, 1998), meat production (Bidanel *et al.*, 1998 and Rattink *et al.*, 1998, Miller *et al.*, 1998, Cepica *et al.*, 1998, Tepas *et al.*, 1998, Maak *et al.*, 1998. Finally, using markers flanking the IGF-1 and GH locus, Casas-Carrillo *et al.* (1997) found a potential association of a QTL, located in the interval between the markers and IGF-1, with average daily gain in one family. No association was found for the GH locus.

4. Genetic Markers Identified in Chicken

4.1. THE CANDIDATE GENE APPROACH

In chicken, the GH gene was also studied. Selection for feed conversion, egg production or growth have been shown to affect GH and GH-receptor levels (Anthony *et al.*, 1990; Vanderpooten *et al.*, 1993). The GH gene is highly polymorphic in meat-type chickens, segregating for at least five different alleles (Fotouhi *et al.*, 1993). In their results, Kuhnlein *et al.* (1997) presented an association between GH-polymorphism disease resistance and egg production. The GH allele co-selected with resistance was associated with a delayed onset of ovulation but a higher persistency of ovulation a age progressed, resulting in an overall increase of egg production by 15 %. The resistance-associated GH was dominant for the onset of ovulation and recessive for the persistency of egg production.

4.2. THE POSITIONAL CLONING APPROACH

In F1992, 100 RFLP loci were mapped in the chicken (Bumstead and Palyga, 1992). As yet, 700 chicken microsatellite markers are available (NAGRP, 1998). In their results, Hu *et al.* (1997) found a G to A substitution in the NRAMP1 (natural resistance-associated macrophage protein 1) gene that is specific to a susceptible line and not observed in any of the resistant lines. These authors estimate that toghether, NRAMP1 and Tnc explain 33 % of the early differential resistance to infection to *S. typhimurium*. On the contrary, Mariani *et al.* (1999) presented a possible association between resistance to salmonellosis trait with a linkage group on chromosome 5 using backross populations from a cross between susceptible and resistant inbred lines. To refine the location of the resistance gene, additional markers known to lie in this region were selected. To day, the gene on chromosome 5 affecting resistance to salmonellosis in chicken is still not found.

5. Conclusions

Over the last few years, a wealth of QTL has been mapped in different farm animals. They are powerful tools in breeding programmes to produce ideal genotypes. As several authors have underlined, it is essential to confirm the presence of mapped QTL before any further step towards a MAS scheme. These confirmations can be done on a greater number of animals and using different breed samples. The development of performant statistical programs for an exact evaluation of the relationships between the trait under study and gene variabilities is also necessary. Some simulation studies are currently carried out in order to estimate the response to MAS as reviewed by Haley and Visscher (1998) and Spelman and Bovenhuis (1998). Finally, biological systems are complex implicating many genes having different influences. Such interactions between loci should always be considered when a putatively interesting locus is used in MAS.

Acknowledgments

Some researches presented in this review were funded by the Federal Belgian Ministry of Small Enterprises, Traders and Agriculture- DGVI (Brussels, Belgium), Alliance-Boviteq (Sy-Hyacinthe, Québec, Canada) and Tomen corp. (Tokyo, Japan).

References

Aarestrup, F.M., Jensen, N.E. and Ostergard, H. (1995) Analysis of associations between major histocompatibility complex (BoLA) class I haplotypes and subclinical mastitis of dairy cows, *J. Dairy Sci. 78*, 1684-1692.

Andersson, L., Haley, C.S., Ellegren, H., Knott, S.A., Johanson, M., Andersson, K., Andersson-Eklund, L., Edfors-Lilja, A., Fredholm, M., Hansson, I., Hakansson, J. and Lundstrom, K. (1994) Genetic Mapping of quantitative trait loci for growth and fatness in pigs, *Science 263*, 1771-1774.

Andersson, L. (1998). Marker-assisted selection : Issues and applications, *Anim. Genet., 29 Suppl. 1*, 1-7.

Andersson-Eklund, L. and Rendel, J. (1993) Linkage between amylase-1 locus and a major gene for milk fat content in cattle, *Anim. Genet. 24*, 101-103.

Anthony, N.B., Vasilatos-Younken, R., Emerson, D.A., Nestor, K.E. and Bacon, W.L. (1990) Pattern of growth and plasma growth hormone in turkeys selected for increased egg production, *Poult. Sci. 69*, 2057-2063.

Archibald, A.L., Haley, C.S., Brown, J.F., Couperwhite, S., McQueen, H.A., Nicholson, D., Coppieters, W., Van de Weghe, A., Stratil, A., Wintero, A.K., Fredholm, M., Larsen, N.J., Nielsen, V.H., Milan, D., Woloszyn, N., Robic, A., Dalens, M., Riquet, J., Gellin, J., Caritez, J.-C., Burgaud, G., Ollivier, L., Bidanel, J.-P., Vaiman, M., Renard, C., Geldermann, H., Davoli, R., Ruyter, D., Verstege, E.J.M., Groenen, M.A.M., Davies, W., Heyheim, B., Keiserud, A., Anderson, L., Ellegren, H., Johansson, M., Marklund, L., Miller, J.R., Anderson, D.V., Dear, S., Signer, A.J., Moran, C., Le Tissier, P., Muladno, H., Rothschild, M.F., Tuggle, C.K., Vaske, D., Helm, J., Liu, H.-C., Rahman, A., Yu, T.-P., Larson, R.G. and Schmitz, C.B. (1995) The PiGMaP consortium linkage map of the pig (Sus scrofa), *Mamm. Genome 6*, 157-175.

Arranz, J., Coppieters, W., Berzi, P., Cambisano, N., Grisart, B., Karim, L., Marcq, F., Moreau, L., Mezer, C., Riquet, J., Simon, P., Vanmanshoven, D., Wagenaar, D. and Georges, M. (1998) A QTL affecting milk yield and composition maps to bovine chromosome 20: a confirmation, *Anim Genet 29*, 107-115.

Ashwell, M.S., Da, Y., Vabraden, P.M., Rexroad, C.E. and Miller, R.H. (1998) Detection of putative loci affecting conformational type traits in an elite population of United States Holstein using microsatellites, *J. Dairy Sci. 81*, 1120-1125.

Ashwell, M.S., Rexroad, C.E., Miller, R.H. and van Raden, P.M. (1996) Mapping economic trait loci for somatic cell score in Holstein cattle using microsatellite markers and selective genotyping, *Anim. Genet. 27*, 235-242.

Baldi, A. (1999) Manipulation of milk production and quality by use of somatotropin in dairy ruminants other than cow, *Dom. Anim. Endocrinol. 17*, 131-137.

Barendse, W., Armitage, S.M., Kossarck, L.M., Shalom, A., Kirkpatrick, B.W., Ryan, A.M., Clayton, D., Li, L., Neibergs, H.L., Zhang, N., Grosse, W.H., Weiss, J., Creighton, P., McCarthy, F., Ron, M., Teale, A.J., Frics, R., McGraw, R.A., Moore, S.S, Georges, M., Soller, M., Womack, J.E. and Hetzel, D.J.S. (1994) A genetic linkage map of the bovine genome, *Nature Genet. 6*, 227-235.

Bauman, D.E. (1999) Bovine somatotropin and lactation: from basic science to commercial application, *Dom. Anim. Endocrinol. 17*, 101-116.

Bidanel, J.P., Milan, D., Woloszyn, N., Bourgeois, F., Caritez, J.C., Gruand, J., Le Roy, P., Bonneau, M., Lefaucheur, L., Mourot, J., Prunier, A., Desautes, C., Mormede, P., Renard, C., Vaiman, M., Robic, A., Ollivier, L., Gellin, J. and Chevalet, C. (1998) Mapping of QTLs in F2 crosses between Meishan and Large White pig breeds, *Anim. Genet. 29 Suppl. 1*, 60-74.

Bishop, M.D., Kappes, S.M., Keele, J.W., Stone, R.T., Hawkins, G.A., Solinas Toldo, S., Fries, R., Grosz, M.D., Yoo, J. and Beattie, C.W. (1994) A genetic linkage map for cattle, *Genetics 136*, 619-639.

Bumstead, N. and Palyga, J. (1992) A prelimonary linkage map of the chicken genome, *Genomics 13*, 690-697.

Casas, E., Shackerlford, S.D., Keele, J.W., Stone, R.T., Kappes, S.M. and Koohmaraie, M. (2000) Quantitative trait loci affecting growth and carcass composition of cattle segregating alternate forms of myostatin, *J. Anim. Sci. 78*, 560-569.

Casas-Carrillo, E., Prill-Adams, A., Price, S.G., Clutter, A.C. and Kirkpatrick, B.W. (1997) Relationship of growth hormone and insulin-like growth factor I and carcass trait in swine, *Anim. Genet. 28*, 88-93.

Cepica, S., Moser, G., Schröffel, J., Müller, E. and Geldermann, H. (1998) Scan of porcine chromosome 4 for QTLs controlling carcass composition and meat quality, *Anim. Genet. 29 Suppl. 1*, 60-74.

Charlier, C., Coppieters, W., Farnir, F., Grobet, L., Leroy, P.L., Michaux, C., Mni, M., Schwers, A., Vanmanshoven, P. and Hanset, R. (1995) The *mh* gene causing double-muscling in cattle maps to bovine chromosome 2, *Mamm. Genome 6*, 788-792.

Chen, R., Ingraham, M.A., Treacy, M.N., Albert, V.R., Wilson, L. and Rosenfeld, M.G. (1990) Autoregulation of Pit-1 gene expression mediated by two cis-active promoter elements, *Nature 346*, 583-589.

Connor, E.E., Ashwell, M.S., Kappes, S.M. and Dahl, G.E. (1998) Characterization of a parial bovine cDNA encoding pituitary growth hormone-releasing hormone receptor (GHRH-R) and mapping of the GH-R to chromosome 4 using a novel PCR-RFLP, *BASE 2 Special issue*, 8.

Coppieters, W., Riquet, J., Arranz, J.J., Berzi, P., Cambisano, N., Grisart, B., Karim, L., Marcq, F., Simon, P., Vanmanshoven, P., Wagenaar, D. and Georges, M. (1998) A QTL with major effect on milk yield and composition maps to bovine chromosome 14 *Mamm. Genome 9*, 540-544.

Cowan, C., Dentine, M. and Coyle, T. (1992) Chromosome substitution effects associated with κ-caseine and β-lactoglobulin in Holstein cattle, *J. Dairy Sci. 75*, 1097-1104.

Crewshaw, E.B., Kalla, K., Simmons, D.M., Swanson, L.W. and Rosenfeld, M.G. (1989) Cell-specific expression of the prolactin gene in transgenic mice is controlled by synergistic interactions between promoter and enhancer elements, *Genes Dev. 3*, 959-964.

Davis, G.P., Hetzel, D.J.S., Corbet, N.J., Scacheri, S., Lowden, S., Renaud, J., Mayne, C., Stevenson, R., Moore, S.S. and Byrne, K. (1998) The mapping of quantitative trait loci for birth weight in a tropical beef herd, *Proc. 6th World Congr. Genet. Appl. Livest. Prod. 26*, 441-444.

Dietz, A.B., Cohen, N.D., Timms, L. and Kehrli, M.E. (1997) Bovine lymphocyte antigen class II alleles as risk factors for high somatic cell counts in milk of lactating dairy cows, *J. Dairy Sci. 80*, 406-412.

Edfors-Lilja, I., Gustafsson, Y., Duval-Iflah, Y., Ellegren, H., Johansson, M., Juneja, R.K., Marklund, L. and Andersson, L. (1995) The porcine intestinal receptor for Escherichia Coli K88ab, K88ac : Regional localization on chromosome 13 and influence of IgG response to the k88 antigen, *Anim. Gent. 26*, 237-242.

Ellegren, H., Chowdhary, B.P., Johansson, M., Marklund, L., Fredholm, M., Gustavsson, I. and Andersson, L. (1994) A primary linkage map of the porcine genome reveals a low rate of genetic recombinaition, *Genetics 137*, 1089-1100.

Eppard, P.J., Bentle, L.A., Violand, B.N.,, Ganguli, S., Hintz, R.L., Kung, L.J., Krivi, G.G. and Lanza, GM. (1992) Comparison of the galactopoetic response to pituitary-derived and recombinant-derived variants of the bovine growth hormone, *J. Endocrinol. 132*, 47-56.

Falaki, M., Gengler, N., Sneyers, M., Prandi, A., Massart, S., Formigoni, A., Burny, A., Portetelle, D. and Renaville, R. (1996) Relationships of polymorphism for growth hormone and growth hormone receptor gene with production traits for Holstein Friesian bulls, *J. Dairy Sci. 79*, 1446-1453.

Falaki, M., Prandi, A., Corradini, C., Sneyers, M., Gengler, N., Massart, S., Fazzini, U., Burny, A., Portetelle, D. and Renaville, R. (1997) Relationships of growth hormone gene and milk protein polymorphisms to milk production traits in Simmental cattle, *J. Dairy Res. 64*, 47-56.

Fotouhi, N., Karatzaq, C.N., Kuhnlein, U. and Zadworny, D. (1993) Identification of growth hormone polymorphisms which respond to divergent selection for abdominal fat content in chickens, *Theor. Appl. Genet. 85*, 931-936.

Fujii, J., Otsu, K., Zorzota, F., de Leon, S., Khanna, V.K., Weiler, J.E., O'Brien, P.J. and McLennan, D.H. (1991) Identification of a mutation in porcine ryanodin receptor associated with malignant hyperthermia, *Science 253*, 448-451.

Furu, L.M., Kazmer, G.W., Zinn, S.A. and Rycroft, H. (1988) Somatotropin MspI and AluI polymorphisms relative to indicators of the genetic merit of Holstein AI sires, *J. Anim. Sci. 76 Suppl 1*, 75.

Gahne, B. and Juneja, R.K. (1985) Prediction of the halothane (Hal) genotypes of pigs by deducing Hal, Phi, PO2, PGD haplotypes of parents and offspring : results of a large-scale practice in Swedish breeds, *Anim. Blood Groups Biochem, Genet. 16*, 265-283.

Ge, W., Davis, M.E. and Hines, H.C. (1997) Two SSCP alleles detected in the 5'-flanking region of bovine IGF1 gene, *Anim. Genet. 28*, 155-156.

Georges, M., Nielsen, D. and Mckinnon, M. (1995) Mapping quantitative trait loci controlling milk production by exploiting progeny testing, *Genetics 139*, 907-920.

Gomez-Raya, L., Vage, D.I. and Olsaker, I. (1996) Mapping QTL affecting traits of economical importance in Norwegian cattle, *Proc. 47th Ann. Meet. Europ. Assoc. Anim. Prod.*, 44.

Grisart, B., Coppieters, W., Arranz, J.-J., Riquet, J., Wagenaar, D., Berzi, P., Cambisano, N., Karim, L., Simon, P., Vanmanshoven, P. and Georges, M. (1998) QTL mapping on candidate chromosomes 6, 14 and 20 using a red and white cattle grand-daughter design, *Anim. Genet. 29 Suppl.1*, 62.

Grobet, L., Royo Martin, L.J., Poncelet, D., Pirottin, D., Browers, B., Riquet, J., Schoeberlein, A., Dunner, S., Menissier, F., Massanbanda, J., Fries, R., Hanset, R. and Georges, M. (1997) A deletion in the myostatin gene causes double-muscling in cattle, *Nat. Genet. 17*, 71-74.

Haley, C.S. and Vissher, P.M. (1998) Strategies to utilise marker-quantitative trait loci associations, *J. Dairy Sci. 8*, 85-97.

Hallerman, E.M., Nave, A., Kashy, Y., Holzer, Z., Soller, M. and Beckmann, J.S. (1987) Restriction fragment length polymorphism in dairy and beef cattle at the growth hormone and the prolactin loci, *Anim. Genet. 18*,213-222.

Haugen, B.R., Wood, W.M., Gordon, D.F. and Ridgeway, E.D. (1993) A thyrotrope-specific variant of Pit-1 transactivates the tyrotropin beta promoter, *J. Biol. Chem. 268*, 20818-20823.

Hauser, S.D., McGrath, M.F., Collier, R.J. and Krivi, G.G. (1990) Cloning and in vivo expression of the bovine growth hormone receptor mRNA, *Mol. Cell. Endocrinol. 72*,187-200.

Hecht, C. and Geldermann, H. (1996) Variants with the 5'-flanking region and the intron I of the bovine growth hormone gene, *Anim. Genet. 27*, 329-332.

Hediger, R., Johnson, S.E., Barendse, W., Drinkwater, R.D., Moore, S.S. and Hetzel, J. (1990) Assignment of the growth hormone gene locus to 19q26-qter in cattle and 11q25-qter in sheep by *in situ* hybridization, *Genomics 8*, 171-174.

Herr, W., Sturm, R.A., Clerc, R.G., Corcoran, L.M., Baltimore, D., Sharp, P.A., Ingraham, H.A., Rosenfeld, M.G., Finney, M., Ruvkun, G. and Harvitz, H.R. (1988) The POU domain: a large conserved region in the mammalian pit-1, oct-1, oct-2 and Caenorhabditis elegans unc-86 gene products, *Genes Dev. 2*, 1513-1516.

Heyen, D.W., Weller, J.I., Ron, M., Band, M., Feldmesser, E., Da, Y., Wiggans, G.R., Van Raden, P.M. and Lewin, H.A. (1998) Genome scan for QTL influencing milk production and health traits in dairy cattle, *Anim. Genet. 29 Suppl. 1*, 61.

Hines, H.C., Ge, W., Zhao, Q. and Davis, M.E. (1998) Association of genetic markers in growth hormone and insulin-like growth factor-I loci with lactation traits in Holsteins, *Anim. Genet. 29 Suppl. 1*, 63.

Hines, H.C. (1990) Genetic markers for quantitative trait loci in dairy cattle. *Proc. 4th World Cong. Genet. App. Livest. Prod. 13*, 121-124.

Hoj, S., Fredholm, M., Larsen, N.J. and Nielsen, V.H. (1993) Growth hormone gene polymorphism associated with selection for milk fat production in lines of cattle, *Anim. Genet. 24*, 91-96.

Hu, J., Bumstead, N., Barrow, P., Sebastiani, G., Olien, L., Morgan, K. and Malo, D. (1997) Resistance to salmonellosis in the chicken is linked to NRAMP1 and TNC, *Genome Res. 7*, 693-704.

Kalm, E., Reinsch, N., Xu, N., Thosen, H., Looft, C., Grupe, S., Brockmann, G., Kuehn, C., Schwerin, M., Leyhe, B., Hiendleder, S., Erhardt, G., Medjugorac, I., Russ, I., Förster, M., Brening, B., Reinhard, F., Reents, R. and Averdunk, G. (1998) Mapping quantitative trait loci on cattle chromosome 2, 5, 10, 16, 18 and 23, *Anim. Genet. 29 Suppl.1*, 62.

Knorr, C., Moser, G., Muller, E. and Geldermann, H. (1997) Associations of GH gene variants with performance traits in F2 generations of European Wild Boar, Pietrain and Meishan pigs, *Anim.. Genet. 28*, 124-128.

Kühn, C., Weikard, R., Goldammer, T., Grupe, S., Olsaker, I. and Schwerin, M. (1996) Isolation and application of chromosome 6 specific microsatellite markers for detection of QTL for milk-production traits in cattle, *J. Anim. Breed. Genet. 113*, 355-362.

Kuhnlein, U., Ni, L., Weigend, S., Gavora, J.S., Fairfull, W. and Zadworny, D. (1997) DNA polymorphisms in the chicken growth hormone gene : response to selection for disease resistance and association with egg production, *Anim. Genet. 28*, 116-123.

Lagziel, A., Lipkin, E., Ezra, E., Soller, M. and Weller, J.I. (1999) An MspI polymorphism at the bovine growth hormone (bGH) gene is linked to a locus affecting milk protein percentage, *Anim. Genet. 30*, 296-299.

Larsen, N.J., Ellegren, H., Bräuner Nielsen, P. and Andersson, L. (1995) Genetic variation at the growth hormone locus in a wild pig intercross : test of association to phenotypic traits and linkage to the blood group D locus, *Theor. Appl. Genet. 91*, 1074-1077.

Lee, B.K, Crooker, B.A., Hansen, L.B. and Chester-Jones, H. (1994) Polymorphism in the third intron of somatotropin (bST) gene and its association with selection for milk yield in Holstein cows, *J. Anim. Sci. 72*, 316.

Lee, B.K., Lin, G.F., Crooker, B.A., Murtaugh, M.P., Hansen, L.B. and Chester-Jones, H. (1993) Association of somatotropin (bST) gene polymorphism with selection for milk yield in Holstein cows, *J. Dairy Sci. 76* Suppl.1, 149.

Leung, D.W., Spencer, S.A., Cachianes, G., Hammonds, R.G., Collins, C. and Henzel, W.J. (1987) Growth hormone receptor and serum binding protein: purification, cloning and expression, *Nature 330*, 537-543.

Lin, C., Lin, S.C., Chang, C.P. and Rosenfeld, M.G. (1992) Pit-1-dependent expression of the receptor for growth hormone releasing factor mediates pituitary cell growth, *Nature 360*, 765-771.

Lira, S.A., Crewshaw, E.B., Glass, C.K., Swanson, L.W. and Rosenfeld, M.G. (1988) Identification of the rat growth hormone genomic sequences targeting pituitary expression in transgenic mice, *Proc. Natl. Acad. Sci. U.S.A. 85*, 4755-4761.

Lira, S.A., Kalla, K.A., Glass, C.K., Drolet, D.W. and Rosenfeld, M.G. (1993) Synergistic interactions between Pit-1 and other elements are required for effective somatotropic rat growth hormone gene expression in transgenic mice, *Mol. Endocrinol. 7*, 694-699.

Lucy, M.C., Hauser, S.D., Eppard, P.J., Krivi, G.G., Clark, J.H., Bauman, D.E. and Collier, R.J. (1993) Variants of somatotropin in cattle: gene frequencies in major dairy breeds and associated milk production, *Domest. Anim. Endocrinol. 10*, 325-333.

Maak, S., Petersen, K. and Von Lengerken, G. (1998) Association of polymorphisms in the porcine HSP 70.2 gene promoter with performance traits, *Anim. Genet. 29 Suppl. 1*, 60-74.

Mariani, P., Lundstrom, K., Gustafsson, U., Enfalt, A.C., Juneja, R.K. and Andersson, L. (1996) A major locus (RN) affecting muscle glycogen content is located on pig chromosome 15, Mamm. Genome 7, 52-54.

Mariani, P. (1999) QTL identification and Marker Assisted Selection: new solutions to old problems, *Proc. A.S.P.A. XIII Congress, Piacenza*, 109-122.

Matise, T.C., Perlin, M. and Chakravarti, A. (1994) Automated construction of genetic linkage maps using an expert system (MultiMap) : A human genome linkage map, *Nature Genet. 6*, 384-390.

Milan, D., Woloszyn, N., Yerle, M., Le Roy, P., Bonnet, M., Riquet, J., Lahbib-Mansais, Y., Caritez, J.C., Robic, A., Sellier, P., Elsen, P. and Gellin, J. (1996) Accurate mapping of the "acid meat" RN gene on genetic and physical maps of pigs chromosome 15, *Mamm. Genome. 7*, 47-51.

Miller, J.R., Thomsen, P.D., Dixon, S.C., Tucker, E.M., Konfortov, B.A. and Harbitz, I. (1991) Synteny mapping of the bovine IGHG2, CRC and IGFI genes *Anim. Genet 23*, 51-58.

Miller, L.M., Paszek, A.A., Schook, L.B. and Louis, C.F. (1998) Denser marker coverage provides greater evidence for carcass trait QTL on SSCP, *Anim. Genet. 29 Suppl. 1*, 60-74.

Moisio, S., Elo, K., Kantanen, J. and Vilki, J. (1998) Polymorphism within the 3' flanking region of the bovine growth hormone receptor gene, *Anim. Genet. 29,* 55-57.

Moody, D.E., Pomp, D., Barendse, W. and Womack, J.E. (1995c) Short communication: assignment of the growth hormone receptor gene to bovine chromosome 20 using linkage analysis and somatic cell mapping, *Anim. Genet. 26,*341-343.

Moody, D.E., Pomp, D. and Barendse, W. (1995a) Short communication: restriction fragment length polymorphism in amplification products of the bovine growth hormone-releasing hormone gene, *J. Anim. Sci. 73,* 3789.

Moody, D.E., Pomp, D. and Barendse, W. (1995b) Restriction fragment length polymorphism in amplification products of the bovine Pit-1 gene and assignment of Pit-1 to bovine chromosome 1, *Anim. Genet. 26,* 45-47.

Moody, D.E., Pomp, D. and Barendse, W. (1996a) Linkage mapping of the bovine insulin-like growth factor-I receptor gene, *Mamm.. Genome 7,* 168-169.

Moody, D.E., Pomp, D., Newman, S. and McNeil, M.D. (1996b) Characterization of DNA polymorphisms in three populations of Hereford cattle and their associations with growth and maternal EPD in line 1 Herefords, *J. Anim. Sci. 74*, 1784-1793.

Nadesalingam, J., Plante, Y., Gibson, J.P., Atchinson, J.L. and Wu, X. (1998) Multiple marker mapping of quantitative trait loci (QTL) for milk production traits on Chromosome 1 in Canadian Holstein bulls, *J. Dairy Sci. 81 suppl. 1,* 72.

NAGRP. (1998) Annual Report, *The National Animal Genome Research Program,* USA.

Neimann-Sorensen, A. and Robertson, A. (1961) The association between blood groups and several production characters in three Danish cattle breeds *Acta Agra. Scand. 11*, 163-196.

Nielsen, V.H., Larsen, N.J. and Agergaard, N. (1995) Associaiton of DNA-polymorphism in the growth-hormone gene with basal-plasma growth-hormone concentration and production traits in pigs, *J. Anim. Breed. Genet. 112*, 205-212.

Paszek, A.A., Dyas, P.A., Finn, S.T., Chen, M., Buttram, S.T. and Didion, B.A. (1998) Allelic effects of microsatellites from putative QTL for reproductive traits, *Anim. Genet. 29 Suppl. 1*, 60-74.

Paterson, A.H. (1995) Molecular dissection of quantitative traits: progress and prospects, *Genome Res. 5,* 321-333.

Pfäffle, R.W., Dimattia, G.E., Parks, J.S., Brown, M.R., Wit, J.M., Jansen, M., Van der Nat, H., Van den Brande, J.L., Rosenfeld, M.G. and Ingraham, H.A. (1992) Mutation of the POU-specific domain of Pit-1 and hypopituitarism without pituitary hypoplasia, *Science 257,* 1118-1121.

Radovick, S., Nations, M., Du, Y., Berg, L.A., Weintraub, D.B. and Wondisford, F.E. (1992) A mutation in the POU-homeodomain of Pit-1 responsible for combined pituitary hormone deficiency, *Science 257,* 1115-1118.

Rattink, A.P., De Koning, D.J., Janss, L.L.G., Vanoers, P.A.M., De Vries, B.J., Groenen, M.A.M., Van Der Poel, J.J., De Groot, P., Brascamp, E.W. and Van Arendonck, J.A.M. (1998) Detection of QTL controlling intramuscular fat and backfat in pigs, *Anim. Genet. 29 Suppl. 1*, 60-74.

Renaville, R., Gengler, N., Vrech, E., Prandi, A., Massart, S., Corradini, C., Bertozzi, C., Mortiaux, F., Burny, A. and Portetelle, D. (1997) Pit-1 gene polymorphism, milk yield and conformation traits for Italian Holstein Friesian bulls, *J. Dairy Sci. 80*, 3431-3438.

Rhodes, S.J., Chen, R., DiMattia, G.E., Scully, K.M., Kalla, K.A., Lin, S.C., Yu, V.C. and Rosenfeld, M.G. (1993) A tissue-specific enhancer confers Pit-1-dependent morphogen inductibility and autoregulation on the Pit-1 gene, *Genes Dev. 7*, 913-919.

Rocha, J.L., Baker, J.F., Womack, J.E., Sanders, J.O. and Taylor J.F. (1992) Statistical associations between restriction fragment length polymorphisms and quantitative traits in beef cattle, *J. Anim. Sci. 70*, 3360-3370.

Rodrigues, C.V., Guimaraes, S.E., Neto, E.D. and Pinheiro, L.E. (1998) Identification of a novel polymorphism in the promoter region of the bovine growth hormone gene, *Anim. Genet. 29*, 63-72.

Rohrer, G., Alexander, L., Keele, J., Smith, T. and Beattie, C. (1994) A microsatellite linkage map of the porcine genome, *Genetics 136*, 231-245.

Ron, M., Band, M., Yanai, A. and Weller, J.I. (1994) Mapping quantitative trait loci with DNA microsatellites in a commercial dairy cattle population, *Anim. Genet. 25*, 259-264.

Ron, M., Heyen, D.W. and Weller, J.I. (1998) Detection and analysis of a locus affecting milk concentration in the US and Israeli dairy cattle populations. *Proc. 6th World Cong. Genet. Appl. Livest. Prod.*, 11-16.

Rothschild, M.F., Vincent, A.L., Tuggle, C.K., Evans, G., Short, T.M., Southwood, O.I., Wales, R. and Platsow, G.S. (1998) A mutation in the prolactin receptor gene is associated with increased litter size in pigs, *Anim. Genet. 29 Suppl. 1,* 60-74.

Schlee, P., Graml, R., Rottmann, O. and Pirchner, F. (1994a) Influence of growth-hormone genotypes on breeding values of Simmental bulls, *J. Anim. Breed. Genet. 111*, 253-256.

Schlee, P., Graml, R., Schallenberger, E., Schams, D. and Rottmann, D. (1994b) Growth hormone and insulin-like growth factor I concentrations in bulls of various growth hormone genotypes, *Theor. Appl. Genet. 88*, 497-500.

Sneyers, M., Renaville, R., Falaki, M., Massart, S., Devolder, A., Boonen, F., Marchand, E., Prandi, A., Burny, A. and Portetelle, D. (1994) TaqI restriction fragment length polymorphisms for growth hormone in bovine breeds and their association with quantitative traits, *Growth Regul. 4*, 108-112.

Soller, M. and Genizi, A. (1978) The efficiency of experimental designs for the detection of linkage between a marker locus and a locus affecting a quantitative trait in segregating populations, *Biometrics 34*, 47-55.

Spelman, R., Coppieters, W., Karim, L., van Arendonk, J.A.M. and Bovenhuis, H. (1996) Quantitative trait loci analysis for five milk production traits on chromosome six in the Dutch Holstein-Friesian population, *Genetics 144*, 1799-1808.

Spelman, R.J. and Bovenhuis, H. (1998) Moving from QTL experimental results to the utilization of QTL in breeding programmes, *Anim. Genet. 29*, 77-84.

Stancekova, K., Vasicek, D., Peskovicova, D., Bulla, J. and Kubeck, A. (1999) Effect of genetic variability of the porcine pituitary-specific transription factor (Pit-1) on carcass traits in pigs, *Anim. Genet. 30*, 313-315.

Starkenburg, R.J., Hansen, L.B., Kehrli, M.E. and Chester-Jones, H. (1997) Frequencies and effects of alternative DRB 3.2 alleles of BoLA for Holstein in milk selection and control lines, *J. Dairy Sci. 80*, 3411-3419.

Steinfelder, H.J., Hausser, P., Nakayama, Y., Radovick, S., McClaskey, J.H., Taylor, T., Weintraub, B.D. and Wondisford, F.E. (1991) Thyrotropin-releasing hormone regulation of human TSHβ expression: role of a pituitary-specific transcription factor (Pit-1/GHF-1) and potential interaction with a thyroid hormone-inhibitory element, *Proc. Natl. Acad. Sci. USA 88*, 3130-3134.

Stone, R., Keele, J.W., Shackerlford, S.D., Kappes, S.M. and Koohmaraie, M. (1999) A primary screen of the bovine genome for quantitative trait loci affecting carcass growth traits, *J. Anim.. Sci. 77*, 1379-1384.

Taylor, J.F., Coutinho, L.L., Herring, K.L., Gallagher Jr, D.S., Brenneman, R.A., Burney, N., Sanders, J.O., Turner, J.W., Smith, S.B., Miller, R.K., Savel, J.W. and Davis, S.K. (1998) Candidate gene analysis of GH1 for effects on growth and carcass composition of cattle, *Anim. Genet. 29*, 194-201.

te Pas, M.F.W., Harders, F.H., Soumillion, A., Verburg, F.J. and Meuwissen, T.H.E. (1998) The Myo D gene family influences birth weight, postnatal growth rate and carcass meat content of pigs, *Anim. Genet. 29 Suppl. 1*, 60-74.

Tornsten, A., Leon, J.T., Alexander, L.J., Andersson, L. and Chowdhary, B.P. (1998) Physical ordering of six YACs from the RN region in pigs, *Anim. Genet. 29*, 319-321.

Tuggle, C.K. and Trenkle, A. (1996) Control of growth hormone synthesis. *Dom. Anim. Endocrinol. 13*, 1-33.

Van der Werf, J.H.J., Verburg, F.J., Harders, F.L., Garssen, G.J. and te Pas, M.F.W. (1998) Influence of genetic variants of bovine growth hormone on milk productivity, *BASE 2 Special issue*, 3.

Van Eijk, M.J.T., Stewart-Haynes, J.A. and Lewin, H.A. (1992) Extensive polymoprhism of the BoLA-DRB 3 gene distinguished by PCR-based DRB3 typing, *Anim. Genet. 23*, 483-496.

Vanderpooten, A., Janssen, W. and Buyse, J. (1993) Study of the hepatic growth hormone receptor at different ages in chickens selected for a good feed conversion and a fast weight gain, *Domest. Animal. Endocrinol. 10*, 199-206.

Velmala, R., Vilkki, J., Elo, K., de Koning, D.J. and Maki-Tanila, A. (1997) QTL for milk production traits on chromosome 6 in Finnish Ayrshire cattle, *48th Ann. Meet. Europ. Assoc. Anim. Prod.* 2.

Vilkki, J., de Koning, D.J., Elo, K., Velmala, R. and Maki-Tanila, A. (1997) Multiple marker mapping of quantitative trait loci of Finnish dairy cattle by regression, *J. Dairy Sci. 73*, 2525-2537.

Wang, Y.D. and Wood, W.I. (1995) Amino acids of the human growth hormone receptor that are required for proliferation and Jak-Stat signaling, *Mol. Endocrinol. 9*, 303-311.

Weber, J.L. and May, P.E. (1989) Abundant class of human DNA polymorphisms which can be typed using the polymerase chain reaction, *Am. J. Hum. Genet. 44*, 388-396.

Weigler, B.J., Hird, D.W., Sischo, W.M., Holmes, J.C., Danaye-Elmi, C., Palmer, C.W. and Utterback, W.W. (1990) Veterinary and nonveterinary costs of disease in 29 California dairies participating in the National Animal Health Monitoring System from 1988 to 1989, *JAVMA 196*, 1945.

Weller, G.R., Kashi, Y. and Soller, M. (1990) Power of 'daughter and 'Granddaughter' designs for cattle using genetic markers, *J. Dairy Sci. 73*, 2525-2537.

Wood, W.M., Dowding, J.M., Haugen, B.R., Bright, T.M. and Ridgway, E.C. (1994) Structural and functional characterisation of the genomic locus encoding the murine beta-2 thyroid hormone receptor, *Mol. Endocrinol. 8*, 1605-1614.

Woollard, J., Schmitz, C.B., Freeman, A.E. and Tuggle, C.K. (1994) Rapid communication: Hinfl polymorphism at the bovine Pit1 locus, *J. Anim. Sci. 72*, 3267.

Yao, J., Aggrey, S.E., Zadworny, D., Hayes, J.F. and Kühnlein, U. (1996) Sequence variations in the bovine growth hormone gene characterized by single-strand conformational polymorphism (SSCP) analysis and their association with milk production traits in Holsteins, *Genetics 144*, 1908-1816.

Yu, T.P., Tuggle, C.K., Schmitz, C.B. and Rothschild, M.F. (1995) Association of Pit-1 polymorphism with growth and carcass traits in pigs, *J. Anim. Sci. 73*, 1282-1288.

Zhang, H.M., Brown, D.R., DeNise, S.K. and Ax, R.L. (1993) Rapid communication: polymerase chain reaction-restriction fragment length polymorphism analysis of the bovine somatotropin gene, *J. Anim. Sci. 71*, 2276.

RECOMBINANT GROWTH HORMONE: POTENTIAL INTEREST AND RISKS OF ITS USE IN BOVINE MILK PRODUCTION[a]

CHILLIARD Y.[1], LERONDELLE C.[2], DISENHAUS C.[3], MOUCHET C.[4] AND PARIS A.[5]

[1]INRA Unité de Recherches sur les Herbivores, Theix, 63122 St-Genès-Champanelle, France; [2]INRA Laboratoire Lentivirus chez les petits ruminants, ENVL, 1 avenue Bourgelat, BP 83, 69280 Marcy l'Etoile, France; [3]ENSA Sciences Animales, 65 rue de St-Brieuc, 35042 Rennes Cedex, France; [4]ENSA Systèmes de Production et Développement Rural,65 rue de St-Brieuc, 35042 Rennes Cedex, France; [5]INRA Laboratoire des Xénobiotiques, 180 chemin de Tournefeuille, St Martin du Touch, 31931 Toulouse Cedex, France.

Abstract

The administration of recombinant bovine growth hormone (rbGH) increases milk yield by 2-6 kg/d, according to cow age, lactation stage and nutritional status, injected dose of GH and prolonged-release formulation of the hormone. Effects on milk composition and body reserves depend on the duration of the trials and on the kind of diets fed. GH increases the activity and/or longevity of mammary secretory cells, probably *via* IGF-I produced by the liver and/or the mammary gland. Simultaneously, GH directs adipose tissue and muscle metabolism towards increased fatty acid mobilisation and oxidation, and towards glucose sparing. The administration of rbGH has few direct effects on the reproductive function, but indirectly tends to delay it when it begins before fertilisation, due to the transient decrease in the energy balance of the cows. Milk processability does not seem to be modified by the administration of rbGH, nor does the increase in milk IGF-I secretion appear to pose a risk to consumers. GH may stimulate immune responses in animals, thereby increasing the milk cell count. On the other hand, the trend towards an increase in the occurrence of mastitis seems to result indirectly from the increased milk yield. High concentrations of GH can stimulate viral yield in some *in vitro* models in monogastric species, but there is currently very few data for dairy ruminants. Furthermore, the elimination rate of xenobiotics (antibiotics, etc...) by the liver could be reduced. Should rbGH be used commercially, this would decrease the efficiency (or increase the cost) of genetic selection, and would be of limited economic

---

[a] translated in part from INRA Prod. Anim. 11, 15-32 (1998)

R. Renaville and A. Burny (eds.), Biotechnology in Animal Husbandry, 65–97.

value for farmers working within a milk quota system. However, its use in other countries without quotas could decrease milk production costs and increase competition in international dairy product market, despite the risks of decreasing the image of these products for consumers.

1. Introduction

Growth hormone can be produced industrially by genetically modified bacteria. The use of GH to increase milk production has been the focus of a large number of studies which made it possible to define its effects and its mode of action. However, GH intervenes in the regulation of numerous physiological factors and its effect on certain mechanisms, especially the immune system, needs to be specified. In addition, the increase in the volume of milk produced resulting from the use of recombinant GH in animal husbandry would not be without consequence on the organisation of milk production on the European scale. This article gives a summary of the animal production results that one can expect to obtain from administering recombinant GH, presents the mechanisms of action of this hormone, and analyses the potential socio-economical consequences should it be used.

For over 70 years, scientists have been aware of the stimulating effects of injecting hypophyseal extracts on tissue growth (Ewans and Long, 1921) and on milk secretion (Stricker and Grueter, 1928). These effects are mainly due to the action of growth hormone (GH), also referred to as somatotropin (ST). The GH naturally produced by the hypophysis has a sequence of 190 or 191 amino acids with four major variants which vary in proportion according to the dairy cow breed, and genetic polymorphism which seems to be related to the production level and/or composition of the milk. Recombinant bovine GH (rbGH or rbST), produced by protein biotechnology, contains between 191 and 199 amino acids, according to the production processes of the different manufacturers (Bauman and Vernon, 1993; Falaki *et al.,* 1996a, 1996b, 1997).

Research on the galactopoietic effects of GH in dairy cows was greatly intensified during the 80's because of the semi-industrial production of rbGH, and then the development of prolonged action forms (prolonged release rbGH) which made it possible for a very large number of long term studies on a great number of cows in experimental and commercial farms to be initiated (Bauman *et al.*, 1985, 1989; Chilliard, 1988b). In addition to these effects on growth and lactation, GH often intervenes directly or indirectly in the physiological regulation of the majority of the organism's tissues, including resistance to diseases and aging, which explains the significant progress which has been made by the studies also being carried out on human GH.

This article focuses on the studies concerning dairy cows. Indeed, these studies illustrate well the advances being made and the theoretical and practical problems concerning an important agro-food sector whose stakes have increased since the commercialisation of rbGH was authorised in the USA in 1993, whereas it is forbidden in the European community. The effects of GH on the performances of dairy cows are presented, followed by the mechanisms of action and, finally, the consequences on the reproduction and health of cows, the quality of milk, dairy herd management and, more generally, the economy of the bovine dairy sector.

2. Effects on Dairy Cow Performance

2.1 SHORT TERM EFFECTS

The daily administration of rbGH over a period of less than four weeks and beginning after the lactation peak causes an increase in milk production of about 4 kg/day, without increasing the food intake level, which in turn greatly reduces the energy balance of the animals (Chilliard, 1998a). The effects on milk composition depend mostly on the nutritional balance of the treated cows (Chilliard *et al.*, 1989): the fat content increases and the protein content decreases when the balance becomes negative, whereas these contents vary little when the balance remains positive (despite its decrease). These effects, which vary depending on the energy balance, are similar to those observed in cows receiving no GH.

2.2. LONG TERM EFFECTS

The trends observed since the end of the 80's in the USA and in Europe (Chilliard, 1988b) and in France (Chilliard *et al.*, 1989) have been confirmed by more recent studies (Bruneau *et al.*, 1991; McGuffey *et al.*, 1991; Thomas *et al.*, 1991; Figure 1).

1.2.1. Milk Production and Composition

With daily injections of increasing doses of rbGH, the response varies from +2.8 to +5.9 kg milk/day (Figure 1), but varies greatly between trials (from 0 to +11 kg/day). Injecting prolonged-release rbGH every 14 or 28 days gives lower results.

However, the injection frequency (14 versus 28 days) modifies the result only slightly if the total injected dose is the same (Figure 1). The variability of responses between herds and trials (ranging between +1 and +8 kg/day) is here again very high. The effect of prolonged-release rbGH on the lactation curve is illustrated in Figure 2, where the cows received the same ration based on preserved fodder during the entire lactation period. The very rapid response of milk production (+4.5 kg/day) decreases over the course of the treatment period to reach +2.5 kg/day during the last weeks. Overall, annual milk production increases by 10 to 30%. The effects of rbGH on milk production were confirmed when cows were treated over several successive lactations (up to eight lactations).

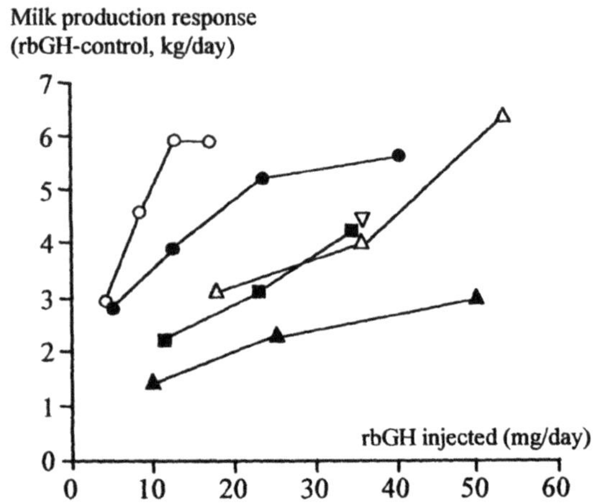

Figure 1. *Long term effects of the dosage and nature of the recombinant growth hormone (rbGH) on the milk production of cows. [● = daily injections of rbGH, 969 cows (Chilliard, 1988b); O = daily injections of Somavubove, 598 cows (Stanisiewski* et al., *1994); ∇ = injection every 14 days of Sometribove, 881 cows (Chilliard, 1988b); = injection every 14 days of Sometribove, 881 cows (Hartnell* et al., *1991); ▲ = injections every 14 days of rbGh-retard, 229 days (Downer* et al., *1993); ■ = injections every 28 days of Somidobove, 468 cows (Chilliard, 1988b)]*

The composition of milk does not vary on average over long periods. However, besides the variations observed at the beginning of the treatment, more or less significant fluctuations can be observed during the time interval separating two injections of prolonged-release rbGH. In fact, in the case of monthly injections, cyclic responses in milk production and composition, generally with maximum milk production and fat content values occurring between days 5 and 10 after each injection, whereas the protein contents reach a minimum on the following day which lasts for approximately one week, then increases during the course of the following week (Figure 3). The increase in fat content follows that in milk production. It is accompanied by an enrichment of the milk fat in long-chain fatty acids derived from the mobilisation of body lipids. The decrease in the protein content which precedes the increase in milk production is difficult to explain by a dilution or undernourishment effect. It could be related to the nitrogen sparing effect of GH on body proteins and the oxidation of amino acids, which would be compatible with the increase in protein content at the end of the cycle, when milk production decreases rapidly.

The same cyclic phenomena exist but at smaller amplitudes in the case of half-dose injections every 14 days, that is for a comparable total monthly dose (Vérité *et al.*, 1989, Rémond *et al.*, 1991; Figure 2).

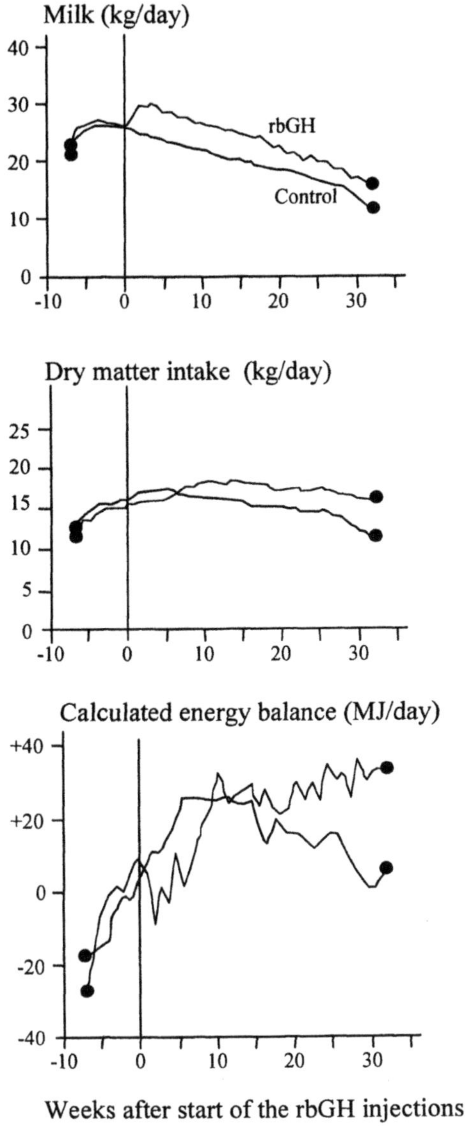

Figure 2. *Effects of rbGH-retard (500 mg every 14 days) on the production, consumption, and energy balance of cows receiving a complete ration* ad libitum. *According to Phipps (1988).*

1.2.2. Feed Intake and Energy Balance of the Animals

Feed intake increases significantly after a period of 6 to 8 weeks (see Figure 2). The increase is of about +1.5 kg DM/day in trials which last 32 weeks, and of about only +0.8 kg DM/day in trials which last only 18 weeks, as the first 6-8 weeks have more weight in the average response.

As a result of these respective variations in milk production and dry matter intake, treated cows initially have a lower energy balance than the control cows, as if a minilactation was beginning, superimposing these effects onto those of the lactation in progress (see Figure 2). However, the balance can become higher again when the cows have access to a highly ingestible ration of high nutritional value (good forage and high proportion of concentrate). In 8 trials on 315 treated cows receiving this type of diet, the energy balance in relation to the control group was 8.5 MJ net energy for lactation (NEl)/day lower during the first 12 weeks of treatment, then 5.0 MJ NEl/day higher during the last 22 weeks (Chilliard 1988b). This theoretically corresponds to a decrease in the deposits of body lipids of 19 kg in 12 weeks, with recovery during the following 22 weeks. This calculation is confirmed, on average for all the available results, by variations in body composition (Figure 4), live weight, body condition score (Bauman *et al.*, 1989; Pell *et al.*, 1992) and blood metabolites. However, it should be pointed out that when the ration has a lower energy content, or when the treatment lasts less than 32 weeks, the body condition of the treated cows at the end of the experiment is not as good (Chilliard *et al.*, 1991) and they should subsequently receive an energy supplement to achieve a comparable degree of fat cover for the following calving.

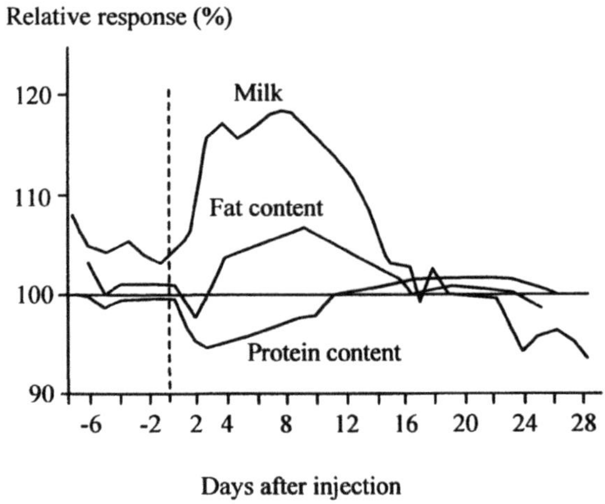

Figure 3. *Daily evolution of the milk production and composition after monthly injections of rbGH-retard (according to Vérité* et al., *1989). The 100 value represents the average of the values observed during days: -7 to -1 and +22 to +28 in the course of 5 injection periods.*

1.2.3. Efficiency of Ration Utilisation

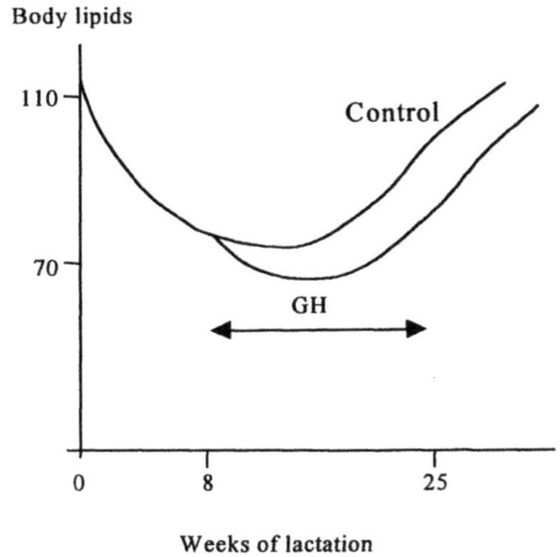

Figure 4. *Variations of body lipids in the dairy cow during lactation and during a course of treatment by growth hormone (GH). According to Chilliard (1999).*

The studies carried out in respiratory chambers and the animal production trials show that GH modifies neither the maintenance requirements nor the efficiency of the digestive and metabolic utilisation of energy ingested when the live weight and body condition of the cows are kept constant (Kirchgessner *et al.*, 1989). Therefore the total energy requirements increase normally with the extra milk yield, and it can be predicted using the usual INRA standards.

The "milk yield/net energy intake" ratio increases as a function of the decrease in the maintenance requirements proportion of the total requirements, *i.e.* +2.6% or +4.4% for responses in milk of +2 or +4 kg/day, respectively. When the apparent efficiency increases to a greater extent, it is due to a decrease in the body condition of the treated cows as compared to the controls which should be compensated for, to the cost of lower efficiency during recovery after the GH treatment.

1.2.4. Responses According to Ration Type

Some trials have aimed to analyse the interaction between GH treatment and the type of ration provided. In some of these trials the highest responses to GH injection were observed with the rations which were the richest in energy and/or protein concentrates but, in most of the other trials, the response was independent of ration type, perhaps due to existence of other limiting factors than those potentially raised by the nutritional factors studied in these trials (Chilliard *et al.*, 1989).

Some indirect comparisons made in a great number of trials (reviews by Chilliard, 1990 and Bruneau *et al.,* 1991) also suggest that the responses are higher (about +5 kg/day) with complete rations based on corn silage and rich in concentrates, than with rations based on grass silage or pasture grass and/or with a separate distribution of concentrates in limited quantities (+2.5 to +3.5 kg milk/day). This confirms that the quantity and quality of nutrients available for satisfying the increased demands of the udder are partly responsible for the milk yield response, particularly in long-term trials.

1.2.5. Responses According to Lactation Number and Stage, Breed, and Milk Potential

The response of primiparous cows (in kg milk per day) has been comparable to that of multiparous cows in several trials, but lower in a number of others. Some systematic comparisons made in a great number of trials (Table 1) show 10 to 25% lower responses in the primiparous cows. Greater differences (response 45% lower in primiparous cows) have been reported in 9 trials with daily injections of rbGH (Stanisiewski *et al.*, 1994). Primiparous cows, when responding well in terms of milk yield, are particularly sensitive to the negative effects of GH on their body condition (Rémond *et al.*, 1991), probably because they simultaneously have a low intake capacity, high maintenance of milk production and a priority for muscular protein deposition, which is energy consuming (Vérité and Chilliard, 1992).

In addition, the responses are higher when the treatment takes place from the third to the sixth lactation month (Table 1). This is probably due to the fragility of the nutritional balance of the cows before the third month, and to the reduction in the secretion potential (number and activity of mammary cells, pregnancy hormones) after the sixth lactation month.

Table 1. *Effects of the lactation number and of the lactation stage on the response (kg of milk/day) to an administration of rbGH-retard (500 mg every 14 days during 12 weeks). (1) Bruneau,* et al. *(1991) - 420 cows treated in 24 French farms; (2) Thomas* et al. *(1991) - 445 cows treated, in 15 American farms.*

Lactation stage at the start of the treatment	Primiparous		Multiparous	
	(1)	(2)	(1)	(2)
60-100 d	2.1	3.2	2.7	4.0
100-140 d	2.5	5.2	4.1	6.6
140-185 d	3.9	4.2	4.8	6.0
185-225 d	3.8	-	4.3	-

The different dairy breeds studied (Holstein, Montbeliarde, Normande, Jersey) give comparable milk yield responses. The response does not seem to vary as a function of herd potential or of cows taken individually (Disenhaus, 1996), although certain studies suggest that high producers may respond less to GH (Michel *et al.,* 1990), perhaps because the secretion of the endogenous hormone would be less limiting in these animals.

2. Mechanisms of Action (Figure 5)

2.1 MAMMARY GLAND, LIVER, AND IGF-I

The liver is a major GH target tissue in ruminants as well as in other species. High affinity receptors have been demonstrated in bovine (Hung and Moore, 1984), ovine (Gluckman *et al.*, 1983), and caprine (Disenhaus *et al.*, 1992) liver. In endocrine terms, GH-receptor binding induces an increase in the concentration of both circulating IGF-I and its type 3 binding-protein (BP3), which is due to greater hepatic synthesis and a decrease in the type 2 binding-protein (BP2) (Sharma *et al.*, 1994), which increases the bioavailability of IGF-I for peripheral tissues, particularly for the mammary gland. In growing ruminants the number of high affinity receptors for GH is increased by GH, glucose and probably insulin, and is reduced by undernutrition (Gluckman and Breier, 1989). This GH induced increase has not been observed in the lactating cow or goat (Hard *et al.*, 1992; Jammes *et al., 1996*). However, the occupancy rate of hepatic receptors increases consistently with the plasma IGF-I response. The effect of undernutrition on hepatic sensitivity to GH action during lactation has not been studied.

The reality of direct GH action on the mammary gland remains controversial. The sensitivity of the udder to GH is probably low because the standard methods used have not made it possible to identify GH receptors (Keys and Djiane, 1998), although the detection of messenger RNAs encoding for the receptors in the cow udder (Glimm *et al.*, 1990) and ewe udder (Jammes *et al.*, 1991) shows that their synthesis can occur there. However, the local infusion of GH does not increase dairy cow yield (McDowell *et al.*, 1987b) and, *in vitro*, the metabolism of bovine mammary explants in culture is only very slightly affected by GH.

Many *in vivo* trials have made it possible to describe the effects of GH on the udder, but without making it possible to specify whether these actions are direct. Thus, the administration of GH rapidly increases mammary blood flow in the ruminant in lactation (Davis *et al.*, 1988) as well as during the dry period (Rulquin and Vérité, 1993), by modifying the gas exchanges in the vascular cells (Nielsen *et al.*, 1995). In addition, this treatment slows down the decrease in the number of mammary epithelial cells during the course of lactation (Knight *et al.*, 1990) and mammary involution (Politis *et al.*, 1990). Lastly, in rodents, GH has a direct action *in vitro* on mammogenesis and on lobuloalveolar development (Feldman *et al.*, 1993). On the other hand, in goats, three-weeks treatment with rbGH only increased lipogenic enzyme activity in the case of three milkings per day, and this effect disappeared during a 22-week treatment (Knight *et al.*, 1990). In cows, a two-month treatment with rbGH did not change the mRNA abundance of lipoprotein lipase and stearoyl-CoA desaturase in mammary gland (Beswick and Kennelly, 2000). This suggests that the long-term galactopoietic effect is exerted mainly *via* longevity, rather than *via* the metabolic activity of secretory cells. Moreover, this activity is not modified by four-weeks treatment with rbGH in cows (Knight *et al.*, 1992). However, a metabolic potential value measured *in vitro* may only reflect part of the actual activity *in vivo*.

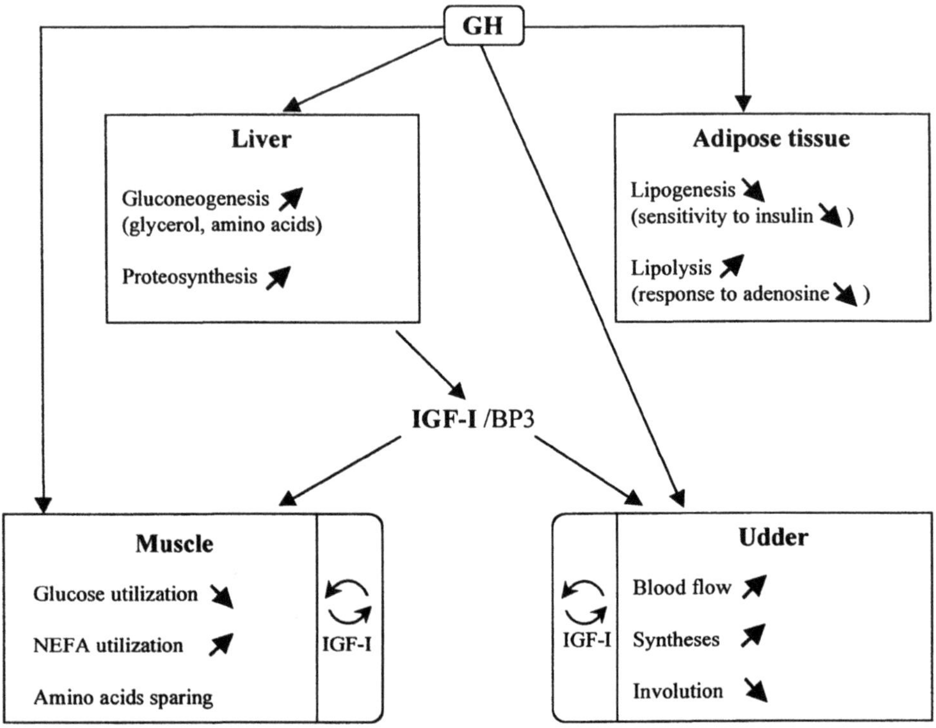

Figure 5. *Mechanisms of short term action of growth hormone (GH) in ruminants during lactation. According to Chilliard (1998a, 1999), Bauman and Vernon (1999) and Burton et al. (1994). (NEFA = nonesterified fatty acids; BP-3 binding protein of type 3; IGF-I = Insulion-like Growth Factor-I)*

In fact, the only decisive argument in favour of direct GH action on the mammary tissue of lactating ruminants is the increase in IGF-I synthesis by this tissue, obtained *in vitro* (Glimm *et al.*, 1992), a result confirmed by the increase *in vivo* of IGF-I concentrations in the udder and milk of treated goats and cows (Prosser *et al.*, 1991; Sharma *et al.*, 1994). Thus IGF-I, synthesised in the liver and/or in the mammary gland, is considered to be the main mediator of the galactopoietic action of GH. In small ruminants, IGF-I infusion at the mammary level makes it possible to increase milk production (Prosser *et al.*, 1990). However, this additional yield is always lower than that observed during GH injection, probably due to the absence of associated BP3 synthesis. *In vitro*, IGF-I induces an increase in DNA synthesis and stimulates the proliferation of bovine mammary cells (Shamay *et al.*, 1988). Furthermore, IGF-I stimulates *in vitro* glucose and α-lactalbumin transport, casein synthesis and lactose production in bovine mammary acini (Shamay *et al.*, 1988; Baumrucker and Stemberger, 1989), but this has not been confirmed by studies using tissue cocultures (Keys *et al.*, 1997).

If the galactopoietic effect of IGF-I is known today, the location of active IGF-I synthesis in the mammary gland has only been partly identified. In fact, although specific IGF-I receptors have been characterised in the mammary tissue of the ewe (Disenhaus *et al.*, 1988), cow (Collier *et al.*, 1993), and goat (Disenhaus *et al.*, 1992), GH treatment modifies neither their number nor their affinity. Moreover, a trial carried out by Prosser *et al.* (1991) made it possible to verify the lack of correlation between the increase in plasma IGF-I content and the galactopoietic response. On the other hand, the latter has been proved to be well correlated with the increase in milk IGF-I content.

The respective modes of action of GH, and of IGF-I produced either by the liver or by the mammary gland, are beginning to be elucidated in rodents. In mice transgenic for IGF-I, the latter induces hypertrophy of the canals and delays the involution of the mammary glands (Hadsell *et al.*, 1996). Thus, the increase in the mammary production of this growth factor by GH could contribute to the local establishment of a hormonal balance favourable to mammary gland development and to the stimulation of milk production during treatment with GH. Provided that they are confirmed in lactating ruminants, these different results illustrate the double mammogenic and galactopoietic role of GH.

2.2 NUTRIENT PARTITIONING BETWEEN DIFFERENT TISSUES

Until now the metabolic adaptation of extramammary tissues has been studied almost exclusively during short-term trials, therefore in situations where the animal has to satisfy greater mammary requirements, without modifying its intake level. Whereas the administration of GH does not seem to change the ration utilisation efficiency, the partitioning of nutrients between different tissues is, to the contrary, profoundly modified.

The glucose flow toward the udder increases, primarily as a consequence of a large decrease in extramammary oxidation, especially in the muscle (Zhao *et al.*, 1996), which can explain up to 30% of the increase in lactose secretion. Stimulation of the intestinal absorption of glucose has also been observed (Bird *et al.*, 1996). In addition, hepatic gluconeogenesis increases, probably due to endogenous precursors such as glycerol, arising from lipomobilisation (27% of supplementary requirements), lactate or amino acids. Muscle proteins may be mobilised but to a limited extent because GH administration stimulates protein anabolism in growing animals, so the protein content of milk decreases when the treated cows are undernourished. Moreover, GH administration reduces amino acid oxidation and the excretion of urinary nitrogen, which explains an increase in protein deposition in cows treated towards the end of lactation, *i.e.* when mammary requirements are low (Chilliard *et al.*, 1991; Binelli *et al.*, 1995).

The effects on muscle and bone metabolism are probably partly due to IGF-I, the synthesis of which is increased by GH, particularly in the liver but also in peripheral tissues (muscles, bones, etc...). However, circulating IGF-I contents are reduced at the beginning of lactation and in undernourished animals, even though the GH contents may be high (Ronge *et al.*, 1988; Cisse *et al.*, 1991). The response of circulating IGF-I to GH injections is greater at the end of lactation or after drying off than at the beginning of lactation when the cows have a very negative energy balance (McDowell *et al.*, 1987a; Ronge and Blum 1989; Vicini *et al.,* 1991), and it is modulated by feed intake (McGuire *et al.*, 1992) and insulinemia (Leonard and Block, 1997). This lack of response (resistance) of IGF-I to GH explains why anabolic effects of GH (due to IGF-I) are not observed in undernourished animals. This improves our understanding of the interactions between GH treatment, lactation stage, and the nutritional state of cows. In the case of undernutrition, the survival of the animal is therefore a priority over the lactation function.

GH administration *in vivo* reduces lipogenic activity and glucose oxidation in adipose tissue in pigs, sheep, and dairy cows. This was confirmed *in vitro*, where physiological concentrations of GH reduced the sensitivity of lipogenesis to insulin (Vernon and Flint 1989; Donkin *et al.*, 1996). Furthermore, GH administration to cows sharply decreased the mRNA abundance of lipoprotein lipase and stearoyl-CoA desaturase in adipose tissue (Beswick and Kennelly, 2000). Generally, GH has no direct lipolytic effect on adipose tissue, and GH administration *in vivo* does not modify the adipose tissue response of cow to the lipolytic effect of catecholamines *in vitro* (French *et al.*, 1990; Lanna *et al.*, 1995). On the other hand, the addition of GH to cultures of ovine adipose tissue tends to increase the beta-adrenergic lipolytic response (Bauman and Vernon, 1993).

The increase in circulating non-esterified fatty acid content (NEFA), which reflects lipomobilisation, in the lactating cow receiving GH mainly depends on the short-term (meal effect) and long-term (physiological stage effect) variations in nutritional state (Sechen *et al.*, 1991). However, Gallo and Block (1990) report in cows a chronic increase in the positive energy balance. The administration of GH increases the response of the circulating NEFA content to adrenaline injection in the lactating cow with a reduced energy balance (Sechen *et al.,* 1990), but has no effect on the dry non-pregnant cow, whatever it is undernourished or overnourished (Ferlay *et al.,* 1996). The greater effect in the lactating cow seems to result mainly from a decrease in the antilipolytic effect of adenosine caused by GH (Lanna *et al.*, 1995; Doris *et al.*, 1996). This is consistent with the lack of effect in dry cows, since their adipose tissue presents lower sensitivity to the antilipolytic effects of adenosine in comparison with the lactating animal.

GH can therefore act directly on adipose tissue (which contains receptors for this hormone) and modify its responses to the homeostatic regulation of lipogenesis and lipolysis, but without abolishing these types of regulation. The effects of GH treatment on the metabolism of adipose tissue are variably increased (increased mobilisation or increased deposition) in the dairy cow *in vivo*, according to its physiological and nutritional state. The increase in lipomobilisation under the GH effect, at the beginning of lactation or at the beginning of rbGH treatment, may, furthermore, contribute to limiting the rate of the increase in feed intake in these two situations. This would decrease the digestive and metabolic risks which usually accompany any sharp increase in the ruminal, intestinal, or hepatic flows of nutrients (Bareille *et al.*, 1997). The subsequent increase in the intake level, which occurs after several weeks of rbGH treatment, would arise from long-term physiological adaptation, thus avoiding any excessive depletion of body reserves. However, this increase does not seem to be affected by going below a minimum threshold for the body condition score of the animals.

Lipid metabolism in the liver of dairy cows treated with GH does not seem to be significantly altered. Neither the liver content in lipids or triglycerides (Gallo and Block, 1990), nor its ketogenic potential *in vitro*, nor the contents of circulating ketone bodies, phospholipids, or cholesterol are modified. The NEFA from adipose tissues are more widely oxidised (which favours gluconeogenesis) without increasing the esterification and secretion of triglycerides by the liver. Moreover this increased NEFA flow is mostly used by the extra-hepatic tissues either directly for mammary lipogenesis or as an energy source substituting for glucose and amino acids, which are thus protected from oxidation.

In summary, the first effect of GH in ruminants is the stimulation (directly or indirectly *via* IGF-I or by other messengers) of the secretory activity and/or life span of mammary cells. In addition, GH exerts short-term diabetogenic type effects (resistance to the peripheral action of insulin and increase in its secretion, reduction in the extramammary utilisation of glucose, increased oxidation of fatty acids...) which increase glucose availability and enables lactose synthesis by the udder to increase significantly. After a few weeks, during which the animal mobilises its body lipids or slows down their deposition, the organism adapts itself through an increase in the intake level which progressively provides the nutrients necessary to the udder and to hepatic gluconeogenesis, then to the re-establishment of the body reserves. Such an outline arises from animal husbandry observations but requires experimental confirmation. Indeed, in the only long-term metabolic study ever carried out, Vernon *et al.* (1995) observed in the goat receiving a diet rich in concentrates, that the lipogenesis and consumption of glucose of the adipose tissue were still low after 22 weeks of GH treatment, without any apparent increase in hepatic gluconeogenesis.

Therefore, GH facilitates the short-term coordinated adaptation ("teleophoresis", Chilliard 1986, 1999) of the metabolism of different tissues and organs (liver, adipose tissue, muscle, bone ...) to satisfy the nutritional requirements of the udder, while maintaining normal homeostatic regulation (see Figure 5). Unfortunately, there is a lack of data on the metabolic adaptation of tissue after long-term GH treatment.

3. Potential Consequences for Herd Management and the Milk Industry

3.1 REPRODUCTION

The diversity of factors influencing reproduction performance makes it difficult to bring to light the effect of GH injections on it. Few specific trials have been carried out, with reproduction performances generally being recorded as additional data, frequently concerning insufficient numbers of animals.

Initiated before conception, GH injections clearly reduce the fecundity and/or fertility of animals. Treatment before reproduction leads to a systematic decrease (15%)

in the pregnancy rate (Figure 6) which is even greater the earlier the treatment and the higher the dose. Primiparous cows appear to be the most sensitive: the percentage of cows producing a calf drops from 91 (control) to 64 after GH in primiparous cows and from 81 to 71 in multiparous cows (J.F. Roche, cited by Chilliard, 1990). An increase in days to first service is frequently observed (from +5 to +45 days, +15 on average), which seems to be due to the difficulty of detecting heats in treated cows (Cole *et al.*, 1991; Lefebvre and Block, 1992). Moreover the success rate for in the first insemination appears to be decreased in variable proportions (from -5 to –30 %) according to the dose injected, the date the treatment was begun, parity, and the galactopoietic response (Bruneau and DeKerchove, 1988). The number of inseminations necessary for conception is generally higher (+0.3 on average, reviews by Chilliard *et al.* (1989) and Burton *et al.* (1994)). The days from calving to conception increase by about two weeks when the GH treatment starts between the fifth and ninth week of lactation (review by Chilliard *et al.* (1989)). This delay can be as high as four weeks in primiparous cows (Roche, op. cit.). When the treatment begins after conception, neither embryonic mortality, abortion, calf weight modification, nor any particular difficulty in the following calving are observed (Leonard *et al.*, 1990; Rajamahendran *et al.*, 1991, review by Burton *et al.* (1994)).

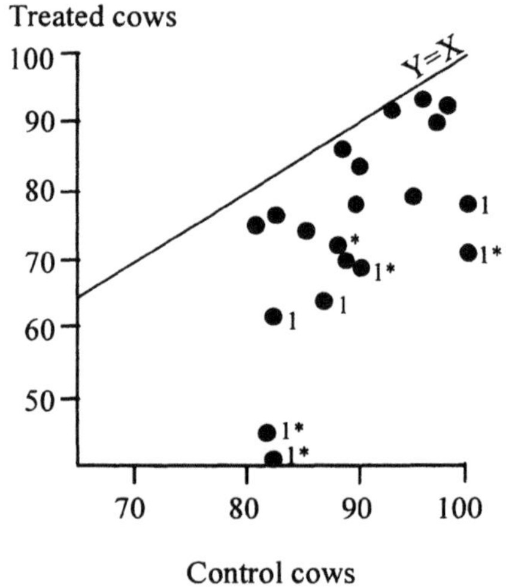

Figure 6. *Effect of recombinant bovine growth hormone on the pregnancy rate (%). Literature review (20 trials, 1589 dairy cows)(1 = first injection before 14 days post-partum; * = daily dose higher than 35 mg).*

The lowest reproduction performances of treated cows are currently difficult to explain. The duration of the different cycle phases (luteal and follicular), and the plasma levels of gonadotropic hormones are not significantly modified by the treatment, even if it starts as of the fourteen day post-partum in the cow (Schemm *et al.*, 1990; Gallo and Block, 1991).

The effect of GH on sexual steroids was studied *in vivo*, *in vitro* following injections *in vivo*, and lastlty in purely *in vitro* systems. GH administered *in vivo* or added to a culture medium *in vitro* stimulates the production of progesterone by the corpus luteum (cyclic or during pregnancy) (Webb *et al.*, 1994) which is consistent with the existence of specific receptors for GH in luteal cells (Tanner and Hauser, 1989). Therefore, the luteal response appears to be favourable to the initiation of pregnancy.

Numerous effects of GH and/or IGF-I on follicular growth have been observed in growing or dry adult females (Table 2). However, in spite of good impregnation by progesterone at the beginning of pregnancy, the increase in the ovulation rate after ovarian stimulation by PMSG or FSH does not occur *via* an increase in the number of transferable embryos in heifers (Gong *et al.*, 1991b). Therefore, using GH to improve superovulation results does not appear to be effective.

Table 2. *Effects of growth hormone (GH) or of IGF-I on the circulating steroids and the ovary function in ruminants during growth (heifer) or maintenance (ewe, cow). According to Adashi* et al. *(1985), Schams* et al. *(1987), Webb* et al. *(1990), Giong* et al. *(1991a, 1991b, 1996), De La Sota* et al. *(1993).*

	Blood	Theca	Granulosa	Follicles
GH *in vivo*	Estrogens ↗ Androgens ↗			
GH *in vitro*		Steroids ↗	No modification	
IGF-I *in vivo* or *in vitro*	Estrogens		Steroids ↗ Evidence for receptors	
GH *in vivo* + ovarian stimulation				Recruiting of small antral follicles ↗ Ovulations ↗

In accordance with the stimulation of follicular growth by GH, a certainly variable but sometimes significant increase in twinning rate has been observed in dairy cows (Roche *op. cit.*, Burton *et al.* (1994)). However, in lactating cows (Spicer *et al.* 1990; Gallo and Block, 1991; De la Sota *et al.*, 1993), neither the circulating follicular steroids modification, nor the stimulation of *in vitro* synthesis under the action of GH and/or IGF-I, described in Table 2 in nonlactating animals, have been observed. To our knowledge, only two experiments have specifically explored follicular growth *in vivo*. In the lactating goat (Driancourt and Disenhaus, 1997), GH had no effect either on the number or proportion of growing follicles in the different classes, or on the ovulation rate after ovarian stimulation. In the lactating cow (De La Sota *et al.*, 1993) it was the class of medium size follicles (6-9 mm), whose number was increased by the treatment. This could explain the decrease in fertility observed in inseminated treated cows and/or the difficulty in detecting heats. In conclusion, the results seem to indicate that follicular growth is directly stimulated by GH in nonlactating ruminants but not when they are in lactation.

In the light of the present knowledge, the direct action of GH on the reproductive function *stricto sensu* appears to be modest and rather favourable. The unfavourable effects on reproduction performance, observed when GH administration is initiated before conception, seem to be indirect and associated with the increase in milk production but, above all, with the decrease in energy balance during the first weeks of treatment (review by Chilliard *et al.* (1989)), in agreement with the already known effects of undernutrition on the reproduction function. Likewise, the increased sensitivity of primiparous cows may be connected with the pecularities of nutrient flows in these animals. In practice, it is therefore preferable to initiate the GH injections after conception. Nevertheless, when cows are treated during the entire lactation period, the standard diets in Europe do not always enable the cows to return to a body condition score sufficient to allow for good subsequent reproduction.

3.2 HEALTH STATE

3.2.1. Metabolic Diseases

In spite of the increase in lipomobilisation frequently observed at the beginning of GH treatment, no increase in acetonemia or lipid infiltration of the liver was observed (Gallow and Block, 1990), which can be explained by the marked orientation of metabolism toward the oxidation of fatty acids and glucose sparing. Treatment with rbGH even appears to improve fetus survival in the ewe with pregnancy toxaemia (Andrews *et al.,* 1996). Likewise, GH treatment appears to decrease the frequency of milk fever, perhaps because of improved calcium homeostasis (C. Baile and G. Montsallier, cited by Chilliard (1990)). However, the metabolic profile of the cows shows a slight tendency towards acidosis and a decrease in albuminemia, as in all high producing cows which mobilise their body reserves (Burton *et al.,* 1994). Furthermore, an increased incidence of foot and leg disorders (by a factor of 2.2) has been reported recently, from an analysis of the post approval monitoring programme on 523 GH-treated cows in the USA , with a potentially negative effect on animal welfare (Broom *et al.*, 1999).

The lack of marked metabolic deviation, *i.e.*, the maintenance of animal homeostasis can be explained, on the one hand, by the coordinated effects (teleophoresis) of GH on overall metabolism and, on the other, by the lower response of the udder in the case of prolonged undernutrition. A careful examination of lameness data in a great number of treated cows is, however, required to validate this partial conclusion.

3.2.2. Immune Responses and Organism Defense

Numerous medical studies, *in vivo* and *in vitro*, show that GH is an important hormone for the development and the functioning of the immune system. In rodents, GH deficiency induces atrophy of the thymus and the spleen, and wasting away. A return to a normal GH level in these animals significantly increases the synthesis of antibodies and the rejection of skin grafts (Kelley, 1989). In a large number of species, the administration of GH and of prolactin restores humoral and cellular immune responses in hypophysectomised animals (Gala, 1991).

The different cells involved in immunity respond to GH and to IGF-I and may even produce these hormones, which suggests that they have an autocrine or paracrine action on the immune system (Burton *et al.*, 1994). The administration of rbGH causes a sharp and permanent increase in the concentration of circulating GH (Cisse *et al.*, 1991). These modifications are capable of increasing the production of IGF-I by immune cells, or of performing a negative retrocontrol on the local production of GH and IGF-I, thus modulating the inflammatory responses.

Numerous cells, particularly lymphocytes, splenocytes, leucocytes (polymorphonuclear and mononuclear), tissue macrophages of various animal species at different physiological states have GH, prolactin, and IGF-I receptors (Burton *et al.*, 1994). The first two belong to a family of cytokine receptors (Cosman *et al.*, 1990). These three hormones stimulate the proliferation and/or numerous functions of lymphocytes B and T, NK (natural killer) cells, and myeloid cells (Kelley, 1989; Weigent and Blalock, 1989; Geffner *et al.*, 1990) which have an important role in immunity.

GH, prolactin, and interferon stimulate phagocytic cells (macrophages and polymorphonuclear) *in vivo* and *in vitro* (Fu *et al.*, 1991, 1994; Kappel *et al.*, 1993), thereby increasing the resistance of rats to lethal infection by *S. Typhimurium* and *L. monocytogenes* (Edwards *et al.*, 1991, 1992).

In bovine animals, the administration of rbGH does not modify immunoglobulin concentrations in the blood (Burton *et al.*, 1991), nor the titers of hemagglutinating antibodies or cutaneous sensitivity (Burton *et al.*, 1992). In addition, there are few published data concerning the effect of rbGH on cellular immunity, but the available results tend to confirm the immunostimulating effect described in humans and rodents (Burton *et al.*, 1994). However, they do not provide precise knowledge on whether rbGH increases resistance to infectious diseases, or not, in highly productive dairy cows receiving this hormone over a long period.

3.2.3. Infectious Diseases

During numerous animal production studies set up to examine the effects of rbGH on the performances of dairy cows, the frequency of infectious diseases, listed in terms of the number of cows affected once or several times during lactation, does not seem to be modified by the treatment (Burton *et al.*, 1994). However, under certain circumstances, GH may modify several infectious processes.

3.2.3.1. Effect of GH on Mammary Infections. The administration of rbGH to dairy cows increases the incidence of clinical mastitis from 15 to 45% in the meta-analyses of the literature cited by Willeberg (1993). This effect is largely related to an increase in the milk production level (Bauman, 1992) as it exists in the genetically superior cows not receiving GH (Barnouin and Karaman, 1986). The frequency of mastitis also tends to increase when cows receive large doses of rbGH and/or are treated over several lactations (Burton *et al.*, 1994). This increase in mastitis can contribute to decreasing the well-being of cows receiving GH (Willeberg, 1993).

In several experiments (Kronfeld, 1988; Burton *et al.,* 1990b, 1994; Lissemore *et al.*, 1991), the administration of rbGH to dairy cows is associated with a slightly higher milk cell number, at least during the first nine weeks of lactation. In the goat, when the number of milk cells is increased by intramammary injection of lipopolysaccharide, the response is greater after administration of rbGH. However, the alteration of the blood-milk barrier is restored more rapidly in the presence of rbGH (Burvenich *et al.,* 1989a). Dairy cows, whose udders have been experimentally infected with *E. coli*, return to normal milk production and a normal lactose content more rapidly when they receive rbGH (Burvenich *et al.*, 1989b).

In conclusion, the administration of rbGH increases the incidence of mastitis, but this may result rather from the increase in milk production. Indeed, there is no observation suggesting that rbGH decreases the immune competence of cows (GH increases them in humans and in rats), and the increase in the cell counts observed at the beginning of lactation could even have a protective role against mammary infections. In fact, the early influx of polymorphonuclear neutrophiles (phagocytic cells) of blood is the main defense mechanism of the udder against the pathogenic agents responsible for mastitis. Indeed this increase constitutes the basis for the diagnosis of subclinical mammary infections in the dairy cow receiving no GH, and the cell counts of milk are taken into account for the milk payment. Considering the issues at stake (improvement in the efficiency of mastitis treatment, but higher frequency of mastitis and a decrease in the price of milk in the case of high cell numbers), some complementary studies using experimental infections under well controlled conditions are necessary to improve our understanding of the rbGH effect on the sanitary state of the udder as well as of the immune response mechanisms that may be involved.

3.2.3.2. Retroviruses. Recombinant human GH (rhGH), according to the dosage used, increases the replication of lentivirus HIV-1 (Human Immunodeficiency Virus - type 1) in vitro from 2 to 20 fold (Laurence *et al.*, 1992). The IGF-I effect is not known, but insulin stimulates the expression of HIV-1 *in vitro* (Spandidos *et al.*, 1990).

It is wise to remain cautious when interpreting these results obtained *in vitro* with human GH. Unfortunately, there are only very few known studies in ruminants, but preliminary reports suggest that prolonged-release rbGH (injections of 500 mg at 14 day intervals) may increase and prolong the production of the visna Maedi virus in the macrophages of milk in the seropositive ewe and may induce the expression of the CAEV virus (caprine arthritis encephalitis virus) in goats (Pascalon and Asso, personal communication). This uncertainty can also be found in a recent study by Lerondelle *et al.* (unpublished results) on four other goats, in which the administration of 5 mg/day of rbGH tends to increase CAEV virus expression in milk cells.

At the molecular level, GH and its main mediator, IGF-I, regulate the expression of some genes. In the nucleus, the transcription factors will bind to the genomic DNA at the level of the target sequences, to induce the transcription of cellular genes. These transcription factors can also become fixed on viral DNA and induce the transcription of viral genes (Le Cam and Legraverend, 1993). This could enable GH to have a direct effect, which remains to be proven, on viral replication. In addition, the indirect action of hormones remains a possibility. Indeed, hormones (including GH), by stimulating the differentiation and maturing process of cells, particularly of mammary epithelial cells at the end of pregnancy, may simultaneously favour viral replication.

Moreover, it should be pointed out that infection with the BIV lentivirus (bovine immunodeficiency virus), of which still little is known, especially from the viewpoints of epidemiology and diagnosis, exists in certain cattle farms. Provided that the preliminary results quoted above on the retrovirus of small ruminant are confirmed , the effects of rbGH administration should be studied in cattle infected by BIV, in order to avoid the risk of an increased propagation of this virus within herds.

3.2.3.3. Other Unconventional Transmissible Viruses and Agents. GH by itself or in association with progesterone, estrogens, and corticoids stimulates (increase by a factor of 2 to 10) the replication of murin cytomegalovirus *in vitro* (Chong and Mims, 1984). Moreover, the addition of rhGH or IGF-I can induce the expression of PrP (prion protein associated with spongiform encephalopathies) messenger RNA in the cells of rat pheochromocytoma *in vitro*. Nevertheless, rhGH and IGF-I act only at very high, nonphysiological, indeed suprapharmacological doses (10 mg/ml for GH), and even then only incompletely (Lasmezas *et al.*, 1993).

In conclusion, when added it in some *in vitro* models, at generally pharmacological doses and in the presence of other hormones, GH can stimulate the production of pathogenic agents exhibiting intracellular development such as viruses and prions. Therefore, it would be logical to say that long-term complementary studies must be carried out *in vivo* with rbGH administered in physiological doses to ruminants before being able to dismiss the possibility of risks to the sanitary state of herds and the safety of consumers.

Then these studies should be carried out on experimental animals during incubation of subacute spongiform encephalopathy, or infected with BIV, or presenting subclinical infection by intracellular bacteria such as Salmonella, Listeria, and Chlamydia as well as by intracellular (protozoon) parasites (Aynaud, personal communication).

3.3 MILK PROCESSABILITY AND CONSUMER SAFETY

The potential advantages of using rbGH to increase milk yield should not be detrimental to milk quality. Otherwise there will be negative consequences concerning milk processing in the dairy industry as well as the consumption of dairy products.

The changes in milk composition already mentioned are mainly observed during the first weeks of rbGH administration. It is explained by a transient decrease in the nutritional balance of treated animals (Kinstedt *et al.*, 1991; Barbano *et al.*, 1992; Lynch *et al.*, 1992). Therefore, these transient changes affect milk yield and the technological processability of milk only slightly. Thus, no study has been able to clearly demonstrate any significant difference in the technological behaviour or organoleptic traits of products processed from milk (milk for consumption, fermented or powdered milk, soft or semi-hard cheeses, or cooked, pressed cheeses) from rbGH-treated cows (Maubois, 1990; Mietton *et al.*, 1990; Laurent *et al.*, 1992).

According to experts from the different federal and international control and regulatory agencies (US Food and Drug Administration, European Commission), rbGH administration to dairy cows to increase milk production is and should be submitted to the same safety evaluation criteria as other xenobiotic products before being commercially authorised. In fact, in terms of residues, no study has been able to establish a net increase in GH concentration in the milk of rbGH-treated cows (Schams, 1988; Butenandt, 1995). Besides, the limitation of the biological activity of bGH to the bovine species (inactivity of rbGH in man), its thermolability and its supposedly almost complete degradation during digestion, should be sufficient to dismiss the possibility of any risk of residual hormonal activity in the consumer (Butenandt, 1995; Groenewegen *et al.*, 1990; Juskewich and Gruyer, 1990a). In addition, no immunological data has been able to demonstrate any allergenic properties of the different recombinant hormones that may be linked to minor differences between their primary structures and the natural hormone (Juskevich and Gruyer, 1990b; Kronfeld, 1990b).

When compared to untreated cows, the IGF-I content of milk from rbGH-treated cows is higher by a factor ranging from 1.5 to 5 (Schams *et al.*, 1991; Breier *et al.*, 1993; Burton *et al.*, 1994). When prepared for nursing infants, the technological processing of milk using high temperatures should reduce this growth factor and consequently eliminate any risk of a significant increase in IGF-I concentrations which could affect the infants (Juskevich and Gruyer, 1990a; Collier *et al.*, 1991). Yet, according to Burton *et al.* (1994), pasteurised milk may still contain IGF-I (or other proteins) which may still be bioactive and which may hypothetically interact with specific targets inside the digestive tract linked to the growth or immune functions of the nursing child (Heinz-Erian *et al.*, 1991). Nevertheless, the IGF-I content of milk from treated dairy cows still remains within the range of values measured in human milk.

In addition, milk composition can be also modified in terms of its contents in residues of xenobiotic agents and their metabolites. Indeed, such xenobiotic compounds can be administered accidentally *via* food (mycotoxins) or intentionally (antibiotics), and their elimination depends mainly on the hepatic detoxication capabilities of the animal. In murine and porcine species, most detoxication activities are reduced by GH (Paris and Bories, 1991; Shapiro *et al.*, 1995). Moreover, Witkamp *et al.* (1993) have shown, in goat, a decrease in the plasma clearance of a veterinary drug, sulphamidine, by increasing the GH injection frequency from one to three times per day. Yet, this effect had not previously been observed in dairy cows treated with implants or daily injections (Witkamp *et al.*, 1989).

At the beginning of the 90's, various American governmental organisations, joined by the international organisations (FAO, WHO), have repeatedly concluded that dairy products derived from cows treated with rbGH are harmless to consumers. However, in a context of intensified milk production by the use of rbGH, the risks of increasing diseases related to lactation (mastitis, etc.), or to latent viral infections, are not nil. If the elimination of xenobiotics is slower in animals receiving rbGH, the rules for using veterinary drugs, especially antibiotics, should probably be modified. Thus, whether it be for milk marketing or carcass marketing purposes, particular attention should be paid to the waiting period so as to respect the maximum residue limits (MRL) authorised for veterinary drugs.

3.4 GENETIC IMPROVEMENT

rbGH is potentially a powerful tool for production, since its considerable flexibility of use (according to animals and periods) is combined with almost-immediate effectiveness. In comparison, the feeding modifications have both a less selective effect, because they concern groups of animals, and are less rapid over time. Therefore, it is necessary to predict as accurately as possible the problems which would prevail after a possible authorisation of rbGH in the European Community, based on the simulations already carried out on this subject (Burnside and Meyer, 1988; Frangione and Cady, 1988; Simianer and Wollny, 1989; Colleau, 1990a, 1990b).

***Table 3**. Impact on some selection parameters of the non-correction of the data for the GH effect. In a situation where GH is used during 8 months, contributing on average 1000 kg more milk, in 10 % of the cows. According to Colleau (1990a, 1990b).*

	Treated cows			
	By chance	The best	The less good	Concentrated in some herds
On the true genetic superiority [(1)]	- 9 %	- 5 %	- 2 %	- 2 %
of elite mothers	- 10 %	- 2 %	- 2 %	- 1 %
of elite fathers	- 2 %	- 2 %	- 3 %	- 2 %
of service fathers	+ 7 %	+ 26 %	- 7 %	0 %
On the average estimation bias of the superiority of these three categories [(2)]				

(1) (b - a/a) x 100, with a = genetic superiority of the selected animals when there is no treatment with GH and b = genetic superiority of the selected animals when GH is used in the population subjected to the selection.
(2) (c - b/b) x 100, with b = genetic superiority of the selected animals when GH is used in the population submitted to selection and c = estimated genetic superiority of the selected animals when GH is used in the population.

If one did not take the possible utilisation of rbGH into account for genetic evaluation, the average genetic value of selected animals (elite cows and bulls, and service bulls) would be overestimated. In a theoretical situation, where one supposes that 10% of the cows are treated for 8 months of their lactation period with a 1000 kg milk increase (Table 3), the efficiency of bull selection is affected as much as that of cow selection, since bulls are judged according to their female descendents. One may be surprised to discover that the results are lower when rbGH is used randomly in the herd. Indeed it would be better if only low-producing cows were treated because this would not enable them to outperform the best cows. From the selection viewpoint, the ideal situation is where the treatments are concentrated in certain herds, because the corrections carried out for the "herd " effect also integrate the rbGH effect in these herds: the loss in genetic progress is practically zero in this case.

The second type of consequences of not taking rbGH into consideration is the existence of bias in genetic evaluation. In ideal indexing, the actual genetic value of a breeder is an unknown value whose average is the calculated index, and which is less variable around this average when the accuracy of the index is greater. In biased indexing, the actual genetic value fluctuates around a value which is lower (overestimated index) or higher (underestimated index) than the index. Table 3 is an illustration of the amplitude of the estimation errors for the indices (+ 26 to -7) when the use of rbGH is not taken into account. The highest biases occur when rbGH is used on the best cows because their apparent superiority is increased further.

The values already mentioned are the results of simulation studies. In reality, the user's confidence with respect to selection programmes would probably be diminished, which could affect their fundamental structures and induce additional losses in genetic progress.

In the longer-term, geneticists and breeders throughout the world have two choices:

- to continue perfecting the genetic evaluation methods already being used, and completely ignore the new variation factor, *i.e.* rbGH ;
- to organise the collection of monthly information regarding the treatment of cows and use it in a genetic evaluation model based not on total lactation but on the different monthly controls (Colleau, 1996).

In the second situation, numerous theoretical studies should be undertaken because variations in response to rbGH are complex. The most limiting factor by far is, however, the collection of exhaustive information from dairy farms and its cost.

The USA is the first country with a considerable milk production industry to have authorised rbGH and, for the time being, the first of these two approaches prevails there. It is not known what the consequences will be for selection efficiency in this country; this should depend on the extent and mode of rbGH utilisation in dairy farms: whether regular and concentrated for all the cows in certain herds, or cyclical, or only utilised for some of the cows in the treated herds. Likewise, it is still too early to say how the world dairy selection organisations will react if rbGH is authorised everywhere, which is what as the present economic (liberal) logic seems to be leading to.

3.5. SOCIAL AND ECONOMIC CONSEQUENCES

3.5.1. Dairy Farm Economics and the Environment

From the viewpoint of farm economics, the use of rbGH appears to be a tool for instantaneous intensification of milk production: the treatment of the animal, which has a cost, entails extra production provided that the supplementary feed requirements are sufficiently met.

The problem, which seems simple since it is a question of comparing the supplementary revenue with the costs, is in fact a complex one, because numerous hypotheses have to be formulated first before turning to the calculations (Mouchet, 1987, 1988; Cordonnier and Bonnafous, 1989; Melet and Mouchet, 1989) :

- maintaining the milk price at the same level as if rbGH was not being used. This hypothesis is acceptable in the short-term but not in the long-term, since adopting more productive techniques is known to entail a decrease in the selling price of the product;
- the possibility of marketing the extra amounts produced without any problems. The present quota system prevents this, unless some significant restructuring is carried out *via* the redistribution of the rights of farms who have ceased their activity;
- modification of herd management, or not, particularly in terms of stocking rate per ha and farming strategy for replacement animals.

In the absence of quota (or any other form of quantitative limitation of production) the use of rbGH is obviously of interest when the additional production is greater than the cost of rbGH and feed, not taking the environment or working conditions into consideration. The simulation results show that, from a purely accounting point of view, the use of rbGH is therefore frequently interesting, which explains its diffusion in the countries where its marketing is authorised.

The situation is entirely different when the farm is subjected, as in France, to a production quota. Here, the farmer, faced with two solutions, should, in both cases, reduce the number of dairy cows: maintain the same forage surface area *i.e* decrease the stocking rate or, conversely, work at the same stocking rate and reduce the forage surface area. The simulation results underline the fact that the solution of reducing the stocking rate, without otherwise modifying the technical and economic management of the herd, is only very rarely profitable.

In the case of reducing the forage surface area with a constant stocking rate, the operation is profitable only if the farmer can use the land thereby made available for production purposes which will make it possible to obtain a return on the fixed factors, excluding productions with low or medium gross margins, and returns to intensive animal or plant production. This is theoretically possible, but faces two obstacles: first, the difficulty of finding such activities whose markets are not saturated and, then, the constraints involved in a change of activity, such as the need for new skills and equipment, the greater or lesser agronomic suitability of the soil, etc. The economic advantage of using rbGH should therefore be established case by case, all the more so as other factors can also vary, such as rbGH administration by the farmer or by the veterinarian, the strategy of applying it to all or some of the herd, the control of the technique, etc. This had led to an interest in using rbGH on a more punctual or occasional basis, for example, to compensate for a deficit in production due to a sick animal, or to exactly adjust the quantity produced to a quota allocated to the farm. It seems that, in this kind of situation, rbGH could be an interesting tool for technical and economic management, due to the flexibility of the conditions in which it can be used (period and duration of utilisation chosen by the farmer, no investment necessary).

In terms of the environment, the consequences of using rbGH can be diverse. The reduction in the number of cows would obviously entail a reduction in the amount of manure, particularly nitrogenous manure, which would be a positive factor if the amount of manure thus reduced was spread out over a surface area kept constant. However, conversely, the restructuring of the dairy farms could lead to a higher concentration of herds and therefore to the risk of punctual surpluses of nitrogen supply. Moreover, the total national and EC livestock would be reduced, all the more if the technique proves to be more efficient and if its penetration rate in dairy farms is high. The studies conducted in this subject area show that this reduction would not be very considerable (most of the observers estimate it would be lower than 10% compared to livestock not treated with rbGH), but it would reinforce a very significant trend towards a reduction following the implementation of quotas and productivity gains. In France, for example, the number of dairy cows decreased from 7.2 million in 1984 to 4.5 million in 1995. This regression, compensated only in part by the increase in meat livestock, usually results in fodder areas being abandoned and cultivated areas being reduced for want of favourable conditions for expansion. Then rbGH, an intensifying tool, would probably reinforce this tendency toward agricultural decline in numerous regions in France. Yet a large part of the society currently considers this to be harmful, which leads to the reaffirmation that, besides food production, agriculture has other goals to fulfill, especially that of land management.

3.5.2. National and International Dairy Product Market

In the medium term, the reduction in production costs induced by the productivity gain would probably entail a decrease in milk prices, all the more so the more widespread this technique becomes (Mouchet, 1987; Cordier, 1988). On EC territory characterised by the single market, the use of this technique, if authorised, will have to begin simultaneously for all the member states, in order to avoid distorting competition. The CAP mechanism guarantees a constant milk price level to meet external competition, relying heavily on the quota system. If the member states are to continue respecting the quotas, and this will no doubt be the case for several more years, there will be no overproduction on the global level. However, the use of rbGH could incite some farmers to produce more, which would be possible only if other producers give up their right to produce. This will be a source of tension in the agricultural sector, to add to those which already exist.

Some conflicts could also appear on the international market, from the moment when big producers, such as the United States of America, authorised the use of rbGH. A "price war" led by American exporters could lead to the reconsideration of the ban on rbGH in the European Union.

In addition to the problems of price and of amounts produced, there is also the problem of product quality and further still that of the image of milk products for the consumers. The milk produced by cows treated with rbGH, beyond any kind of scientific consideration, could appear as disqualified, even dangerous whereas, for numerous historical and sociological reasons, milk is seen as a noble product, particularly because of its role in children's food. This modification in consumer appraisal, judged as being irrational by certain operators in the milk sector or by certain scientists, is nevertheless possible in a context where problems of public health and ethics are frequently associated, as in the case of bovine spongiform encephalopathy.

One solution would be to reserve the milk obtained by using rbGH for certain uses, or to leave it up to the consumer to choose. This assumes that one can easily and undoubtedly, identify the origin of milk, whether marketed in its unprocessed or processed state. What is conventionally termed product "traceability" should then give rise to long and complex studies in different scientific, economic, administrative, and judicial fields.

It is also possible that, in the dairy sector as well as in others, taking into consideration the different goals of the agriculture sector and the revolution in relations between the agricultural sector and society, will lead consumer-citizens to regard the price of an agricultural product as representative not only of the food production function but also of the different aspects of its quality. They could then accept to pay a higher price in return for the assurance that the milk (or another agricultural product) has been obtained using a production technique they approve of (as it happens, without rbGH) and can control.

4. Conclusions

The long history of the use of GH for milk production today enables us to make use of a significant volume of knowledge on the effects and mechanisms of action of this hormone, as well as on the context and the stakes involved in the economic development of one of the first hormones produced on a large scale by genetic recombination.

The spectacular effects of the hormone on animal production are presently described with sufficient precision. Its utilisation is efficient and free of any major problems if applied to healthy animals which are safe from latent infection, properly fed and receiving no medical treatment, and as long as the treatment starts after conception. The treatment of primiparous cows requires particular attention in terms of nutrition and reproduction.

Knowledge of GH action mechanisms and of the physiology of "normal" lactation has also made spectacular progress. However, uncertainty remains concerning the respective direct or indirect roles of GH and of its mediators (mainly IGF-I) on mammary metabolism. In addition, the longterm effects of GH treatment on tissue metabolism are still not totally known.

Most of the technical questions concerning the use of rbGH in dairy production have received favourable responses, whether in terms of productive efficiency, the technologic quality of dairy products or nutritional safety for the consumer. However, four points remain to be clarified: the possible direct or indirect effects of GH on the occurrence of lameness and mastitis, on the rate of xenobiotic elimination by the liver, and its possible effects on the reactivation of latent infections by viruses or other pathogenic infectious agents. The technical risks appear to be sufficiently low to the American authorities for them to authorise the commercial utilisation of rbGH in the USA. However, the systematic sanitary monitoring of treated cows and of their milk will be necessary and could be insisted on by consumers, as well as the traceability of milk produced at a lower cost with the aid of rbGH. However, at the present time there are no technical means to control this traceability.

This authorisation has not been granted in western Europe for several social and economic reasons since, in the dairy quota system, the profitability of using rbGH is limited, and its use in the context of free production would, as is the case for other new technologies, accelerate the decrease in the number of farmers, the desertion of certain rural zones and an the increase in unemployment. Moreover, any damage to the positive image of dairy products in the eyes of the consumers, in a context of general anxiety about the "artificialisation" of agriculture, could reinforce the problems of the whole dairy sector. Finally, the questions which remain on the subject of the effects on viral infections in herds could justify the application of a precautionary principle by the EU authorities if these concerns are supported by new experimental results. However, it is necessary to be aware that, in the case of the complete liberalisation of world trade, the American producers authorised to use hormonal treatment to produce milk at a lower cost would benefit from being at a significant advantage when competing with European producers.

Acknowledgments

We thank J.J. Colleau for writing the section on "genetic improvement" in the French manuscript as well as J.M. Aynaud, F. Bocquier, M.H. Farce, F. Grosclaude, H. Jammes, J.P. Lafont, B. Remond, J. Robelin and H. Rulquin for critical and constructive remarks, A.M. Wall from the translation Unit of INRA, for helping to correct the English, and M. Borel and P. Béraud for the preparing of the final draft.

References

Adashi, E.Y., Resnik, E.C., D'Ercole, A.J., Svoboda, M.E. and Van Wyk, J.J. (1985) Insulin-like growth factors as regulators of granulosa cell growth and function, *Endocrinol. Rev. 6*, 400-417.

Andrews, A.H., Holland-Howes, V.E. and Wilkinson, J.I.D. (1996) Naturally occurring pregnancy toxaemia in the ewe and treatment with recombinant bovine somatotropin, *Small Rum. Res. 23,* 191-197.

Barbano, D.M., Lynch, J.M., Bauman, D.E., Hartnell, G.F., Hintz, R.L. and Nemeth, M.A. (1992) Effect of a prolonged-release formulation of N-methionyl bovine somatotropin (Sometribove) on milk composition, *J. Dairy Sci. 75,* 1775-1793.

Bareille, N., Faverdin, P. and Hay, M. (1997) Modification of feed intake response to a beta (2)-agonist by bovine somatotropin in lactating or dry dairy cows, *J. Dairy Sci. 80,* 52-66.

Barnouin, J. and Karaman, Z. (1986) Enquête écopathologique continue : 9. influence du niveau de production sur la pathologie de la vache laitière, *Ann. Rech. Vét. 17,* 331-346.

Bauman, D.E. (1992) Bovine somatotropin : review of an emerging animal technology, *J. Dairy Sci. 75,* 3432-3451.

Bauman, D.E., Eppard, P.J., DeGeeter, M.J. and Lanza, G.M. (1985) Responses of high-producing dairy cows to long-term treatment with pituitary somatotropin and recombinant somatotropin, *J. Dairy Sci. 68,* 1352-1362.

Bauman, D.E., Hard, D.L., Crocker, B.A., Partridge, M.S., Barric, K., Sandles, L.D., Erb, H.N., Franson, S.E., Hartnell, G.F. and Hintz, R.L. (1989) Long-term evaluation of a prolonged-release formulation of N-methionyl bovine somatrotropin in lactating dairy cow, *Dairy Sci. 72,* 642-651.

Bauman, D.E. and Vernon, R.G. (1993) Effects of exogenous bovine somatotropin on lactation, *Annu. Rev. Nutr. 13*, 437-461.

Baumrucker, C.R. and Stemberger, B.H. (1989) Insulin and insulin-like growth factor-I stimulate DNA synthesis in bovine mammary tissue *in vitro, J. Anim. Sci. 67*, 3503-3514.

Beswick, N.S., Kennely, J.J., (2000) Influence of bovine hormone and growth hormone-releasing factor on messenger RNA abundance of lipoprotein lipase and stearoyl-CoA desaturase in the bovine mammary gland and adipose tissue, *J Anim. Sci.* 78, 412-419.

Binelli, M., Vanderkooi, W.K., Chapin, L.T., Vandehaar, M.J., Turner, J.D., Moseley, W.M. and Tucker, H.A. (1995) Comparison of growth hormone-releasing factor and somatotropin : body growth and lactation of primiparous cows, *J. Dairy Sci.* 78, 2129-2139.

Bird, A.R., Croom, W.J. Jr., McBride, B.W., Fan, Y.K., Daniel, L.R. and Taylor, I.L. (1996) Recombinant bovine somatotropin increases nutrient absorption by the proximal small intestine in sheep, Can. *J. Anim. Sci. 76*, 343-350.

Breier, B.H., Milsom, S.R., Blum, W.F., Schwander, J., Gallaher, B.W. and Gluckman, P.D. (1993) Insulin-like growth factors and their binding proteins in plasma and milk after growth hormone-stimulated galactopoiesis in normally lactating women, *Acta Endocrinol. 129*, 427-435.

Broom, D., Dantzer, R., Willeberg, P., Mepham, B. and Noordhuizen-Stassen, E. (1999) Animal welfare aspects of the use of bovine somatotropin, in *Report of the scientific committee on animal health and welfare*, Directorate General XXIV of the European Commission, Brussels, 10 March 1999, 90 pp.

Bruneau, P., Bertrand, G., Chilliard, Y. and De Kerchove, G. (1991) Somatotropine bovine (Sometribove) et production laitière. Efficacité et innocuité chez 448 paires de vaches dans les conditions de 24 exploitations françaises, *Rec. Méd. Vét. 167*, 859-872.

Bruneau, P. and DeKerchove, G. (1988) Somatotropine, reproduction et génétique chez la vache laitière, *Elevage et insémination 228*, 3-14.

Burnside, E.B. and Meyer, K. (1988) Potential impact of bovine somatotropin on dairy sire evaluation, *J. Dairy Sci. 71*, 2210-2219.

Burton, J.H., Mc Leod, G.K., Mc Bride, B.W., Burton, J.L., Bateman, K., Mc Millan, I. and Eggert, R.G. (1990b) Overall efficacy of chronically administrated recombinant bovine somatotropin to lactating dairy cows, *J. Dairy Sci. 73*, 2157-2167.

Burton, J.L., Mc Bride, B.W., Block, E., Glimm, D.R. and Kenelly, J.J. (1994) A review of bovine growth hormone, Can. *J. Anim. Sci. 74*, 167-201.

Burton, J.L., Mc Bride, B.W., Burton, J.H. and Eggert, R.G. (1990a) Health and reproductive performance of dairy cows treated for up to two consecutive lactations with bovine somatotropin, *J. Dairy Sci. 73*, 3258-3265.

Burton, J.L., Mc Bride, B.W., Kennedy, B.W., Burton, J.H., Elsasser, T.H. and Woodward, B. (1992) Contact sensitivity and systemic antibody responses in dairy cows treated with recombinant bovine somatotropin, *J. Dairy Sci. 75*, 747-755.

Burton, J.L., Mc Bride, B.W., Kennedy, B.W., Burton, J.H., Elsasser, T.H. and Woodward, B. (1991) Serum immunoglobulin profiles of dairy cows chronically treated with recombinant bovine somatotropin, *J. Dairy Sci. 74*, 1589-1598.

Burvenich, C., Massart-Leen, A.M., Vandeputte-van Messon, G., Roets, E. and Kiss, G. (1989a) Effect of recombinant bovine somatotropin on endotoxin induced mastitis in lactating goats, *Arch. Intern. Phys. Bioch. 97*, 91-95.

Burvenich, C., Vandeputte-van Messon, G., Roets, E., Fabry, J. and Massart-Leen, A.M. (1989b) Effect of bovine somatotropin on milk yield and milk composition in periparturient cows experimentally infected with Escherichia coli, in K. Sejrsen, M. Vestergaard, A. Neimann-Sorensen (eds.), *Somatotropin in Livestock Production.* Elsevier Applied Science, London, pp. 277-280.

Butenandt, O. (1995) Growth hormone in meat production: are there any risks for the human consumer ?, in *Scientific Conference on Growth Promotion in Meat Production,* Directorate General VI Agriculture of the European Commission, Brussels, 29 November -1 December 1995.

Chilliard, Y. (1986) Revue bibliographique : Variations quantitatives et métabolisme des lipides dans les tissus adipeux et le foie au cours du cycle gestation-lactation, 1ère Partie : chez la ratte, *Reprod. Nutr. Dev. 26*, 1057-1103.

Chilliard, Y. (1988a) Rôles et mécanismes d'action de la somatotropine (hormone de croissance) chez le ruminant en lactation, *Reprod. Nutr. Dev. 28*, 39-59.

Chilliard, Y. (1988b) Review. Long term effects of recombinant bovine somatotropin (r BST) on dairy cow performances, *Ann. Zootech. 37,* 159-180.

Chilliard, Y. (1990) Somatotropine et lactation. Données récentes présentées au séminaire Monsanto (Telfs, Autriche, 9-11 mars 1990), *Rec. Med. Vet. 166,* 513-516.

Chilliard, Y. (1999) Metabolic adaptations and nutrient partitioning in the lactating animal, in J. Martinet, L.M. Houdebine and H.H Head (eds), *Biology of lactation.* INRA Ed., Paris, pp. 503-552,.

Chilliard, Y., Cissé, M., Lefaivre, R. and Rémond, B. (1991) Body composition of dairy cows according to lactation stage, somatotropin and concentrate supplementation, *J. Dairy Sci. 74,* 3103-3116.

Chilliard, Y., Vérité, R. and Pflimlin, A. (1989) Effets de la somatotropine bovine sur les performances des vaches laitières dans les conditions françaises d'élevage, *INRA Prod. Anim. 2,* 301-312.

Chong, K.T. and Mims, C.A. (1984) Effects of pregnant mouse serum and pregnancy hormones on the replication *in vitro* of murine cytomegalovirus, *Arch. Virol. 82,* 223-231.

Cissé, M., Chilliard, Y., Coxam, V., Davicco, M.J. and Rémond, B. (1991) Slow release somatotropin in dairy heifers and cows fed two levels of energy concentrate. II- Plasma hormones and metabolites, *J. Dairy Sci. 74,* 1382-1394.

Cole, W.J., Madsen, K.S., Hintz, R.L. and Collier, R.J. (1991) Effect of recombinantly derived bovine somatotropin on reproductive performance of dairy cattle, *Therionology 36,* 573-595.

Collie, J.J. (1990a) Impact of the use of bovine somatotropin (BST) on dairy cattle selection, *Génét. Sél. Evol, 21,* 479-491.

Collie, J.J. (1990b) Effet de la somatotropine bovine (BST) sur l'efficacité des programmes de sélection laitière, *INRA Prod. Anim. 3,* 93-102.

Colleau, J.J. (1996) Evaluation génétique des animaux d'élevage, *INRA Prod. Anim. Hors série,* 27-40.

Collier, R.J., Miller, M.A., Hildebrandt, J.R., Torkelson, A.R., White, T.C., Madsen, K.S., Vicini, J.L., Eppard, P.J. and Lanza, G.M. (1991) Factors affecting insulin-like growth factor I (IGF-I) concentration in bovine milk; *J. Dairy Sci.* 4, 2905-2911.

Cordier, J. (1988) Influence économique de la B.S.T. en France, *Economie et gestion agro-alimentaire n°8,* 6-12.

Cordonnier, P. and Bonnafous, P. (1989) Essai d'appréciation de l'impact économique des utilisations de la somatotropine bovine dans les systèmes laitiers français, *INRA Economie et Sociologie Rurale,* Notes et Documents, 138 p.

Cosman, D., Layman, S.D., Idzerda, R.L., Beckmann, M.P., Park, L.S., Goodwind, R.G. and March, C.J. (1990) A new cytokine receptor superfamily, *T.I.B.S. 15,* 265-270.

Davis, S.R., Collier, R.J., McNamara, J.P., Head, H.H. and Sussman, W. (1988) Effects of thyroxine and growth hormone treatment of dairy cows on milk yield, cardiac output and mammary blood flow, *J. Anim. Sci. 66,* 70-79.

De La Sota, R.L., Lucy, M.C., Staples, C.R. and Thatcher, W.W. (1993) Effect of recombinant bovine somatotropin (sometribove) on ovarian function in lactating and non lactating dairy cows, *J. Dairy Sci. 76,* 1002-1013.

Disenhaus, C., Belair, L. and Djiane, J. (1988) Caractérisation et évolution physiologique des récepteurs pour les insulin-like growth factors I et II (IGFs) dans la glande mammaire de brebis, *Reprod. Nut. Dév. 28,* 241-252.

Disenhaus, C., Jammes, H., Belair, L., Kann, G. and Djiane, J. (1992) Récepteurs hépatiques à la somatotropine et récepteurs mammaires à la prolactine et à l'IGF1 après injections de somatotropine bovine recombinée chez la chèvre en lactation, *Ann. Zootech. 41,* 111-112.

Disenhaus, C. (1996) La somatotropine bovine recombinée : réponses des ruminants en lactation. Thèse Institut National Agronomique Paris-Grignon (171p).

Donkin, S.S., Chiu, P.Y., Yin, D., Louveau, I., Swencki, B., Vockroth, J., Evock-Clover, C.M., Peters, J.L. and Etherton, T.D. (1996) Porcine somatotropin differentially down-regulates expression of the GLUT4 and fatty acid synthase genes in pig adipose tissue, *J. Nutr. 126,* 2568-2577.

Doris, R.A., Thompson, G.E., Finley, E., Kilgour, E., Houslay, M.D. and Vernon, R.G. (1996) Chronic effects of somatotropin treatment on response of subcutaneous adipose tissue lipolysis to acutely acting factors *in vivo* and *in vitro, J. Anim. Sci. 74,* 562-568.

Downer, J.V., Patterson, D.L., Rock, D.W., Chalupa, W.V., Cleale, R.M., Firkins, J.L., Lynch, G.L., Clark, J.H., Brodie, B.O., Jenny, B.F. and De Gregorio, R. (1993) Dose titration of sustained-release recombinant bovine somatotropin in lactating dairy cows, *J. Dairy Sci. 76,* 1125-1136.

Driancourt, M.A. and Disenhaus, C. (1997) Lack of effect of somatotropin administration on ovarian function of lactating goats, *Anim. Reprod. Sci. 46*, 123-132.

Edwards, C.K., Ghiasuddin, S.M., Yunger, L.M, Lorence, R.M., Arkins, S., Dantzer, R. and Kelley, K.W. (1992) *In vivo* administration of recombinant growth hormone or gamma interferon activates macrophages : enhanced resistance to experimental *Salmonella typhimurium* infection is correlated with generation of reactive oxygen intermediates, *Infect. Immun. 60*, 2514-2521.

Edwards, C.K., Yunger, L.M., Lorence, R.M. and Dantzer, R. (1991) The pituitary gland is required for protection against lethal effects of *Salmonella typhimurium*, *Proc. Natl. Acad. Sci. USA 88*, 2274-2277.

Evans, H.M. and Long, J.A. (1921) Characteristic effects upon growth, oestrus and ovulation induced by the intraperitoneal administration of fresh anterior hypophyseal substance, *Proc. Natl. Acad. Sci. USA 8*, 38-39.

Falaki, M., Gengler, N., Sneyers, M., Prandi, A., Massart, S., Formigoni, A., Burny, A., Portetelle, D. and Renaville, R. (1996a) Relationships of polymorphisms for growth hormone and growth hormone receptor genes with milk production traits for italian holstein-friesian bulls, *J. Dairy Sci. 79*, 1446-1453.

Falaki, M., Sneyers, M., Prandi, A., Massart, S., Corradini, C., Formigoni, A., Burny, A., Portetelle, D. and Renaville, R. (1996b) *Taq* I growth hormone gene polymorphism and milk production traits in holstein-friesian cattle, *Anim. Sci. 63*, 175-181.

Falaki, M., Prandi, A., Corradini, C., Sneyers, M., Gengler, N., Massart, S., Fazzini, U., Burny, A., Portetelle, D. and Renaville, R. (1997) Relationships of growth hormone gene and milk protein polymorphisms to milk production traits in Simmental cattle, *J. Dairy Res. 64*, 47-56.

Feldman, M., Ruan, W., Cunningham, B.C., Wells, J.A. and Kleinberg, D.L. (1993) Evidence that growth hormone receptor mediates differenciation and development of the mammary gland, *Endocrinology 133*, 1602-1608.

Ferlay, A., Chilliard, Y., Sala, A.M., Durier, C. and Bocquier, F. (1996) Somatotropin treatment does not affect nonesterified fatty acid response to adrenergic injections in underfed or overfed nonlactating cows, *J. Nutr. 126*, 945-954.

Frangione, T. and Cady, R.A. (1988) Effects of bovine somatotropin on sire summaries for milk production and milk yield heritabilities, *J. Dairy Sci. 71, (suppl. 1)*, 239.

French, N., De Boer, G. and Kennelly, J.J. (1990) Effects of feeding frequency and exogenous somatotropin on lipolysis, hormone profiles, and milk production in dairy cows, *J. Dairy Sci. 73*, 1552-1559.

Fu, Y.K., Arkins, S., Li, Y.M., Dantzer, R. and Kelley, K.W. (1994) Reduction in superoxide anion secretion and bactericidal activity of neutrophils from aged rats : reversal by the combination of gamma interferon and growth hormone, *Infect. Immun. 62*, 1-8.

Fu, Y.K., Arkins, S., Wang, B.S. and Kelley, K.W. (1991) A novel role of growth hormone and insulin-like growth factor 1. Priming neutrophils for superoxide anion secretion, *J. Immunol. 146*, 1602-1608.

Gala, R.R. (1991) Prolactin and growth hormone in the regulation of the immune system, *Proc. Soc. Exp. Biol. Med. 198*, 513-527.

Gallo, G.F. and Block, E. (1990) Effect of recombinant bovine somatotropin on nutritional status and liver function of lactating dairy cows, *J. Dairy Sci. 73*, 3276-3286.

Gallo, G.F. and Block, E. (1991) Effects of recombinant bovine somatotropin on hypophyseal and ovarian functions of lactating cows, Can. *J. Anim. Sci. 71*, 343-353

Geffner, M.E., Bersch, N., Lippe, B.M., Rosenfeld, R.G., Hintz, R.L. and Golde, D.W. (1990) Growth hormone mediates the growth of T-lymphoblast cell lines via locally generated insulin-like growth factor 1, *J. Clin. Endocrinol. Metabol. 71*, 464-469.

Glimm, D.R., Baracos, V.E. and Kennelly, J.J. (1990) Molecular evidence for the presence of growth hormone receptors in the bovine mammary gland, *J. Endocrinol. 126*, R5-R8.

Glimm, D.R., Baracos, V.E. and Kennelly, J.J. (1992) Northern and in situ hybridation analyses of the effect of somatotropin on bovine mammary gene expression, *J. Dairy Sci. 75*, 2687-2705.

Gluckman, P.D. and Breier, B.H. (1989) The regulation of the growth hormone receptor, Pages 27-33 in R.B. Heap, C.G. Prosser, G.E. Lamming (eds.) *Biotechnology in growth regulation*. Butterworths, Toronto, ON.

Gluckman, P.D., Butlet, J.H. and Elliott, T.B. (1983) The ontogeny of somatotropic binding sites in ovine hepatic membranes, *Endocrinology 113*, 1697-1702.

Gong, J.G., Bramley, T. and Webb, R. (1991) The effect of recombinant bovine somatotropin on ovarian function in heifers. Follicular populations and peripheral hormones, *Biol. Reprod. 45*, 941-949

Gong, J.G., Bramley, T., Wilmut, I. and Webb, R. (1991) Effect of recombinant bovine somatotropin on the superovulatory response to pregnant mare serum gonadotropin in heifers, *Biol. Reprod. 48,* 1141-1149.

Gong, J.G., Campbell, B.K., Bramley, T. and Webb, R. (1996) Treament with recombinant bovine somatotropin enhances ovarian follicle development and increases the secretion of insulin-like growth factor I by ovarian follicle in ewes, *Animal Reprod. Sci. 41,* 13-26.

Groenewegen, P.P., McBride, B.W., Burton, J.H. and Elsasser, T.H. (1990) Bioactivity of milk from bST-treated cows, *J. Nutr. 120,* 514-520.

Hadsell, D.L., Greenberg, N.M., Fligger, J.M., Baumrucker, C.R. and Rosen, J.M. (1996) Targeted expression of des (1-3) human insulin-like growth factor-I (IGF I) in transgenic mice influences mammary gland development and IGF-binding protein expression, *Endocrinology 137,* 321-330.

Hard, D.L., Adriens, F., Prosser, C.G., Newbold, J.A., Heap, R.B., Phipps, R.H. and Hannah, M.J. (1992) Effect of recombinant bovine somatotropin (bST) and nutrition on galactopoiesis, hepatic bST receptors and plasma hormone concentrations in the dairy cow, *J. Physiol. 446,* Abstract n°271P.

Hartnell, G.F., Franson, S.E., Bauman, D.E., Head, H.H., Huber, J.T., Lamb, R.C., Madsen, K.S., Cole, W.J. and Hintz, R.L. (1991) Evaluation of Sometribove in a prolonged-release system in lactating dairy cows. Production responses, *J. Dairy Sci. 74,* 2645-2663.

Heinz-Erian, P., Kessler, U., Funk, B., Gais, P. and Kiess, W. (1991) Identification and *in situ* localization of insulin-like growth factor II/Mannose-6-phosphate (IGF-II/M6P) receptor in the gastrointestinal tract: comparison with the IGF-I receptor, *Endocrinologica 129,* 1769-1778.

Hung, C.H. and Moore, W.V. (1984) Binding of bovine growth hormone to bovine liver membranes, *Mol. Cell. Endocrinol 35,* 151-157.

Jammes, H., Disenhaus, C., Ouriet, V., Kayser, C., Postel-Vinay, M.C. and Djiane, J. (1996) Growth hormone-binding protein in the goat : characterization, evolution under exogenous growth hormone treatment, and correlation with liver growth hormone receptor levels, *Domest. Anim. Endocrinol. 13,* 477-489.

Jammes, H., Gaye, P., Belair, L. and Djiane, J. (1991) Identification and characterization of GH receptor mRNA in the mammary gland, *Mol. Cell. Endocr. 75,* 27-35.

Juskevich, J.C. and Guyer, C.G. (1990a) Bovine growth hormone : human food safety evaluation, *Science 249,* 875-884.

Juskevich, J.C. and Guyer, C.G. (1990b) Safety of bovine growth hormone, *Reply. Science 251,* 256-257.

Kappel, M., Hansen, M.B., Diamant, M., Jorgensen, J.O.L., Gyhrs, A. and Pedersen, B.K. (1993) Effects of an acute bolus growth hormone infusion on the human immune system, *Horm. Metabol. Res. 25,* 579-585.

Kelley, K.W. (1989) Growth hormone, lymphocytes and macrophages, *Biochem. Pharmacol. 38,* 705-713.

Keys, J.E. and Djiane, J. (1988) Prolactin and growth hormone binding in mammary gland and liver of lactating cows, *J. Recept. Res. 8,* 731-739.

Keys, J.E., Van Zyl, J.P. and Farrell, H.M. Jr. (1997) Effect of somatotropin and insulin-like growth factor-I on milk lipid and protein synthesis *in vitro, J. Dairy Sci. 80,* 37-45.

Kindstedt, P.S., Pell, A.N., Rippe, J.K., Tsang, D.S. and Hartnell, G.F. (1991) Effect of long-term bovine somatotropin (Sometribove) treatment on nitrogen (protein) distribution in Jersey milk, *J. Dairy Sci. 74,* 72-80.

Kirchgessner, M., Schwab, W. and Mueller, H.L. (1989) Effect of bovine growth hormone on energy metabolism of lactating cows in long-term administration. Proc. 11th Symp. Energy Metabol. (Lunteren, Netherlands, September 1988), EAAP Publ. n°43, 143-146.

Knight, C.H., Fowler, P.A. and Wilde, C.J. (1990) Galactopoietic and mammogenic effects of long-term treatment with bovine growth hormone and thrice daily milking in goats, *J. Endocrinol. 127,* 129-138.

Knight, C.H., Hillerton, J.E., Kerr, M.A., Teverson, R.M., Turvey, A. and Wilde, C.J. (1992) Separate and additive stimulation of bovine milk yield by the local and systemic galactopoietic stimuli of frequent milking and growth hormone, *J. Dairy Res. 59,* 243-252.

Kronfeld, D.S. (1988) Biologic and economic risks associated with use of bovine somatotropins, *J. Am. Vet. Med. Assoc. 192,* 1693-1696.

Kronfeld, D.S. (1990) Safety of bovine growth hormone, *Science 251,* 256.

Lanna, D.P.D., Houseknecht, K.L., Harris, D.M. and Bauman, D.E. (1995) Effect of somatotropin treatment on lipogenesis, lipolysis, and related cellular mechanisms in adipose tissue of lactating cows, *J. Dairy Sci., 78,* 1703-1712.

Lasmezas, C., Deslys, J.P. and Dormont, D. (1993) Recombinant human growth hormone and insulin-like growth factor 1 induce PrP gene expression in PC-12 cells, *Biochem. Biophys. Res. Com. 196*, 1163-1169.

Laurence, J., Grimison, B. and Gonenne, A. (1992) Effect of recombinant human growth hormone on acute and chronic human immunodeficiency virus infection *in vitro*, *Blood 79*, 467-472.

Laurent, F., Vignon, B., Coomans, D., Wilkinson, J. and Bonnel, A. (1992) Influence of bovine somatotropin on the composition and manufacturing properties of milk, *J. Dairy Sci. 75*, 2226-2234.

Le Cam, A. and Legraverend, C. (1993) Mode d'action de l'hormone de croissance, *Med. Sci. 9*, 1352-1361.

Lefebvre, D.M. and Block, E. (1992) Effect of recombinant bovine somatotropin on estradiol-induced estrous behaviour in ovariectomized heifers, *J. Dairy Sci. 75*, 1461-1464.

Léonard, M., Gallo, M., Gallo, G. and Block, E. (1990) Effects of a 28-day sustained-release formulation of recombinant bovine somatotropin (rbST) administered to cows over two consecutive lactations, Can. *J. Anim. Sci. 70*, 795-809.

Léonard, M. and Block, E. (1997) Effects on nutrient and hormonal profile of long-term infusions of glucose or insulin plus glucose in cows treated with recombinant bovine somatotropin before peak milk yield, *J. Dairy Sci. 80*, 127-143.

Lissemore, K.D., Leslie, K.E., Mc Bride, B.W., Burton, J.H., Willan, A.R. and Batenma, K.G. (1991) Observations on intra mammary infection and somatic cell count in cows treated with recombinant bovine somatotropin, Can. *J. Vet. Res. 55*, 196-198.

Lynch, J.M., Barbano, D.M., Bauman, D.E., Hartnell, G.F. and Nemeth, M.A. (1992) Effect of a prolonged-release formulation of N-methionyl bovine somatotropin (Sometribove) on milk fat, *J. Dairy Sci. 75*, 1794-1809.

Maubois, J.L. (1990) Incidence de l'utilisation du Sometribove, somatotropine bovine méthionylée recombinée, sur les propriétés technologiques du lait de vache et sur les qualités organoleptiques des produits résultants, *Lait 70*, 369-382.

McDowell, G.H., Hart, I.C., Bines, J.A., Lindsay, D.B. and Kirby, A.C. (1987a) Effects of pituitary-derived bovine growth hormone on production parameters and biokinetics of key metabolites in lactating dairy cows at peak and mid lactation, Aust. *J. Biol. Sci. 40*, 191-202.

McDowell, G.H., Hart, I.C. and Kirby, A.C. (1987b) Local intra-arterial infusion of growth hormone into the mammary glands of sheeps and goats : effects on milk yield and composition, plasma hormones and metabolites, Aust. *J. Biol. Sci. 40*, 181-190.

McGuffey, R.K., Basson, R.P., Snyder, D.L., Block, E., Harrison, J.H., Rakes, A.H., Emery, R.S. and Muller, L.D. (1991) Effect of Somidobove sustained release administration on the lactation performance of dairy cows, *J. Dairy Sci. 74*, 1263-1276.

McGuire, M.A., Bauman, D.E., Miller, M.A. and Hartnell, G.F. (1992) Response of somatomedins (IGF-I and IGF-II) in lactating cows to variations in dietary energy and protein and treatment with recombinant n-methionyl bovine somatotropin, *J. Nutr. 122*, 128-136.

Melet, J. and Mouchet, C. (1989) Impact microéconomique de la BST. Journée d'information sur la somatotropine bovine. Institut Technique de l'Elevage Bovin, Syndicat de l'Industrie du Médicament Vétérinaire. Paris, 18 Avril 1989, pp. 43-56.

Michel, A., McCutcheon, S.N., Mackenzie, D.D.S., Tait, R.M. and Wickham, B.W. (1990) Effects of exogenous bovine somatotropin on milk yield and pasture intake in dairy cows of low or high genetic merit, *Anim. Prod. 51*, 229-234.

Mietton, B., Gauzere, Y. and Groualle, F. (1990) Etude comparative en technologie gruyère thermisé de la supplémentation de vaches laitières en somatotropine bovine (BST), *Revue des ENIL 144*, 14-24.

Mouchet, C. (1987) Conséquences économiques de l'emploi de la B.S.T. dans la production laitière en France, *Sciences Agronomiques (ENSA Rennes), Numéro spécial thématique*, 80 p.

Mouchet, C. (1988) Conséquences économiques de l'emploi de la B.S.T. dans la production laitière en France, *Economie Rurale n° 192-193*, 49-53.

Nielsen, M.O., Schleisner, C., Jacobsen, K. and Andersen, P.H. (1995) The effect of mammary O2 uptake, CO2 and H+ production on mammary blood flow during pregnancy, lactation and somatotropin treatment in goats, *Comp. Bioch. Physiol. 112A*, 591-599.

Paris, A. and Bories, G. (1991) Modifications des activités hépatiques de detoxification induites chez le porc par le traitement à l'hormone de croissance recombinante. Rapport interne, INRA, Laboratoire des Xénobiotiques, Toulouse.

Phipps, R.H. (1988) The use of prolonged release bovine somatotropin in milk production, Int. *Dairy Fed. Bull. n°228.*

Politis, I., Block, E. and Turner, J.D. (1990) Effect of somatotropin on the plasminogen and plasmin system in the mammary gland : proposed mechanism of action for somatotropin on the mammary gland, *J. Dairy Sci. 73*, 1494-1499.

Prosser, C.G., Fleet, I.R., Corps, A.N., Froesch, E.R. and Heap, R.B. (1990) Increase in milk secretion and mammary blood flow by intra-arterial infusion of insulin-like growth factor-I into the mammary gland in a goat, *J. Endocrinol. 126*, 437-452.

Prosser, C.G., Royle, C., Fleet, I.R. and Mepham, T.B. (1991) The galactopoietic effect of bovine growth hormone in goats is associated with increased concentrations of insulin-like growth factor-I in milk and mammary tissues, *J. Endocrinol. 128*, 457-463.

Rajamahendran, R., Shelford, J.A., Ravindran, V. and Kennelly, J.J. (1991) Effect of a single lactation treatment with recombinant bovine growth hormone on reproductive performance in dairy cows, *Anim. Reprod. Sci. 24*, 211-216.

Rémond, B., Cissé, M., Ollier, A. and Chilliard, Y. (1991) Slow release somatotropin in dairy heifers and cows fed two levels of energy concentrate. I. Performance and body condition, *J. Dairy Sci. 74*, 1370-1381.

Ronge, H., Blum, J., Clement, C., Jans, F., Leuenberger, H. and Binder, H. (1988) Somatomedin C in dairy cows related to energy and protein supply and to milk production, *Anim. Prod. 47*, 165-183.

Ronge, H. and Blum, J.W. (1989) Insulin-like growth factor I responses to growth hormone in dry and lactating dairy cows, *J. Anim. Physiol. 62*, 280-288.

Rulquin, H. and Vérité, R. (1993) The effects of bovine somatotropin injections on mammary blood flow in dry and lactating cows, *Ann. Zootech. 42*, 209.

Schams, D. (1988) Analytik des endogenen und exogenen Wachstumshormons beim Rind, BST-Symposium, Landbauforschung Völkenrode 88, 211-219.

Schams, D., Graf, F., Meyer, J., Graule, B., Mauthner, M. and Wollny, C. (1991) Changes in hormones, metabolites, and milk after treatment with sometribove (recombinant methionyl bST) in Deutches Fleckvieh and German Black and White cows, *J. Anim. Sci. 69*, 1583-1592.

Schams, D., Koll, R. and Li, C.H. (1987) Insulin-like growth factor-I stimulatesq oxytocin and progesterone production by bovine granumosa cells in culture, *J. Endocrinol. 116*, 97-100.

Schemm, S.R., Deaver, D.R., Griel, L.C Jr. and Muller, L.D. (1990) Effects of recombinant bovine somatotropin on luteinizing hormone and ovarian function in lactating dairy cows, *Biol. Reprod. 42*, 815-821.

Sechen, S.J., Dunshea, F.R. and Bauman, D.E. (1990) Somatotropin in lactating cows : effect on response to epinephrine and insulin, *Am. J. Physiol. 258*, E582-588.

Shamay, A., Cohen, N., Niwa, M. and Gertler, A. (1988) Effect of insulin-like growth factor I on deoxyribonucleic acid synthesis and galactopoiesis in bovine undifferenciated and lactating mammary tissue *in vitro*, *Endocrinology 123*, 804-809.

Shapiro, B.H., Agrawal, A.K. and Pampori, N.A. (1995) Gender differencces in drug metabolism regulated by growth hormone, *Int. J. Biochem. Cell. Biol. 27*, 9-20.

Sharma, B.K., Vendehaar, M.J. and Ames, N.K. (1994) Expression of insulin-like growth factor I in cows at different stages of lactation and in late lactation cows treated with somatropin, *J. Dairy Sci. 77*, 2232-2241.

Simianer, H. and Wollny, C. (1989) Impact of the potential use of Bovine Somatotropin (BST) on the efficiency of a conventional dairy cattle breeding scheme, *Livest. Prod. Sci. 22*, 31-47.

Spandidos, D.A., Zoumpourlis, V., Kotsinas, A., Tsiriyotis, and Sekeris, C.E. (1990) Response of human immunodeficiency virus long terminal repeat to growth factors and hormones, *Anticancer Res. 10*, 1241-1246.

Spicer, L.J., Tucker, W.C. and Adam, G.D. (1990) Insulin-like growth factor I in dairy cows: relationships among energy balance, body condition, ovarian activity and oestrous behavior, *J. Dairy Sci. 73*, 929-937.

Stanisiewski, E.P., McAllister, J.F., Ash, K.A., Taylor, V.N., Kratzer, D.D. and Lauderdale, J.W. (1994) Production performance of dairy cattle administered recombinant derived bovine somatotropin (USAN, Somavubove) daily : a dose range study, *Domest. Anim. Endocrinol. 11*, 239-260.

Stricker, P. and Grueter, F. (1928) Action du lobe antérieur de l'hypophyse sur la montée laiteuse, *C.R. Acad. Sci. Paris 99*, 1978-1980.

Tanner, J.W. and Hauser, S.D. (1989) Molecular evidence for the presence of the somatotrophin receptor in the bovine ovary, *J. Anim. Sci. 67 (Suppl. 1),* 413 (Abstr).

Thomas, J.W., Erdman, R.A., Galton, D.M., Lamb, R.C., Arambel, M.J., Olson, J.D., Madsen, K.S., Samuels, W.A., Peel, C.J. and Green, G.A. (1991) Responses by lactating cows in commercial dairy herds to recombinant bovine somatotropin, *J. Dairy Sci. 74*, 945-964.

Vérité, R. and Chilliard, Y. (1992) Effect of age of dairy cows on body composition changes throughout the lactation cycle as measured with deuteriated water, *Ann. Zootech. 41*, 118.

Vérité, R., Rulquin, H. and Faverdin, Ph. (1989) Effects of slow-released somatotropin on dairy cow performances, In Sejrsen, K., Vestergaard, M., Neimann-Sorensen, A., (Eds). Use of Somatotropin in *Livestock Production, 269-273*, Elsevier Applied Science, London.

Vernon, R.G., Faulkner, A., Finley, E., Watt, P.W. and Zammit, V.A. (1995) Effects of prolonged treatment of lactating goats with bovine somatotropin on aspects of adipose tissue and liver metabolism, *J. Dairy Res. 62,* 237-248.

Vernon, R.G. and Flint, D.J. (1989) Role of growth hormone in the regulation of adipocyte growth and function. In RB Heap, CG Proser, GE Lamming (eds) : *Biotechnology in growth regulation.* Butterworths, London, pp. 57-71.

Vicini, J.L., Buonomo, F.C., Veenhuizen, J.J., Miller, M.A., Clemmons, D.R. and Collier, R.J. (1991) Nutrient balance and stage of lactation affect responses of insulin, insulin-like growth factors I and II, and insulin-like growth factor-binding protein 2 to somatotropin administration in dairy cows, *J. Nutr. 121*, 1656-1664.

Webb, R., Gong, J.G. and Bramley, T.A. (1994) Role of growth hormone and intrafollicular peptides in the control of follicle development in cattle, *Therionology 41*, 25-30.

Weigent, D.A. and Blalock, J.E. (1989) Structural and functional relationship between the immune and neuroendocrine systems, *Bull. Inst. Pasteur 87*, 61-92.

Willeberg, P. (1993) Bovine somatotropin and clinical mastitis : epidemiological assessment of the welfare risk, *Livest. Prod. Sci. 36*, 55-66.

Witkamp, R.F., Nijmeijer, S.M., van Duin, C.T.M., Bevers, M.M., Wensing, T.H., van Gogh, H. and van Miert A.S.J.P.A.M. (1989) Has somatotropin (bST) an effect upon drug disposition ? Comparative studies in goats and cattle with sulphamidine and antipyrine after parenteral administration of bST, zeranol and proligestone, *J. Vet. Pharmacol. Therap. 12*, 163-178.

Witkamp, R.F., Nijmeijer, S.M., van Duin, C.T.M., Noordhoek, J. and van Miert, A.S.J.P.A.M. (1993) The regulation of oxidative drug metabolism by growth hormone in the dwarf goat: differences from and similarities to the mechanisms in rats, *J. Endocrinol. 136*, 313-317.

Zhao, F.Q., Moseley, W.M., Tucker, H.A. and Kennelly, J.J. (1996) Regulation of glucose transporter gene expression in mammary gland, muscle, and fat of lactating cows by administration of bovine growth hormone and bovine growth hormone-releasing factor, *J. Anim. Sci.* 74, 183-189.

Zhao, X., Burton, J.H. and Mc Bride, B.W. (1992) Lactation, health and reproduction of dairy cows receiving daily injectable or substained-release, *J. Dairy Sci. 75*, 3122-3130.

IS-IT POSSIBLE TO DETECT BOVINE TREATED WITH BST ?

BERTOZZI C., PORTETELLE D., PIRARD M., PARMENTIER I., HAEZEBROECK V. AND RENAVILLE, R.
Animal and Microbial Biology Unit, Gembloux Agricultural University, 5030 Gembloux, Belgium

Abstract

Biotechnology has been heralded as a science that will have a revolutionary impact on agriculture development. Joint efforts among industry, universities, and authorities will result in new and novel products that establish new frontiers in the livestock industries. One of the first biotechnology products is somatotropin.

The galactopoietic effect of bovine somatotropin in dairy cows has been firmly established. Also, administration of exogenous somatotropin markedly improves productive efficiency in lactating cows. However, in the international network, a control strategy of treated animals is required not only because BST use is not authorised in Europe and in Canada, but also because rbST treatment interferes with the genetic selection scheme of reproductive animals. Besides studies on eventual adverse effects on animal health, ethical aspects as well as consumer protection have to be considered. Therefore, the present review tends to describe different possible ways to develop strategies for identification of BST treated cattle.

1. Introduction

Sixty years ago, Russian scientists were the first to report that bovine pituitary extracts stimulated milk production in dairy cows (Azimov and Krouze, 1937). During the 1940s, scientists in the United Kingdom further refined these extracts and established that somatotropin (ST) was the component with a galactopoietic effect (Young, 1947). During World War II, attempts by British scientists to use bST to offset milk shortage were hampered by the limited amount of bovine ST (bST), which must be purified from each pituitary gland. With the advent of biotechnology, it has been possible to clone in bacteria the gene for bST and to induce the bacteria to produce higher quantities of the recombinant protein. In late 1981, the first experiment utilising recombinant methionyl bST indicated that it was as effective as pituitary derived ST in enhancing milk production (Bauman *et al.*, 1982). Studies with recombinant bST (rbST) have

R. Renaville and A. Burny (eds.), Biotechnology in Animal Husbandry, 99–110.

consistently demonstrated that treatment resulted in an unprecedented increase in milk secretion. The magnitude of the gain in efficiency of milk production was equal to that normally achieved over a 10 to 20 year period with artificial insemination and genetic selection technologies (Bauman, 1992). Moreover, because this biotechnology increases the food output per unit of feed resource input, the benefits of bST include effects on environmental impact through reductions in animal waste products.

Despite the advantages afforded by this molecule, rbST is the subject of hot controversy over health effects on the consumer or the target animal, specially between U.S. Food and Drug Administration and the European Committee for Veterinary and Medical Products scientists. Based on 40 years of studies on the hormone by its manufacturers and independent scientists, the FDA has concluded that rbST presents « no increased health risk to consumers ». On the contrary, European researchers believe that the safety of bST has not been established with adequate scientific rigour (European Report, 1999), because of possible adverse effects of substantially increased concentrations of insulin-like growth factor-I (IGF-I) in the milk of rbST-treated cows. At concentrations of IGF-I found in milk, it is unlikely that systemic effects would be induced, but there is already evidence for local effects on gut tissues (Mepham, 1992; Playford *et al.*, 1993; Olanrewaju *et al.*, 1992). Risk characterisations were also pointed out by different authors like i.e. an association between circulating insulin-like growth factor-I (IGF-I) levels which are increased by BST treatment and a possible effect on breast and prostate cancer development (Epstein, 1996) or an increase of antibiotic residues in milk after the BST treatment in relation with mastitis problems (Willeberg, 1993).

Recently, Health Canada Office also announced that rbST would not be approved for use in Canada because the hormone increases the likelihood of health problems in dairy cattle (Nolen, 1999). The decision was based on findings of review panels examining the effects of ST on human and animal health. One report that indicated cows treated with rbST had increased frequencies of mastitis, infertility, and lameness.

Finally, because rbST is not authorized in all countries, it has been speculated about the possible alteration by bST of genetic improvement programs. However, Weigel *et al.* (1998) have concluded their study that bST treatment had little impact on the genetic selection decisions.

2. Structure and Chemical Identity of Somatotropin

Pituitary bST exists as 4 variants comprised of 190 or 191 amino acids with heterogeneity at the amino terminus (Phe or Ala-Phe) and at position 126 (Val or Leu) of the molecule (Santome *et al.*, 1976). Synthesis of recombinant bST variants has allowed exploration of the effect of various amino acid substitutions on the biopotency of the bST molecule (Bauman *et al.*, 1985). Substitution of the amino-terminal alanyl residue of rbST with methionine did not affect milk response in dairy cows but deletion of the first four amino acids lowered the response (Eppard *et al.*, 1992). [Ala^1]-bST variants with valine at position 127 elicited a greater milk response than Leu^{127} variants even though immunoreactive bST concentrations in blood were statistically equivalent

(Eppard *et al.*, 1992). Other molecular bST variants are also available, harboring either three additional amino acids (Burton *et al.*, 1991a ; Juskevich and Guyer, 1990) or nine additional amino acids (Bick *et al.*, 1990; Juskevich and Guyer, 1990; Zwickl *et al.*, 1990) at the amino terminus.

The most popular biosynthetic form of rbST was developed by Monsanto (designated as Sometribove®) and has the same number and sequence of amino acids as except for the substitution of methionine for alanine at the amino terminus.

3. Strategies Developed for Identification of rbST Treated Animals

It is important to exactly identify authorized or banned exogenous hormones to meet with consumer expectations if systems and procedures to assess the safety of food components for consumption are to be considered as efficient.

Because recombinant preparations are chemically similar to pituitary GH, analytical methods are presently available to identify natural and recombinant bST in biological fluids. It is however impossible to distinguish GH from natural or recombinant origin. This point is one of the major arguments of the consumer associations to ban bST use in dairy industry. To resolve this problem, numerous scientists develop alternative indirect methods to discriminate control and treated animals. These methods include measurement of bST, IGF-I, IGFBP-3 and/or IGFBP-2 levels in blood or milk or evaluation of immune responses to exogenously administered GH.

3.1. ALTERATION OF ENDOGENOUS HORMONAL OR PROTEIN FACTORS BY BST TREATMENT

3.1.1. bST Variations

Exogenous bST treatment increases considerably plasma ST concentrations (by 500%) (Vanderkooi *et al.*, 1995). Slaba *et al.* (1994) observed that mean plasma bST concentration in Somidobove® (Eli Lilly / Elanco rbST) or Sometribove® treated cows increased rapidly during the first 24 h following hormone administration and returned to pre-treatment levels by day 10 post-injection. This is not in agreement with the results obtained by Schams *et al.* (1991) who reported a gradual increase of plasma bST concentration until day 3 after the injection of Somidobove and with the results reported by Cisse *et al.* (1991) who found higher plasma bST concentrations at day 10 after administration of Sometribove. Moreover, Zhao *et al.* (1994) observed a significant increase ($P<0.05$) in plasma bST concentrations for all early, mid- and late lactation periods irrespective of the mode of GH administration, daily injections or sustained-release bST treatment. Recently, we observed that plasma bST levels measured by an ELISA sandwich assay using two monoclonal antibodies against rbST Sometribove®, increased immediately after rbST treatment according to the rhythm of injections. Highly significant differences ($P<0.01$) between control and treated animals were observed after the first injection until the end of the trial. Animal response was maximum 5 to 9 days after the injection.

Because milk samples are easily available in food quality controls, some studies aimed to identify bST-treated cows via the induced increase of hormone concentrations in milk. Daily or sustained administration of exogenous bST to dairy cattle has been shown to raise milk concentrations of bST (Zhao *et al.*, 1994). However, the daily injection group showed constant hormone profiles, while the sustained release group showed a cyclic pattern with high concentrations early in the injection cycle followed by declining concentrations thereafter.

The main problem of bST quantification in milk is the lower hormone concentrations (ten time less than in plasma) which are near the limit of detection of numerous published RIA or ELISA methods. Using an electrochemiluminescent (ECL) detection system and our monoclonal antibodies, we quantified bST in milk (Baronheid *et al.*, 2000) with a detection limit of 5 pg bST/ml. In our study, the ECL method appeared as a reliable alternative to RIA and ELISA, characterized by a high specificity of antibodies used and an improved assay sensitivity. Milk hormone levels were significantly higher ($P<0.05$) after the first bST treatment, and remained significantly higher ($P<0.05$) than in the control group until the end of the experiment. After validation on a sufficient number of biological samples, the ECL-based assay could be used to identify bST treated dairy cows.

3.1.2. IGF-I Quantification

The mechanisms inducing galactopoiesis in ruminants are complex and involve a multiplicity of events. Nevertheless, observations have suggested that the effects of bST on the mammary gland itself are mediated in part by IGF-I, and that the availability of IGF-I to mammary tissue is an important component of the overall galactopoietic response to bST (Cohick *et al.*, 1989; Mielke *et al.*, 1990; Prosser *et al.*, 1992; Zhao *et al.*, 1994; Bauman, 1999). Following Sometribove® administration, the IGF-I concentrations 24 h post-injection were elevated to nearly 2-fold and by 48 h to nearly 4-fold values. In the next few days, the mean IGF-I concentration increased and reached the maximum value at day 10. Then, it decreased but still remained higher on day 14 than those in placebo-treated cows (Slaba *et al.*,1994). This period corresponded to the lowest activity of the implants and imposed a new injection. More recently, heifers treated with bST presented a significant increase of plasma IGF-I on day 10 after the first injection and the IGF-I level remained significantly higher than the control group until the end of the trial (150 days). IGF-I concentrations decreased and returned to pre-injection or control levels within 30 days after the last injection (Bertozzi *et al.*, 1998).

This increase in serum IGF-I was accompanied by increases of ≈140% in the abundance of IGF-I mRNA in the liver (Vanderkooi *et al.*, 1995). Because IGF-I synthesis is regulated, in large part, at the level of mRNA abundance (Sharma *et al.*, 1994), these results suggest that increased synthesis of IGF-I in the liver is largely responsible for the increased concentration of IGF-I in serum of cows treated with bST. Moreover, Vanderkooi *et al.* (1995) found that continuous bST treatement nearly abolished pulsatility of endogenous ST by pituitary gland. These data indicate that steady concentrations of ST may stimulate the IGF axis more than do pulsatile concentrations. This hypothesis is consistent with data from rats where continuous infusions of ST increased serum IGF-I in hypophysectomized rats more than did intermittent injections of ST at the same daily dose (Bick *et al.*, 1992).

The concentration of immunoreactive IGF-I in bovine milk varies depending on the stage of lactation. Malven *et al.* (1987) reported levels of 20 ± 2 nmol/l for IGF-I in *pre-partum* milk, declining to 3.2 ± 0.3 nmol/l by d 6 *post-partum*. Prosser *et al.* (1989) showed that after peak lactation (35-47 weeks) only trace amounts of IGF-I (0.44 ± 0.04 nmol/l) are present in cows'milk. Of particular interest was the 3.6-fold increase in the concentrations of IGF-I to a maximum of 1.6 ± 0.2 nmol/l following 7 d of treatment with rbST. There was a parallel increase in milk yield over this period such that the output of IGF-I into milk of one udder half increased 6-fold. Daxenberger *et al.* (1998) demonstrated that IGF-I measurements in milk might be useful to monitor for bST treatment in milk samples, especially as a screening procedure prior to further confirmatory testing, for which commonly accepted methods remain to day inefficient.

3.1.3. Determination of IGFBP-3/IGFBP-2 Ratio

In serum of different species, over 99% of IGF molecules circulate complexed to at least six distinctly different binding proteins (IGFBP), suggesting that the IGFBP play a pivotal role in IGF bioactivity (Davis *et al.*, 1989; McGuire *et al.*, 1992, Bauman, 1999). Association with IGFBPs increases IGF half-life dramatically, as compared to free IGF, and IGFBP are thought to regulate bioavailability of IGF in the target tissues. The major IGFBP species is IGFBP-3, which contains more than 75% of the circulating IGF and forms a ternary complex with IGF and the acid labile subunit (ALS) (Jones and Clemmons, 1995).

Numerous studies with IGFBP data have been published, but usually measurements were at that time obtained by western ligand blot. Based on this method, Sharma *et al.* (1994) and Vanderkooi *et al.* (1995) observed that bST treatments increased the amount of IGFBP-3 in serum by ≈ 60 % and decreased the amount of IGFBP-2 by 25 %. This is consistent with observations in humans and pigs after ST administration (Walton and Etherton, 1989; Bauman, 1999). In confirmation of results of Cohick *et al.* (1992) and Vicini *et al.* (1991), Bertozzi *et al.* (1998) observed with Western Ligand blotting (WLB) that plasma IGFBP-3 was positively ($P<0.05$) and IGFBP-2 negatively ($P<0.05$) correlated with IGF-I which increased significantly during bST treatment. Results showed that the calculated IGFBP-3/IGFBP-2 ratio was drastically changed by bST treatment and amplifies significant differences between treated and untreated animals.

However, this WLB technique is both more time-consuming and less precise than the RIA method. Recently, a RIA method for IGFBP-2 measurement in bovine species has been published (Vleurick *et al.*, 2000). These authors reported that treated groups with bST or somatocrinine (GRF) both have diminished IGFBP-2 levels as compared to the control group, corresponding to our present data. However, no difference between the bST and GRF groups could observed using WLB, while the RIA was superior in this respect. Recently, we produced antibodies against bovine IGFBP-3 and developed a RIA method (Bertozzi *et al.*, 2000). Using this new strategy, we confirmed that IGFBP-3 synthesis and IGFBP-3/IGFBP-2 ratio were increased by bST treatment (unpublished data). We concluded that as in cattle little circadian variation occurred in plasma levels of IGFBP-2 and IGFBP-3 as well, and it is tempting to speculate on the use of bovine IGFBP-2 and IGFBP-3 as indicators for bST treated animal.

3.1.4. Hormonal Approach in bST-Treated Animal Strategy

As mentioned above, bST treatment in cattle significantly affect endogenous somatotropin, IGF-I, IGFBP-2 and IGFBP-3 plasma and milk levels. Also, quantitative assessment of these factors, particularly IGFBPs since RIA methods are available, could be used to identify treated animal.

However, because the somatotropic axis is known to be modulated by different factors other than bST treatment, the detection strategy of bST treated animals by this approach requires numerous validation tests including the effects of genetic potential, nutritional plan, stage of lactation, health status, ... For example, based on the difference in concentrations of IGFBPs between early lactation and the dry period, it seems that IGFBP-3 may be lower during negative nutrient balance; on the other hand, IGFBP-2 may be higher (McGuire *et al.*, 1992; Formigoni *et al.*, 1996). Conversely, basal concentrations of IGFBP-2 were highest during early lactation when ST was high, and lowest during the dry period when ST was the lowest or when IGF-I was high, and was reduced during bST administration when IGF-I concentrations were elevated (McGuire *et al.*, 1992). Similar results were described by Sharma *et al.* (1997) for hepatic IGFBP mRNA expression after diet restriction or bST treatement. These different parameters could affect the cut-off values for each factors used to discriminate treated and untreated animals.

Finally, considering bibliographic information, milk appears as the best mistake biological fluid for detection bST-treated animal. Considering the relative nycthemeral stability and the recent RIA quantification, IGFBPs ratio could be the most interesting parameter for bST treatment detection in dairy cattle.

3.2. HUMORAL IMMUNITY APPROACH

The actions of ST and IGF-I on immune cell activity suggest the existence of a specific neuro-endocrine-immune axis. This concept was proposed more than 50 years ago. ST, a central component of this axis, has many functions at both a molecular and cellular level, including thymocyte proliferation, stimulation of the cytotoxic activity of natural killer cells and induction of lymphocyte proliferation. Exogenous ST given by injection has been shown to increase thymic, lymph node, and spleen mass in ST-deficient animals (reviewed by Berczi, 1997). ST also was shown to regenerate a variety of T cell-dependent immune responses in ST-deficient animals (Berczi, 1986; Kelley, 1989). Furthermore, exogenous ST has been implicated in the function of T-helper cells (Arrenbrecht and Sorkin, 1973) and in the maintenance of peripheral blood leukocyte numbers in immunodeficient dwarf mice (Fabris *et al.*, 1971a , 1971b). Further evidence for the participation of ST in the immune system was gained by studies that demonstrated the presence of ST receptors on lymphocytes (Hattori *et al.*, 1996). Many of the growth-promoting and galactopoietic effects of ST are mediated by IGF-I. It is possible that IGF-I also participates in immunomodulatory phenomena such as those reported above for exogenous ST. Binding of ST to its receptors on lymphocytes stimulates the production of IGF-I, which mediates the effects of ST on cell proliferation (Geffner, 1997).

Exogenous hormones elicit mild immunological responses in human subjects. Therapeutic administration of human pituitary or recombinant somatotropin (hST) to humans resulted in the production of circulating anti-hST antibodies (Preece, 1986). Approximately 20 to 40% of patients treated with recombinant methionyl hST developed measurable antibody levels (Rougeot *et al.*, 1991). Most cows treated with bST developed antibody against bST (Zwickl *et al.*, 1990), but not adverse health or performance effects were observed. Burton *et al.* (1991b) also reported no detrimental effect of daily injection of bST on blood concentrations of immunoglobulin (Ig). Even if rbST preparations can be chemically similar to pituitary bST (Langley *et al.*, 1987), antigenic differences may be expected between specific molecular variants that differ structurally depending on the source of rbST. Eppard *et al.* (1992) also observed that only bST-treated cows developed antibodies and an increase in relative anti-bST binding compared with their individual pre-treatment binding percentages.

Bertozzi *et al.* (1998) demonstrated by ELISA methods that bST treated animals developed circulating anti-bST antibodies which were detected after the second injection and during the rest of the experimental period (Figure 1). Ten days after the first bST injection and until 20 days after the last one, mean optical densities in treated group were significantly ($P<0.01$) over the maximum individual value within the control group during the whole experimental period.

3.3. CELLULAR IMMUNITY APPROACH

Exogenous ST has been shown to reconstitute deficiencies in the lymphocytic component of the bone marrow and the thymus, numbers of peripheral blood lymphocytes (PBL), and polymorphonuclear cells, and cell-mediated immune responses (e.g., contact sensitivity induced by dinitrochlorobenzene, skin graft rejection time, and proliferative responsiveness of PBL to mitogen stimulation) in ST-deficient animals (Fabris *et al.*, 1971a, 1971b; Roth *et al.*, 1984; Berczi *et al.*, 1986; Kelley, 1989). Discovery of ST receptors on human lymphocytes and on human peripheral blood mononuclear cells (Gagnerault *et al.*, 1996 ; Mertani *et al.*, 1996) suggests an additional direct peripheral role of ST on lymphocytic cells. Moreover, it has also been proposed that ST has an autocrine or paracrine mode of action in the immune system. Hattori *et al.* (1990) reported that short-term cultures of human B lymphocytes and T lymphocytes secreted immunoreactive ST.

Burton *et al.* (1991b) showed that lymphocytes from the blood of cows treated with rbST responded to mitogen with higher proliferative responsiveness than cells from control cows, but that this effect required long-term treatment and adequate mitogen concentrations to be detected. Recombinant bST treatment during the non lactating period enhanced interleukin-2 (IL-2) secretion by mononuclear cells isolated from cows treated also with rbST during lactation. (Torre *et al.*, 1993). However, rbST treatment during the nonlactating period had little effect on IL-2 secretion by mononuclear cells from cows not treated with rbST during lactation. An another indicator of cell-mediated immunity was the contact sensitivity induced by dinitrochlorobenzene. Burton *et al.* (1992) observed that the cutaneous sensitivity response was similarly not affected by bST. These results were in contrast with their previous finding of augmented proliferative responsiveness of mitogen-stimulated peripheral blood lymphocytes in bST-treated versus control cows. It is possible that exogenous bST affects some immune processes and lymphocyte subsets and not others. That contradiction demonstrates some limits of this approach. Limits could also be induced by the great individual variability observed with that kind of analytical method (e.g. cell proliferation).

4. Conclusions

Some parameters of the somatotropic axis (bST, IGF-I and specially IGFBP-3/IGFBP-2 ratio) and the detection of rbST antibodies could provide good indirect methods to screen rbST treated animals. But further studies are needed. It will be necessary to study the transmission of antibodies between rbST cows and their calves during the nursing stage in order to avoid possible false positive animals. It wil be also interesting studying the individual variability of antigenic activity against rbST in a wide population and the effects of nutritional, genetic, health and environmental factors on hormone and protein levels in order to determine a threshold value for treated animal. An interesting perspective is offered by the ECL method due to its high sensitivity. Moreover, stability of reagents used in this technique is an advantage for standardisation of the protocol.

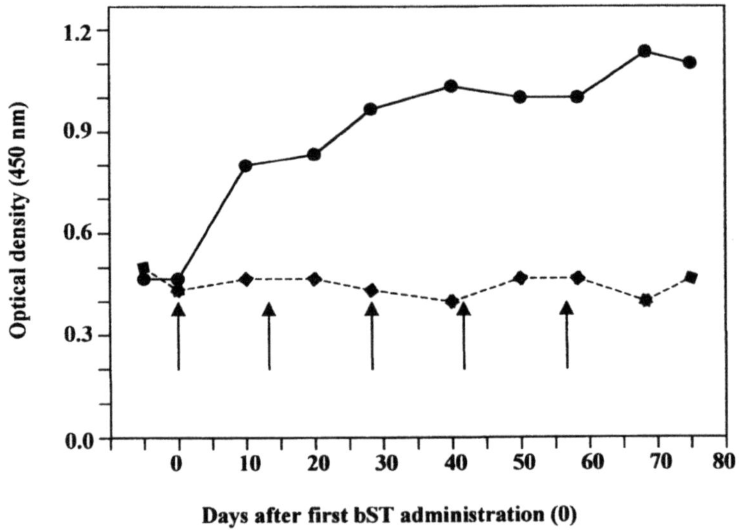

Figure 1. Mean plasma bST antibody estimated by ELISA (Optical Density) in not treated (n=6) (-----------) or bST treated (n=6) (————) Belgian White Blue bST Somitrobove ® treated heifers (↑ = day of treatment).

Acknowledgements

Some researches presented in this review were funded by the Federal Belgian Ministry of Small Enterprises, Traders and Agriculture - DGVI (Brussels, Belgium),

References

Arrenbrecht, S. and Sorkin, E. (1973) Growth hormone induced T cell differentiation, *Eur. J. Immunol. 3*, 601-605.

Azimov, G.J. and Krouze, N.K. (1937) The lactogenic preparations from the anterior pituitary and the increase of milk yield in cows, *J. Dairy Sci. 20*, 289-306.

Baronheid, C., Bertozzi, C., Renaville, R., Portetelle, D. and Deaver, D. (2000) Identification of animals treated with bovine somatotropin : a new non-isotopic detection system for bST, *Base 4*, 28.

Bauman D.E., Eppard P.J., DeGeeter M.J. and Lanza G.M. (1985) Responses of high-producing dairy cows to long-term treatment with pituitary somatotropin and recombinant somatotropin, *J. Dairy Sci. 68*, 1352-1362.

Bauman, D.E. (1992) Bovine somatotropin: review of an emerging animal technology, *J. Dairy Sci. 75*, 3432-3451.

Bauman, D.E. (1999) Bovine somatotropin and lactation : from basic science to commercial application, *Dom. Anim. Endocrinol. 17*, 101-116.

Bauman, D.E., DeGeeter, M.J., Peel, C.J., Lanza, G.M., Gorewit, R.C., and Hammond, R.W. (1982) Effects of recombinantly derived bovine growth hormone (bGH) on lactational performance of high yielding dairy cows, *J. Dairy Sci. 65(Suppl.1)*, 121.

Berczi, I. (1986) The effects of growth hormone and related hormones on the immune system, in *Pituitary Function and Immunity*, CRC Press Inc Boca Raton, FL, pp. 133.

Berczi, I. (1997) Pituitary hormones and immune function, *Acta Paediatr. (Suppl.) 423*, 70-75

Bertozzi, C., Portetelle, D., Massart, S., Deaver, D., Vleurick, L., Parmentier, I., Mortiaux, F. and Renaville, R (1998) Antibody formation and plasma profiles of insulin-like growth factor-I (IGF-I) and IGF-binding proteins in growing heifers after sustained-release exogenous bST administration, *Base 2 (Special Issue)*, 15.

Bertozzi, C., Portetelle, D., Parmentier, I., Haezebroeck, V., Pirard, M. and Renaville, R (2000) A homologous radioimmunoassay for insulin-like growth factor binding protein-3 in cattle, *Endocrinology, submitted.*

Bick, P.H., Brown, A.C.G., Smith, H.W., Tamura, R.N. and Zwickl, C.M. (1990) Antibody responses to recombinant bovine somatotropin in dairy cows, *Anim. Biotechnol. 1*, 61-73.

Brief, S., Kelley, K.W., Davila, D.R., Simon J., Hammer, R.E. and Brinster, R.L. (1987) Effects of high levels of growth hormone on lectin-induced proliferation of splenocytes from aged, trangenic and nude rodents, *Fed. Proc. 46*, 4508.

Burton, J.L., Mc Bride, B.W., Kennedy, B.W., Burton, J.H. Elsasser, T.H. and Woodward, B. (1991a) Serum immunoglobulin profiles of dairy cows chronically treated with recombinant bovine somatotropin, *J. Dairy Sci. 74*, 1589-1598.

Burton, J.L., Mc Bride, B.W., Kennedy, B.W., Burton, J.H. Elsasser, T.H. and Woodward, B. (1991b) Influence of exogenous bovine somatotropin on the responsiveness of peripheral blood lymphocytes to mitogen, *J. Dairy Sci. 74*, 916-928.

Burton, J.L., Mc Bride, B.W., Kennedy, B.W., Burton, J.H. Elsasser, T.H. and Woodward, B. (1992) Contact sensitivity and systemic antibody responses in dairy cows treated with recombinant bovine somatotropin, *J. Dairy Sci. 75*, 747-755.

Cisse, M., Chilliard,Y., Coxam,V., Davicco, M.J. and Remond, B. (1991) Slow release somatotropin in dairy heifers and cows fed two levels of energy concentrate. 2. Plasma hormones and metabolites, *J. Dairy Sci. 74*, 1382-1394.

Cohick, W.S., Plaut, K., Sechen, S.J. and Bauman, D.E. (1989) Temporal pattern of insulin-like growth factor-I response to exogenous bovine somatotropin in lactating cows, *Domest. Anim. Endocrinol. 6*, 263-274.

Cohick, W.S., McGuire, M.A., Clemmons, D.R. and Bauman, D.E. (1992) Regulation of insulin-like growth factor-binding proteins in serum and lymph of lactating cows by somatotropin, *Endocrinology 130,* 1508-1514.

Davis, S.R., Hodgkinson, S.C., Moore, L.G. and Gluckman, P.D. (1989) Improved estimates clearence of 125I-labelled insulin-like growth factor-I carrier protein complexes from blood plasma of sheep, *J. Endocrinol. 123,* 469-475.

Daxenberger, A., Breier, B.H. and Sauerwein, H. (1998) Increased milk levels of insulin-like growth factor 1 (IGF-1) for the identification of bovine somatotropin (bST) treated cows, *Analyst. 123,* 2429-2435.

Eppard, P.J., Hudson, S., Cole, W.J., Hintz, R.L., Hartnell, G.F., Hunter, T.W., Metzger, L.E., Torkelson, A.R., Hammond, B.G. and Collier, R.J. (1991) Response of dairy cows to high doses of a sustained-release bovine somatotropin administered during two lactations. 1. Production response, *J. Dairy Sci. 74,*3807-3821.

Eppard, P.J., Bentle, L.A., Violand, B.N., Ganguli, S., Hintz, R.L., Kung, L. Krivi, G.G. and Lanza, G.M. (1992) Comparison of the galactopoietic response to pituitary-derived and recombinant-derived variants of bovine growth hormone, *J. Endocrinol. 132,* 47-56.

Epstein, S.S. (1996) Unlabeled milk from cows treated with biosynthetic growth hormones: a case of regulatory abdication, *Int. J. Health Serv. 16,* 173-185.

European Report (1999) Public health aspects on the use of bovine somatotropin, DGXXIV of the European Commission, http://europa.eu.int/comm/dg24/health/sc/scv/out19_en.html.

Fabris, N., Pierpaoli, W. and Sorkin, E. (1971a) Hormone and the immunological capacity. III. The immunodeficiency disease of the hypopituitary Snell Bagg dwarf mouse, *Clin Exp. Immunol. 9,* 209-226.

Fabris, N., Pierpaoli, W. and Sorkin, E. (1971b) Hormone and the immunological capacity. IV. Restorative effects of developmental hormones or of lymphocytes on the immunodeficiency syndrome of the dwarf mouse, *Clin Exp. Immunol. 9,* 227-241.

Formigoni, A., Cornil, M.C., Prandi, A., Mordenti, A., Rossi, A., Portetelle, D. and Renaville, R. (1996) Effect of propylene glycol supplementation around parturition on milk yield, reproduction performances and some hormonal and metabolic parameters in dairy cows, *J. Dairy Res. 63,* 11-24.

Gagnerault, M.C., Postel-Vinay, M.C. and Dardenne, M. (1996) Expression of growth hormone receptors in murine lymphoid cells analyzed by flow cytofluorometry, *Endocrinology 137,* 1719-1726.

Geffner, M. (1997) Effects of growth homone and insulin-like growth factor I on T- and B-lymphocytes and immune function, *Acta Paediatr. 423,* 76-79.

Hattori, N., Shimatsu, A., Sugita, M., Kumagai, S. and Imura, H. (1990) Immunoreactive growth hormone (GH) secretion by human lymphocytes: augmented release by exogenous GH, *Biochem. Biophys. Res. Commun. 168,* 396-401.

Hattori, Y., Vera, J.C., Rivas, C.I., Bersch, N., Bailey, R.C. and Geffner, M.E. (1996) Decreased insulin-like growth factor I (IGF-I) receptor expression and function in immortalized African Pygmy T-cells, *J. Clin. Endocrinol. Metab. 81,* 2257-2263.

Jones, I.J. and Clemmons, D.R. (1995) Insulin-like growth factors and their binding proteins: biological actions, *Endoc. Rev. 16,* 3-34.

Juskevich, J.C. and Guyer, C.G. (1990) Bovine growth hormone : human food safety evaluation . *Science 249,* 875-884.

Kelley, K.W. (1989) Commentary : growth hormone, lymphocytes and macrophages, *Biochem. Pharmacol. 38,* 705-715.

Langley, K.E., Lai, P.H., Wypych, J., Everett, R.R., Berg, T.F., Krabill, L.F., Davis J.M. and Souza, L.M. (1987) Recombinant-DNA-derived bovine growth hormone from Escherichia coli 12. Biochemical, biophysical, immunological and biological comparison with the pituitary hormone, *Eur. J. Biochem. 163,* 323-328.

Malven, P.V., Head, H.H., Collier, R.J. and Buonomo, F.C. (1987) Periparturient changes in secretion and mammary uptake of insulin and in concentration of insulin-like growth factors in milk of dairy cows, *J. Dairy Sci. 70,* 2254-2265.

McGuire, M.A., and Bauman, D.E. (1997) Regulation of nutrient use by bovine somatotropin : the key to animal performance and well-being, *Proc. Ixth Int. Conf. Prod. Dis.,*125-137.

McGuire, M.A., Vicini, J.L., Bauman D.E. and Veenhuizen, J.J. (1992) Insulin-like growth factors and binding proteins in ruminants and their nutritional regulation, *J. Anim. Sci. 70,* 2901-2908.

Mepham, T.B. (1992) Public health implications of bovine somatotropin use in dairying: discussion paper, *J. R. Soc. Med. 85*, 736-739.

Mertani, H.C., Delehaye-Zervas, M.C., Martini, M., Postel-Vinay, M.C. and Morel, G. (1996) Localization of growth hormone receptor messenger RNA in human tissues, *Endocrine 3*, 135-142.

Mielke, H. Lochmann, G. Simon, I. Bier, H. Jahreis, G. Hesse, V. and Gallowitsch, F. (1990) bSTH and IGF-I levels in blood and milk of black and white dairy cows before and after application of recombinantly derived bovine somatotropin, *J. Vet Med. A 37*, 554-557.

Nolen, S. (1999) US and Canada draw different conclusions on rbST, *J. Am. Vet. Med. Assoc. 214*, 985-986.

Olanrewaju, H., Patel, L. and Seidel, E.R. (1992) Trophic action of local intraileal infusion of insulin-like growth factor I: polyamine dependence, *Am. J. Physiol. 263*, E282-E286.

Playford, R.J., Woodman, A.C. and Clark, P (1993) Effect of luminal growth factor preservation on intestinal growth, *Lancet 341*, 843-848.

Preece, M.A., (1986) Experience of treatment with pituitary derived human growth hormone with special reference to immunological aspects, in R.D.G. Milner and H. Flodh (eds.), *Med. Educ. Serv.*, Oxford, p 9.

Prosser, C.G., Fleet, I.R., and Corps, A.N. (1989) Increased secretion of insulin-like growth factor I into milk of cows treated with recombinantly derived bovine growth hormone, *J. Dairy Res. 56*, 17-23.

Prosser, C.G., Royle, C., Fleet, I.R. and Mepham,T.B. (1992) The galactopoietic effect of bovine growth hormone in goats is associated with increased concentrations of insulin-like growth factor- I in milk and mammary tissue, *J. Encrinol. 128*, 457-463.

Ross, M., Francis, G.L., Szabo, L., Wallace, J.C. and Ballard, F.J. (1989) Insulin-like growth factor (IGF)-binding proteins inhibit the biological activities of IGF-I and IGF-2 but not des-(1-3)-IGF-I, *Biochem. J. 258*, 267-273.

Roth, J.A., Kaeberle, M.L., Grier, R.L. Hopper, J.G., Spiegel, R.E. and McAllister, H.A. (1984) Improvement in clinical condition and thymus morphologic features associated with growth hormone treatment of immunodeficient dwarf dogs, *Am. J. Vet. Res. 45*, 1151-1156.

Rougeot, C., Marchand, P., Dray, F., Girard, F., Job, J.C., Pierson, M., Ponte, C., Rochiccioli, P. and Rappaport, R. (1991) Comparative study of biosynthetic human growth hormone immunogenicity in growth hormone deficient children, *Horm. Res. 35*, 76-81.

Santome, J.A., Dellacha J.M. and Paladini, A.C. (1976) Chemistry of somatotropin, *Pharmacol. Ther. B2*, 571-590.

Schams, D., Graf, F., Meyer, J., Graule, B., Mauthner, M. and Wollny, C. (1991) Changes in hormones, metabolites, and milk after treatment with sometribove (recombinant methionyl bST) in Deutsches Fleckvieh and German black and white cows, *J. Anim. Sci. 69*, 1583-1592.

Sharma, B.K., VandeHaar, M.J. and Ames, N.K. (1994) Expression of insulin-like growth factor-I in cows at different stages of lactation and in late lactation cows treated with somatotropin, *J. Dairy Sci. 77*, 2232-2241.

Sharma, B.K., VandeHaar, M.J., Tucker, H.A., Renaville, R., Portetelle, D., Radcliff, R.P., VanderKooi, W.K., Binelli, M. and Liesman, J.S. (1997) Effect of diet and bST on mRNA for IGF-I-binding proteins (BP) -2 and -3 in dairy cattle, *J. Dairy Sci. 80(Suppl 1)*, 176.

Slaba, J., Krejci, P., Skarda, J., Huybrechts, L.M., Decuypere, E. and Herrmann, H. (1994) Plasma profiles of somatotropin and IGF-I in dairy cows following application of two preparations of recombinant bovine somatotropin in a sustained release vehicle, *Physiol. Res. 43*, 37-43.

Torre, P.M., Lewis, M.J., Ingle, T.L. and Olivier, S.P. (1993) Influence of recombinant bovine somatotropin (Sometribove) on mononuclear cells during the nonlactating period, *J. Dairy Sci. 76*, 983-991.

Vanderkooi, W.K., VandeHaar, M.J., Sharma, B.K., Binelli, M., Tucker, H.A., Akers, R.M. and Moseley, W.M. (1995) Comparison of growth hormone-releasing factor and somatotropin: the somatotropic axis in lactating primiparous cows, *J. Dairy Sci. 78*, 2140-2149.

Vicini, J.L., Buonomo, F.C., Veenhuizen, J.J., Miller, M.A., Clemmons, D.R. and Collier, R.J. (1991) Nutrient balance and stage of lactation affect responses of insulin, insulin-like growth factors I and II, and insulin-like growth factor-binding protein 2 to somatotropin administration in dairy cows, *J. Nutr. 121*, 1656-1664.

Vleurick, L., Renaville, R., Massart, S., VandeHaar, M., Istasse, L., Parmentier, I., Bertozzi, C., Van Eenaeme, C. and Portetelle, D. (2000) A homologous radioimmunoassay for plasma Insulin-like growth factor binding protein-2 in cattle, *J. Dairy Sci. 83*, 452-458.

Walton, P.E. and Etherton, T.D. (1989) Effects of porcine growth hormone and insulin-like growth factor-I (IGF-I) on immunoreactive IGF-binding protein concentration in pig, *J. Endocrinol. 120*, 153-160.
Weigel, K.A., Fisher, T.M., Van der Linde, C., Gianola, D. and Rekaya, R. (1998) Impact of bovine somatotropin on genetic evaluation of dairy sires and cows, *J. Dairy Sci. 81*, 2045-2051.
Willeberg, P. (1993) Bovine somatotropin and clinical mastitis: epidemiological assessment of the welfare risk, *Liv. Prod. Sci. 36*, 55-66.
Young, F.G. (1947) Experimental stimulation (galactopoiesis) of lactation, *Br. Med. Bull 5*, 155-160.
Zhao, X., Mc Bride, B.W., Trouten-Radfor, L.M., Golfman L. and Burton, J.H. (1994) Somatotropin and insulin-like growth factor-I concentrations in plasma and milk after daily or sustained-release exogenous somatotropin administrations, *Dom. Anim. Endocrinol. 11*, 209-216.
Zwickl, C.M., Smith, H.W., Tamura, R.N. and Bick, P.H. (1990) Somatotropin antibody formation in cows treated with a recombinant bovine somatotropin over two lactations, *J.Dairy Sci.* 73, 2888-2897.

BIOLOGY AND ACTIONS OF SOMATOTROPIN IN THE PIG

LOUVEAU I. AND BONNEAU M.
INRA, Unité Mixte de Recherches sur le Veau et le Porc, 35590 Saint Gilles, France

Abstract

The availability of recombinant somatotropin (ST) in the eighties resulted in an exponential increase in investigations of the effects of this molecule. Numerous studies have been performed in the pig which is very responsive to porcine ST (pST) treatment. The aim of the present paper is to review the latest data concerning the biology of ST and the actions of pST in pigs. The effects of ST are mediated at least in part by insulin-like growth factor-I. Changes in hormones and binding proteins in plasma as well as receptors expression in target tissues are considered during development and in response to nutritional status. Somatotropin is involved in the regulation of many physiological processes. The actions of pST are presented with a special emphasis on the effects in adipose tissue and skeletal muscle. In the growing pig, the reduction in lipid deposition results primarily from a decrease in the rate of lipogenesis. The increase in muscle mass in response to pST treatment involves at least an increase in protein synthesis. Somatotropin also stimulates longitudinal growth and affects the metabolism of all nutrients classes. The effects on performance and meat quality in growing pigs and on the reproductive system are then presented. It has been shown that treatment of growing pigs with pST consistently improved growth performance including average daily gain and feed efficiency without adversely affecting reproductive functions and meat quality.

1. Introduction

The involvement of the anterior pituitary gland in growth was first demonstrated in the 1920s. Following these initial observations, the specific protein in the pituitary extract was identified as somatotropin (ST). The amino acid sequence was determined in the 1970s for different species. The development of biotechnologies has led to the availability of large quantities of recombinant ST. During the past 15 years, a considerable number of experiments have been carried out to investigate the role of ST in growth and development and to examine the possibility of enhancing growth in farm

R. Renaville and A. Burny (eds.), Biotechnology in Animal Husbandry, 111–131.

livestock and other species. Several strategies have been used to evaluate the effects of ST. They involved an increase in circulating and/or tissue hormones concentrations. One approach is the administration of the hypothalamic peptide Growth Hormone Releasing Factor (GRF) that stimulates endogenous ST synthesis and secretion (Etherton *et al.*, 1986; Dubreuil *et al.*, 1987). Another approach is the production of transgenic animals that overexpress ST (Pursel *et al.*, 1997). The widely used approach is the administration of porcine somatotropin (pST).

The aim of the present paper is to review the latest data concerning the biology of ST and the actions of pST in pigs. The effects of ST are mediated at least in part by insulin-like growth factors (IGFs) and primarily by IGF-I. Changes in hormones, binding proteins and receptors in plasma and/or target tissues are considered during development and in response to changes in nutritional status. The review will then focus on the actions of ST evaluated in vitro or after exogenous administration of pST.

2. The Somatotropic Axis

2.1. HYPOTHALAMIC FACTORS

Growth Hormone Releasing Factor (also known as GHRH - Growth Hormone Releasing Hormone, somatocrinin or somatoliberin) is secreted by specialised neurons located in the arcuate nucleus of the hypothalamus (Daikoku *et al.*, 1988). The 44 amino acid sequence of porcine GRF is similar to that of human GRF, with only 3 amino acids differing at positions 34, 38 and 42 (Böhlen *et al.*, 1983). GRF1-29 (the first 29 amino acid portion of GRF) is as efficient as GRF1-44 in stimulating pST secretion (Petitclerc *et al.*, 1987).

Somatostatin (also known as SRIF - Somatotropin Release Inhibiting Factor) is secreted in various tissues and organs particularly the hypothalamus (Daikoku *et al.*., 1988). It has numerous biological actions (Hall *et al.*, 1988) including the inhibition of ST secretion. Different forms of somatostatin have been described, particularly somatostatin-14 (predominant in brain) and somatostatin-28. Amino acid sequences of both forms of somatostatin are identical in the mammalian species so far studied (Brazeau, *et al.*, 1973; Schally *et al.*, 1976; 1980; Pradayrol *et al.*, 1980).

2.2. SOMATOTROPIN, RECEPTORS AND BINDING PROTEINS

Somatotropin is a 22 kDa polypeptide hormone (Chen *et al.*, 1970) released primarily from the anterior pituitary gland under the influence of GRF and somatostatin. Both the synthesis and release of ST are pulsatile. Porcine ST shares a high degree of homology with bovine ST (bST). There are 18 positions between pST and bST that have different amino acid residues whereas there are 59 residues that differ between pST and human ST. The three-dimensional structure of recombinant pST has been reported (Abdel-Meguid *et al.*, 1987).

The first step in ST action is binding to a specific cell surface receptor (STR). The STR was first purified, sequenced and cloned from rabbit liver (Leung *et al.*, 1987).

Subsequently, the cDNA for the STR has been cloned in several other species including pigs (Cioffi *et al.*, 1990). This cDNA sequence shares 89% sequence identity with the human and rabbit STR cDNAs. STR is a transmembrane protein encoded by nine exons (Edens and Talamantes, 1998). Receptors are highly expressed in liver but are also found in adipose tissue, kidney, lung, skeletal muscle, ovary and pancreas (Chung and Etherton, 1986; Louveau and Etherton, 1992; Sǿrensen *et al.*, 1992; Lee *et al.*, 1993; Peng *et al.*, 1998; Quesnel, 1999). The STR is a member of the cytokine receptor superfamily, a group of receptors characterised by a single transmembrane domain and defined sequence homologies in the extracellular and cytoplasmic regions (Kelly *et al.*, 1991). The STR is able to associate with and activate the tyrosine kinase JAK2 that in turn activates a number of intracellular pathways (Wojcik and Postel-Vinay, 1999).

A ST binding protein (STBP), which essentially corresponds to the extracellular domain of the STR (Leung *et al.*, 1987) has been identified in the serum of many species including pigs (Davis *et al.*, 1992). STBP is derived from the same gene as STR and is generated by mechanisms that differ between species (Edens and Talamantes, 1998). Little is known about the physiological role of STBP. It has been shown to enhance the growth-promoting effects of ST in vivo probably by increasing its half-life in the circulating compartment.

2.3. IGFS, RECEPTORS AND BINDING PROTEINS

IGF-I and IGF-II are ~7.5 kDa single chain polypeptides that retain 70% amino acid homology with each other and 50% homology with proinsulin. The IGF-I molecule contains 70 amino acid residues and IGF-II has 67 amino acid residues. The sequence is highly conserved across species. Porcine IGF-I is identical to human or bovine IGF-I (Tavakkol *et al.*, 1988). Porcine IGF-II differs from human IGF-II only by one amino acid residue (Francis *et al.*, 1989). IGF-I is predominantly synthesised by the liver under the influence of GH. Local production of both IGF-I and IGF-II has also been demonstrated in numerous other tissues and organs.

IGF-I and IGF-II interact with two types of receptors, type I and type II, that differ in their amino acid sequence, secondary structure and ligand-binding specificity (Rechler and Nissley, 1985; Jones and Clemmons, 1995). The type I receptor has a heterotetrameric structure which is homologous to that of the insulin receptor. The type II receptor is a monomeric protein. This receptor is identical to the cation-independent mannose 6-phosphate receptor (M-6PR) involved in lysosomal enzyme transport whose major, if not sole, function is IGF-II uptake for purposes of internalisation and degradation. The IGF-II/M-6P receptor has very weak affinity for IGF-I and no affinity for insulin. IGF-I receptors (IGF-IRs) are widely distributed (Jones and Clemmons, 1995). In the pig, IGF-IRs are expressed in adipose tissue, intestine, kidney, liver, lung, skeletal muscle, ovary, pancreas and uterus (Schober *et al.*, 1990; Hofig *et al.*, 1991; Lee *et al.*, 1993; Louveau *et al.*, 1996; Peng *et al.*, 1996; 1998; Gerfault *et al.*, 1999; Quesnel, 1999).

The IGFs are present in all biological fluids almost entirely (95 to 99%) bound to a family of structurally related binding proteins (IGFBPs). To date, six genetically distinct IGFBPs (IGFBP-1 to IGFBP-6) have been cloned and sequenced (Jones and

Clemmons, 1995; Rajaram *et al.*, 1997) and five of them have been identified in pigs (McCusker *et al.*, 1985a; Coleman *et al.*, 1991). Like IGFs, they are synthesised ubiquitously. In adults, most of the bound serum IGFs are present in a 150 kDa complex that contains IGFBP-3 and an additional 85 kDa protein known as the acid-labile subunit (ALS). The IGFBPs differ in their regulation and their affinities for IGF-I and IGF-II. Structural alteration elicited by proteolysis and other posttranslational modifications such as phosphorylation can modify IGFBP affinities for IGFs. In addition to serving as carrier proteins (150 kDa complex), there is increasing evidence that they act as potentiators or modulators of several complex physiological activities of IGFs.

3. The Somatotropic Axis during Growth and Development

Numerous changes occur during development at many levels of the somatotropic axis. In a number of species including pigs, ST concentrations are considerably higher in foetuses than in adults (Figure 1). Plasma pST concentrations increase to very high levels (80-150 ng/ml) until 90-110 days of gestation and then remain stable (Klindt and Stone, 1984; Macdonald *et al.*, 1985; Spencer *et al.*, 1989) or decline (Herbein *et al.*, 1977; DeHoff *et al.*, 1986; Bauer and Parvizi, 1996). Foetal secretion is pulsatile at least between d89 and d113 of gestation (Bauer and Parvizi, 1996). During the first 2-3 days postnatally, plasma GH concentrations decrease sharply and then decline slowly (Klindt, 1986; Scanes *et al.*, 1987; Carroll *et al.*, 1998) at least until 5-6 months of age (Siers and Swiger, 1971; Chappel and Dunkin, 1975; Althen and Gerrits, 1976; Dubreuil *et al.*, 1987; Scanes *et al.*, 1987; Louveau *et al.*, 1991).

In contrast to GH, plasma IGF-I concentrations increase with age in pigs like in other species (Figure 1). Plasma IGF-I concentration is low in foetuses. It increases during the latter half of foetal life. It further increases postnatally (Lee *et al.*, 1991; 1993; Owens *et al.*, 1991; Louveau *et al.*, 1996) and decreases after 180 days of age (Weiler *et al.*, 1998). Plasma IGF-II concentration remains high throughout the prenatal period (Hausman *et al.*, 1991; Lee *et al.*, 1991; Lee *et al.*, 1993). It then increases postnatally, but the change in IGF-II levels is not as pronounced as that of IGF-I levels (Lee *et al.*, 1991; 1993; Owens *et al.*, 1991). The preponderance of IGF-II over IGF-I in foetal serum and the postnatal increase in serum levels of both IGFs support the view that IGF-II is both a foetal and a postnatal growth factor, whereas IGF-I may be primarily a postnatal growth factor in pigs. In the kidney, liver and skeletal muscle, IGF-II mRNA levels are high during the foetal and early postnatal life and decrease thereafter (Kampman *et al.*, 1993; Lee *et al.*, 1993; Peng *et al.*, 1996; Gerrard *et al.*, 1998). IGF-II mRNA levels are highest at 59 days of gestation in muscle (Gerrard *et al.*, 1998). In the liver, IGF-I mRNA levels are the highest at 21 days of age (Lee *et al.*, 1993; Peng *et al.*, 1996). In kidney and skeletal muscle, IGF-I mRNA declines after birth (Lee *et al.*, 1993; Peng *et al.*, 1996; Gerrard *et al.*, 1998).

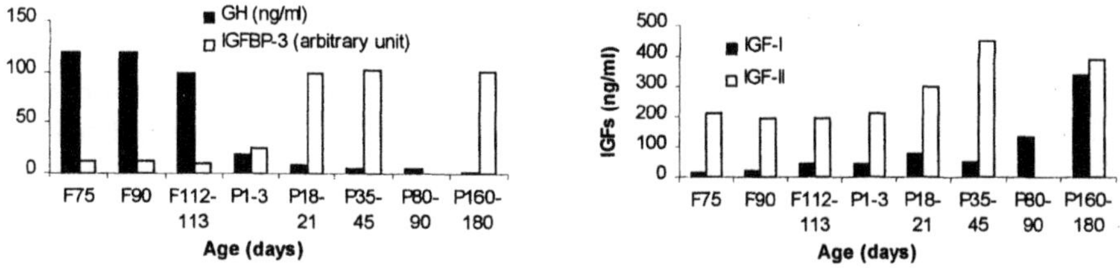

Figure 1. *Developmental changes in plasma GH, IGFBP-3, IGF-I and IGF-II from 75 days of gestation to 180 days of age in pigs (adapted from Klindt and Stone, 1984; Klindt, 1986; Lee* et al., *1991; Louveau* et al., *1996).*

Marked changes in the relative proportions of circulating IGFBPs occur with age. IGFBP-2 is 2- to 3-fold more abundant in foetal than in postnatal serum (McCusker *et al.*, 1991; Lee *et al.*, 1993). After birth, the IGFBP profile changes towards a pattern that is more similar to a peripubertal animal (McCusker *et al.*, 1985a). Plasma IGFBP-3 predominates and increases with age (Lee *et al.*, 1991; Figure 1). After birth, IGFBP-3 mRNA level does not change in liver but decreases in kidney, pancreas and skeletal muscle (Peng *et al.*, 1996; 1998). Hepatic IGFBP-2 mRNA levels decline postnatally (Kampman *et al.*, 1993; Lee *et al.*, 1993).

The expression of STR is tissue- and perhaps cell-specific. The STR is expressed at a much earlier stage of development in muscle than in liver, and the ontogenic profiles differ between these two tissues (Breier *et al.*, 1989; Duchamp *et al.*, 1996; Schnoebelen-Combes *et al.*, 1996). After birth, STR increases in liver whereas there is no age-related change in skeletal muscle (Figure 2). Plasma STBP levels also increases with age (Mullins and Davis, 1992; Schnoebelen-Combes *et al.*, 1996). The specific binding of IGF-I is detected in liver and skeletal muscle from at least 75 days of gestation until the adult stage (Figure 2). It increases between 75 and 90 days of gestation (Louveau *et al.*, 1996). After birth, the levels of IGF-IR decline markedly in kidney, liver, pancreas and skeletal muscle (Lee *et al.*, 1993; Louveau *et al.*, 1996; Peng *et al.*, 1996; 1998). In the small intestine, IGF-I binding is highest at birth, declines by day 3 and increases by day 21 (Schober *et al.*, 1990).

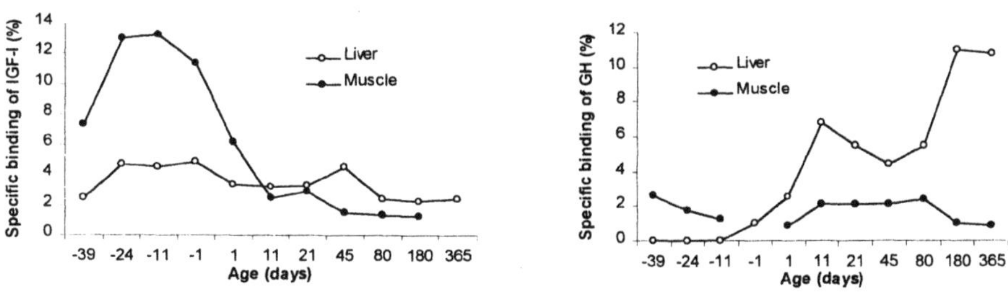

Figure 2. *Developmental changes in IGF-I and GH binding to liver and skeletal muscle (longissimus) from 75 days of gestation to 365 days of age in pigs (adapted from Louveau* et al., *1996; Schnoebelen-Combes* et al., *1996).*

4. Influence of Nutritional Status on the Somatotropic Axis

It is widely accepted that nutritional status plays a major role in regulating circulating levels of ST, IGFs and IGFBPs (Table 1). Serum ST concentrations are higher in non-suckled than in suckled piglets (Campion *et al.*, 1986). Prolonged fasting (at least 24 hours) induces an elevation of plasma pST concentration in young piglets as well as in older animals (Kasser *et al.*, 1981; McCusker *et al.*, 1985b; Buonomo and Baile, 1991) and a decrease of plasma IGF-I levels (Campion *et al.*, 1986; Buonomo and Baile, 1991). It also causes a decline in serum IGFBP-3, IGFBP-2 and 24 kDa IGFBP-4 (McCusker *et al.*, 1985a). The decrease of IGFBP-2 is related to degradation of intact IGFBP-2 in serum (McCusker *et al.*, 1991).

Refeeding after a 48-hour fast is associated with a decline in plasma pST levels within 2 hours whereas it takes longer for IGF-I to return to prefast levels (Buonomo and Baile, 1991). In pigs submitted to limited feed restriction, circulating pST levels are not altered (Kirkwood *et al.*, 1987). Both moderate and severe food restrictions decrease plasma IGF-I concentrations and IGFBP-3 levels (Dauncey *et al.*, 1993; 1994a; Louveau and Le Dividich, unpublished data). Weaning results also in a reduction in serum IGF-I and IGF-II levels (Carroll *et al.*, 1998). There is increasing evidence that circulating IGF-I is related directly to energy intake in neonatal like in older pigs (Dauncey *et al.*, 1994a; Ritacco *et al.*, 1997; Louveau and Le Dividich, unpublished data). Pigs fed a protein deficient diet present lower circulating IGF-I levels than controls (Caperna *et al.*, 1990). In contrast, Grant *et al.* (1991) report no response of plasma IGF-I to dietary protein level. Porcine ST secretion is not altered in pigs receiving a tryptophan deficient diet, which induces a significant decrease in feed intake (Montgomery *et al.*, 1980).

Table 1*. Influence of fasting and undernutrition on the somatotropic axis.*

		Fasting	Undernutrition	
			Moderate	Severe
Plasma	ST	↓	↔	
	STBP		↓	
	IGF-I	↓	↓	↓
	IGFBP-3	↓	↓	↓
	IGFBP-2	↓		
Liver	STR		↔ or ↑	↓
	IGF-I mRNA			↓
	IGF-IR		↓	↔
Skeletal muscle	STR		↔	↔ or ↑
	IGF-I mRNA			↓ or ↔
	IGF-IR		↔	↑

The effects on IGF and ST receptor levels have been studied less extensively, although our recent investigations are starting to elucidate some of the responses involved. Whereas moderate food restriction induced by a larger litter size does not significantly affect STR, it decreases IGF-IR in liver but not in skeletal muscle (Louveau, unpublished data). A moderate but long term food restriction in older pigs induces an increase in hepatic STR but does not affect STR in muscle (Combes *et al.*, 1997a). Severe food restriction from birth to 7 days of age decreases STR in liver but not in skeletal muscle and increases IGF-IR in skeletal muscle but not in liver (Louveau and Le Dividich, unpublished data). A low food intake in pigs aged between 3 and 7 weeks results in down-regulation of hepatic STR mRNA but an upregulation in muscle, which is reflected in a marked decrease in hepatic IGF-I mRNA and growth rate (Dauncey *et al.*, 1994b; Weller *et al.*, 1994). Altogether, these data clearly indicate that the regulation of these receptors is tissue-specific and dependent on the type of undernutrition and/or the age and stage of development.

5. Biological Effects of pST

Research conducted during the past 15 years has established the diversity of the biological effects of ST. Somatotropin not only stimulates longitudinal growth but has numerous other effects in a variety of tissues and regulatory systems. The effects on kidney (Feld and Hirschberg, 1996), muscle (Florini *et al.*, 1996), adipose tissue (Etherton and Bauman, 1998) and bone (Ohlsson *et al.*, 1998) have been reviewed. The metabolism of all nutrients classes is affected by pST treatment: carbohydrate, lipid protein, minerals. The effects on the somatotropic axis and on general metabolism will be developed. Because the major effect of pST is a reduction in adipose tissue deposition and an increase in skeletal muscle growth in the growing pig, the review will focus on the effects of pST in these two tissues.

5.1. EFFECTS OF PST ON THE SOMATOTROPIC AXIS

Exogenous administration of pST consistently induces an increase in the circulating concentrations of IGF-I in growing pigs (Buonomo *et al.*, 1987; 1988; Walton and Etherton, 1989; Caperna *et al.*, 1990; Ambler *et al.*, 1992), in gilts and sows during gestation (Sterle *et al.*, 1995) and in lactating sows (Toner *et al.*, 1996). This IGF-I rise is dose-dependent (Etherton *et al.*, 1987; Sillence and Etherton, 1987; Evock *et al.*, 1988). It also occurs in piglets (Matteri *et al.*, 1997; Wester *et al.*, 1998) for high dose but not for a dose commonly used in growing pig (Dunshea *et al.*, 1999). A significant increase in IGF-I mRNA abundance has been reported in liver, in subcutaneous adipose tissue and in semitendinosus muscle following pST administration (Grant *et al.*, 1991; Wolverton *et al.*, 1992; Coleman *et al.*, 1994, Brameld *et al.*, 1996). In contrast, IGF-I mRNA is not affected by pST in longissimus muscle (Grant *et al.*, 1991; Coleman *et al.*, 1994; Brameld *et al.*, 1996). These data suggest that the response to pST administration is muscle-specific (Brameld *et al.*, 1996). In contrast to IGF-I, the effect of pST on plasma IGF-II concentration is not consistent. Plasma IGF-II has been shown to increase or to remain unchanged in growing pigs (Buonomo *et al.*, 1988; Klindt *et al.*, 1992) but to decrease in pregnant gilt (Sterle *et al.*, 1995). Serum IGFBP-3 levels are increased whereas serum IGFBP-2 levels are decreased by pST (Walton and Etherton, 1989; Coleman and Etherton, 1991). IGFBP-3 mRNA levels also increase in response to ST in both liver and kidney but not in longissimus muscle, stomach or jejunum (Dunaiski *et al.*, 1999). Administration of pST increases STR level in liver (Chung and Etherton, 1986; Ambler *et al.*, 1992; Combes *et al.*, 1997b) but not in adipose tissue (Sørensen *et al.*, 1992) or skeletal muscle (Combes *et al.*, 1997b). The rise in ST binding in liver is associated with an increase (Mullins and Davis, 1992) or no change (Combes *et al.*, 1997b) in plasma STBP.

5.2. EFFECTS OF PST ON GENERAL METABOLISM

There is evidence for altered insulin action in vivo during pST treatment. Numerous studies have demonstrated that exogenous administration of pST increases plasma insulin and glucose concentrations (Chung *et al.*, 1985; Etherton *et al.*, 1986; Kveragas *et al.*, 1986; Gopinath and Etherton, 1989a; Mikel *et al.*, 1993). It has been shown that the reduced insulin sensitivity is expressed within 21 hours of the initial pST treatment (Kerber *et al.*, 1998). The ability to clear glucose from the circulation in response to a glucose tolerance test is reduced in pST-treated pigs (Gopinath and Etherton, 1989b; Wray-Cahen *et al.*, 1993). Insulin resistance after pST administration in pigs is due to 1) the diminution of glucose clearance because of decreased glucose uptake by peripheral tissues and 2) the augmentation of glucose output from the liver (Gopinath and Etherton, 1989b; Wray-Cahen *et al.*, 1993).

Circulating free fatty acids are elevated and blood urea nitrogen is reduced by pST (Chung *et al.*, 1985; Etherton *et al..*, 1986; Kveragas *et al.*, 1986; Gopinath and Etherton, 1989a; Mikel *et al.*, 1993). These changes must be the result of increased lipolysis and reduced amino acid catabolism in the liver respectively. However, the mechanisms governing these metabolic changes in pigs are still poorly understood.

5.3. SOMATOTROPIN ACTIONS ON ADIPOSE TISSUE

Exogenous pST treatment consistently decreases lipid deposition in pigs regardless of sex, genotype, age or nutrient intake. The reduction in lipid deposition results primarily from a decrease in the rate of lipogenesis as assessed in vivo (Dunshea *et al.*, 1992a) and in vitro (Walton and Etherton, 1986; Walton *et al.*, 1987; Magri *et al.*, 1990; Wolverton *et al.*, 1992; Harris *et al.*, 1993; Wang *et al.*, 1999). The reduction in lipogenesis results from a decrease in both mRNA expression and activities of several lipogenic enzymes (Magri *et al.*, 1990; Mildner and Clarke, 1991; Harris *et al.*, 1993; Liu *et al.*, 1994; Donkin *et al.*, 1996). Glucose transport may also be involved. The mRNA abundance of GLUT4, the major insulin-regulated glucose transporter gene in adipose tissue, is decreased (Donkin *et al.*, 1996). However, the magnitude of the decrease is much lower than the one observed for expression of the fatty acid synthase gene. Somatotropin may also regulate lipid deposition through affecting exogenous lipid uptake by adipose tissue. It has been shown that ST decreases adipose tissue lipoprotein lipase activity (Wang *et al.*, 1999). An increase in the rate of lipolysis could also be involved in the reduction in adipose tissue. Although indices of fat mobilisation tended to be higher in pST-treated pigs, the changes are too small to account for a significant decrease in lipid accretion (Dunshea *et al.*, 1992b).

Besides its effects on lipid metabolism, ST exerts an effect on adipose tissue cell formation. This effect has been mostly studied in preadipose cell lines. In the pig, ST inhibits preadipocytes differentiation in primary culture (Hausman and Martin, 1989; Gerfault *et al.*, 1999). It has also been shown that ST reduces proliferation of stromal-vascular cells in primary culture (Gerfault *et al.*, 1999).

5.4. SOMATOTROPIN ACTIONS ON SKELETAL MUSCLE

The action of pST in skeletal muscle has not been studied as extensively as in adipose tissue in pigs. The administration of exogenous pST to pigs is known to increase muscle mass and to slightly alter muscle composition. Moisture and protein contents are higher whereas the lipid content decreases or tends to decrease in pST-treated pigs (Beerman *et al.*, 1990; Bidanel *et al.*, 1991; Fabry *et al.*, 1991; McPhee *et al.*, 1991; Nieuwhof *et al.*, 1991; Mourot *et al.*, 1992). These changes involve at least an increase in protein synthesis in pigs fed an adequate level of protein (Seve *et al.*, 1993). Several histochemical studies have been conducted to determine the effects of pST on muscle fibre morphology and fibre types (Solomon *et al.*, 1988; 1990; 1991; 1994; Beerman *et al.*, 1990; Lefaucheur *et al.*, 1992; Whipple *et al.*, 1992; Rehfeldt and Ender, 1993; Oksbjerg *et al.*, 1995; Ono *et al.*, 1995). Most studies indicate that the administration of pST increases muscle fibre size resulting in a concomitant increase in cross-sectional area of muscle. The effects of pST on fibre types are more conflicting. Ono *et al.* (1995) indicate that most of the muscle that respond to pST treatment are located in hindlimb region suggesting a relationship between pST and muscle function and/or metabolism. The discrepancies between the studies may be related to duration or onset of pST treatment, nutritional status or the dose of pST used.

6. Effects of Exogenous Administration of pST on Performances

6.1. SOMATOTROPIN ADMINISTRATION IN SUCKLING PIGS

Although the somatotropic axis is considered to be essential for postnatal growth, it is widely believed that its role in neonatal growth is relatively limited. Few studies have examined the effect of pST in the suckling pig. Whereas Matteri *et al.* (1997) and Dunshea *et al.* (1999) report no effect on growth rate, Wester *et al.* (1998) indicate that body weight is 10% greater in pST-treated than in control pigs using high doses of pST (1 mg.kg-1.d-1). Taken together, these recent studies indicate that the sensitivity of the neonatal pig to exogenous pST is reduced compared to older pigs.

6.2. SOMATOTROPIN ADMINISTRATION IN GROWING PIGS

Beneficial effects of administration of exogenous pST on growth performance and carcass characteristics of pigs are well documented (Bonneau, 1991). In ad libitum fed pigs administered pST during the finishing period, daily feed intake is reduced (2-22%), growth is generally accelerated (up to 47%) and feed efficiency is dramatically improved (3-38%), in connection with a sharp reduction in fat deposition (7-44%). Lean content of the carcass is augmented (2-23%). Due to the increased weight of some organs (liver, heart, etc.), dressing percentage is reduced (1-4%).

It is well known that performance response to pST administration varies extensively between studies. We performed a statistical analysis of available literature data, as an attempt to estimate the effects of pST dose, animal's potential and dietary lysine content, on performance response to pST administration. Data from several experiments were selected on the basis of availability of relevant data and reliable description of experimental procedures (Azain *et al.*, 1991, 1992; Becker *et al.*, 1992; Bidanel *et al.*, 1991; Bonneau *et al..*, unpublished data; Etherton *et al..*, 1986; Evock *et al.*, 1988, 1991; Evock-Clover *et al.*, 1992; Fabry *et al.*, 1991; Goodband *et al.*, 1993; Hacker *et al.*, 1993; Hagen *et al.*, 1991; Klindt *et al.*, 1992; Knight *et al.*, 1991; Krick *et al.*, 1992; McLaren *et al.*, 1990; McNamara *et al.*, 1991; Quiniou *et al.*, 1993; Smith and Kasson, 1991; Thiel *et al.*, 1993; Verstegen *et al.*, 1991; Weeden *et al.*, 1993a, 1993b).

6.2.1. Effect of pST Dose on Performance Response to pST Administration

For all criteria, the best coefficients of determinations are obtained with negative exponential regressions of performance on pST dose. The pST-induced decrease in daily feed intake (DFI) or backfat thickness (BFT) is almost linear (Figure 3). However, the exponential estimates are significantly better than the linear ones, indicating that the marginal reduction in DFI or BFT, obtained from any increase in pST dose, tends to decline with higher doses. Only part of the maximum theoretical response (calculated as the response for an infinite dose of pST) for DFI (60%) or BFT (65%) is achieved with a dose of 100 $\mu g.kg^{-1}.d^{-1}$.This means that the maximum theoretical responses for DFI and BFT to pST cannot be practically achieved, because high doses of pST are associated with sanitary problems, particularly oesophagogastric ulcers, eliciting some mortality (Smith and Kasson, 1990; McNamara *et al.*, 1991; Smith *et al.*, 1991; Lefaucheur *et al.*, 1992). The responses of average daily gain (ADG) and feed conversion ratio (FCR) to pST are clearly not linear. Most of the maximum theoretical response is achieved for a dose of 100 $\mu g.kg^{-1}.d^{-1}$ (94% for ADG and 81% for FCR), and responses to pST tend to reach a plateau for high doses.

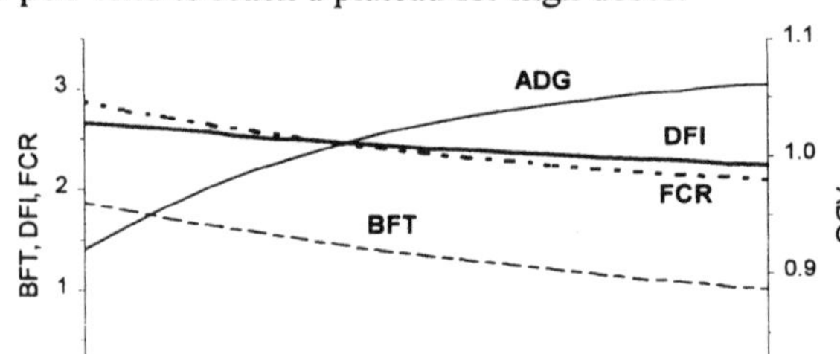

Figure 3. *The influence of pST dose (µg.kg-1.d-1) on daily feed intake (DFI), average daily gain (ADG), feed conversion ratio (FCR) and backfat thickness (BFT) (results obtained from an overall analysis of literature data).*

6.2.2. Effect of Animal's Potential on Performance Response to pST Administration

Four performance level groups were formed on the basis of FCR and BFT measured in control animals (Figure 4). The effect of pST administration on ADG is similar in all performance level groups. On the other hand, the pST-induced reductions in DFI, FCR and BFT are significantly larger in low performance than in high performance pigs.

In high performance pigs, the limited reduction in energy requirements for fat deposition is balanced by the increase in energy requirements for both maintenance and lean tissue deposition (Verstegen *et al.*, 1991; Noblet *et al.*, 1992), so that total energy requirements, and consequently feed intake, are poorly affected. In contrast, in pigs with low performance, the elevation in energy requirement due to increases in lean tissue growth rate and maintenance is largely over-compensated by the sharp reduction in energy requirement for fat deposition, resulting in a reduction of feed intake.

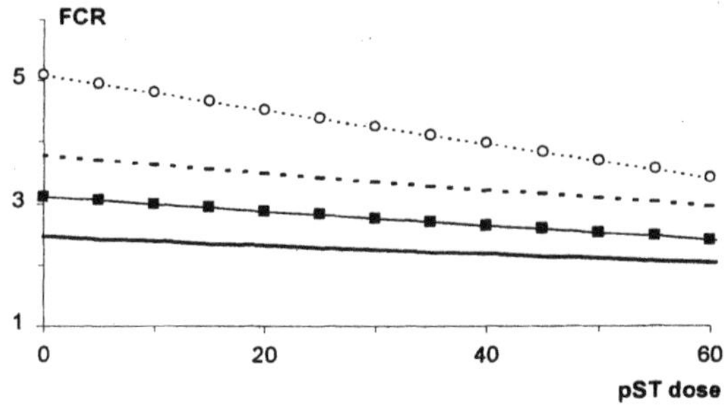

Figure 4. *The influence of pST dose ($\mu g.kg^{-1}.d^{-1}$) on feed conversion ratio (FCR) in 4 performance groups (results obtained from an overall analysis of literature data; performance groups formed on the basis of FCR and backfat thickness in untreated animals).*

6.2.3. Effect of Dietary Lysine Level on Performance Response to pST Administration

With the exception of DFI, the effects of pST on performance are affected by dietary lysine. When dietary lysine levels are too low (0.7%), pST administration has a small effect on FCR and negatively affects ADG. In connection with the pST-elicited reduction of DFI, daily lysine supply is more limiting in pST-treated than in control animals, despite the fact that pST improves the efficiency of nitrogen (Verstegen *et al.*, 1990; Noblet *et al.*, 1993) and amino acid (Goodband *et al.*, 1993) utilisations for lean tissue deposition.

With dietary lysine levels equal to or higher than 0.8%, the improvements in ADG and FCR, obtained from any elevation of pST dose, increase with dietary lysine. The increase in protein deposition elicited by high doses of pST is not fully compensated by the better efficiency of nitrogen deposition, so that essential amino acid requirements are elevated. High lysine levels are thus required to compensate for the pST-elicited decrease in daily lysine supply resulting from the reduced feed intake.

7. Meat Quality in pST-Treated Pigs

In contrast to performance, the effects of pST on meat quality have not been studied extensively. Despite the changes induced in muscle composition by pST, the influence of pST on physical measurements of meat quality is rather low (Beerman *et al.*, 1990; Bidanel *et al.*, 1991; Fabry *et al.*, 1991; McPhee *et al.*, 1991; Nieuwhof *et al.*, 1991; Mourot *et al.*, 1992). Somatotropin treatment has no significant effect on muscle pH fall after slaughter according to Bidanel *et al.* (1991). However, it has been reported in other studies that ultimate pH (24 hours after slaughter) is elevated after pST administration (Beerman *et al.*, 1990; Fabry *et al.*, 1991). Shear force, drip and cooking losses and reflectance of meat are generally not significantly affected by pST treatment (Beermann *et al.*, 1990; Bidanel *et al.*, 1991; Mourot *et al.*, 1992). Administration of pST has been shown to reduce the incidence of boar taint in entire male pigs (Hagen *et al.*, 1991; Bonneau *et al.*, 1992).

Few studies have evaluated the consumer acceptance of pork from pST-treated pigs (Prusa *et al.*, 1993; Fox *et al.*, 1995). Sensory characteristics of pig meat are generally unaffected by pST treatment (Beermann *et al.*, 1990). However, with high pST dosage, a slight but significant reduction in tenderness, juiciness or flavour can be observed (Evock *et al.*, 1988; Beermann *et al.*, 1990).

8. Effects of Exogenous Administration of pST on Reproductive Functions

8.1. SOMATOTROPIN ADMINISTRATION TO PREPUBERTAL GILTS

Porcine ST has been administered to prepubertal gilts to determine whether this treatment alters the onset of puberty and subsequent reproductive performance. Whereas one study indicates that the onset of puberty is delayed (Bryan *et al.*, 1989), other studies fail to show any effect of pST on age at puberty (Andres *et al.*, 1991; Terlouw *et al.*, 1991). Genital tract development of prepubertal gilts is not affected by pST treatment (Bryan *et al.*, 1989, 1990). Moreover, length of oestrus, oestrus cycle length, ovulation rate, percentage pregnant females and embryo survival are not significantly affected by pST treatment (Andres *et al.*, 1991; Terlouw *et al.*, 1991). The concentration of IGF-I is consistently increased by pST in follicular fluid (Bryan *et al.*, 1989, 1992; Beltranema *et al.*, 1994; Echternkamp *et al.*, 1994). However, the significance of this change remains unclear with respect to follicular development and maturation.

8.2. EFFECTS OF PST ADMINISTRATION ON PREGNANT GILTS AND ITS EFFECTS ON THE PROGENY

Administration of pST from d30 to 43 of gestation results in greater foetal and placental weights (Sterle *et al.*., 1995). Kelley *et al.* (1995) reported that pST administration from d28 to 39 of gestation increases foetal survival and crown rump lengths but not foetal weight. The effect of pST administration in the pregnant gilt may be dependent on the gestational period of treatment (Rehfeldt *et al.*, 1993). It has been shown that pST increases muscle fibre number by 27% at birth when sows are treated early in gestation (d10 to d24) (Rehfeldt *et al.*, 1993). The same authors observe an increase in birth weight of piglets from sows given pST in late gestation (d80 to d94) whereas they report no effect for other periods of treatment. A lack of significant effect on birth and weaning weights of piglets has been also reported after pST administration to sows during d28 to d40 of gestation (Kelley *et al.*, 1995) or during the last 3 weeks of gestation (Kveragas *et al.*, 1986). However, body lipid content at birth as well as fasting blood glucose were higher in piglets born from pST treated than from control sows (Kveragas *et al.*, 1986).

8.3. SOMATOTROPIN ADMINISTRATION IN THE LACTATING SOW

In contrast to dairy cattle, very few studies have been conducted to evaluate the effects of pST in lactating sows. Sows treated with pST during late gestation and/or lactation consistently consume less feed and lose more backfat than untreated sows (Harkins *et al.*, 1989; Cromwell *et al.*, 1992; Toner *et al.*, 1996). The effects on milk production are more variable. According to Harkins *et al.* (1989), exogenous pST administration during lactation (d12 through d29) induces a large increase in milk production, with no alteration in milk composition. Piglets suckling a pST treated sow weigh 6% more at weaning. Subsequent studies failed to show any increase in milk production in sows treated with pST (Cromwell *et al.*, 1992; Toner *et al.*, 1996). The failure of pST to affect milk production may be related to the period of treatment or to the litter size (Toner *et al.*, 1996).

9. Conclusions

Exogenous pST administration markedly enhances pig growth performance and affects body composition. It stimulates muscle growth and decreases lipid deposition. The research conducted during the past 15 years has increased our knowledge on the biology of ST including in pigs. Nevertheless, various questions remain to be resolved. Although no adverse effects have been reported so far, the impact of pST administration on meat quality and reproductive functions need to be further evaluated. The mechanism by which ST regulates muscle growth need to be investigated further. The signalling pathways that are responsible for the numerous effects of ST and the mechanisms that are involved in the tissue-specific effects of ST are still unknown.

References

Abdel-Meguid, S.S., Shieh, H.S., Smith, W.W, Dayringer, H.E., Violand, B.N. and Bentle, L.A. (1987) Three-dimensional structure of a genetically engineered variant of porcine growth hormone, *Proc. Natl. Acad. Sci. USA* 85, 6434-6437.

Althen, T.G. and Gerrits, R.J. (1976) Pituitary and serum growth hormone levels in Duroc and Yorkshire swine genetically selected for high and low backfat, *J. Anim. Sci.* 42, 1490-1497.

Ambler, G.R., Breier, B.H., Blair, A.S.H., McCutcheon, S.N., Silbergeld, A. and Gluckman, P.D. (1992) The interrelationship between and the regulation of hepatic growth hormone receptors and circulating GH binding protein in the pig, *Acta Endocrinol. (Copenh)* 126, 155-161.

Andres, C.J., Green, M.L., Clapper, J.A., Cline, T.R. and Dickman, M.A. (1991) Influence of daily injections of porcine somatotropin on growth puberty, and reproduction in gilts, *J. Anim. Sci.* 69, 3754-3761.

Azain, M.J., Seerley, R.W., Reagan, J.O. and Anderson, M.K. (1991) Effect of a high fat diet on the performance response to porcine somatotropin (pST) in finishing pigs, *J. Anim. Sci.* 69, 153-161.

Azain, M.J., Bullock, K.D., Kasser, T.R. and Veenhuizen, J.J. (1992) Relationship of mode of porcine somatotropin administration and dietary fat to the growth performance and carcass characteristics of finishing pigs, *J. Anim. Sci.* 70, 3086-3095.

Bauer, M. and Parvizi N. (1996) Pulsatile and diurnal secretion of GH and IGF-I in the chronically catheterized pig fetus, *J. Endocrinol.* 149, 125-133.

Becker, B.A., Knight, C.D., Buonomo, F.C., Jesse, G.W., Hedrick, H.B. and Baile, C.A. (1992) Effect of a hot environment on performance, carcass characteristics, and blood hormones and metabolites of pigs treated with porcine somatotropin, *J. Anim. Sci.*, 70, 2732-2740.

Beermann, D.H., Fishell, V.K., Roneker, K., Boyd, R.D., Armbruster, G. and Souza, L. (1990) Dose-response relationships between porcine somatotropin, muscle composition, muscle fiber characteristics and pork quality, *J. Anim. Sci.* 68, 2690-2697.

Beltranema, E., Schaefer, A.L., Aherne, F.X. and Foxcroft G.R. (1994) Recombinant porcine somatotropin effects on sexual development and metabolic status of gilts, *Can. J. Anim. Sci.* 74, 265-271.

Bidanel, P., Bonneau, M., Pointillart, A., Gruand, J., Mourot, J. and Demade, I. (1991) Effects of exogenous porcine somatotropin (pST) administration on growth performance, carcass traits, and pork meat quality of Meishan, Pietrain, and crossbred gilts, *J. Anim. Sci.* 69, 3511-3522.

Böhlen, P., Esch, F. and Brazeau, P. (1983) Isolation and characterization of the porcine hypothalamic growth hormone releasing factor, *Biochem. Biophys. Res. Comm.* 116, 726-734.

Bonneau, M. (1991) Regulation of pig growth by somatotropic hormones: II. The effect of exogenous GRF or pST administration on performance and meat quality, *Pig News Inf.* 12, 39-45.

Bonneau, M., Meadus,, W.J. and Squires, E.J. (1992) Effects of exogenous porcine somatotropin on performance, testicular steroid production and fat levels of boar-taint-related compounds in young boars, *Can. J. Anim. Sci.* 72, 537-545.

Brameld, J.M., Atkinson, J.L., Saunders, J.C., Pell, J.M., Buttery, P.J. and Gilmour, R.S. (1996) Effects of growth hormone administration and dietary protein intake on insulin-like growth factor I and growth hormone receptor mRNA expression in porcine liver, skeletal muscle, and adipose tissue, *J. Anim. Sci.* 74, 1832-1841.

Brazeau, P., Vale, W., Burgus, R., Ling, N., Butcher, M., Rivier, J. and Guillemin, R. (1973) Hypothalamic polypeptide that inhibits the secretion of immunoreactive pituitary growth hormone, *Science* 179, 77-79.

Breier, B.H., Gluckman, P.D., Blair, H.T. and McCutcheon, S.N. (1989) Somatotrophic receptors in hepatic tissue of the developing pig, *J. Endocrinol.* 123, 25-31.

Bryan, K.A., Hammond, J.M., Canning, S., Mondschein, J., Carbaugh, D.E., Clark, A.M. and Hagen, D.R. (1989) Reproductive and growth responses of gilts to exogenous porcine pituitary growth hormone, *J. Anim. Sci.* 67, 196-205.

Bryan, K.A., Clark, A.M. and Hagen, D.R. (1990) Effect of treatment with and subsequent withdrawal of exogenous porcine somatotropin on growth and reproductive characteristics of gilts, *J. Anim. Sci.* 68, 2357-2361.

Bryan, K.A., Hagen, D.R. and Hammond, J.M. (1992) Effect of frequency of administration of exogenous porcine growth hormone on growth and carcass traits and ovarian function of prepubertal gilts, *J. Anim. Sci.* 70, 1454-1463.

Buonomo, F.C. and Baile, C.A. (1991) Influence of nutritional deprivation on insulin-like growth factors I, somatotropin, and metabolic hormones in swine, *J. Anim. Sci.* 69, 755-760.

Buonomo, F.C., Lauterio, T.J., Baile, C.A. and Campion, D.R. (1987) Determination of insulin-like growth factor 1 (IGF-1) and IGF binding protein levels in swine, *Domest. Anim. Endocrinol.* 4, 23-31.

Buonomo, F.C., Grohs, D.L., Baile, C.A. and Campion, D.R. (1988) Determination of circulating levels of insulin-like growth factor II (IGF-II) in swine, *Domest. Anim. Endocrinol.* 5, 323-329.

Campion, D.R., McCusker, R.H., Buonomo, F.C. and Jones, W.K., Jr (1986) Effect of fasting neonatal piglets on blood hormone and metabolite profiles and on skeletal muscle metabolism, *J. Anim. Sci.* 63, 1418-1427.

Caperna, T.J., Steele, N.C., Komarek, D.R., McMurtry, J.P., Rosebrough, R.W., Solomon, M.B. and Mitchell, A.D. (1990) Influence of dietary protein and recombinant porcine somatotropin administration in young pigs: growth, body composition and hormone status, *J. Anim. Sci.* 68, 4243-4252.

Carroll, J.A., Veum, T.L. and Matteri, R.L. (1998) Endocrine responses to weaning and changes in post-weaning diet in the young pig, *Domest. Anim. Endocrinol.* 15, 183-194.

Chappel, R.J. and Dunkin, A.C. (1975) Relation of concentration of growth hormone in blood plasma to growth rate and carcass characteristics in the pig, *Anim. Prod.* 20, 51-61.

Chen, H.C., Wilhelmi, A.E. and Howard, S.C. (1970) Purification and characteristics of porcine growth hormone, *J. Biol. Chem.* 245, 3402-3406.

Chung, C.S. and Etherton, T.D. (1986) Characterization of porcine growth hormone (pGH) binding to porcine liver microsomes: chronic administration of pGH induces pGH binding, *Endocrinology* 119, 780-786.

Chung, C.S., Etherton, T.D. and Wiggins, J.P. (1985) Stimulation of swine growth by porcine growth hormone, *J. Anim. Sci.* 60, 118-130.

Cioffi, J.A., Wang, X. and Kopchick, J.J. (1990) Porcine growth hormone cDNA sequence, *Nucleic Acids Res.* 18, 6451.

Coleman, M.E. and Etherton, T.D. (1991) Effects of exogenous porcine growth hormone on serum insulin-like growth factor-binding proteins in growing pigs, *J. Endocrinol.* 128, 175-180.

Coleman, M.E., Pan, Y.C. and Etherton, T.D. (1991) Identification and NH2-terminal amino acid sequence of three insulin-like growth factor-binding proteins in porcine serum, *Biochem. Biophys. Res. Commun.* 181, 1131-1136.

Coleman, M.E., Russell, L. and Etherton, T.D. (1994) Porcine somatotropin (pST) increases IGF-I mRNA abundance in liver and subcutaneous adipose tissue but not in skeletal muscle of growing pigs, *J. Anim. Sci.* 72, 918-924.

Combes, S., Louveau, I. and Bonneau, M. (1997a) Moderate food restriction affects skeletal muscle and liver growth hormone receptors differently in pigs, *J. Nutr.* 127, 1944-1949.

Combes, S., Louveau, I. and Bonneau, M. (1997b) Effect of GH administration on GH and IGF-I receptors in porcine skeletal muscle and liver in relation to plasma GH-binding protein, *J. Endocrinol.* 155, 19-26.

Cromwell, G.L., Stahly, T.S., Edgerton, L.A., Monegue, H.J., Burnell, T.W., Schenck, B.C. and Schricker B.R. (1992) Recombinant porcine somatotropin for sows during late gestation throughout lactation, *J. Anim. Sci.* 70, 1404-1416.

Daikoku, S., Hisano, S., Hitoshi, K., Chikamori-Aoyama, M., Katogani, Y., Zhang, R. and Chihara, K. (1988) Ultrastructural evidence for neuronal regulation of growth hormone secretion, *Neuroendocrinology* 47, 405-415.

Dauncey, M.J., Rudd, B.T., White, D.A. and Shakespear, R.A. (1993) Regulation of insulin-like growth factor binding proteins in young growing animals by alteration of energy status, *Growth Regul.* 3, 198-207.

Dauncey, M.J., Burton, K.A. and Tivey, D.R. (1994a) Nutritional modulation of insulin-like growth factor-I expression in early postnatal piglets, *Pediatr. Res.* 36, 77-84.

Dauncey, M.J., Burton, K.A., White, P., Harrison, A.P., Gilmour, R.S., Duchamp, C. and Cattaneo, D. (1994b) Nutritional regulation of growth hormone receptor gene expression, *FASEB J.* 8, 81-88.

Davis S.L., Graf, M., Morrisson, C.A., Hall, T.R. and Swift, P.J. (1992) Identification and partial purification of serum growth hormone binding protein in domestic animal species, *J. Anim. Sci.* 70, 773-780.

DeHoff, M.H., Stoner, C.S., Bazer, F.W., Collier, R.J., Kraeling, R.R. and Buonomo, F.C. (1986) Temporal changes in steroids, prolactin and growth hormone in pregnant and pseudopregnant gilts during mammogenesis and lactogenesis, *Domest. Anim. Endocrinol.* 3, 95-105.

Donkin, S.S., Chiu, P.Y., Yin, D., Louveau I., Swencki, B., Vockroth, J., Evock-Clover, C.M., Peters, J.L. and Etherton, T.D. (1996) Porcine somatotropin differentially down-regulates expression of the GLUT4 and fatty acid synthase genes in pig adipose tissue, *J. Nutr.* 126, 2568-2577.

Dubreuil, P., Pelletier, G., Petitclerc, D., Lapierre, H., Couture, Y., Brazeau, P., Gaudreau, P. and Morisset, J. (1987) Influence of age and sex on basal secretion of growth hormone (GH) and on GH-induced release by porcine GH-releasing factor (pGRF(1-29)NH2) in growing pigs, *Domest. Anim. Endocrinol.* 4, 299-307.

Duchamp, C., Burton, K.A., Herpin, P. and Dauncey, M.J. (1996) Perinatal ontogeny of porcine growth hormone receptor gene expression is modulated by thyroid status, *Eur. J. Endocrinol.* 134, 524-532.

Dunaiski, V., Dunshea, F.R., Walton, P.E. and Goddard, C. (1999) Effect of growth hormone administration on IGF binding protein-3 mRNA levels in porcine tissues, *J. Mol. Endocrinol.* 22, 261-272.

Dunshea, F.R., Harris, D.M., Bauman, D.E., Boyd, R.D. and Bell, A.W. (1992a) Effect of porcine somatotropin on in vivo glucose kinetics and lipogenesis in growing pigs, *J. Anim. Sci.* 70, 141-151.

Dunshea, F.R., Harris, D.M., Bauman, D.E., Boyd, R.D. and Bell, A.W. (1992b) Effect of somatotropin on nonesterified fatty acid and glycerol metabolism in growing pigs, *J. Anim. Sci.* 70, 132-140.

Dunshea, F.R., King R.H., Owens, P.C. and Walton, P.E. (1999) Moderate doses of porcine somatotropin do not increase plasma insulin-like growth factor-I (IGF-I) or IGF binding protein-3, *Domest. Anim. Endocrinol.* 16, 149-157.

Echternkamp, S.E., Spicer, L.J., Klindt, J., Vernon, R.K., Yen, J.T. and Buonomo, F. (1994) Administration of porcine somatotropin by a sustained-release implant: effects on follicular growth, concentrations of steroids and insulin-like growth factor I, and insulin-like growth factor binding protein activity in follicular fluid of control, lean and obese gilts, *J. Anim. Sci.* 72, 2431-2440.

Edens, A. and Talamantes F. (1998) Alternative processing of growth hormone receptor transcripts, *Endocr. Rev.* 19, 559-582.

Etherton, T.D. and Bauman, D.E. (1998) Biology of somatotropin in growth and lactation of domestic animals, *Physiol. Rev.* 78, 745-761.

Etherton, T.D., Wiggins, J.P., Chung, C.S., Evock, C.M., Rebhun, J.F. and Walton, P.E. (1986) Stimulation of pig growth performance by porcine growth hormone and growth hormone-releasing factor, *J. Anim. Sci.* 63, 1389-1399.

Etherton, T.D., Wiggins, J.P., Evock, C.M., Chung, C.S., Rebhun, J.F., Walton, P.E. and Steele, N.C. (1987) Stimulation of pig growth performance by porcine growth hormone: determination of the dose-response relationship, *J. Anim. Sci.* 64, 433-443.

Evock, M., Etherton, T.D., Chung, C.S. and Ivy, R.E. (1988) Pituitary porcine growth hormone (pGH) and a recombinant pGH analog stimulate pig growth performance in a similar manner, *J. Anim. Sci.* 66, 1928-1941.

Evock, C.M., Caperna, T.J., Steele, N.C., McMurtry, J.P. and Rosenbrough, R.W. (1991) Influence of time of injection of recombinant porcine somatotropin (rpST) relative to time of feeding on growth performance, hormone and metabolite status, and muscle RNA, DNA, and protein in pigs, *J. Anim. Sci.* 69, 2443-2451.

Evock-Clover, C.M., Steele, N.C., Caperna, T.J. and Solomon, M.B. (1992) Effects of frequency of recombinant porcine somatotropin administration on growth performance, tissue accretion rates, and hormone and metabolite concentrations in pigs, *J. Anim. Sci.* 70, 3709-3720.

Fabry, J., Demeyer, D., Thielemans, M.F., Deroanne, C., Van de Voorde, G., Deroover, E. and Dalrymple, R.H. (1991) Evaluation of recombinant porcine somatotropin on growth performance, carcass characteristics, meat quality, and muscle biochemical properties of Belgian Landrace pigs, *J. Anim. Sci.* 69, 4007-4018.

Feld S. and Hirschberg, R. (1996) Growth hormone, the insulin-like growth factor system, and the kidney, *Endocr. Rev.* 17, 423-480.

Florini, J.R., Ewton, D.Z. and Coolican, S.A. (1996) Growth hormone and the insulin-like growth factor system in myogenesis, *Endocr. Rev.* 17, 481-517.

Francis, G.L., Owens, P.C., McNeil, K.A., Wallace, J.C. and Ballard, F.J. (1989) Purification, amino acid sequences and assay cross-reactivities of porcine insulin-like growth factor-I and –II, *J. Endocrinol.* 122, 681-687.

Gerfault, V., Louveau, I. and Mourot, J. (1999) The effect of GH and IGF-I on preadipocytes from Large White and Meishan pigs in primary culture, *Gen. Comp. Endocrinol.* 114, 396-404.

Gerrard, D.E., Okamura, C.S., Ranalletta, M.A.M. and Grant, A.L. (1998) Developmental expression and location of IGF-I and IGF-II mRNA and protein in skeletal muscle, , *J. Anim. Sci.* 76, 1004-1011.

Goodband, R.D., Nelssen, J.L., Hines, R.H., Kropf, D.H., Stoner, G.R., Thaler, R.C., Lewis, A.J. and Schricker, B.R. (1993) Interrelationships between porcine somatotropin and dietary lysine on growth performance and carcass characteristics of finishing swine, *J. Anim. Sci.* 71, 663-672.

Gopinath, R. and Etherton, T.D. (1989a) Effects of porcine growth hormone on glucose metabolism of pigs: I. Acute and chronic effects on plasma glucose and insulin status, *J. Anim. Sci.* 67, 682-688.

Gopinath, R. and Etherton, T.D. (1989b) Effects of porcine growth hormone on glucose metabolism of pigs: II. Glucose tolerance, peripheral tissue insulin sensitivity and glucose kinetics, *J. Anim. Sci.* 67, 689-697.

Grant, A.L., Helferich, W.G., Kramer, S.A., Merkel, R.A. and Bergel, W.G. (1991) Administration of growth hormone to pigs alters the relative amount of insulin-like growth factor-I mRNA in liver and skeletal muscle, *J. Endocrinol.* 130, 331-338.

Hacker, R.R., Deschutter, A., Adeola, O. and Kasser, T.R. (1993) Evaluation of long-term somatotropin implants in finishing pigs, *J. Anim. Sci.* 71, 564-570.

Hagen, D.R., Mills, E.W., Bryan, K.A. and Clark, A.M. (1991) Effects of exogenous porcine growth hormone (pGH) on growth carcass traits, reproductive characteristics, and meat sensory attributes of young boars, *J. Anim. Sci.* 69, 2472-2479.

Harkins, M., Boyd, R.D. and Bauman, D.E. (1989) Effect of recombinant porcine somatotropin on lactational performance and metabolite patterns in sows and growth of nursing pigs, *J. Anim. Sci.* 67, 1997-2008.

Hall, R., Page, M.D., Dieguez, C. and Scanion, M.F. (1988) Somatostatin: a historical perspective, *Horm. Res.* 29, 50-53.

Harris, D.M., Dunshea, F.R., Bauman, D.E., Boyd, R.D., Wang, S.-Y., Johnson, P.A. and Clarke, S.D. (1993) Effect of in vivo somatotropin treatment of growing pigs on adipose tissue lipogenesis, *J. Anim. Sci.* 71, 3293-3300.

Hausman, G.J. and Martin, R.J. (1989) The influence of human growth hormone on preadipocyte development in serum-supplemented and serum-free cultures of stromal-vascular cells from pig adipose tissue, *Domest. Anim. Endocrinol.* 6, 331-337.

Hausman, G.J., Campion, D.R. and Buonomo F.C. (1991) Concentration of insulin-like growth factors (IGF-I and IGF-II) in tissues of developing lean and obese pig fetuses, *Growth Dev. Aging* 55, 43-52.

Herbein, J.H., Martin, R.J., Griel, L.C. and Kavanaugh, J.F. (1977) Serum hormones in the perinatal pig and the effect of exogenous insulin on blood sugars, *Growth* 41, 277-283.

Hofig, A., Michel, F.J., Simmen, F.A. and Simmen, R.C.M. (1991) Constitutive expression of uterine receptors for insulin-like growth factor-I during the peri-implantation period in the pig, *Biol. Reprod.* 45, 533-539.

Jones, J.I. and Clemmons, D.R. (1995) Insulin-like growth factors and their binding proteins: biological actions, *Endocr. Rev.* 16, 3-34.

Kampman, K.A., Ramsay, T.G. and White, M.E. (1993) Developmental changes in hepatic IGF-II and IGFBP-2 mRNA levels in intrauterine growth retarded and control swine, *Comp. Biochem. Physiol.* 104B, 415-421.

Kasser, T.R., Martin, R.J., Gahagan, J.H. and Wangness, P.J. (1981) Fasting plasma hormones and metabolites in fetal and domestic newborn pigs, *J. Anim. Sci.* 53, 420-426.

Kelley, R.L., Jungst, S.B., Spencer, T.E., Owsley, W.F., Rahe, C.H. and Mulvaney, D.R. (1995) Maternal treatment with somatotropin alters embryonic development and early postnatal growth of pigs, *Domest. Anim. Endocrinol.* 12, 83-94.

Kelly, P.A., Djiane, J., Postel-Vinay, M.-C. and Edery, M. (1991) The prolactin / growth hormone receptor family, *Endocr. Rev.* 12, 235-251.

Kerber, J.A., Wray-Cahen, D., Boyd, R.D. and Bauman, D.E. (1998) Decreased glucose response to insulin is màximal within 24 hours of somatotropin injection in growing pigs, *Domest. Anim. Endocrinol.* 15, 267-270.

Kirkwood, R.N., Evans, F.D. and Aherne, F.X. (1987) Influence of age, weight and growth rate on basal LH, growth hormone and cortisol, and estrogen-induced LH release in prepubertal gilts, *Can. J. Anim. Sci.* 67, 1001-1010.

Klindt, J. (1986) Porcine growth hormone and prolactin secretion: the first month of postnatal life, *Growth* 50, 516-525.

Klindt, J. and Stone, R.T. (1984) Porcine growth hormone and prolactin concentrations in the fetus and secretory patterns in the growing pig, *Growth* 48, 1-15.

Klindt, J., Buonomo, F.C. and Yen, J.T. (1992) Administration of porcine somatotropin by sustained-release implant: growth and endocrine responses in genetically lean and obese barrows and gilts, *J. Anim. Sci.* 70: 3721-3733.

Knight, C.D., Kasser, T.R., Swenson, G.H., Hintz, R.L., Azain, M.J., Bates, R.O., Cline, T.R., Crenshaw, J.D., Cromwell, J.L., Hedrick, H.B., Jones, S.J., Kropf, D.H., Lewis, A.J., Mahan, D.C., McKeith, F.M., McLaughlin, C.L., Nelssen, J.L., Novakofski, J.E., Orcutt, M.W. and Parrett, N.A. (1991) The performance and carcass composition responses of finishing swine to a range of porcine somatotropin doses in a 1-week delivery system, *J. Anim. Sci.* 69, 4678-4689.

Krick, B.J., Ronecker, K.R., Boyd, R.D., Beermann, D.H., David, P.J. and Meisinger, D.J. (1992) Influence of genotype and sex on the response of growing pigs to recombinant porcine somatotropin, *J. Anim. Sci.* 70, 3024-3034.

Kveragas, C.L., Seerley, R.W., Martin, R.J. and Vandergrift, W.L. (1986) Influence of exogenous growth hormone and gestational diet on sow blood and milk characteristics and on baby pig blood, body composition and performance, *J. Anim. Sci.* 63, 1877-1887.

Lee, C.Y., Bazer, F.W., Etherton, T.D. and Simmen, F.A. (1991) Ontogeny of insulin-like growth factors (IGF-I and IGF-II) and IGF-binding proteins in porcine serum during fetal and postnatal development, *Endocrinology* 128, 2336-2344.

Lee, C.Y., Chung, C.S. and Simmen, F.A. (1993) Ontogeny of the porcine insulin-like growth factor system, *Mol. Cell. Endocrinol.* 63, 71-80.

Lefaucheur, L., Missohou, A., Ecolan, P., Monin, G. and Bonneau, M. (1992) Performance, plasma hormones, histochemical and biochemical muscle traits, and meat quality of pigs administered exogenous somatotropin between 30 or 60 kilograms and 100 kilograms body weight, *J. Anim. Sci.* 70, 3401-3411.

Leung, D.W., Spencer, S.A., Cachianes, G., Hammonds, R.G., Collins, C; Henzel, W.J., Barnard, R., Waters, M.J. and Wood, W.I. (1987) Growth hormone receptor and serum binding protein : purification, cloning and expression, *Nature* 330, 537-543.

Liu, C.Y., Grant, A.L., Kim, K.H. and Mills, S.E. (1994) Porcine somatotropin decreases acetyl-CoA carboxylase gene expression in adipose tissue, *Domest. Anim. Endocrinol.* 11, 125-132.

Louveau, I. and Etherton, T.D. (1992) Characterization of somatotropin binding sites in pig skeletal muscle, *J. Anim. Sci.* 70, 1801-1805.

Louveau, I., Bonneau, M. and Salter, D. (1991) Age-related changes in plasma growth hormone (pGH) profiles and insulin-like growth factor-I (IGF-I) concentrations in Large White and Meishan pigs, *Reprod. Nutr. Dev.* 31, 205-216.

Louveau, I., Combes, S., Cochard, A. and Bonneau, M. (1996) Developmental changes in insulin-like growth factor-I (IGF-I) receptor levels and plasma IGF-I concentrations in Large White and Meishan pigs, *Gen. Comp. Endocrinol.* 104, 29-36.

Macdonald, A.A.; Spencer, G.S.G. and Hallett, K.G. (1985) Plasma growth hormone and thyroxine levels in the chronically catheterised pig foetus and the effect of thyrotropin releasing hormone, *Acta Endocrinol. (Copenh)* 109, 126-129.

Magri, K.A., Adamo, M., Leroith, D. and Etherton, T.D. (1990) The inhibition of insulin action and glucose metabolism by porcine growth hormone in porcine adipocytes is not the result of any decrease in insulin binding or insulin receptor kinase activity, *Biochem. J.* 266, 107-113.

Matteri, R.L., Becker, B.A., Carroll, J.A. and Buonomo, F.C. (1997) Suppression of somatotroph function induced by growth hormone treatment in neonatal pigs, *Domest. Anim. Endocrinol.* 14, 109-118.

McCusker, R.H., Campion, D.R., Jones, W.K. and Clemmons, D.R. (1985a) The insulin-like growth factor-binding proteins of porcine serum: endocrine and nutritional regulation, *Endocrinology* 125, 501-509.

McCusker, R.H., Wangness, P.J., Griel, L.C., jr. and Kavanaugh, J.F. (1985b) Effects of feeding, fasting and refeeding on growth hormone and insulin in obese pigs, *Physiol. Behavior* 35, 383-388.

McCusker, R.H., Cohick, W.S., Busby, W.H. and Clemmons, D.R. (1991) Evaluation of the developmental and nutritional changes in porcine insulin-like growth factor-binding protein-1 and -2 serum levels by immunoassay, *Endocrinology* 129, 2631-2638.

McLaren, D.G., Bechtel, P.J., Grebner, G.L., Novakofski, J.E., McKeith, F.K., Jones, R.W., Dalrymple, R.H. and Easter R.A. (1990) Dose-response in growth of pigs injected daily with porcine somatotropin from 57 to 103 kilograms, *J. Anim. Sci.* 68, 640-651.

McNamara, J.P., Brekke, C.J., Jones, R.W. and Dalrymple, R.H. (1991) Recombinant porcine somatotropin alters performance and carcass characteristics of heavyweight swine and swine fed alternative feedstuffs, *J. Anim. Sci.* 69, 2273-2281.

McPhee, C.P., Thornton, R.F., Trappett, P.C., Biggs, J.S., Shorthose, W.R. and Ferguson, D.M. (1991) A comparison of the effects of porcine somatotropin, genetic selection and sex on performance, carcass and meat quality traits of pigs fed ad libitum, *Livest. Prod. Sci.* 28, 151-162.

Mikel, W.B., Althen, T.G., Rogers, R.W., Moore, A.B., Miller, H.W. and Miller, L.F. (1993) Effects of exogenous porcine somatotropin on the carcass composition, hormonal and metabolic profiles, lipogenic capacity, and binding of insulin to erythrocyte receptors of fast- versus slow-growing swine, *J. Anim. Sci.* 71, 1786-1795.

Mildner, A.M. and Clarke, S.D. (1991) Porcine fatty acid synthase: cloning of a complementary DNA, tissue distribution of its mRNA and suppression of expression by somatotropin and dietary protein, *J. Nutr.* 121, 900-907.

Montgomery, G.W., Flux, D.S. and Greenway, R.M. (1980) Tryptophan deficiency in pigs: changes in food intake and plasma levels of glucose, amino acids, insulin and growth hormone, *Horm. Metab. Res.* 12, 304-309.

Mourot, J., Bonneau, M., Charlotin, P. and Lefaucheur, L. (1992) Effects of exogenous porcine somatotropin (pST) administration on pork meat quality, *Meat Sci.* 31, 219-227.

Mullins, T.M. and Davis, S.L. (1992) Assessment of factors regulating serum growth hormone binding proteins in pigs, *J. Anim. Sci.* 70, 2741-2745.

Nieuwhof, G.J., Kanis, E., van der Hel, W., Verstegen, M.W.A., Huisman, J., van der Wal, P. (1991) Effect of recombinant porcine somatotropin on body composition and meat quality in growing pigs; interactions with genotype, sex and slaughter weight, *Meat Sci.* 30, 265-278.

Noblet, J., Herpin, P. and Dubois, S. (1992) Effect of recombinant porcine somatotropin on energy and protein utilization by growing pigs: interactions with capacity form lean tissue growth, *J. Anim. Sci.* 70, 2471-2484.

Noblet, J., Dourmad, J.Y. and Dubois, S. (1993). In: M.W.A. Verstegen, L.A. den Hartog, G.J.M. van Kempen and J.H.M. Metz, (Editors), Nitrogen flow in pig production and environmental consequences. Pudoc Scientific Publishers, Wageningen, 189-193

Ohlsson, C., Bengtsson, B.-A., Isaksson, O.G..P., Andreassen, T.T. and Slootweg, M.C. (1998) Growth hormone and bone, *Endocr. Rev.* 19, 55-79.

Oksbjerg, N., Peterson, J.S., Sorensen, M.T., Henckel, P. and Agergaard, N. (1995) The influence of porcine growth hormone on muscle fibre characteristics, metabolic potential and meat quality, *Meat Sci.* 39, 375-385.

Ono, Y., Solomon, M.B., Evock-Clover, C.M., Steele, N.C. and Maruyama, K. (1995) Effects of porcine somatotropin administration on porcine muscles located within different regions of the body, *J. Anim. Sci.* 73, 2282-2288.

Owens, P.C., Conlon, M.A., Campbell, R.G., Johnson, R.J., King, R. and Ballard F.J., (1991) Developmental changes in growth hormone, insulin-like growth factors (IGF-I and IGF-II) and IGF-binding proteins in plasma of young growing pigs, *J. Endocrinol.* 128, 439-447.

Peng, M., Pelletier, G., Palin, M.F., Véronneau, S., LeBel, D. and Abribat, T. (1996) Ontogeny of IGFs and IGFBPs mRNA levels and tissue concentrations in liver, kidney and skeletal muscle of pig, *Growth, Dev. Aging* 60, 171-187.

Peng, M., Abribat, T., Calvo, E., LeBel, D., Palin, M-F., Bernatchez, G., Morisset, J. and Pelletier, G. (1998) Ontogeny of insulin-like growth factors (IGF), IGF binding proteins, IGF receptors, and growth hormone receptor mRNA levels in porcine pancreas, *J. Anim. Sci.* 76, 1178-1188.

Petitclerc, D., Pelletier, G., Lapierre, H., Gaudreau, P., Couture, Y., Dubreuil, P., Morisset, J. and Brazeau, P. (1987) Dose response of two synthetic human growth hormone-releasing factors on growth hormone release in heifers and pigs, *J. Anim. Sci.* 65, 996-1005.

Pradayrol, L., Jornvall, M., Mutt, V. and Ribet, A. (1980) N-terminally extended somatostatin: the primary structure of somatostatin-28, *FEBS Lett.* 209, 55-58.

Prusa, K.J., Fedler, C.A. and Miller, L.F. (1993) National in-home consumer evaluation of pork roasts from pigs administered porcine somatotropin (pST), *J. Food Sci.* 58, 480-481.

Pursel, V.G., Wall, R.J., Solomon, M.B., Bolt, D.J., Murray, J.D. and Ward, K.A. (1997) Transfer of an ovine metallothionein-ovine growth hormone fusion gene into swine, *J. Anim. Sci.* 75, 2208-2214.

Quesnel, H. (1999) Localization of binding sites for IGF-I, insulin and GH in the sow ovary, *J. Endocrinol.* 163, 363-372.

Quiniou, N., Noblet, J. and Dourmad, J.Y. (1993) Influence de l'administration de somatotropine porcine et d'une réduction de taux protéique du régime sur les rejets d'azote et de phosphore chez le porc, *Journées Rech. Porcine en France* 25, 287-294.

Rajaram, S., Baylink, D.J. and Mohan, S. (1997) Insulin-like growth factor-binding proteins in serum and other biological fluids: regulation and functions, *Endocr. Rev.* 18, 801-831.

Rechler, M.M. and Nissley, S.P. (1985) The nature and regulation of the receptors for insulin-like growth factors, *Annu. Rev. Physiol.* 47, 425-442.

Rehfeldt, C. and Ender, K. (1993) Skeletal muscle cellularity and histochemistry in response to porcine somatotropin in finishing pigs, *Meat Sci.* 34, 107-118.

Rehfeldt, C., Fiedler, I., Weikard, R., Kanitz, E. and Ender, K. (1993) It is possible to increase skeletal muscle fibre number in utero, *Biosci. Rep.* 13, 213-220.

Ritacco, G., Radecki, S.V. and Schoknecht, P.A. (1997) Compensatory growth in runt pigs is not mediated by insulin-like growth factor-I, *J. Anim. Sci.* 75, 1237-1243.

S¢rensen, M.T., Chaudhuri, S., Louveau, I., Coleman, M.E. and Etherton, T.D. (1992) Growth hormone binding proteins in pig adipose tissue: number, size and effects of pGH treatment on pGH and bGH binding, *Domest. Anim. Endocrinol.* 9, 13-24.

Scanes, C.G., Lazarus, D., Bowen, S., Buonomo, F.C. and Gilbreath, R.L. (1987) Postnatal changes in circulating concentrations of growth hormone, somatomedin C and thyroid hormones in pigs, *Domest. Anim. Endocrinol.* 4, 252-257.

Schally, A.V., Dupont, A., Arimura, A., Redding, T.W., Niski, N., Linthieum, G.L. and Schlesinger, D.H. (1976) Isolation and structure of somatostatin from porcine hypothalamus, *Biochemistry* 15, 509-514.

Schally, A., Huang, W. and Change, R. (1980) Isolation and structure of pro-somatostatin: a putative somatostatin precursor from pig hypothalamus, *Proc. Natl. Acad. Sci. USA* 77, 4489-4493.

Schnoebelen-Combes, S., Louveau, I., Postel-Vinay, M-C. and Bonneau, M. (1996) Ontogeny of GH receptor and GH-binding protein in the pig, *J. Endocrinol.* 148, 249-255.

Schober, D.A., Simmen, F.A., Hadsell, D.L. and Baumrucker, C.R. (1990) Perinatal expression of type I IGF receptors in porcine small intestine, *Endocrinology* 126, 1125-1132.

Sève, B., Ballèvre, O., Ganier, P., Noblet, J., Prugnaud, J. and Obled C. (1993) Recombinant porcine somatotropin and dietary protein enhance protein synthesis in growing pigs, *J. Nutr.* 123, 529-540.

Siers, D.G. and Swiger, L.A. (1971) Influence of live weight, age and sex on circulating growth hormone levels in swine, *J. Anim. Sci.* 32, 1229-1232.

Sillence, M.N. and Etherton, T.D. (1987) Determination of the temporal relationship between porcine growth hormone, serum IGF-1 and cortisol concentrations in pigs, *J. Anim. Sci.* 64, 1019-1023.

Smith, V.G. and Kasson, C.W. (1990) Growth performance and carcass characteristics of pigs administered recombinant porcine somatotropin during 30 to 110 kilogram live weight, *J. Anim. Sci.* 68, 4109-4116.

Smith, V.G. and Kasson, C.W. (1991) The interrelationship between crude protein and exogenous porcine somatotropin on growth, feed and carcass measurements of pigs, *J. Anim. Sci.* 69, 571-577.

Smith, V.G., Leman, A.D., Seaman, W.J. and VanRavensway, F. (1991) Pig weaning weight and changes in hematology and blood chemistry of sows injected with recombinant porcine somatotropin during lactation, *J. Anim. Sci.* 69, 3501-3510.

Solomon, M.B., Campbell, R.G., Steele, N.C., Caperna, T.J. and McMurtry, J.P. (1988) Effect of feed intake and exogenous porcine somatotropin on longissimus muscle fiber characteristics of pigs weighing 55 kilograms live weight, *J. Anim. Sci.* 66, 3279-3284.

Solomon, M.B., Campbell, R.G. and Steele, N.C. (1990) Effect of sex and exogenous porcine somatotropin on longissimus muscle fiber characteristics of growing pigs, *J. Anim. Sci.* 68, 1179-1181.

Solomon, M.B., Campbell, R.G., Steele, N.C. and Caperna, T.J. (1991) Effects of exogenous porcine somatotropin administration between 30 and 60 kilograms on longissimus muscle fiber morphology and meat tenderness of pigs grown to 90 kg, *J. Anim. Sci.* 69, 641-645.

Solomon, M.B., Caperna, T.J., Mroz, R.J. and Steele, N.C. (1994) Influence of dietary protein and recombinant porcine somatotropin administration in young pigs: III. Muscle fiber morphology and shear force, *J. Anim. Sci.* 72, 615-621.

Spencer, G.S.G., Hallett, K.G., Beermann, U. and Macdonald, A.A. (1989) Changes in the levels of growth hormone, insulin, cortisol, thyroxine and somatomedin-C/IGF-1, with increasing gestational .age in the fetal pig, and the effect of thyroidectomy in utero, *Comp. Biochem. Physiol.* 93A, 467-472.

Sterle, J.A., Cantley, T.C., Lamberson, W.R., Lucy, M.C., Gerrard, D.E., Matteri, R.L. and Day, B.N. (1995). Effects of recombinant porcine somatotropin on placental size, fetal growth, and IGF-I and IGF-II concentrations in pigs, *J. Anim. Sci.* 73, 2980-2985.

Tavakkol, A., Simmen, F.A. and Simmen, R.C.M. (1988) Porcine insulin-like growth factor-I (pIGF-I): complementary deoxyribonucleic acid cloning and uterine expression of messenger ribonucleic acid encoding evolutionarily conserved IGF-I peptides, *Mol. Endocrinol.* 2, 674-681.

Terlouw, S.L., Rieke, A.R., Cantley, T.C., Miller, L.F. and Day, B.N. (1991) The effects of recombinant porcine somatotropin on reproductive function in gilts treated during the finishing phase, *J. Anim. Sci.* 69, 4294-4298.

Thiel, L.F., Beermann, D.H., Krick, B.J. and Boyd, R.D. (1993) Dose-dependent effects of exogenous porcine somatotropin on the yield, distribution and proximate composition of carcass tissues in growing pigs, *J. Anim. Sci.* 71, 827-835.

Toner, M.S., King, R.H., Dunshea, F.R., Dove, H. and Atwood, C.S. (1996) The effect of exogenous somatotropin on lactation performance of first-litter sows, *J. Anim. Sci.* 74, 167-172.

Verstegen, M.W., Van der Hel, W., Huisman, J., Kanis, E., van der Wal, P. and van Weerden, E.J (1990) Effect of exogenous porcine somatotropin administration on nitrogen and energy metabolism in three genotypes of pigs, *J. Anim. Sci.* 68, 1008-1016.

Verstegen, M.W., Van der Wal, W., Brandsma, H.A., Henken, A.M., Kanis, E. and Van der Wal, P. (1991) Effects of recombinant porcine somatotropin on metabolic rate in growing pigs, *J. Anim. Sci.* 69, 2961-2970.

Walton, P.E. and Etherton, T.D. (1986) Stimulation of lipogenesis by insulin in swine adipose tissue: antagonism by porcine growth hormone, *J. Anim. Sci.* 62, 1584-1595.

Walton, P.E., Etherton, T.D. and Evock, C.M. (1986) Antagonism of insulin action in cultured pig adipose tissue by pituitary and recombinant porcine growth hormone: potentiation by hydrocortisone, *Endocrinology* 118, 2577-2581.

Walton, P.E. and Etherton, T.D. (1989) Effects of porcine growth hormone and insulin-like growth factor-1 (IGF-I) on immunoreactive IGF-binding protein concentration in pigs, *J. Endocrinol.* 120, 153-160.

Walton, P.E., Etherton, T.D. and Chung, C.S. (1987) Exogenous pituitary and recombinant growth hormones induce insulin and insulin-like growth factor I resistance in pig adipose tissue, *Domest. Anim. Endocrinol.* 4, 183-189.

Wang Y., Fried, S.K., Petersen, R.N. and Schoknecht, P.A. (1999) Somatotropin regulates adipose tissue metabolism in neonatal swine, *J. Nutr.* 129, 139-145.

Weeden, T.L., Nelssen, J.L., Goodband, R.D., Hansen, J.A., Fitzner, G.E., Friesen, K.G. and Laurin, J.L. (1993a) Effects of porcine somatotropin and dietary phophorus on growth performancee and bone properties of gilts, *J. Anim. Sci.*, 71: 2674-2682.

Weeden, T.L., Nelssen, J.L., Goodband, R.D., Hansen, J.A., Friesen, K.G. and Richert, B.T. (1993b) The interrelationship of porcine somatotropin administration and dietary phosphorus on growth performance and bone properties in developing gilts, *J. Anim. Sci.* 71, 2683-2692.

Weiler, U., Claus, R., Schnoebelen-Combes, S. and Louveau, I. (1998) Influence of age and genotype on endocrine parameters and growth performance: a comparative study in Wild boars, Meishan, and Large White boars, *Livest. Prod. Sci.* 54, 21-31.

Weller, P.A., Dauncey, M.J., Bates, P.C., Brameld, J.M., Buttery, P.J. and Gilmour, R.S. (1994) Regulation of porcine insulin-like growth factor I and growth hormone receptor mRNA expression by energy status, *Am. J. Physiol.* 266, E776-E785.

Wester, T.J., Davis, T.A., Fiorotto, M.L. and Burrin, D.G. (1998) Exogenous growth hormone stimulates somatotropic axis and function and growth in neonatal pigs, *Am. J. Physiol.* 274, E29-E37.

Whipple, G., Hunt, M.C., Klemm, R.D., Kropf, D.H., Goodman, R.D. and Schricker, B.R. (1992) Effects of porcine somatotropin and supplemental lysine on porcine muscle histochemistry, *J. Muscle Foods* 3, 217-227.

Wojcik, J. and Postel-Vinay, M.C. (1999) Signal transduction of the growth hormone (GH) receptor, and GH-binding protein, *Growth Horm. IGF Res.* 9, 51-55.

Wolverton, C.K., Azain, M.J., Duffy, J.Y., White, M.E. and Ramsay T.G. (1992) Influence of somatotropin on lipid metabolism and IGF gene expression in porcine adipose tissue, *Am. J. Physiol.* 263: E637-E645.

Wray-Cahen, D., Boyd, R.D., Bauman, D.E. and Ross, D.A. (1993) Effect of porcine somatotropin on the response of growing pigs to acute challenges of glucose, insulin and epinephrine and during a hyperinsulinin emic-euglycemic clamp, *Domest. Anim. Endocrinol.* 10, 103-115.

MODULATION OF THE GROWTH HORMONE (GH) AXIS BY THE USE OF ANTIBODIES TO HORMONES, RECEPTORS AND BINDING PROTEINS

BEATTIE J., ALLAN G.J., SHAND J.H. AND FLINT D.J.
Hannah Research Institute, Ayr KA6 5HL, UK

Abstract

Hormonal growth promoters (growth hormone (GH), ß-adrenergic agonists, steroids) which improve growth rate and/or lean:fat ratios in the carcass have received considerable adverse publicity and are either banned or have no licence for use in the EC. This has led to the development of a number of techniques, involving the use of antibodies, aimed at regulating metabolic processes involved in determining growth and body composition. In this article we will discuss the immunoneutralization of somatostatin to increase GH secretion, the use of antibodies to enhance GH action, anti-idiotypic antibodies as GH mimics, antibodies to the GH receptor and the possibilities for manipulating the activity of IGF-I, a mediator of GH actions. Whether these new approaches will be perceived as acceptable to the general public remains a serious concern and a potential limitation to their development as many would-be sponsors cut back their support for research in these areas.

1. Introduction

Animal production systems have seen improvements in the efficiency of growth through improved nutrition and genetic selection programmes. Further improvements have been achieved by use of antibiotics and hormone-based approaches, involving steroids which have been used world-wide. Opposition to the use of steroids, however led to a ban on their use in the EC, a ban that was also extended to growth hormone (GH), otherwise known as bST, although GH is licensed for use in a number of countries world-wide, including the United States. The European ban served to encourage research into alternative approaches which might be more "consumer acceptable" as well as taking full account of increasing concern for the welfare of the animals which would be treated with such preparations. Some of these alternative techniques are considered in this review.

Any novel approaches to manipulate growth and body composition need to fulfill a number of important criteria in addition to being effective in promoting increased

R. Renaville and A. Burny (eds.), Biotechnology in Animal Husbandry, 133–141.

growth rates, improved body composition or improved efficiency of conversion of feed intake into meat. They must also be able to produce long-term effects without the need to use frequent treatment (a practical requirement) and without leaving potentially hazardous residues in the meat of animals destined for human consumption (consumer safety). The use of bovine (b) GH for improvement of milk yield in dairy herds has been in operation in the US for several years now with a relatively high initial uptake by dairy farmers. Within the European Union, however, the use of bGH has been banned. Such a policy has been viewed as reflecting the view of consumers, although the scientific evidence to support this position is somewhat lacking. More recently, however, a number of concerns regarding insulin-like growth factor –1 (IGF-1) in milk and its potential effects upon the gastrointestinal tract of the consumer has become a subject of considerable debate. The EU moratorium on the use of GH ends in December 1999 and the member states are currently reconsidering this issue. Aside from the safety aspect to the consumer, attention has also turned to the concerns for animal welfare, including such considerations as the effects of GH treatment on metabolic stress, on reproductive problems on the incidence of mastitis and problems of reactions at injection sites in animals treated with GH. It is not within the scope of this review to discuss these topics but they are mentioned here to explain the current background to the use of potential agents for increasing animal productivity. They will no doubt continue to be areas of animated debate within Europe for some time to come. Accepting these general concerns, it is questionable whether immunomodulating approaches as described in this article indicate a possible direction forward, or whether this will be considered to be no more than a subtle modification of a technology which many consider unacceptable. Again, it is the belief of the authors that, irrespective of the eventual commercialisation of such approaches, the fundamental science underpinning these potential vaccines is aiding our understanding of the biological actions of GH. It is also useful to remember that, although this technology has been developed for use in agricultural species, its utility, admittedly in the longer-term, in humans or companion animals may influence judgements on the acceptability of these strategies. This is particularly pertinent to the situation of obesity, which is now recognised as a disease rather than a cosmetic problem and which clearly would involve some adjustment of the "risk-benefit" analysis when it is compared with an approach which is being developed purely as an agricultural "production tool". The approaches we will discuss include using antibodies to neutralize, mimic or enhance the actions of certain hormones involved in regulating growth and body composition.

Although all of the areas discussed in this review are amenable to molecular biological approaches, particularly transgenics, the major distinction to be made is that immunization techniques offer flexibility and speed of response to changing demands although they suffer the drawback of the necessity to treat all animals rather than benefit from transmission through the germline.

2. Immunoneutralization of Somatostatin

The effectiveness of GH in stimulating growth and milk production provided the opportunity to consider various aspects of the GH axis as potential sites for immunomodulation. One of the first to be assessed was immunization against somatostatin because it inhibits GH secretion from the pituitary. Thus, inhibition of somatostatin action should result in elevation of endogenous GH secretion. The initial studies showed considerable increases in liveweight gain in sheep (Spencer *et al.*, 1983a, 1983b). Subsequent studies, however, supported or refuted these findings in terms of growth (for review see Flint, 1990). Similar differences in results have also been described for milk production, one study in goats demonstrated that immunization against somatostatin could increase milk yield in early lactation (Spencer and Garssen, 1985) whereas a study using the same approach in sheep failed to show any effect on milk production, despite the fact that the same study did show an increase in weight gain of female lambs immunized in the same way (Deligeorgis *et al.*, 1998). More recently a study in sheep has described an increase in milk yield after immunization with somatostatin (Sun *et al.*, 1990). The reason for these discrepant results has been proposed to be due to differences in antibody titres, although no relationship between titre and biological effect has been described. The fact that the original hypothesis was too simplistic has also been proposed as a possible explanation for variable results, since somatostatin plays a role in regulating a number of hormones and affects nutrient absorption and digesta flow (Elsaesser and Drath, 1995; Flint, 1990; Reichlin, 1987).

3. Antibodies as Enhancers of GH Activity

The biological function of antibodies is most widely perceived as the binding and neutralisation of antigenic molecules. For example, administration of polyclonal antiserum raised against endogenous GH neutralises hormone activity and this kind of approach can provide useful information in relation to the endocrine regulation of physiological processes such as lactation and growth (Flint *et al.*, 1992; Palmer *et al.*, 1994). This approach, however, attracted commercial interest when certain monoclonal antibodies were shown to increase, rather than inhibit GH action. There are many potential benefits both for industry and consumer in devising methods whereby activity of GH is augmented rather than neutralised. This approach received experimental support over a decade ago with the demonstration that the somatogenic activity of human hGH could be enhanced by pre-complexing it with monoclonal antibodies (Mab) prior to administration in a suitable animal model (Aston *et al.*, 1986, 1989). *In vitro* assays using [35S]-sulphate uptake into costal cartilage of dwarf mice indicated enhanced somatogenic activity of Mab-hGH complexes. This antibody-mediated enhancement of hGH activity confirmed earlier experiments which had demonstrated the enhancement of epidermal growth factor (EGF), insulin and other polypeptide hormones by antibodies (reviewed in Aston *et al.*, 1987).

These observations of Mab-mediated augmentation of hGH activity, were followed by studies which demonstrated that the same approach could be used to enhance bGH activity (Aston *et al.*, 1991). This finding had obvious implications for the field of animal performance and productivity. Clearly the injection or infusion of selected anti-bGH Mabs into livestock is not currently feasible both on grounds of cost and practicality. Such reagents would be quickly recognised as foreign protein and an anti-antibody response raised in species receiving such foreign antibodies. Considerations such as these led to the development of the concept of antisera directed towards specific sites or epitopes on the GH molecule as alternatives to Mabs. In its ultimate form, this strategy would employ peptide fragments of the bGH molecule as vaccines to be used in active immunisation procedures to generate antibodies, which would bind to, and enhance, endogenous GH activity. The feasibility of this approach gained support from several reports that antisera raised against fragments of both bovine and porcine GH enhanced hormone activity (Bomford and Aston, 1990; Tainer *et al.*, 1984; Wang *et al.*, 1990, 1995). However some of these studies still employed dwarf mice or hypophysectomised rats as a GH-deficient animal model and enhancement of GH activity was demonstrated by complexing the hormone with anti-peptide antisera prior to administration to animals. Under such conditions these complexes did show greater activity than GH administered alone when parameters such as increases in body weight or [35S]-sulphate uptake into costal cartilage were used.

In an attempt to improve the prospects for the development of a peptide vaccination strategy, epitope mapping sudies of the GH molecule were undertaken. Previous GH peptides had largely been selected on the basis of computer algorithms which predicted immunogenic regions within the GH molecule. This typically resulted in the selection of peptides which were part of loop structures connecting helices within the GH molecule and which, with reference to the published crystal structure of the protein, were accessible on the surface of the molecule. Antisera generated from such peptides may be more likely to bind to native protein than antisera raised to peptides which represent elements of secondary structure within a protein (Beattie *et al.*, 1992). Indeed, initial epitope mapping studies, using a technique which identifies linear epitopes, showed that polyclonal anti-sera raised against bGH in rabbits and guinea pigs contained major continuous epitopes which were present in all of the loops joining the 4 α-helices as well as in the non-ordered C-terminus of the protein protein (Beattie *et al.*, 1992). Although many monoclonal antibodies could not be epitope-mapped using this technique, presumably because they recognised discontinuous epitopes, one anti-ovine GH enhancing antibody, Mab OA15, was identified as defining a functionally continuous epitope lying between residues 91-102 of bGH in a loop region which joins helices 2 and 3 of the molecule (Beattie and Holder, 1994). In agreement with earlier studies, the peptide sequence defined by this enhancing Mab could also be used to elicit a polyclonal antiserum which was capable of enhancing ovine (o)GH activity when administered as the hormone-antibody complex in the hypopituitary dwarf mouse model (Mockridge, 1997; Mockridge *et al.*, 1998). However, in active immunisation experiments in lambs, which involved immunisation with this peptide conjugated to Keyhole Limpet Haemocyanin, no increase in weight gain was observed compared with control animals (Mockridge, 1997). These results suggest that peptide fragments of growth hormone(s) can be used to generate polyclonal anti-sera, which enhance GH activity when complexed with the hormone *in vitro* and administered to a GH-deficient animal model but active immunisation leading to *in vivo* complexes of GH and antibody are incapable, at least for the moment, of reproducing this growth- enhancing effect in our hands. There is however one report (Pell and Aston, 1991) which describes the successful active immunisation of lambs with a modified fragment 134-154 of porcine (p) GH with resulting increases in carcass weight, although no other studies have repeated this success. This may be related to the typically low anti-GH titres induced in those animals which are immunised with peptide fragments of the hormone. This may not be too surprising since this is effectively an attempt to induce an autoimmune response against an endogenous protein. Clearly, a re-evaluation of this strategy is required including the influence of such parameters as peptide structure, formulation and presentation to vaccinated livestock. In one such trial, involving synthesis and presentation of oGH fragment 91-102 in the form of a multiple antigenic peptide (MAP) structure (Tam, 1988), data have shown no increase in body and individual organ weights when this MAP peptide was used as immunogen. This correlated with a failure of anti-serum from treated lambs to bind bGH *in vitro* (unpublished observations - J Mockridge). Clearly, this area requires further investigation to aid understanding of some of the current limitations. One aspect which has been developed to a limited extent is an investigation into the mechanism of antibody-mediated hormone-enhancement. Most theories have concentrated on specific *in vivo* effects such as increased half-life of the hormone-antibody complex compared with the hormone alone, increased GH receptor occupancy time by hormone-antibody complexes and the targetting of hormone by antibody into specific target tissues (eg liver). Recently evidence has been provided to support some of these hypotheses (Massart *et al.*, 1993; Tans *et al.*, 1994; Wang *et al.*, 1992) and clearly a greater understanding of the mechanistic aspects of this phenomenon would assist in the development of this technology.

4. Anti-idiotypic Antibodies as Mimics of Hormone Action

Other strategies have been used to manipulate the GH axis with antibodies. An anti-idiotypic approach (Jerne, 1974) has been attempted using both monoclonal and polyclonal anti-GH antibodies. According to idiotypic network theory the use of such reagents should allow the generation of anti- (anti-GH) antibodies, a proportion of which may exist as structural and functional mimics of the original immunogen (GH). Such approaches have met with varying degrees of success. For example polyclonal anti-idiotype antibodies raised in sheep to polyclonal anti-rGH anti-serum were able to stimulate growth in hypophysectomised rats (Gardner *et al.*, 1990). Similarly an anti-idiotypic Mab (2A6) raised to an anti-pGH Mab (PS-7.6) also promoted growth in hypophysectomised rats (Wang *et al.*, 1994). In another study it was shown that administration of a purified polyclonal rabbit anti-bGH anti-serum to lactating cows resulted in an increase in "bGH-like" immunoreactivity in the sera of immunised animals. However, there was no effect of immunisation on either milk yield or serum IGF-I concentrations (Schalla *et al.*, 1994), two biological responses which are stimulated by bGH. Together with these conflicting data are the inherent limitations associated with the use of the anti-idiotypic network in this way, ie the fact that such approaches may still rely on the administration of antibody and also the ability of recipient animals, over a period of time, to generate specific neutralising anti- (anti-idiotypic) antibodies (Ab3)

5. Antibodies to the GH Receptor as Receptor-Agonists

GH, like other class I cytokine ligands activates its receptor by a homodimerisation mechanism involving separate sites on a single hormone molecule engaging largely similar areas of two different receptor proteins resulting in a 1:2 (hormone: receptor) stoichiometry (De Vos *et al.*, 1992). Such a mechanism of receptor binding suggested that activation may be achieved by the direct binding to receptor of appropriate bivalent anti-receptor antibodies. Such receptor activating antibodies have already been described for the prolactin receptor. Many considerations apply in the case of anti-receptor antibodies. For example, although some reports have demonstrated a stimulatory effect of anti-prolactin receptor antibodies on mammary gland casein synthesis *in vitro* and *in vivo* (Dusanter-Fourt *et al.*, 1983, 1984) such antibodies also have the ability to inhibit prolactin activity and on their own may show inhibitory or stimulatory activity depending on dosage (Djiane *et al.*, 1981). Much the same considerations would apply to anti-growth hormone receptor antibodies although to date no reports of anti-GH receptor antibodies which stimulate the GH axis *in vitro* or *in vivo* have been published. In addition, as indicated above, administration of anti-GHR antibodies is scientifically and commercially unattractive and therefore active immunisation with receptor peptide fragments is an option. However, the difficulties of this approach were clearly illustrated in the context of anti-GH enhancing antibodies and these same problems of peptide selection and presentation apply. The 3-dimensional structure of the extra-cellular domain of the hGHR has been reported (De Vos *et al.*, 1992) and together with published epitope mapping studies of the bGHR extra-cellular domain (Beattie *et al.*, 1996) may assist in the selection of candidate receptor peptides for immunisation. In the case of direct binding anti-receptor antibodies, however, it must be borne in mind that the appropriate spatial geometry of the receptors which is induced by ligand binding may not be replicated by antibody-receptor engagement and under these circumstances only a weak (if any) signal may be propagated. Alternatively, such antibodies may prove to be antagonists at the GHR and may form the basis of an approach to treatment of GH over-activity syndromes (eg acromegaly). In a refinement of this technique we have adopted a strategy of chimeric vaccine production where putative peptide epitopes are presented within the framework of a homologous protein. We believe that the use of such recombinant proteins as vaccines may overcome problems associated with the appropriate presentation of peptides during vaccination. Our initial studies using chimeric rat/ovine GH binding protein molecules have indicated the feasability of this approach (Allan *et al.*, 1999).

6. Manipulation of the Insulin-like Growth Factor (IGF) Axis

IGF-I is believed to mediate many of the actions of GH and the actions of IGF-I are in turn modulated by a series of 6 IGF binding proteins (IGFBPs) (reviewed in [38]). The interactions of IGFBPs with IGF-I are complex and can involve both inhibition or enhancement of IGF action. At least one of the functions of the IGFBPs involves transport of IGFs within the blood and it is interesting to note that increased activity of IGF-I has been achieved using antibodies to IGF-I, which might act like a pseudo-binding protein (Hill and Pell, 1998). It is probably significant to note that the affinity of such an antibody was 10-8M, one order of magnitude lower than the affinity of the IGF-receptor, and thus the antibody would not be expected to inhibit IGF interaction with its receptor. By contrast some of the IGFBPs have affinities for IGF-I which are 10-10M and these IGFBPs are believed to act as inhibitors of IGF-action under such circumstances, acting to sequester IGF-I away from cell surface receptors. Clearly, if antibodies to IGF-I could be created with such high affinities they could be used to inhibit IGF-action.

An alternative strategy might involve the use of antibodies to the IGFBPs, antibodies which bind at, or close, to, the site at which IGF-I binds. This would effectively prevent the inhibitory action of the IGFBPs on IGF-I action. In a development of this approach small molecules have been produced which bind to IGFBPs and prevent IGF binding and these have been shown to enhance IGF-action (Lowman *et al.*, 1998).

7. Conclusions

The practical and political limitations on the use of hormonal growth promoters such as anabolic steroids, ß-adrenergic compounds and GH have led to the search for alternative strategies which are more "consumer acceptable".

The use of antibodies to manipulate biological processes which lead to improvements in body composition of agricultural animals has clear potential advantages in terms of safety to the consumer. Antibodies are proteins and thus are not considered to be orally active; they do not survive high temperatures involved in food preparation. In addition, these antibodies are generally species-specific and would thus not have biological activities even if injected into humans. Despite these considerable safety aspects, it is impossible to predict whether they will be deemed acceptable particularly by the European consumer. Inevitably, political considerations will come to the fore as consumer concerns are balanced against the approval of such techniques elsewhere in the world and enforced entry of such products into the European Market under the auspices of the GATT agreement. Equally interesting will be the public reaction to these same technologies when they are developed for clinical use, for example, to treat obesity.

References

Allan, G.J., Shand, J.H., Beattie, J. and Flint, D.J. (1999) Identification of novel sites in the ovine growth hormone receptor involved in binding hormone and conferring species specificity, *Eur J Biochem.* 261, 555-562.

Aston, R., Cowden, W.B. and Ada, G.L. (1989) Antibody-mediated enhancement of hormone activity, *Mol. Immunol. 26*, 435-446.

Aston, R., Holder, A.T., Ivanyi, J. and Bomford, R. (1987) Enhancement of bovine growth hormone activity *in vivo* by monoclonal antibodies, *Mol. Immunol. 24*, 143-150.

Aston, R., Holder, A.T., Preece, M.A. and Ivanyi, J. (1986) Potentiation of the somatogenic and lactogenic activity of human growth hormone with monoclonal antibodies, *J. Endocrinol. 110*, 381-388.

Aston, R., Rathjen, D.A., Holder, A.T., Bender, V., Trigg, T.E., Cowan, K., Edwards, J.A. and Cowden, W.B. (1991) Antigenic structure of bovine growth hormone: location of a growth enhancing region, *Mol. Immunol. 28*, 41-50.

Beattie, J., Fawcett, H.A.C. and Flint, D.J. (1992) The use of multiple pin peptide synthesis in an analysis of the continuous epitopes recognised by various anti-recombinant bovine (rb) growth hormone anti-sera: comparison with predicted regions of immunogenicity and location within the 3-dimensional structure of the molecule, *Eur. J. Biochem. 210*, 59-66.

Beattie, J. and Holder, A.T. (1994) Location of an epitope defined by an enhancing monoclonal antibody to growth hormone: some structural details and biological implications, *Mol. Endocrinol. 8*, 1103-1110.

Beattie, J., Shand, J.H. and Flint, D.J. (1996) An immobilised peptide array identifies antibodies to a discontinuous epitope in the extra-cellular domain of the bovine growth hormone receptor, *Eur. J. Biochem. 239*, 479-486.

Bomford, R. and Aston, R. (1990) Enhancement of bovine growth hormone activity by antibodies against growth hormone peptides, *J. Endocrinol. 125*, 31-38.

Clemmons, D.R. (1997) Insulin-like growth factor binding proteins and their role in controlling IGF actions, *Cyt.Growth Fact. Rev. 8*, 45-62

De Vos, A.M., Ultsch, M. and Kossiakoff, A.A. (1992) Human growth hormone and extra cellular domain of its receptor: crystal structure of the complex, *Science 255*, 306-312.

Deligeorgis, S.G., Rogdakis, E., Mantzios, A., Nikolaou, E., Evengelatos, G. and Sotiriadis-Viahos, C. (1998) A note on the effect of active immunization against somatostatin on milk production and growth in sheep, *Anim. Prod. 46*, 304-308.

Djiane, J., Houdebine, L.M. and Kelly, P.A. (1981) Prolactin-like activity of anti-prolactin receptor antibodies on casein and DNA synthesis in the mammary gland, *Proc. Nat. Acad. Sci. 78*, 7445-7448.

Dusanter-Fourt, I., Djiane, J., Houdebine, L.M. and Kelly, P.A. (1983) *In vivo* lactogenic effects of anti prolactin receptor antibodies in pseudopregnant rabbits, *Life Sci. 32*, 407-412.

Dusanter-Fourt, I., Djiane, J., Kelly, P.A., Houdebine, L.M. and Teyssot, B. (1984) Differential biologial activities between mono- and bivalent fragments of anti-prolactin receptor antibodies, *Endocrinology 14*, 1021-1027.

Elsaesser, F. and Drath, S. (1995) The potential of immunoneutralization of somatostatin for improving pig performance, *Liv. Prod. Sci. 42*, 255-263.

Flint, D.J. (1990) Immunological approaches for manipulation of animal growth, body composition and fecundity, *Biotech. Genet. Eng. Rev. 8*, 97-131.

Flint, D.J., Tonner, E., Beattie, J. and Panton, D. (1992) Investigation of the mechanism of action of growth hormone in stimulating lactation in the rat, *J. Endocrinol. 134*, 377-383.

Gardner, M.J., Morrison, C.A., Stevenson, L.Q. and Flint, D.J. (1990) Production of anti-idiotypic anti-sera to rat GH antibodies capable of binding to GH receptors and increasing body weight gain in hypophysectomised rats, *J. Endocrinol. 125*, 53-59.

Hill, R.A. and Pell, J.M. (1998) Regulation of insulin-like growth factor I (IGF-I) bioactivity *in vivo*: further characterization of an IGF-I-enhancing antibody, *Endocrinology 139*, 1278-1287.

Jerne, N.K. (1974) Toward a network theory of the immune system, *Ann.Immunol. (Paris) 125C*, 373-389.

Lowman, H.B., Chen, Y.M., Skelton, N.J., Mortensen, D.L., Tomlinson, E.E., Sadick, M.D., Robinson, I.C.A.F. and Clark, R.C. (1998) Molecular mimics of insulin-like growth factor 1 (IGF-1) for inhibiting IGF-1: IGF-binding protein interactions, *Biochem. 37*, 8870-8878.

Massart, S., Maiter, D., Portetelle, D., Adam, E., Renaville, R. and Ketelslegers, J.M. (1993) Monoclonal antibodies to bovine growth hormone potentiate hormonal activity *In vivo* by enhancing growth hormone binding to hepatic somatogenic receptors, *J. Endocrinol. 139*, 383-393.

Mockridge, J.W. (1997) Immunomodulation of GH activity: application and potential mechanisms. PhD Thesis (University of Cambridge).

Mockridge, T.W., Holder, A.T. and Beattie, J. (1998) Enhancement of growth hormone (GH) activity by anti-sera prepared against analogues of an epitope peptide defined by a GH enhancing monoclonal antibody, *Liv. Prod. Sci. 55*, 1-11.

Palmer, R.M., Loveridge, N., Thomason, B.M., Mackie, S.C., Tonner, E. and Flint, D.J. (1994) Effects of a polyclonal antiserum to rat growth hormone on circulating insulin-like growth factor-I (IGF-I) and IGF-binding protein concentrations and the growth of muscle and bone, *J. Endocrinol. 142*, 85-91.

Pell, J.M. and Aston, R. (1991) Active immunisation with a synthetic peptide region of growth hormone: increased lean tissue growth, *J. Endocrinol. 131*, R1-R4.

Reichlin, S. (1987) Secretion of somatostatin and its physiological function, *J. Lab. Clin. Med. 109*, 320-326.

Schalla, C.L., Roberge, S., Wilkie, B.N., McBride, B.W. and Walton, J.S. (1994) Production of anti-idiotypic antibodies resembling bovine somatotrophin by active immunisation of lactating cows. *J. Endocrinol. 141*, 203-208.

Spencer, G.S.G. and Garssen, G.J. (1985) Increased milk production by goats following immunization against somatostatin, *Anim. Prod. 40*, 572-573.

Spencer, G.S.G., Garssen, G.J. and Hart, I.C. (1983a) A novel approach to growth promotion using autoimmunization against somatostatin. I. Effects on growth and hormone levels in lambs, *Liv. Prod. Sci. 10*, 25-37.

Spencer, G.S.G., Garssen, G.J. and Hart, I.C. (1983b) A novel approach to growth promotion using autoimmunization against somatostatin. II. Effects on appetite, carcass composition and food utilization in lamb, *Liv. Prod. Sci. 10*, 469-477.

Sun, Y.X., Sinclair, S.E., Wynn, P.C. and McDowell, G.H. (1990) Immunization against somatotropin release inhibiting factor increases milk yield in ewes, *Austr. J. Agricul. Res. 41*, 393-400.

Tainer, J.A., Getzoff, E.D., Alexander, H., Houghten R.A., Olson A.J., Lerner R.A. and Hendrickson W.A. (1984) The reactivity of anti-peptide antibodies is a function of the atomic mobility of sites in a protein, *Nature 312*, 127-134.

Tam, J.P. (1988) Synthetic peptide vaccine design: synthesis and properties of a high density multiple antigenic peptide system, *Proc. Nat. Acad. Sci. 85*, 5409-5413.

Tans, C., Dubois, F., Zhong, Z-D., Jadot, M., Wattiaux, R. and Wattiaux-de Conincks, S. (1994) Uptake by rat liver of bovine growth hormone free or unbound to a monoclonal antibody, *Biol.Cell 82*, 45-49.

Wang, B.S., Corbett, M.J., Shieh, H-M. and Sadeghi, H. (1995) Augmentation of the effectiveness of porcine GH with swine antibodies induced by a synthetic GH peptide, *J. Endocrinol. 145*, 163-167.

Wang, B.S., Lumanglas, A.L., Szewczyke, E., McWilliams, W., Loullis, C.C. and Hart, I.C. (1992) A proposed mechanism of action of a growth hormone-specific monoclonal antibody in the enhancement of hormonal activity, *Mol. Immunol. 29*, 313-317.

Wang, B.S., Szewczyk, E., Shieh, H-M. and Hart, I.C. (1990) Potentiation of the growth-promoting activity of porcine growth hormone (pGH) with an antibody generated in rabbits to the peptide sequence pGH 110-118, *J. Endocrinol. 127*, 481-485.

Wang, B.S., Zhang, R.J., Bona, C.A. and Moran, T.M. (1994) Promotion of animal growth with a monoclonal anti-idiotype specific to anti-porcine growth hormone antibody. *Mol. Immunol. 1*, 651-656.

GENE THERAPEUTIC ENHANCEMENT OF ANIMAL HEALTH AND PERFORMANCES

RUXANDRA DRAGHIA-AKLI R.
Center for Cell and Gene Therapy, Department of Cellular and Molecular Biology, Baylor College of Medicine, One Baylor Plaza, Houston, Texas, 77030, USA

Abstract

The last 10 years have seen substantial progress in the development of gene therapy. Gene therapy is a new modality with the potential for treating or preventing a variety of diseases, inherited or acquired. New viral and nonviral vectors, and applications are being developed at a rapid pace. The ability to reversibly turn genes on and off through gene therapy is a powerful tool in the control of gene expression. Until recently, gene therapy focused mainly on studying disease mechanisms, whereas recent research has put potential clinical and agricultural applications to the forefront of attention. Thus, significant progress has been made in the continued development of viral systems, including retro- and adenoviruses, as well as in the exploration of novel tools such as plasmid DNA, herpes virus-based systems or liposomes. New vaccines using DNA vectors are used to induce a long-lasting immune response. Gene therapeutic approaches were used to direct the ectopic production of growth factors for chronic periods. The success of these projects has broad ramifications: in addition to being potentially more physiological and economical that the traditional ones they could be adopted for the delivery of genes that encode a wide variety of peptides, such as other releasing hormones, neurotrophic factors or enzymes. The application therefore can be extended to enhance the economic efficiency of animal food production, and overall animal health and performances.

1. Introduction

Gene therapy promises to revolutionize agriculture as well as medicine. We are nearing the end of the first decade of gene therapy, and important recent developments were made. The early results on the clinical efficacy of gene therapies were disappointing, largely because the available gene-transfer vectors proved to be inadequate. Great progress was made recently in selecting and improving vectors, and subsequently first

R. Renaville and A. Burny (eds.), Biotechnology in Animal Husbandry, 143–157.

positive results were reported (Mountain, 2000). The use of molecular and cellular tools to genetically modify and improve food supply is playing a leading role in the continuing concern to produce more food and of better quality for the increasingly demanding world population. Many of the approaches supplement or enhance conventional breeding, address environmental concerns or stabilize food production (Kasha, 1999). Genetically manipulated animals generated by transgenic and gene-targeting (knockout) technology contributed tremendously to our understanding of gene function and regulation at the molecular level in the context of the whole organism. The development of new biotechnologies offers the potential to improve productive efficiency of animal agriculture, increase meat and milk production and the efficiency of these processes (Etherton, 1999). Consumers' risk perceptions toward the food-related biotechnology, as involuntary risk exposure, unnatural product characteristics, lack of trust in FDA regulator's ability to protect consumers in the marketplace, and consumers' inability to distinguish treated products compared to untreated is a parallel increasing problem. Studies indicate that often the public perceives no consumer benefits from farmers' use of recombinant proteins or transgenic animals (Grobe *et al.*, 1999). On the other hand, the Food and Agriculture Organization of the United Nations (FAO) underlines the fact that the world population more than doubled in the last 40 years, and it is expected to double again in the next 40-50 years, while the food surplus felt below 17%. The question now is how can we better feed the next generations (Kasha, 1999). Very recently investigators focused on developing a new type of technology in the agricultural field, approaches of gene therapies rather than transgenesis for improving growth and lactation in farm animals by enhancement of endogenous animal performances or for vaccine production.

2. Gene Therapy Vectors and their Applications

2.1. VIRAL VECTORS

The efficient delivery of therapeutic genes and appropriate gene expression are the crucial issues for clinically relevant gene therapy. Viruses are vehicles that efficiently transfer their genes into host cells. This ability made them desirable for engineering virus vector systems for the delivery of therapeutic genes. The viral vectors currently used in research are based on RNA and DNA viruses processing very different genomic structures and host ranges (Walther and Stein, 2000). Particular viruses were selected as gene delivery vehicles because of their capacities to carry foreign genes, efficiently deliver these genes associated with efficient gene expression. These are the major reasons why viral vectors derived from retroviruses, adenovirus, adeno-associated virus, herpesvirus and poxvirus are employed in more than 70% of clinical gene therapy trials worldwide - several applications are mostly correlated with the viral vectors like vaccines.

2.1.1. Retrovirus Vectors

Retrovirus vectors are an efficient transfer systems characterized by their reverse-transcriptase activity that allows their viral RNA genome to be transcribed into a double stranded DNA that stably inserts into the host DNA. The advantages of retroviral vectors are the stable integration into the host genome, generation of titers that allow efficient gene transfer into a broad variety of target cells and the ability to carry foreign genes of up to 8kb (Wu and Ataai, 2000). Unfortunately, retrovirus vectors have also disadvantages, as instability of some viral vectors, with possibility of recombination, possible insertional mutagenesis by integration at random sites in the cell host genome and the ability of vectors based on Moloney murine leukemia virus (MuLV) to transduce only dividing cells. Vectors based on lentiviruses such as HIV have the ability to transduce a variety of post-mitotic tissues as heart, muscle or brain, but biosafety remains a major concern in the production of such vectors (Naldini, 1998; Naldini *et al.*, 1996). Despite these problems, retroviruses are the system of choice for some animal gene therapy applications.

2.1.2. Somatic Gene Therapy for Growth

Somatic gene therapy consists in stable expression of a transgene product from an implanted group of cells that could be eventually removed if desired. The vectors used in these cases are retroviruses because of their property of integrating into the transduced cells genome and express the transgene for the sequential generations in the cell-line. Investigators developed stable transfections of porcine myoblasts and fibroblasts isolated from muscle of young pigs first using a green fluorescence protein (GFP) reporter plasmid (Blanton *et al.*, 2000). Porcine cells were transduced with GFP using vesicular stomatitis virus glycoprotein G pseudotyped retrovirus and resulted in efficiencies of 1:1.2 for myoblasts and 1:1.1 for fibroblasts. Retroviral transduction produced stable reporter gene expression in more than 80% of porcine muscle cells. Transduced GFP-positive cells could be separated from negative cells by fluorescence-activated cell sorting and implanted into pigs. Several days after the transplant, implanted muscles were removed and subjected to immunodetection of GFP protein. Fibroblast implantation resulted in limited GFP expression within muscle, whereas myoblast implantation resulted in GFP within muscle fibers.

This experiment suggested that cell-mediated gene transfer is possible in porcine muscle and the technology may be useful as an approach for studying and/ or promoting muscle growth in pigs. As a step further, Cheng *et al.* (1998) used this strategy to deliver long-term growth hormone (GH) to swine. Hormones stimulating growth and lactation are the obvious targets of animal gene therapy. Administration of growth hormone (GH) to farm animals has been used to improve muscling and milk production, while improving feed efficiency. Administration of porcine GH to growing pigs increases average daily weight gain by 10-20%, improves feed efficiency by 10-30%, decreases lipid accretion rates to 70%, and stimulates protein deposition (muscle growth) to 60% (Etheron *et al.*, 1993). The porcine GH (pGH) gene was constructed inside a bicistronic retroviral vector transfected into fibroblasts further encapsulated with immunoprotective microcapsules. The nonautologous microencapsulated fibroblasts carrying the pGH cDNA were implanted into the Tao-Yuan swine. Encapsulation of the pGH with an alginate-poly-L-lysine-alginate membrane did not show any deterioration in their proliferation and survival both *in vitro* and *in vivo*. The pGH gene in encapsulated recombinant fibroblasts was fully expressed after it was transplanted into the peritoneal cavity of the Tao-Yuan swine for at least 1 month as detected by PCR. The treated midget Tao-Yuan swine showed a significant increase in weight gain of 34% compared with age-matched untreated control group or with those with encapsulated recombinant fibroblasts only.

2.1.3. Adenovirus Vectors

Adenoviruses used in gene therapy applications are members of type 2 (Ad2) and type 5 (Ad5) serotypes. Replication incompetent vectors can be created by specific deletions in the wild-type virus, which allows for a packaging capacity of at least 7-8 kb. The great advantage of the Ad vectors is that very high viral titers can be easily achieved, making them suitable for direct in vivo applications. Furthermore, Ad vectors can infect a wide range of cell, including non-dividing cells (Walther and Stein, 2000). Recently, targeted viral vectors localize gene transfer to specific cell types, thus reducing immunogenicity and toxicity, increase safety, and enable the systemic administration of these vectors for multiple indications including cancer, cardiovascular disease, and inflammatory disease (Wickham, 2000). The disadvantage of Ad is that their episomal state within the host cells allows only transient expression of the therapeutic gene. Also, undesired expression from some Ad genes provokes inflammatory reactions and toxicity that limit repeated administration.

2.1.3.1. Gene therapy for the Next Generation? Maybe an alternative to the classical transgenic animal is the introduction of a manipulated gene into the spermatozoa. Farre *et al.* (1999) tested the ability of adenoviral vectors to transfer DNA into boar spermatozoa and to offspring. Exposure of spermatozoa to adenovirus bearing the reporter Eschrichia coli lacZ gene resulted in the transfer of the gene to the head of the spermatozoa. Treatment did not affect either viability or acrosomal integrity of boar sperm. Approximately 22% of cell embryos obtained after *in vitro* fertilization with adenovirus-exposed sperm expressed the LacZ product. Also, 7% of the piglets obtained after artificial insemination with adenovirus-exposed spermatozoa, as well as two of the stillborn piglets showed the presence of the LacZ mRNA in all of the tissues tested. The offspring obtained after mating of the two positive animals did not show LacZ gene presence. Their results indicate that adenovirus could be a feasible mechanism for the delivery of DNA into spermatozoa, even though the transfer of the transgene seams to be limited to the first generation.

2.1.4. Adeno-Associated Virus Vectors

The AAV is a single stranded DNA virus, and a broad range of cell types is susceptible to infection with AAV. The AAV requires an adenovirus or a herpes virus for viral replication. No pathology was linked to this virus. As a gene therapy vehicle the AAV has important characteristics: high level of infection in different cell types, including post-mitotic cells, long-term expression of the transgene for at least up to 2 years (Fisher *et al.*, 1997; Koebert *et al.*, 1997), low immunogenicity and could integrate in a specific site on chromosome 19 (property often lost in AAV gene therapy construct). Unfortunately, the recombinant AAV production requires superinfection with a Ad that results in low-quality viral stocks. Another potential rate-limiting aspect is the relatively small packaging capacity, less than 5 kb (Monaham and Samulski, 2000). Based on impressive success in delivering genes by this method in rodents and primates, many clinical trial using AAV are ongoing for a variety of human inherited diseases, from cystic fibrosis to beta-thalassaemia (Hallek and Wendtner, 1996). In contrast, this type of vector is relatively unused for animal improvement purposes.

2.1.5. Herpes Virus Vectors

HSV-1 is a DNA virus with a large genome of 150 kb, relatively stable, which allows packaging of up to 50 kb of foreign sequence. Vectors can be produced at very high titers. HVS-1 vectors have a broad range of target cells, and it can infect both dividing and non-dividing cells. Because of the natural tropism to neuronal cells the vectors are mostly used for corrective human gene therapies or cancer within the nervous system (Wolfe *et al.*, 1992; Miyanohara *et al.*, 1992).

2.2. PLASMID DNA

Direct gene transfer with naked plasmid DNA into somatic cells that would further function as bioreactors for production of secreted proteins, as hormones or enzymes, or for vaccines is now a routine technique. The organ of choice for plasmid delivery is the skeletal muscle (Wolff *et al.*, 1990), but skin, some tumors and immune cells are successfully transfected using naked DNA (Ulmer *et al.*, 1997).

For farm animals, skeletal muscle possesses properties that make it an attractive target organ for gene therapy. It offers an easily accessible site for injection of DNA, and the post-mitotic nature and longevity of muscle fibers permit sustained expression of genes that are delivered. Studies to date suggest that when plasmids are injected into skeletal muscle they persist as an episome and are not integrated into host chromosomal DNA (Wolff *et al.*, 1992) even though expression was reported to persist for up to 19 months (Manthorpe *et al.*, 1993). The transfer efficiency of plasmid was reported to be superior to adenovirus and retrovirus in rodent muscle (Davis *et al.*, 1993), and expression of recombinant protein from intramuscular injection of plasmid were also reported in primates and humans (Jiao *et al.*, 1992). Comparative with viruses, the nonviral techniques for gene transfer *in vivo*, the direct injection of plasmid DNA is simple to use, easy to produce on large scale, inexpensive, and safe, as it lacks specific immune response.

In addition to expression of reporter enzymes, skeletal muscle is now used as a bioreactor to express therapeutic proteins having either a local (Acsadi *et al.*, 1991) or systemic effect (Draghia-Akli *et al.*, 1997, 1999). Until recently this methodology was limited by the relatively low expression levels due to inefficient DNA uptake into muscle fibers to ensure systemic physiological levels of secreted proteins (Prentice *et al.*, 1994; Wells *et al.*, 1997). Several approaches were developed to enhance the efficiency of gene transfer via naked DNA including gene gun (Mahvi *et al.*, 1997) or electroporation. Electroporation was previously used in un-anesthetized humans to transfect tumor cells after injection of plasmid DNA (Nishi *et al.*, 1996; Rols *et al.*, 1998). The electrotransfer method was first used experimentally in rodents and other small animals (Aihara and Miyazaki, 1998; Mir *et al.*, 1998; Muramatsu *et al.*, 1998) and it was found to be effective with expression levels at 40-100 fold over injection alone bringing the range of expression to physiological realm. Recently, different polymers that weakly interact with DNA such as PVP were shown to improve the efficiency of gene transfer (Anwer *et al.*, 1998).

2.2.1.Growth Enhancement using Naked DNA

Porcine GH treatment, while widely used, could induce insulin resistance of protein metabolism (Etherton *et al.*, 1986), and consequently reduces the theoretical possibility for increased protein synthesis in the fed state. Numerous studies have shown that GH markedly reduces the amount of carcass fat; consequently the quality of products increases (Etherthon *et al.*, 1993). Nevertheless, side effects of continuous GH administration by implants are frequent, as increased glucose or insulin levels (Klindt *et al.*, 1996); the requirement for daily administration is labor intensive. GH transgenic animals developed problems as lethargy, lameness, gastric ulcers and anoestrus (Solomon *et al.*, 1994; Pursel *et al.*, 1990).

The use of growth hormone releasing hormone (GHRH), the upstream stimulator of GH, was considered to be an alternate strategy that may increase not only growth performance or milk production for large species such as pigs or cattle, but also more importantly, the efficiency of production from both practical and metabolic perspectives (Draghia-Akli *et al.*, 1999). GHRH therapy seems to be more physiological than GH therapy, as interaction between GH and its receptor suggests that the molecular heterogeneity of circulating GH may have important implications in growth and homeostasis (Caiozzo *et al.*, 1992; Diffee *et al.*, 1993). It is thought that GH side effects are correlated with the excess GH, which abolishes the natural GH episodic pulses. GHRH therapy is capable of a degree of feed-back, which is totally abolished in the GH therapies, which explains the fact that virtually no side effects have been reported for GHRH therapy (Thorner *et al.*, 1986).

GHRH cDNAs were characterized in porcine, bovine and many other species (Baird *et al.*, 1986; Kato *et al.*, 1990) . It is well established that extracranially secreted GHRH as mature peptide or truncated molecules can be biologically active and even produce acromegaly in humans (Melmed, 1991; Faglia *et al.* 1992) in a wide range of dosage. Thus the conclusion that hypothalamic tissue specific expression of the GHRH gene is not required for activity, as long as the peptide has access to the circulation. The approach for a GHRH gene therapy is complex. This hormone has a short half-life in serum that would theoretically imply that a large quantity of plasmid would be needed to assure that a sufficient number of cells are transfected and produce the hormone. In order to minimize the quantity of plasmid necessary for the injection, the design of the plasmid is critical. In our laboratory we combined several improved elements in the injected constructs: we designed and integrated strong muscle specific promoters (Li *et al.*, 1999), mutated GHRH cDNAs coding for molecules with longer half-life and cloned them on plasmid backbones that don't enhance an immune response.

Several mutated GHRH coding vectors were tested *in vitro* into porcine primary myocytes and used for the stimulation of pig GH release from porcine primary pituitary culture. A shorter form (1-40)OH form of the hormone was used as a base construct in all cases. Growth hormone release rose from baseline values of 200ng/ml up to 1600ng/ml, when stimulated with media from HV-GHRH mutant (figure 1). For the *in vivo* studies the best mutant construct (pSP-HV-GHRH), as well as the wild-type construct (pSP-wt-GHRH) as a positive control, and an *Eschericia coli* beta-galactosidase expressing construct as the negative control (pSP-βgal) was selected

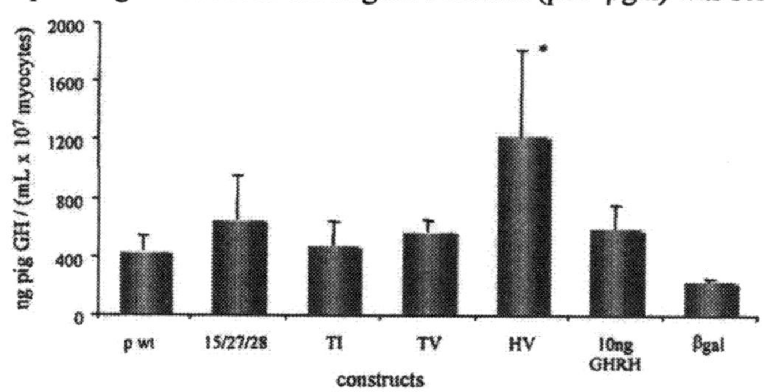

***Figure 1.** Pig GH release in porcine primary pituitary culture is stimulated by media from skeletal muscle cells transfected with porcine GHRH mutant plasmids (media from cells transfected with: pwt - porcine wild-type construct, 15/27/28 - mutation of Gly15 to Ala, Met27 to Leu and Ser28 to Asn, TI - mutations as in 15/27/28 plus change of Ala2 to Ile, TV - mutations as in 15/27/28 plus change of Ala2 to Val, HV - mutations as in 15/27/28 plus change of Tyr1 with His, and Ala2 with Val, βgal - construct coding for E.coli beta-gal, used as a negative control, 10ng GHRH- cells stimulated with 10 ng of recombinant GHRH.*

At day 0, three-week-old pigs were anesthetized and 10mg of pSP-GHRH or pSP-βgal plasmid DNA was injected into intact semitendinosous muscle (figure 2), and then the thigh was placed in between two calipers and electroporated, as described (Aihara and Miyazaki, 1998; Wang *et al.*, 1998).

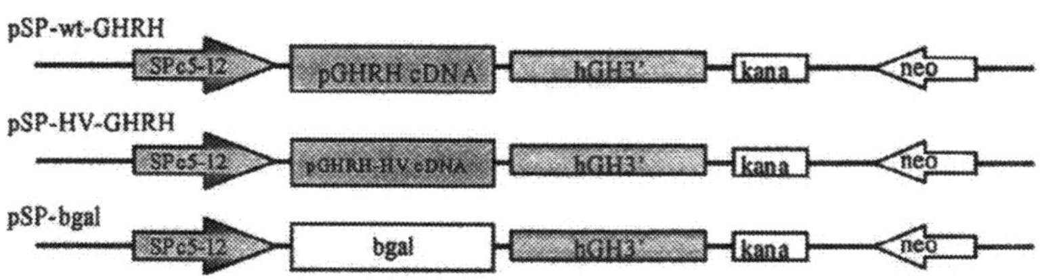

***Figure 2.** GHRH constructs injected* in vivo *in porcine.*

In vivo activity of these myogenic expression vectors was first evaluated by measuring the relative GHRH serum levels {2696}. In pigs injected with pSP-HV-GHRH, GHRH levels increased at 7 days post-injection (figure 3a) and were 150% the βgal control levels at 14 days (652.4±77pg/ml versus 419.6±13pg/ml). pSP-HV-GHRH serum GHRH levels reached a plateau by 60 days that was 2 to 3 fold greater than the injected control values ($p<0.002$). The absolute quantity of GHRH (figure 3b), corrected for the increase in body weight between day 0 and day 60 (blood volume accounts for 8% of total body weight), secreted by the pSP-HV-GHRH injected pigs was 3 fold the βgal control values (1426.49±10.47ng versus 266.84±25.45ng, $p<0.034$). By comparison, the wild-type injected animals required 45 days to demonstrate an increase in GHRH secretion, and had attained only a 2-fold increase at the 60 day time point ($p<0.16$).

Young animals have very high levels of GH, levels gradually decreasing with age. The GHRH injected animals (figure 4) had a less dramatic decline or even an increase in their GH levels evident at day 7 post-injection (delta variation HV = +1.52, wt = -0.73 versus control = -3.2ng/ml) and 14 days post-injection (delta variation HV = +1.09, wt = -4.42 versus control = -6.88ng/ml).

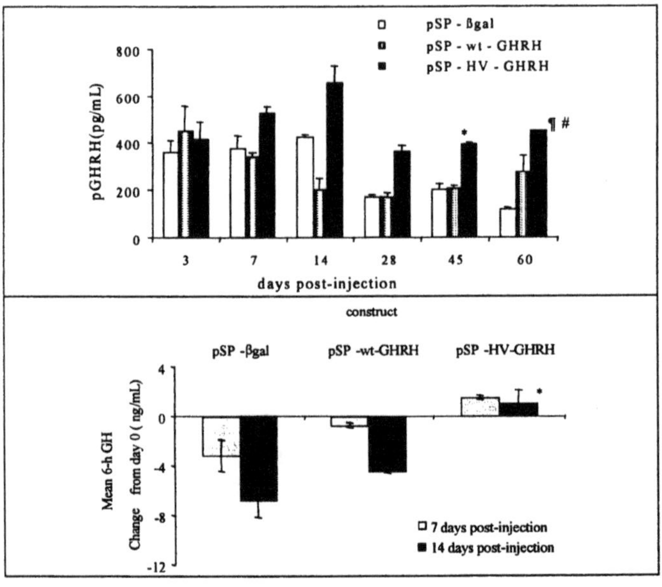

Figure 3. *Serum GHRH levels in pSP-GHRH injected pigs versus control animals.* ***A.*** *Relative levels of serum GHRH levels in pSP-GHRH injected pigs versus placebo injected pigs, *p<0.02; ¶p<0.002. When experimental data from HV-GHRH and controls through the entire experiment were compared by ANOVA, #p<0.002.* ***B.*** *Absolute levels of serum GHRH in pSP-GHRH injected pigs versus control pigs corrected for weight/blood volume increase, *p<0.05 at all marked time points. When data from HV-GHRH and controls were compared throughout the experiment by ANOVA, #p<0.004.*

Plasma IGF-I level started to rise in pig HV-GHRH injected pigs at 3 days post-injection (figure 5). At 21 days, the same animals had in average 3 times more IGF-I, level maintained over 65 days ($p < 0.03$). The pig wild-type GHRH injected animals had in average 40% increase in their circulating IGF-I ($p = 0.39$). IGF-binding protein-3 was slightly increased in treated animals.

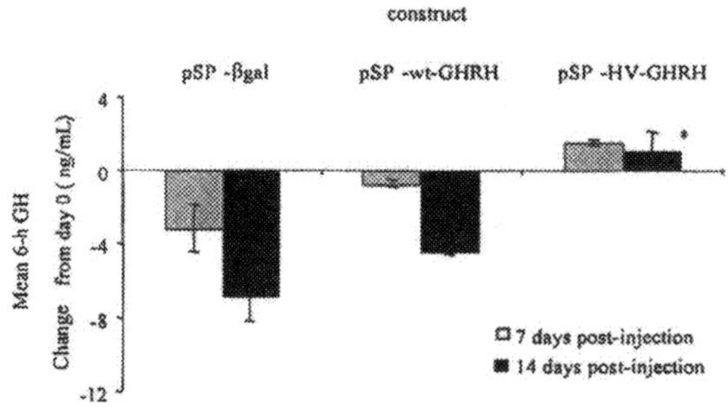

*Figure 4. pSPc5-12-HV-GHRH injected pigs show a positive variation of GH levels. *p<0.033.*

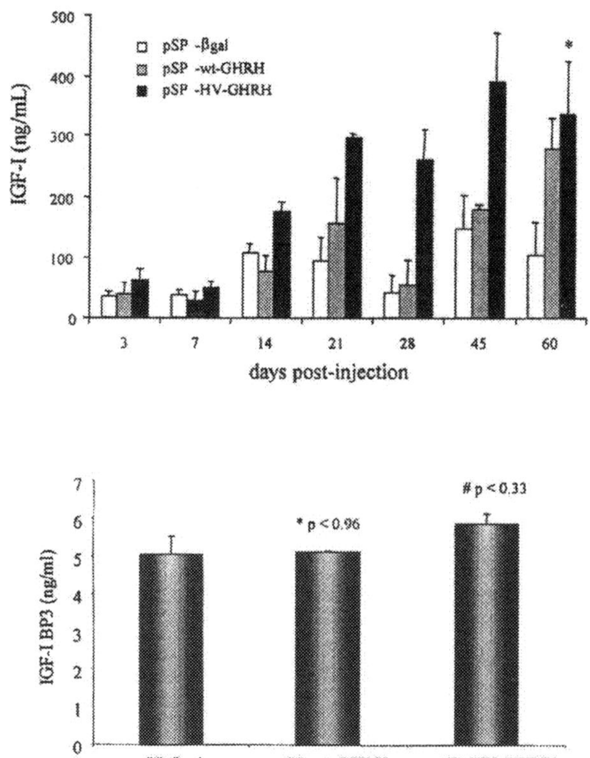

*Figure 5. Plasma IGF-1 and IGF-BP3 levels after direct intramuscular injection of pSP-GHRH constructs. ANOVA for the entire series * p<0.03.*

Sixty-five days post-injection, GHRH injected animals showed a significant increase in weight (figure 6A): pig wild-type GHRH injected animals were in average 21.5% heavier than the controls (37.125kg vs. 29.375kg), while the HV-GHRH injected pigs were 37.8% heavier than the βgal controls (41.775kg) ($p = 0.014$).

Feed efficiency (figure 6B) is increased with 20% in pigs injected with GHRH constructs when compared with controls (0.267 kg of food/day for each kg weight gain in pSP-HV-GHRH, and 0.274 kg in pSP-wt-GHRH, versus 0.334 kg in pSP–βgal injected pigs).

Body composition studies showed a proportional increase of all body components in GHRH injected animals, with no organomegaly, and no sign of associated pathology. A decrease in the relative proportion of body fat (11.35% versus 16% in controls, $p = 0.09$) and an increase in the lean body mass (86.2% versus 81% in control animals, $p < 0.02$) was observed in pigs injected with wt-GHRH. Pigs injected with the mutated HV-GHRH had similar body composition with the control animals (table 1).

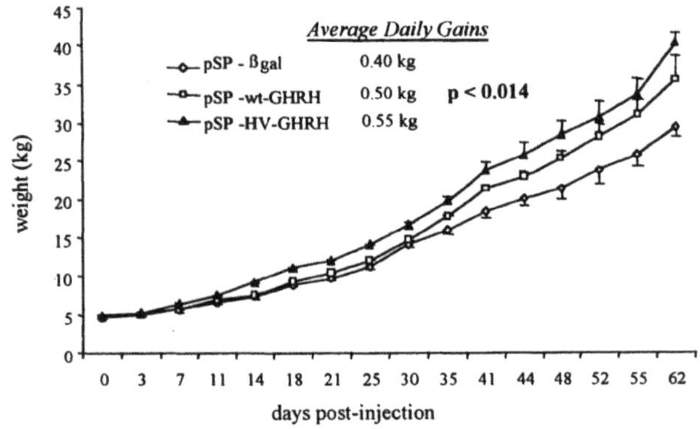

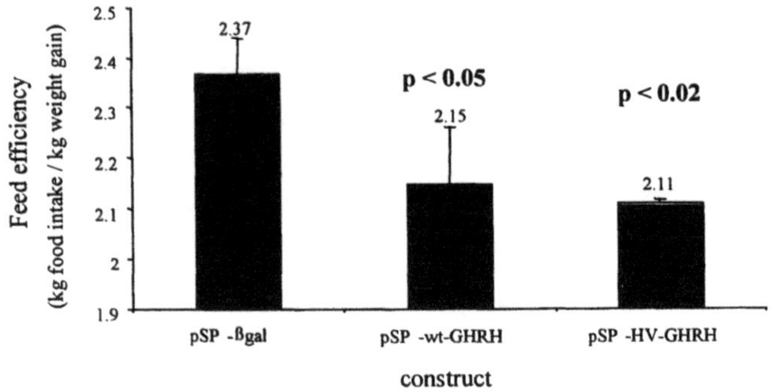

***Figure 6. A.** Average weight increase in injected pigs over 2 month after pSP-GHRH intramuscular injection. **B.** Improved feed efficiency in the pSP-GHRH injected pigs versus controls.*

The metabolic profile of pSP-HV-GHRH injected pigs showed a significant decrease in serum urea level ($p < 0.006$), indicating decreased protein catabolism. Serum glucose level was similar in between the controls and the plasmid GHRH injected pigs. Importantly, no other metabolic changes, including the absence of hyperglycemia, were found (table 2).

Table 1. *Composition of weight gain as determined by dual x-ray absorbitometry (DEXA), compared with IGF-I levels and feed consumption over 63 days in piglets subjected to gene transfer treatment with pSP-ßgal, porcine pSP-wt-GHRH, or pSP-HV-GHRH vectors.*

Gain (0-63d)	**control**	**wt-GHRH**	**HV-GHRH**	**pool SD**	**p**
weight (g/d)	390	493	563	48	**0.06**
fat (g/d)	62	56	103	15	**0.09**
lean (g/d)	316	425	459	24	**0.02**
BMD at 63d (g/cm2)	0.63	0.60	0.70	0.04	**0.15**
IGF-I (ng/ml)	127	231	345	57	**0.07**
feed efficiency (g/g BW)	2.37	2.16	2.11	0.08	**0.04**

Table 2. *Plasma glucose, urea, creatinine, and total protein concentrations in control and GHRH-injected pigs 63 days after receiving the plasmid injection.*

	Glucose (mg/dL)	**Ure (mg/dL)**	**Ceatinin (mg/Dl)**	**Total protein(g/dL)**
pSP ßga	99.2	9.0	0.82	4.6
pSP-wt-GHRH	97.5	8.3	0.83	4.76
pSP-wt-GHRH	104.8	6.9	0.78	4.88

These results demonstrate that the i.m. injection of pSP-GHRH followed by electroporation could be used to produce physiological levels of GHRH in the circulation of farm animals. Thus, it should be possible to enhance endogenous GH and IGF-I secretion in livestock in a more physiological and less expensive manner than traditional methods.

New experiments in our laboratory using injectable electrodes rather than calipers showed that the quantity of plasmid could be reduced to as low as 100 micrograms with similar results.

2.2.2. Plasmid DNA – The Next Generation of Vaccines

Nonviral vectors are particularly suitable for creating vaccines, as parts of the viral antigens can be expressed. The generated immune responses are sufficient to protect animals from a wide variety of live infectious agents, leading to the creation of a new class of therapeutic agents, the DNA vaccines. Different studies in mice showed that

DNA immunization can induce neutralizing antibodies and cytotoxic T-cells against several viral pathogens, including rabies virus (Xiang and Ertl, 1995), herpes simplex, murine cytomegalovirus or papilloma virus (Ertl and Xiang, 1996). In porcine, investigators tested pseudorabies virus (PRV) based on surface glycoproteins B (gB), gC and gD important antigens implicated in protective immunity against PRV infection (van Rooij *et al.*, 2000). As cell-mediated immunity plays a major role in this protective immunity, pigs were vaccinated with plasmid DNA constructs coding for gB, gC or gD and challenged with the virulent strain of pseudorabies virus. Vaccination with gB plasmid DNA induced the strongest cell-mediated immune responses including cytotoxic T cell responses, whereas plasmid DNA coding for gD induced the strongest virus neutralizing antibody responses. Interestingly, vaccination with gB-DNA reduced virus excretion early after challenge infection.

Naked DNA vaccines were developed as candidates for foot-and-mouth disease (FMD), an important disease of domestic animals (Beard *et al.*, 1999). The virus that causes this disease, FMDV, is a member of the picornavirus family, which includes many important human pathogens, such as poliovirus, hepatitis A virus, and rhinovirus. The first DNA vaccine, pP12X3C, encoded for the viral capsid gene (P1) and the processing proteinase (3C). Cells transfected with this DNA produce processed viral antigen, and animals inoculated with this DNA using a gene gun produced detectable antiviral immune responses. The second DNA vaccine candidate, pWRMHX, encoded the entire FMDV genome, permitting the plasmid-encoded viral genomes to undergo amplification in susceptible cells; as a measure of protection the construct encoded also a mutation at the cell-binding site, preventing the replicated genomes from causing disease. Swine inoculated with this vaccine candidate produced viral particles lacking the cell binding site, and neutralizing antibodies that recognize the virus. Comparison of the immune responses elicited by pP12X3C and pWRMHX in swine indicate that the plasmid encoding the replicating genome stimulated a stronger immune response, and swine inoculated with pWRMHX by the intramuscular, intradermal, or gene gun routes were partially protected from a highly virulent FMD challenge. Using this methodology, recombinant swinepox virus vaccines expressing pseudorabies virus antigens were developed and shown to provide protection against challenge (Tripathy, 1999). These studies and evidence of local infection of the oral tract by swinepox virus indicate its potential as a recombinant vector for providing immunity against various swine pathogens including those that infect the respiratory and gastrointestinal tracts. Other investigators (Pirzadeh and Dea, 1998) proved that following a massive intratracheal challenge with the virulent interstitial pneumonitis and broncho-alveolitis virus, DNA-vaccinated pigs were protected from generalized viraemia and the development of typical macroscopic lung lesions that were observed in unvaccinated, virus-challenged controls, as well as in pigs that were immunized with a mock construct. Interstitial pneumonitis and broncho-alveolitis were remarkably milder in DNA-vaccinated animals.

The last ten years have seen substantial progress in the development and application of gene therapy. Many problems remain to be resolved before delivery of genes for improvment of animal health and performances can become a standard practice. The achievement of these goals has broad ramifications, and finally will enhance the economic efficiency of animal food production.

References

Acsadi, G., Dickson, G., Love, D.R., Jani, A., Walsh, F.S., Gurusinghe, A., Wolff, J.A. and Davies, K.E. (1991) Human dystrophin expression in mdx mice after intramuscular injection of DNA constructs [see comments], *Nature 352*, 815-818.

Aihara, H. and Miyazaki, J. (1998) Gene transfer into muscle by electroporation *in vivo*, *Nat.Biotechnol. 16*, 867-870.

Anwer, K., Shi, M., French, M.F., Muller, S.R., Chen, W., Liu, Q.S., Proctor, B.L., Wang, J.J., Mumper, R.J., Singhal, A., Rolland, A.P. and Alila, H.W. (1998) Systemic effect of human growth hormone after intramuscular injection of a single dose of a muscle-specific gene medicine, *Human Gene Therapy 9*, 659-670.

Baird, A., Wehrenberg, W.B. and Ling, N. (1986) Relative potencies of human, rat, bovine/caprine, porcine and ovine hypothalamic growth hormone-releasing factors to release growth hormone by the rat anterior pituitary *in vitro*, *Neuroendocrinology 42*, 273-276.

Beard, C., Ward, G., Rieder, E., Chinsangaram, J., Grubman, M.J. and Mason, P.W. (1999) Development of DNA vaccines for foot-and-mouth disease, evaluation of vaccines encoding replicating and non-replicating nucleic acids in swine, *J. Biotechnol. 73*, 243-249.

Blanton, J.R. Jr., Bidwell, C.A., Sanders, D.A., Sharkey, C.M., McFarland, D.C., Gerrard, D.E. and Grant, A.L. (2000) Plasmid transfection and retroviral transduction of porcine muscle cells for cell-mediated gene transfer, *J. Anim. Sci. 78*, 909-918.

Caiozzo, V.J., Ma, E., McCue, S.A., Smith, E., Herrick, R.E. and Baldwin, K.M. (1992) A new animal model for modulating myosin isoform expression by altered mechanical activity, *J. Appl. Physiol 73*, 1432-1440.

Cheng, W.T., Chen, B.C., Chiou, S.T. and Chen, C.M. (1998) Use of nonautologous microencapsulated fibroblasts in growth hormone gene therapy to improve growth of midget swine, *Hum.Gene Ther. 9*, 1995-2003.

Davis, H.L., Demeneix, B.A., Quantin, B., Coulombe, J. and Whalen, R.G. (1993) Plasmid DNA is superior to viral vectors for direct gene transfer into adult mouse skeletal muscle, *Human Gene Therapy 4*, 733-740.

Diffee, G.M., Caiozzo, V.J., McCue, S.A., Herrick, R.E. and Baldwin, K.M. (1993) Activity-induced regulation of myosin isoform distribution: comparison of two contractile activity programs, *J. Appl. Physiol 74*, 2509-2516.

Draghia-Akli, R., Fiorotto, M.L., Hill, L.A., Malone, P.B., Deaver, D.R. and Schwartz, R.J. (1999) Myogenic expression of an injectable protease-resistant growth hormone-releasing hormone augments long-term growth in pigs, *Nat. Biotechnol. 17*, 1179-1183.

Draghia-Akli, R., Li, X.G. and Schwartz, R.J (1997) Enhanced Growth By Ectopic Expression Of Growth Hormone Releasing Hormone Using An Injectable Myogenic Vector, *Nat. Biotechnol. 15*, 1285-1289.

Ertl, H.C. and Xiang, Z. (1996) Novel vaccine approaches, *J. Immunol. 156*, 3579-3582.

Etherton, T.D. (1999) Emerging strategies for enhancing growth: is there a biotechnology better than somatotropin?, *Domest. Anim. Endocrinol. 17*, 171-179.

Etherton, T.D., Kris-Etherton, P.M. and Mills, E.W. (1993) Recombinant bovine and porcine somatotropin: safety and benefits of these biotechnologies, *J. Amer. Diet. Assoc. 93*, 177-180.

Etherton, T.D., Louveau, I., Sorensen, M.T. and Chaudhuri, S. (1993) Mechanisms by which somatotropin decreases adipose tissue growth, *Am. J. Clin. Nutr. 58*, 287S-295S.

Etherton, T.D., Wiggins, J.P., Chung, C.S., Evock, C.M., Rebhun, J.F. and Walton, P.E. (1986) Stimulation of pig growth performance by porcine growth hormone and growth hormone-releasing factor, *J. Anim. Sci. 63*, 1389-1399.

Faglia, G., Arosio, M., Bazzoni, N. (1992) Ectopic acromegaly. [Review]. *Endocrinology & Metabolism Clinics of North America* 21, 575-595.

Farre, L., Rigau, T., Mogas, T., Garcia-Rocha, M., Canal, M., Gomez-Foix, A.M. and Rodriguez-Gil, J.E. (1999) Adenovirus-mediated introduction of DNA into pig sperm and offspring, *Mol. Reprod. Dev. 53*, 149-158.

Fisher, K.J., Jooss, K., Alston, J., Yang, Y.P., Haecker, S.E., High, K., Pathak, R., Raper, S.E. and Wilson, J.M. (1997) Recombinant adeno-associated virus for muscle directed gene therapy, *Nature Med. 3*, 306-312.

Grobe, D., Douthitt, R. and Zepeda, L. (1999) A model of consumers' risk perceptions toward recombinant bovine growth hormone (rbGH): the impact of risk characteristics, *Risk Anal. 19*, 661-673.

Hallek, M. and Wendtner, C.M. (1996) Recombinant adeno-associated virus (rAAV) vectors for somatic gene therapy: recent advances and potential clinical applications, *Cytokines Mol. Ther. 2*, 69-79.

Jiao, S., Williams, P., Berg, R.K., Hodgeman, B.A., Liu, L., Repetto, G. and Wolff, JA. (1992) Direct gene transfer into nonhuman primate myofibers *in vivo*, *Human Gene Therapy 3*, 21-33.

Kasha, K.J. (1999) Biotechnology and world food supply, *Genome 42*, 642-645.

Kato, Y., Shimokawa, N., Kato, T., Hirai, T., Yoshihama, K., Kawai, H., Hattori, M., Ezashi, T., Shimogori, Y. and Wakabayashi, K. (1990) Porcine growth hormone: molecular cloning of cDNA and expression in bacterial and mammalian cells, *Biochim. Biophys. Acta 1048*, 290-293.

Klindt, J., Buonomo, F.C., Wise, T. and Yen, J.T. (1996) Endocrine and metabolite responses to porcine growth hormone administered by sustained release implant for different lengths of time in male pigs, *Endocrinology 137*, 3689-3695.

Koeberl, D.D., Alexander, I.E., Halbert, C.L., Russell, D.W. and Miller, A.D. (1997) Persistent expression of human clotting factor IX from mouse liver after intravenous injection of adeno-associated virus vectors, *Proc. Natl. Acad. Sci. U.S.A 94*, 1426-1431.

Li, X., Eastman, E. M., Schwartz, R.J. and Draghia-Akli, R. Synthetic muscle promoters: activities exceeding naturally occurring regulatory sequences, *Nat. Biotechnol. 17*, 241-245 1999.

Mahvi, D.M., Sheehy, M.J. and Yang, N.S. (1997) DNA cancer vaccines: a gene gun approach, *Immunol. Cell Biol. 75*, 456-460.

Manthorpe, M., Cornefert-Jensen, F., Hartikka, J., Felgner, J., Rundell, A., Margalith, M. and Dwarki, V. (1993) Gene therapy by intramuscular injection of plasmid DNA: studies on firefly luciferase gene expression in mice, *Human Gene Therapy 4*, 419-431.

Melmed, S. (1991) Extrapituitary Acromegaly. [Review]. *Endocrinology & Metabolism Clinics of North America* 20, 507-518.

Mir, L.M., Bureau, M.F., Rangara, R., Schwartz, B. and Scherman, D. (1998) Long-term, high level in vivo gene expression after electric pulse- mediated gene transfer into skeletal muscle. *C.R.Acad.Sci.III 321*, 893-899.

Miyanohara, A., Johnson, P.A., Elam, R.L., Dai, Y., Witztum, J.L., Verma, I.M. and Friedmann, T. (1992) Direct gene transfer to the liver with herpes simplex virus type 1 vectors: transient production of physiologically relevant levels of circulating factor IX, *New Biol. 4*, 238-246.

Monahan, P.E. and Samulski, R.J. (2000) AAV vectors: is clinical success on the horizon?, *Gene Ther.* 7, 24-30.

Mountain, A. (2000) Gene therapy: the first decade, *Trends Biotechnol. 18*, 119-128.

Muramatsu, T., Nakamura, A. and Park, H.M. (1998) In vivo electroporation: A powerful and convenient means of nonviral gene transfer to tissues of living animals, *Int. J. Mol. Med. 1*, 55-62.

Naldini, L. (1998) Lentiviruses as gene transfer agents for delivery to non-dividing cells, *Curr.Opin.Biotechnol. 9*, 457-463.

Naldini, L., Blomer, U., Gallay, P., Ory, D., Mulligan, R., Gage, F.H., Verma, I. M. and Trono, D. (1996) In vivo gene delivery and stable transduction of nondividing cells by a lentiviral vector , *Science 272*, 263-267.

Nishi, T., Yoshizato, K., Yamashiro, S., Takeshima, H., Sato, K., Hamada, K., Kitamura, I., Yoshimura, T., Saya, H., Kuratsu, J. and Ushio, Y. (1996) High-efficiency in vivo gene transfer using intraarterial plasmid DNA injection following in vivo electroporation, *Cancer Res. 56*, 1050-1055.

Pirzadeh, B. and Dea, S. (1998) Immune response in pigs vaccinated with plasmid DNA encoding ORF5 of porcine reproductive and respiratory syndrome virus, *J. Gen. Virol. 79*, 989-999.

Prentice, H., Kloner, R. A., Prigozy, T., Christensen, T., Newman, L., Li, Y. and Kedes, L. (1994) Tissue restricted gene expression assayed by direct DNA injection into cardiac and skeletal muscle, *J. Mol. Cell. Cardiol. 26*, 1393-1401.

Pursel, V.G., Bolt, D.J., Miller, K.F., Pinkert, C.A., Hammer, R.E., Palmiter, R.D. and Brinster, R.L. (1990) Expression and performance in transgenic pigs, *J. Reprod. Fertil. Suppl 40,* 235-45, 235-245.

Rols, M.P., Delteil, C., Golzio, M., Dumond, P., Cros, S. and Teissie, J. (1998) In vivo electrically mediated protein and gene transfer in murine melanoma, *Nat. Biotechnol. 16*, 168-171.

Solomon, M.B., Pursel, V.G., Paroczay, E.W. and Bolt, D.J. (1994) Lipid composition of carcass tissue from transgenic pigs expressing a bovine growth hormone gene, *J. Anim. Sci. 72*, 1242-1246.

Thorner, M.O., Vance, M.L., Evans, W.S., Blizzard, R.M., Rogol, A.D., Ho, K., Leong, D.A., Borges, J.L., Cronin, M.J. and McLeod, R. M. (1986) Physiological and clinical studies of GRF and GH, *Rec. Prog. Horm. Res. 42*, 589-640.

Tripathy, D.N. (1999) Swinepox virus as a vaccine vector for swine pathogens, *Adv. Vet. Med. 41*, 463-480.

Ulmer, J.B., Deck, R.R., Dewitt, C.M., Fu, T.M., Donnelly, J.J., Caulfield, M.J. and Liu, M.A. (1997) Expression Of A Viral Protein By Muscle Cells In Vivo Induces Protective Cell-Mediated Immunity, *Vaccine 15*, 839-841.

van Rooij, E.M., Haagmans, B.L., Glansbeek, H.L., de Visser, Y.E., de Bruin, M.G., Boersma, W. and Bianchi, A.T. (2000) A DNA vaccine coding for glycoprotein B of pseudorabies virus induces cell-mediated immunity in pigs and reduces virus excretion early after infection, *Vet. Immunol. Immunopathol. 74*, 121-136.

Walther, W. and Stein, U. (2000) Viral vectors for gene transfer: a review of their use in the treatment of human diseases, *Drugs 60*, 249-271.

Wang, L., Ma, W., Markovich, R., Lee, W.L. and Wang, P.H. (1998) Insulin-like growth factor I modulates induction of apoptotic signaling in H9C2 cardiac muscle cells, *Endocrinology 139*, 1354-1360.

Wells, K.E., Maule, J., Kingston, R., Foster, K., McMahon, J., Damien, E., Poole, A. and Wells, D.J. (1997) Immune responses, not promoter inactivation, are responsible for decreased long-term expression following plasmid gene transfer into skeletal muscle, *FEBS Lett. 407*, 164-168.

Wickham, T. J. (2000) Targeting adenovirus, *Gene Ther. 7*, 110-114.

Wolfe, J.H., Deshmane, S.L. and Fraser, N.W. (1992) Herpesvirus vector gene transfer and expression of beta-glucuronidase in the central nervous system of MPS VII mice, *Nat.Genet. 1*, 379-384

Wolff, J.A., Ludtke, J.J., Acsadi, G., Williams, P. and Jani, A. (1992) Long-term persistence of plasmid DNA and foreign gene expression in mouse muscle, *Human Molecular Genetics 1*, 363-369.

Wolff, J.A., Malone, R.W., Williams, P., Chong, W., Acsadi, G., Jani, A. and Felgner, P.L. (1990) Direct gene transfer into mouse muscle *in vivo, Science 247*, 1465-1468.

Wu, N. and Ataai, M.M. (2000) Production of viral vectors for gene therapy applications, *Curr. Opin. Biotechnol. 11*, 205-208.

Xiang, Z. and Ertl, H.C. (1995) Manipulation of the immune response to a plasmid-encoded viral antigen by coinoculation with plasmids expressing cytokines, *Immunity 2*, 129-135.

ANTI-ADIPOCYTE MONOCLONAL ANTIBODIES: A NEW TECHNOLOGY FOR REGULATING ADIPOSE CONVERSION

REMACLE C., DE CLERCQ L., MOUROT* J. AND BOONE C.
*Université catholique de Louvain, Laboratory of Cell Biology, Croix du Sud 5, B1348, Louvain-la-Neuve, Belgium, and * INRA, Station de Recherches Porcines, F35590 Saint-Gilles, France*

Abstract

During the last decades, livestock breeders attempted to develop strategies for decreasing fat content in meat. One of these strategies consists in immunization against adipocytes. Polyclonal antisera have been raised against fat cells or adipocyte plasma membranes of different species, and in vivo reduction of fat depots has been observed in some cases. Facing the problems of specificity and reproducibility inherent to polyclonal antisera, monoclonal antibodies could offer a promising alternative, which has been developed only in the porcine species.

In our laboratory, a mouse monoclonal antibody of the IgG2b sub-class was raised against porcine adipocyte plasma membranes. This antibody did not cross-react in immunocytofluorescence with any tested cell-type or tissue other than adipocytes. Complement-mediated cytotoxicity was demonstrated in primary cultures of porcine stromal-vascular cells. When antibody and complement were added to already differentiated cultures, the treatment resulted in elimination of lipid-filled preadipocytes, whereas an early treatment of cultures prevented the appearance of these cells. Newborn piglets were injected at a time when most fat tissues are developing, except the intramuscular one maturation of which occurs later. The dose of 1 mg/kg of monoclonal antibody on d 2 and 5 of life produced more than a 20% reduction of subcutaneous and leaf fat lipids at 35 d of age, whereas the lipid content of muscle remained unaffected. In animals sacrificed at 50 and 100 kg, the early treatment with antibody led to a decrease of subcutaneous and visceral fat without affecting the growth performance of the animals. These results demonstrate that early systemic treatment of piglets with a specific anti-adipocyte antibody is able to reduce the fat mass, while keeping less affected the intramuscular fat depot.

R. Renaville and A. Burny (eds.), Biotechnology in Animal Husbandry, 159–167.

1. Introduction

Considering the food surplus stocks existing in European countries, increasing meat production is no more needed. By contrast, and in response to consumer requirement, livestock breeders now attempt to develop new strategies for improving the qualitative aspect of meat. The main objective is to develop a meat with high nutritional values as well as good technological (e.g. conservation, consistency, sensitivity to oxidation) and organoleptical (e.g. colour, flavour, texture) qualities (Lebret and Mourot, 1998). Meat is essentially composed of muscular, connective and adipose tissues. Numerous studies attempted to favour the growth of the muscular mass at the expense of fat tissues, considered as useless or even undesirable for health and taste of consumers. For this purpose, hormones and antibiotics treatments were administered but, because of consumer reluctance, development of new approaches was investigated. One of these alternative strategies for reducing body fat of meat-producing animals consists in immunisation against adipocytes.

2. Using Anti-Adipocyte Monoclonal Antibodies for Controlling the Development of Adipose Tissue

Polyclonal antisera have been raised against fat cells or adipocyte plasma membrane proteins of rat (Flint, 1998; Flint *et al.*, 1986; Panton *et al.*, 1990), rabbit (Dulor *et al.*, 1990), cattle (Cryer *et al.*, 1984), sheep (Moloney and Allen, 1989; Nassar and Hu., 1991, 1992), pig (Kestin *et al.*, 1993), and chicken (Butterwith *et al.*, 1989, 1992; Dong *et al.*, 1991). *In vitro* and in the presence of complement, fat cells of rat (Flint *et al.*., 1986), cattle (Cryer *et al.*, 1984) and chicken (Butterwith *et al.*, 1989, 1992) origin could be lysed but the most relevant results are those observed *in vivo*. Indeed, studies on rats showed that injection of anti-adipocytes polyclonal antisera resulted in a long-term reduction of the number of adipocytes in internal fat depots, probably due to adipocyte lysis (Flint, 1998; Flint *et al.*, 1986). Panton *et al.* (1990) also observed reduced body fat depots in antisera-treated rats, which was accompanied by an increased body weight gain suggesting a compensatory increase in lean body mass. Decrease in total body fat as well as of dorsal and perirenal adipose tissues was obtained in experimental rabbits (Dulor *et al.*, 1990). In breeding species, such as sheep, several fat depots (i.e. backfat, subcutaneous, omental, perirenal and kidney pelvic) were reduced after injection of antisera against adipocyte plasma membranes (Moloney and Allen, 1989; Nassar and Hu, 1991, 1992). However, these experiments did not show out obvious improvement of carcass production. In the case of the porcine species however, such antisera led to a long-term reduction in body fat and to an increase in lean carcass development (Kestin *et al.*, 1993). The main drawback of this technique is that these antisera are not specific for fat cells because they recognise a myriad of uncharacterised determinants. To reduce this problem, polyclonal antisera are usually adsorbed with extracts of non-adipose tissues in order to increase the relative specific recognition of adipocyte determinants (Cryer *et al.*, 1984). However, the damages observed in vitro on erythrocytes (Butterwith *et al.*, 1989) clearly indicate that such antisera, even when partially purified, are able to cross-react with tissues other than fat in vivo. Another problem is that antisera administration can be accompanied by anaphylactic reactions, which can be reduced by multiple injections of lower concentrations (Hu *et al.*, 1992). Finally, their lack of reproducibility renders their long-term utilisation impossible.

Consequently, another strategy, developed so far in the porcine species, and consisting in the production of monoclonal antibodies, was adapted. It is characterised by a constant and abundant production of molecules with high specificity for fat cell surface proteins (De clercq *et al.*, 1997; Hausman *et al.*, 1993; Killefer and Hu 1990a,b; Wright and Hausman, 1990). Such an approach avoids the adsorption procedures performed for polyclonal antisera and ensures specific cytotoxicity in vivo. This technology has already found applications in vitro and in vivo. Indeed, by using a monoclonal antibody named AD-3, Yu *et al.* (1997) have depleted committed porcine preadipocytes in stromal-vascular culture by cytotoxic reaction and showed that the remaining cells can then be recruited by a specific treatment. In addition to indicate for the first time that these cultures contain preadipocytes at different stages of differentiation, this study also highlighted the possibility to detect preadipocytes which are at a very early stage of adipose conversion and undetectable with classical markers. AD-3 is now currently used by these authors as an early marker of fat cell differentiation (Hausman and Richardson, 1998). On the other hand, Killefer and Hu (1990a, b) isolated a monoclonal antibody (named LA-1) that specifically binds a 64 kDa antigen present on the adipocyte surface of lean but not obese pigs. This work clearly demonstrates that searching for monoclonal antibodies having such characteristics in humans could reveal antigens playing an important role in the development of avoiding or triggering human obesity. Such studies could open new ways of investigation for treating this quite sperad dysfunction of human fat tissue. Another treatment of this pathology could consist in using anti-adipocyte monoclonal antibodies for their direct cytotoxic effect on adipose tissue. In our laboratory, we are developing this last concept for better controlling fat development in the porcine species.

The different adipose depots (e.g. visceral, subcutaneous, intramuscular) have different breeding and economical importance and the organoleptical quality of meat is mainly dependent on intramuscular fat tissue. Then, if previous strategies led to a marked reduction in carcass fat content in pigs through various techniques (e.g. selection, hormones and antibiotics treatments), that was to the detriment of meat quality. Consequently, a very interesting new approach consists in depressing the development of most fat depots but not the intramuscular adipose tissue. This strategy is made easier by the fact that this fat tissue develops later than the other ones, as demonstrated for porcine (Hauser *et al.*, 1997; Moody *et al.*, 1978) and bovine (Chakrabarty and Romans, 1972; Lin *et al.*, 1992) species. In the particular case of pigs, we showed that during the early postnatal period, numerous lipid inclusions are found inside the muscle fibres, whereas there are very few differentiated intramuscular preadipocytes (Hauser *et al.*, 1997). Thereafter, the number of intramuscular adipose cells increases whereas lipid droplets in muscle fibers disappear, highlighting the asynchronous aspect of intramuscular adipose tissue maturation. In our laboratory, we started to develop monoclonal antibodies raised against fat cells in view of reducing body fat mass in pigs (De clercq *et al.*, 1997). This study has the final and original objective to destroy both preadipocyte and adipocyte pools in very young animals. Such an approach show favour long-term effects and avoid impact on intramuscular adipose tissue.

3. Identification, Production and Characterisation of Monoclonal Antibodies against Porcine Preadipocytes and Adipocytes

We immunised BALB/c mice with purified plasma membranes of adipocytes. However, this technique can lead to the loss of interesting plasma membrane determinants, due to the method of membrane preparation (Wright and Hausman, 1990). Consequently, we also immunised BALB/c mice with entire adipocytes, (known to be better immunogens than plasma membranes), from 1 week piglets. Two protocols of immunisation were performed. In a short-term protocol, immunogens were injected four times (every three days) in the hind soles, the popliteal lymph nodes being dissected, and the plasma cells recovered at the 12th day. In a long-term protocol, immunogens were injected intraperitoneally twice in one month. Then, after four to six months, the spleen was dissected and the splenocytes recovered. The antibodies content in mice serum was checked by ELISA after blood sampling from periocular plexus. Plasma cells or splenocytes were then fused with SP2/O myeloma cells, in the presence of polyethylene glycol and a feed layer from peritoneal cells. These hybridomas were cloned in RPMI culture medium containing 3% hypoxanthine-aminopterin-thymidine supplement and 10% foetal bovine serum, and the supernatants were tested in ELISA. When a supernatant was positive on adipocyte plasma membrane, the clone was developed until obtaining confluent culture wells. Then, these cells were subcloned at least twice by limit dilution and were cultured in RPMI medium until confluence. Supernatants were then tested in ELISA for their ability to react with adipocyte but not hepatocyte antigens. We obtained 44 clones that responded to this criterion but four of them, named 1B5, 2C2, 4G7 and 3B12, all derived from short-term immunizations with entire adipocytes, were still positive after the various cloning and freezing steps.

Additional tests were performed in order to better characterise these hybridomas (Table 1). The clones 1B5 and 3B12 secreted immunoglobulins (Ig) G1k, like the AD-1, AD-2 and AD-3 monoclonal antibodies developed in Hausman's laboratory (Wright and Hausman, 1995). They secreted very low amounts (less than 0.2 μg/ml) of antibody. On the other hand, the clones 2C2 and 4G7 produced IgG2bK at a concentration of 0.2 and 5.1 μg/ml respectively. The fact that these Igs are of G type is an advantage since they are better suited for cytotoxic activity, when compared to IgMs. Additional experiments showed that the supernatant of 4G7 clone, after the addition of fluorescein-coupled rabbit anti-mouse IgGs, led to a strong immunofluorescent reaction with cultured adipocytes and preadipocytes containing at least small intracytoplasmic lipid droplets (De clercq *et al.*, 1997). Supernatant of 2C2 clone did not provide so obvious results. In addition, no or a very weak reaction was observed either with lipid-free (fibroblastoid) cells or endothelial cells from porcine pulmonary artery. Finally, the supernatants were tested on various frozen porcine tissue sections (i.e. adipose tissue, skeletal muscle and kidney) and it appeared that IgGs produced by 4G7 were very specific to adipose tissue (De clercq *et al.*, 1997). Comparatively, polyclonal anti-serum, obtained from immunized mice, led to strong reactions with all the frozen tissue sections tested, demonstrating once again the interest to produce monoclonal antibodies. Finally, the 4G7 monoclonal antibody was purified by affinity chromatography on a protein A-sepharose CL-4B column. Its purity was assessed by SDS-PAGE electrophoresis and its activity by ELISA.

Table 1. *Characteristics of the monoclonal antibodies*

	1B5	3B12	2C2	4G7
Immunoglobulin	IgG1K	IgG1K	IgG2bK	IgG2bK
Production (µg/ml)	< 0.2	< 0.2	0.2	5.1
Reactivity in culture to:				
-lipid-containing preadipocytes	strongly	weakly	weakly	strongly
-fibroblastoid cells	very weakly	very weakly	very weakly	very weakly
-endothelial cells	no signal	no signal	no signal	no signal
Reactivivity on frozen tissues to:				
-adipocytes in adipose tissu	yes	yes	yes	yes
-adipocytes in muscular tissue	yes	yes	yes	yes
-capillaries in muscular tissue	no	no	yes	no
-perivascular smooth muscle cells	no	yes	no	no
-kidney	no	no	no	no

We attempted to produce 4G7 and 2C2 hybridomas in vivo, by injection into NMRI-BALB/c mice and recovering the ascites fluid, but this technique appeared inefficient. Then, we decided to produce the antibodies in vitro by using a batch incubator and a good productivity was then obtained.

4. Cytotoxicity of 4G7 Antibody on Preadipocytes and Adipocytes *in vitro*

A first step for investigating the cytotoxic activity of monoclonal antibodies consisted in testing these proteins on differentiated stromal-vascular cells from adipose tissue in culture.

Wright and Hausman (1995) showed that, in the presence of complement, several monoclonal antibodies (named AD-1, AD-3, SC3, IC3, AB3 and JC3) have cytotoxic activity on porcine fat cell clusters grown in serum-containing medium. By contrast, pools of antibodies were necessary for inducing cell death in serum-free medium. It would be interesting to investigate whether this discrepancy is due or not to the lack of glucocorticoids in the serum-deprived medium. Indeed, it is possible that this lipophilic hormone induces or amplifies the expression of some crucial antigenic determinants on stromal-vascular cells.

In our laboratory, we have shown that the addition of the 4G7 monoclonal antibody alone was sufficient to induce complement-mediated cytotoxicity on lipid-containing cells cultured in a medium containing insulin and hydrocortisone in the presence or not of serum (De clercq *et al.*, 1997). By contrast with AD-3 that has been shown to bind an antigen appearing early in adipose conversion, before the lipid accumulation in cells (Yu *et al.*, 1997), we do not presently know at what precise stage the epitope (92 kDa) recognised by our 4G7 antibody is expressed during adipose conversion. It does not seem to be a very early determinant, since only lipid-accumulating cells were detected by immunocytofluorescence. However, similarly to the observations of Wright and Hausman (1995), cultures treated with 4G7 antibody and complement before lipid accumulation do not differentiate well (De clercq *et al.*, 1997). These observations suggest that the antigen recognised by 4G7 antibody might emerge just before lipid accumulation. Finally, no secondary effect (e.g. nuclear shrinking and membrane protrusions) was detected in the undifferentiated cells remaining in the culture dishes after complement-mediated lysis of the adipocytes (De clercq *et al.*, 1997).

5. Treatment of Young Piglets with 4G7 Antibody *in vivo*

5.1. EFFECTS DURING THE FIRST MONTH

As mentioned above, injections of antisera in rats (Flint, 1998; Flint *et al.*, 1986; Panton *et al.*, 1990), rabbits (Dulor *et al.*, 1990), sheep (Moloney and Allen, 1989; Nassar and Hu, 1991, 1992) and pigs (Kestin *et al.*, 1993), as well as injections of monoclonal antibodies in rats (Wright and Hausman, 1995), led to reduction in body fat. The problem of using polyclonal antisera has already been discussed and the experiments of Wright and Hausman (1995) consisted in injections of pooled monoclonal antibodies.

Because of the high efficiency and specificity of the 4G7 antibody in vitro, we had a good potential tool to selectively reduce porcine adipose tissue in vivo. This had never been performed on this species before. The 4G7 monoclonal antibodies were injected intraperitoneally in young Large White pigs two and five days after birth (De clercq *et al.*, 1997). Pigs were sacrificed at 38 days of age and various tissues were analysed. It appeared that heart, spleen and liver were unaffected in injected piglets whereas all the adipose tissues examined were slightly decreased in weight, expressed as percent of the carcass weight. The backfat and subcutaneous ham adipose tissues reached the most significant reductions. In addition, whereas the percentage of dry matter was only slightly reduced in the carcass, it was significantly decreased in backfat, in which adipocytes were also of smaller sizes than in controls, and leaf fat. Finally, the total amount of lipids in the carcass was significantly decreased, the amount of tissue lipids expressed as percentage of wet tissue weight was decreased by the treatment in backfat, breast, shoulder and leaf fat adipose tissues, but the lipid content of the *longissimus* muscle was not affected. Taken together, these results indicate that the 4G7 monoclonal antibody reduced fat mass without adversely affecting either the average daily gain or the weights of tested organs in young pigs. Another interesting conclusion was that neonatal anti-adipocyte antibody treatment did not modify the lipid content of the *longissimus* muscle, suggesting that intramuscular fat development was not adversely affected.

5.2. LONG-TERM EFFECTS

The 4G7 monoclonal antibody was then tested for its long-term effects after intraperitoneal injection in young Large White x Piétrain pigs, two and five days after birth, that were sacrificed after reaching 50 to 100 kg.

A first series of experiments indicated that treatment led to a lower thickness of subcutaneous adipose tissue in shoulder, backfat and kidney area (Table 2) without affecting growth performance of the animals since the carcass mass remained unaffected (not shown). This was supported by the fact that the content in tissue lipids, expressed as percentage of wet tissue weight, tended to decrease in backfat, shoulder, leaf fat and ham, but not in *longissimus* muscle (Table 3). Furthermore, in treated pigs of 50 and 100 kg, adipocytes were smaller in the tested adipose tissues except in internal ham, an intermuscular fat depot (Table 4).

Table 2. *Ultrasonic measurements of the thickness of subcutaneous adipose tissue (mm) (mean ± s.e.m., n=6)*

Groups	Pig weight (kg)	Shoulder	Backfat	Kidney
Controls	62.5 ± 1.4	10 ± 2	7 ± 1	8 ± 2
Treated	63.5 ± 2	9 ± 1	7 ± 1	8 ± 1
Controls	76.5 ± 2.2	14 ± 1	10 ± 3	11 ± 2
Treated	72.2 ± .09	12 ± 1	8 ± 1	8 ± 1
Controls	85.8 ± 3.1	17 ± 1	11 ± 2	13 ± 2
Treated	85.3 ± 2.4	13 ± 1	10 ± 1	12 ± 1

Table 3. *Tissue lipid content expressed as percentage of wet tissue weight (mean ± s.e.m., n=6)*

Groups	Backfat	Shoulder	Leaf fat	Internal ham	*Longissimus* muscle
Controls	66.9 ± 3.6	71.9 ± 3.8	80.2 ± 4.6	62 ± 4.2	1.8 ± 0.2
Treated	60.5 ± 3.9	65.1 ± 4	71 ± 6	57.2 ± 5.7	2 ± 0.2

Table 4. *Adipocyte diameter (µm) (mean ± s.e.m., n=6)*

Pig weight (kg)	Groups	Backfat	Shoulder	Leaf fat	Internal ham
50	Controls	50.3 ± 4.1	47.9 ± 3.2	50.7 ± 1	41.6 ± 2.1
	Treated	42 ± 3.6	44.3 ± 0.3	40 ± 6.5	39.3 ± 0.3
100	Controls	69.3 ± 3.9	67.4 ± 4.9	73.4 ± 4.3	44.7 ± 7.2
	Treated	63.6 ± 6.1	60.8 ± 3	65.9 ± 6.6	46.1 ± 7.1

6. Conclusions

These *in vivo* results show that the inhibiting effect of 4G7 monoclonal antibody on porcine adipose conversion observed after five weeks is sustained until pigs reach the weight of slaughter (i.e. about 100 kg). This clearly demonstrates that the technology we developed during these experiments is a promising tool for controlling long-term adipose tissue development in this species.

Acknowledgements

This research was subsidised by the "Institut pour la Recherche Scientifique dans l'Industrie et l'Agriculture" and the Belgian Ministry of Agriculture – Administration de la Recherche et du Developpement.

References

Butherwith, S.C., Kestin, S.C., Griffin, H.D. and Flint, D.J. (1989) Cytotoxic antibodies to chicken adipocytes and their precursors: lack of tissue specificity, *Br. Poult. Sci. 30*, 371-378.

Butterwith, S.C., Kestin, S.C., Griffin, H.D., Beattie, J. and Flint, D.J. (1992) Identification of chicken (*Gallus domesticus*) adipocyte plasma membrane and differentiation specific proteins using SDS-PAGE and western blotting, *Comp. Biochem. Biophys. 110B*, 147-151.

Chakrabarty, K. and Romans, J.R. (1972) Lipogenesis in the adipose cells of the bovine as related to their intramuscular fat content, *Comp. Biochem. Physiol. 41B*, 603-607.

Cryer, A., Gray, B.R. and Woodhead, J.S. (1984) Study on the characterisation of bovine adipocyte precursor cells and their differentiation in vitro, using an indirect-labelled-second antibody cellular immunoassay, *J. Dev. Physiol. 6*, 159-176.

De clercq, L., Mourot, J., Genart, C., Davidts, V., Boone, C. and Remacle, C. (1997) An anti-adipocyte monoclonal antibody is cytotoxic to porcine preadipocytes *in vitro* and depresses the development of pig adipose tissue, *J. Anim. Sci. 75*, 1791-1797.

Dong, J., Froman, D.P. and Hu, C.Y. (1991) Development and characterization of polyclonal antibodies against chicken adipocytes, *Comp. Biochem. Physiol. 99A*, 195-198.

Dulor, J., Reyne, Y. and Nougues, J. (1990) In vivo effects of a treatment with antibodies to adipocyte plasma membranes in the rabbit, *Reprod. Nutr. Dev. 30*, 49-58.

Flint, D.J. (1998) Effects of antibodies to adipocytes on body weight, food intake, and adipose tissue cellularity in obese rats, *Biochem. Biophys. Res. Commun. 252*, 263-268.

Flint, D.J., Coggrave, H., Futter, C.E., Gardner, M.J. and Clarke, T.J. (1986) Stimulatory and cytotoxic effects of an anti-serum to adipocyte plasma membranes on adipose tissue metabolism *in vitro* and *in vivo*, *Int. J. Obesity 10*, 69-77.

Hauser, N., Mourot, J., De clercq, L., Genart, C. and Remacle, C. (1997) The cellularity of developing adipose tissues in Pietrain and Meishan pigs, *Reprod. Nutr. Dev.* 37, 617-625

Hausman, G.J. and Richardson, R.L. (1998) Newly recruited and pre-existing preadipocytes in cultures of porcine stromal-vascular cells: morphology, expression of extracellular matrix components, and lipid accretion, *J. Anim. Sci. 76*, 48-60.

Hausman, G.J., Wright, J.T., Dean, R. and Richardson, R.L. (1993) Cellular and molecular aspects of the regulation of adipogenesis, *J. Anim. Sci. 71 (Suppl. 2)*, 33-55.

Hu, C.Y., Suryawan, A., Killefer, J. and Wehr, N.B. (1992) Effect of antiserum to rat adipocytes on growth and body composition of the rat, *Comp. Biochem. Physiol. 101A*, 669-673.

Kestin, S., Kennedy, R., Tonner, E., Kiernan, M., Cryer, A., Griffin, H., Butterwith, S., Rhind, S. and Flint, D. (1993) Decreased fat content and increased lean in pigs treated with antibodies to adipocyte plasma membranes, *J. Anim. Sci. 71*, 1486-1494.

Killefer, J. and Hu, C.Y. (1990a) Expression of a 64 kD adipocyte-specific plasma membrane protein in genetically lean but not obese porcine adipocytes, *J. Cell. Biochem. 44*, 167-175.

Killefer, J. and Hu, C.Y. (1990b) Production of a novel monoclonal antibody to porcine adipocyte plasma membrane, *Proc. Soc. Exp. Biol. Med. 194*, 172-176.

Lebret, B. and Mourot, J. (1998). Caractéristiques et qualité des tissus adipeux chez le porc. Facteurs de variation non génétiques, *INRA Prod. Anim. 11*, 131-143.

Lin, G., Cross, C. and Smith, M.H. (1992) Esterification of fatty acids by bovine intramuscular and subcutaneous adipose tissues, *Lipids 27*, 111-116.

Moloney, A.P. and Allen, P. (1989) Growth and weights of abdominal and carcass fat in sheep immunized against adipose cell membranes, *Proc. Nutr. Soc. 48*, 14A.

Moody, W.G., Enser, M.B., Wood, J.D., Restall, D.J. and Lister, D. (1978) Comparison of fat and muscle development in Piétrain and Large White piglets, *J. Anim. Sci. 46*, 618-633.

Nassar, A.H. and Hu, C.Y. (1991) Growth and carcass characteristics of lambs passively immunized with antibodies developed against ovine adipocyte plasma membrane, *J. Anim. Sci., 69*, 578-586.

Nassar, A.H. and Hu, C.Y. (1992) Characterization of polyclonal antibodies against ovine adipocyte plasma membranes, *Int. J. Biochem. 24*, 599-604.

Panton, D., Futter, C., Kestin, S. and Flint, D. (1990) Increased growth and protein deposition in rats treated with antibodies to adipocytes, *Amer. J. Physiol. 258*, E985-E989.

Wright, J.T. and Hausman, G.J. (1990) Monoclonal antibodies against cell surface antigens expressed during porcine adipocyte differentiation, *Int. J. Obesity 14*, 395-409.

Wright, J.T. and Hausman, G.J. (1995) Monoclonal antibody-mediated cytotoxicity of adipocytes, *Obesity Res. 3*, 265-272.

Yu, Z.K., Wright, J.T. and Hausman, G.J. (1997) Preadipocyte recruitment in stromal vascular cultures after depletion of committed preadipocytes by immunocytotoxicity, *Obesity Res. 5*, 9-15.

IMMUNOCASTRATION OF FARM ANIMALS

PIRARD M., PORTETELLE D., BERTOZZI C., PARMENTIER I., HAEZEBROECK V., FONTAINE, S. AND RENAVILLE, R.
Department of Applied Biochemistry and Biology, Gembloux Agricultural University, Belgium

Abstract

In the farm animal species, the control of fertility and of sexual and aggressive behaviour are important factors of productivity. Traditionally, they have been reduced by surgical approaches. But, castration becomes unacceptable generating stress and decreasing animal efficiency and carcass quality. Therefore, alternative castration method such as immunocastration has been considered. This alternative method involves active immunisation against a hormone or factor, which is implicated in the cascade of events of the hypothalamic-pituitary-gonadal axis. Finally, a last procedure which consists in the application of recombinant DNA technologies to produce viral vector for hormone and other factors is under studies.

1. Introduction

The relation between reproductive function and production characteristics is strong. Secretion of anabolic steroids from the gonads improves the production characteristics of farm or domestic animals such as growth efficiency, growth rate and carcass quality (Renaville *et al.*, 1996). Experiments comparing entire or castrated animals supplemented with anabolic gonadal steroids have confirmed these effects (Schanbacher 1984; Renaville *et al.*, 1988; Sheridan *et al.*, 1990).
However, in feedlot industry, there are difficulties like sexual and aggressive behaviours associated with rearing gonad-intact animals (D'Occhio, 1993). Traditionally, they have been overcome or reduced by surgical/invasive forms of castration, by laborious management practices or, in the case of female, by administration of progestogens. Unfortunately, conventional castration is non-reversible, stressful, and generally decreases animal efficiency and carcass leanness.

It is also being increasingly scrutinised from ethical and environmental points of view. For example, because intact males have better feed efficiency, avoiding castration would significantly reduce the amount of biological pollutants, excreted by animals into

R. Renaville and A. Burny (eds.), Biotechnology in Animal Husbandry, 169–178.

the environment (Bonneau *et al.*, 1995; Thompson, 2000). In the long term, such considerations may have important influences on animal production policy. Therefore, there is a need for a practical alternative approaches to traditional castration.

2. Principle of Immunocastration

Different immunological approaches have been studied for the control of reproduction. In active immunisation, vaccines are composed of peptides that mimicked the epitopes of a target protein. After injection, antibodies against these peptides are produced. The rationale for endocrine-directed vaccines is that binding of a hormone to a specific antibodies will diminish or neutralise the biological activity of the endogenous hormones (Hadley, 1992). There are many factors other than animal variation which determine the success of an active immunisation approach, including species, age, structure of the peptide/hormone, type of adjuvant used, immunisation frequency and their relative timing.

Reproduction function is under control of the hypothalamus, hypophysis and gonads (Figure 1). The hypophysis secreted the luteinising hormone releasing hormone (LHRH) which stimulates the pituitary gland to secrete the luteinising hormone (LH) and the follicle stimulating hormone (FSH). The LH and FSH induce a steroid production by the gonads of progesterone, testosterone, androstenedione and 17beta-oestradiol. Immunisation against LHRH, LH, FSH and human chorionic gonadotrophin playing an important role in the reproduction, these hormones have been studied as targets for immunocastration. Immunisation against other targets as gonadal steroids and sperm and ova surface antigens will be also described.

3. Targets for Contraceptive Vaccines

3.1. IMMUNISATION AGAINST LHRH

LHRH is the first intracellular messager of the control of fertility. Because comparisons of LHRH amino acids sequences have showed that these proteins are highly conserved in all mammalian and between male and female (Flanagan *et al.*, 1997), we may expect that one vaccine kit for one species could be quickly adapted for an another one. Largely studied (Table 1), immunological neutralisation of LHRH would expect to block pituitary secretion of LH and FSH and lead to gonadal quiescience. Experiments of immunisation results in a markedly reduction of steroidogenic and gametogenic functions of the ovaries and the testis. After long-term treatment with LHRH immunization, a decrease of the secretion of FSH and LH is observed in cattle. Bulls animals immunised against this hormone have their testicular function suppressed, plasma testosterone concentrations being similar to those of surgical castrates (Tshewang *et al.*, 1997; D'Occhio, 1993). The individual immune response was sometimes variable and the vaccines effects reversible (Cook *et al.*, 2000). In lambs, immunoneutralisation reduced testis weight and the concentration of testosterone in serum. Measures of sexual behavior (frequency of mounts and ejaculation) were also reduced. A better stimulation was observed when Freund's complete adjuvant was used as carrier protein (Kiyma, 2000). Likewise, immunisation of heifers against LHRH leads to a reduction of fertility equivalent to that of ovariectomised heifers (Adams and Adams, 1990; Adams *et al.*, 1990). Animals treated with this vaccines have also there sexual behaviours reduced (D'Occhio, 1993). The effects were reversible and the subsequent fertility of the vaccinated fillies was normal (Tshewang *et al.*, 1997).

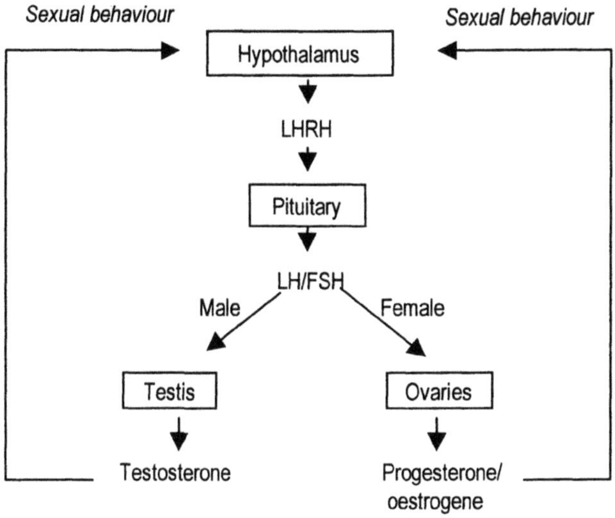

Figure 1. *Sexual axis in mammalian.*

A LHRH vaccine, containing 10 amino acids which are well conserved among species, has been commercialised under the name of Vaxstrate and is available in Australia (Hoskinson *et al.*, 1990).

Administration of Vaxstrate suppresses ovarian cyclicity in post pubertal females for a period of six to eight months. After this time, the effect of the vaccine disappears and ovarian function retains to normal. In production systems, the reversibility of the vaccine is not useful. This problem could be solved by including an additional adjuvant to the vaccine which should extended the immunological response of the animals (Hoskinson *et al.*, 1990).

Administration of Vaxstrate to bulls have plasma testosterone concentrations similar to those of surgical castrates (Jeffcoate *et al.*, 1982; D'Occhio, 1993). They also showed a decreased in testis size for about 25 weeks after which they returned to normal (Hoskinson *et al.*, 1990).

The effects of a newly developed gonadotrophin releasing hormone (GnRH) vaccine, Improvac, on growth performance and incidence of boar taint-related compounds were investigated in male pigs (Dunshea *et al.*, 2000; Mc Cauley *et al.*, 2000). All animals exhibited anti-GnRH titres. Compared with control, testis and bulbourethral gland were reduced and serum testosterone were below 2 nM. Compounds responsible for boar taint include androstenone and skatole in subcutaneous fat were suppressed to low or non detectable levels. Pigs treated with Improvac grew more rapidly. The vaccine was well tolerated by the pigs, and no observable site reactions could be detected at slaughter. Vaccination of pigs with Improvac allows the production of heavy entire boars with improved meat quality through the prevention and control of boar taint (Dunshea *et al.*, 2000).

Male pigs were also used in study to investigate the interactions between Improvac and porcine somatotrophin (pST, Reporcin) regimes. While neither Improvac or Reporcin alone had any effect upon daily gain, the combined treatment increased daily gain by 25% (McCauley *et al.*, 2000).

Table 1. *Immunisation against sexual hormone in male and female animals*

Hormone	Animal	Sex	References
LHRH	Pig	Male	Falvo *et al.* (1986), Esbenshade and Johnson (1987), Awoniyi *et al.* (1988)
		Female	Esbenshade and Britt (1985), Traywick and Esbenshade (1988)
	Horse	Male	Dowsett *et al.* (1996)
		Female	Garza *et al.* (1986, 1988), Safir *et al.* (1987), Dowsett *et al.* (1993), Tshewang *et al.* (1997)
	Sheep	Male	Lincoln and Fraser (1979), Jeffcoate *et al.* (1982), Schanbacher (1982), Chase *et al.* (1982)
		Female	Clarke *et al.* (1978), Fraser and McNeilly (1982), McNeilly *et al.* (1986)
	Cattle	Male	Robertson *et al.* (1979, 1981, 1982), Jeffcoate *et al.* (1982), Goubeau *et al.* (1989)
		Female	Johnson *et al.* (1988), Adams and Adams (1990), Adams *et al.* (1990), Hoskinson *et al.* (1990)
LH	Cattle	Male	Schanbacher (1988)
		Female	De Silva *et al.* (1986), Grieger and Reeves (1990), Johnson *et al.* (1988), Grieger *et al.* (1990), Roberts *et al.* (1990)
	Sheep	Female	Roberts and Reeves (1989)
FSH	Sheep	Male	Mariana *et al.* (1998)
hCG	Cattle	Female	Johnson *et al.* (1988), Talwar *et al.* (1993)

3.2. IMMUNISATION AGAINST GONADOTROPHIC HORMONES

Because LH, FSH and human chorionic gonadotrophin share a common alpha subunit, immunisation against the beta subunit has been carried in cattle and sheep (Table 1). Experiments on LH immunisation have tended to focus on female because of the requirement for LH in follicule maturation and ovulation. Immunisation against LH leads to acyclicity for a period comprised between 42 to 96 week in females and to the suppression of sexual behaviour in both sexes (Johnson *et al.*, 1988; Roberts and Reeves, 1989). Immunisation against FSH is characterised by the suppression of fertility by presumably blocking the preovulatory surge of LH and by the conservation of the sexual behaviour (Mariana *et al.*, 1998). Application of a FSH contraceptive preparation could be interesting when only the infertility aspect is desired.

Human chorionic gonadotrophin (hCG) has been used as target particularly for immunocontraception in women. A vaccine incorporating a synthetic peptide antigen representing the aminoacid sequence of the C-terminal region of the beta subunit of human chorionic gonadotrophin has been tested. The protein has been linked chemically to tetanus toxoid and diphtheria has been tried as vaccine. Potentially contraceptive levels of antibodies to hCG were observed for six months (Jones *et al.*, 1988). Immunisation was well tolerated with no significant changes in endocrine, metabolic and haematological indices. Normal ovulatory cycles were maintained as indicated by menstrual regulation. The vaccine was highly effective in preventing pregnancy and reversible fertility control is feasible (Talwar *et al.*, 1993).

3.3. IMMUNISATION AGAINST GONADAL STEROIDS

Gonadal steroids serve important roles in sexual behaviour. Therefore, it is not surprising that numerous experiments have investigated the reproductive consequences of immunisation against progesterone, testosterone (Table 1). Anti-progesterone monoclonal antibody injected completely blocked pregnancy (Wang *et al.*, 1984; Rider *et al.*, 1987). The action of the antibody in females was associated with a failure to initiate an implantation response. The most pronounced effect, however, was a break of embryonic development at a stage prior to cavitation (Ellis *et al.*, 1988; Wang *et al.*, 1989). The results show that passive immunisation against progesterone shortly after mating interferes with early hormone-dependent steps which are essential for normal embryonic development. In some studies, immunisation against progesterone induces infertility in females (Price *et al.*, 1987) but in others is associated with ovarian abnormalities like cystic follicles (Martin *et al.*, 1978) which is not desirable.

Testosterone has been shown to regulate spermatogenesis, primarily by controlling meiosis, the step responsible for the production of spermatids (Suresh *et al.*, 1995). However, immunisation against testosterone was not characterised by an inhibition of the sexual behaviour in rams and there were no testicular function suppressed (Walker *et al.*, 1984).

3.4. SPERM AND OVA ANTIGENS

Sperm and ova have both antigens on their surfaces. These antigens play an important role during fertilisation and are involved in interactions between sperm and egg. (Wassarman, 1990). An interesting possibility is to made antibodies that interact with gamete antigens. Antibody binding to sperm or ova will block gamete interactions required for fertilisation. This kind of vaccine allows us to block the conception while other reproductive functions still normal. It is particularly relevant when the control of reproduction should not disrupt social hierarchies (Tyndale-Biscoe, 1991).

The zona pellucida (ZP) forms a coating around mammalian eggs. The ZP is comprised essentially of three major glycoproteins, which are involved in sperm binding during their penetration through the zona. The ZP3 was identified as a major sperm receptor and targeted as the most likely female gamete antigen for a contraceptive vaccine (D'Occhio, 1993). In several species, immunisation of females against ZP antigens can lead to the control of fertility (Liu *et al.*, 1989). However, some experiments were associated with abnormal ovarian function and altered reproductive hormone profile (D'Occhio, 1993)

Sperm antigens can be used to immunise males or females while zona pellucida antigen vaccination will be only effective in females (Liu *et al.*, 1989).

3.5. IMMUNISATION WITH TROPHOBLASTS

Early events of pregnancy involve complex interactions between maternal tissues and the developing trophoblast. Unique antigens can be express on the surface of trophoblasts and could be the targets for antibodies. A large number of monoclonal antibodies has been generated against human trophoblast and have been selected. These antibodies could influence the immune interactions between maternal tissues and the conceptus and could be used as vaccine (D'Occhio, 1993).

3.6. RECOMBINANT DNA TECHNOLOGIES

The application of recombinant DNA technologies is under studies. The principle of the technique is to introduce a vector (plasmid, retrovirus, herpes virus, adenovirus) containing the DNA sequence of a gene coding a protein implicated in the reproduction functions. The synthesis of the recombinant protein induces an immunological response and the production of antibodies. We could expect a reject reaction against the protein, a reduction of fertility and an inhibition of the sexual behaviour of the animals.

First experiments of immunisation of mouse result in a decrease of fertility for a period of 28 weeks. After this time, the effect of the vaccine disappears. However, some problems have to be solved before this new approach could be used. The introducing of the DNA sequence was carried out with mouse pox vector which is not acceptable for the population. The virus could mutate, become uncontrollable and spread to other species. Additional works has to be done to find an acceptable vector to introduce the DNA sequence in the animals. The next step is to adapt this method to new species (Norrie, 1997).

4. Perspectives

The best results of immunocastration were obtained with the LHRH and LH vaccines. But some improvements have to be done to obtain a better vaccine. An ideal product would be the one that induces a long-term response (up to 3 years). In this case a single injection is necessary to induce a infertility of the animals. The vaccine should be reversible if necessary. Some improvements could occur through a better understanding of the factors controlling the immune response and the effects of the adjuvant. This might satisfy both the production and reproduction aspects of managing farm animals reserved for meat production

Encouraging results were obtained through the use of recombinant DNA technologies to incorporate either a sexual hormone into vector delivery systems. The main advantage is that application of these technologies would require a single injection that lead to continued expression of the hormone.

References

Adams, T.E. and Adams, B.M. (1990) Reproductive function and feedlot performance of beef heifers actively immunized against GnRH, *J. Anim. Sci. 68*, 2793-2802.

Adams, T.E., Daley, C.A., Adams, B.M. and Sakurai, H. (1993) Testis function and feedlot performance of bulls actively immunized against gonadotropin-releasing hormone : effect of implants containing progesterone and oestradiol benzoate, *J. Anim. Sci. 71*, 811-817.

Adams, T.E., Dunbar, J.R., Berry, S.L., Garrett, W.N., Famula, T.R. and Lee, Y.B. (1990) Feedlot performance of beef heifers implanted with Synovex-H: effect of melengestrol acetate, ovariectomy or active immunization against GnRH, *J Anim. Sci.68*, 3079-3085.

Auclair, D., Sowerbutts, S.F. and Setchell, B.P. (1995) Effect of active immunization against oestradiol in developing ram lambs on plasma gonadotrphin and testosterone concentration, time of onset of puberty and testicular blood flow, *J. Reprod. Fert. 104*, 7-16.

Awoniyi, C.A., Chandrashekar, V., Arthur, R.D., Schanbacher, B.D. and Falvo, R.E. (1988) Changes in testicular morphology in boars actively immunized against gonadotropin hormone-releasing hormone, *J. Androl. 9*, 160-171.

Bonneau, M. and Enright, W.J. (1995) Immunocastration in cattle and pigs, *Lives. Prod. Sci. 42*, 193-200.

Brown, B.W., Mattner, P.E., Carroll, P.A., Holland, E.J., Paull, D.R., Hoskinson, R.M. and Rigby, R.D.G. (1994) Immunization of sheep against GnRH early in life : effects on reproductive function and hormones in raws, *J. Reprod. Fert. 101*, 15-21.

Chase, D.J., Schanbacher, B.D. and Lunstra, D.D. (1988) Effects of pulsatile and continuous luteinizing hormone (LH) infusions on testosterone responses to LH in rams actively immunized against gonadotropin-releasing hormone, *Endocrinology 123*, 816-826.

Clarke, I.J., Fraser, H.M. and McNeilly, A.S. (1978) Active immunization of ewes against luteinizing hormone releasing hormone, and its effects on ovulation and gonadotrophin, prolactin and ovarian steroid secretion, *J. Endocrinol. 78*, 39-47.

Cook, R.B., Popp, J.D., Kastelic, J.P., Robbins, S. and Harland, R. (2000) The effects of active immunization against GnRH on testicular development feedlot performance, and carcass characteristics of beef bulls, *J. Anim. Sci. 78*, 2778-2783.

D'Occhio, M.J. (1993) Immunological suppression of reproductive functions in male and female mammals, *Anim. Reprod. Sci. 33*, 345-372.

De Silva, M., Tjoelker, L, Chang, C.F. and Reeves, J.J. (1986) Ovarian response to active immunization against luteinizing hormone in the cow, *Theriogenology 26*, 89-100.

Dowsett, K.F., Knott, L.M., Tshewang, U., Jackson, A.E., Bodero, D.A. and Trigg, T.E. (1996) Suppression of testicular function using two dose rates of a reversible water soluble gonadotrophin releasing hormone (GnRH) vaccine in colts, *Aust. Vet. J. 74*, 228-235.

Dowsett, K.F., Tshewang, U., Knott, L.M., Jackson, A.E. and Trigg, T.E. (1993) Immunocastration of colts and immunospeying of fillies, *Immunol. Cell. Biol. 71*, 501-508.

Dunshea, F.R., Colantoni, C., Howard, K., Jackson, P., Long, K.A., Lopaticki, S., Nugent, E.A., Simons, J.A., Walker, J. and Hennessy, D.P. (2000) Vaccination of entire boars with Improvac eliminates boar taint and increases growth performance, *J. Anim. Sci. 78 suppl 1*, 138.

Ellis, S.T., Heap, R.B., Butchart, A.R., Rider, V., Richardson, N.E., Wang, M.W. and Taussig, M.J. (1988) Efficacy and specificity of monoclonal antibodies to progesterone in preventing the establishment of pregnancy in the mouse, *J. Endocrinol. 118*, 69-80.

Esbenshade, K.L. and Britt, J.H. (1985) Active immunization of gilts againsts gonadotropin-releasing hormone : effects on secretion of gonadotropins, reproductive function, and responses to agonists of gonadotropin-releasing hormone, *Biol. Reprod. 33*, 569-577.

Esbenshade, K.L. and Johnson, B.H. (1987) Active immunization of boars against gonadotropin releasing hormone. II. Effects on libido and response to testosterone propionate, *Theriogenology 27*, 581-585.

Falvo, R.E., Chandrashekar, V., Arthur, R.D., Kuenstler, A.R., Hasson, T., Awoniyi, C. and Schanbacher, B.D. (1986) Effect of active immunization against LHRH or LH in boars : reproductive consequences and performance traits, *J. Anim. Sci. 64*, 986-694.

Flanagan, C.A., Millar, R.P. and Illing, N. (1997) Advances in understanding gonadotropin-releasing hormone receptor structure and ligand interactions, *Rev. Reprod. 2*, 113-120.

Fraser, H.M. and McNeilly, A.S. (1982) Effect of immunoneutralisation of luteinizing hormone releasing hormone on the estrogen-induced luteinizing hormone and follicle-stimulating hormones surges in the ewe, *Biol. Reprod. 27*, 548-555.

Garza, F., Thompson, D.L., French, D.D., Wiest, J.J., George, R.L., Ashley, K.B., Jones, L.S., Mitchell, P.S. and McNeill, D.R. (1986) Active immunization of intact mares against gonadotropin-releasing hormone : differential effects on secretion of luteinizing hormone and follicle-stimulating hormone, *Biol. Reprod. 35*, 347-352.

Garza, F., Thompson, D.L., Mitchell, P.S and Wiest, J.J. (1988) Effects of active immunization against gonadotropin releasing hormone on gonadotropin secretion after ovariectomy and testosterone propionate administration to mares, *J. Anim. Sci. 66*, 479-486.

Goubau, S., Silversides, D.W., Gonzalez, A., Laarveld, B., Mapletoft, R.J. and Murphy, B.D. (1989) Immunization of cattle against modified peptides of gonadotropin releasing hormone conjugated to carriers : effectiveness of Freund's and alternative adjuvants, *Theriogenology 32*, 557-567.

Grieger, D.M. and Reeves, J.J. (1990) Active immunization of beef heifers against luteinizing hormone : II. Evaluation of conjugation technique on antigenicity of LH, *J. Anim. Sci. 68*, 3747-3754.

Grieger, D.M., Scarborough, R., deAvila, D.M., Johnson, H.E. and Reeves, J.J. (1990) Active immunization of beef heifers against luteinizing hormone. III. Evaluation of dose and longevity, *J. Anim. Sci. 68*, 3755-3764.

Hadley, M.E. (1992) Endocrinology, third edition, Prentice-hall inc., New Jersey, 74 pages.

Hoskinson, R.M., Rigby, R.D.G., Mattner, P.E., Huynh, V.L., D'Occhio, M., Neish, A., Trigg, T.E., Moss, B.A., Lindsey, M.J., Coleman, G.D. and Schwartzkoff, C.L. (1990) Vaxstrate : an anti-reproductive vaccine for cattle, *Aust J. Biotechnol. 4*, 166-170.

Jeffcoate, I.A., Lucas, J.M.S. and Crighton, D.B. (1982) Effects of active immunization of ram lambs and bull calves against synthetic luteinizing hormone releasing hormone, *Theriogenology* 18, 65-77.

Johnson, H.E., deAvila, D.M., Chang, C.F. and Reeves, J.J. (1988) Active immunization of heifers against luteinizing hormone-releasing hormone, human chorionic gonadotropin and bovine luteinizing hormone, *J. Anim. Sci. 66*, 719-726.

Jones, W.R., Bradley, J., Judd, S.J., Denholm, E.H., Ing, R.M.Y., Mueller, U.W., Powell, J., Griffin, P.D. and Stevens, V.C. (1988) Phase 1 clinical trial of a world Health organisation birth control vaccine, *Lancet 1*, 1295-1298.

Kiyma, Z., Adams, T.E., Hess, B.W., Riley, M.L., Murdoch, W.J. and Moss, G.E. (2000) Gonadal function, sexual behavior, feedlot performance, and carcass traits of ram lambs actively immunized against GnRH, *J. Anim. Sci. 78*, 2237-2243.

Lincoln, G.A. and Fraser, H.M. (1979) Blockage of episodic secretion of luteinizing hormone in the ram by the administration of antibodies to luteinizing hormone releasing hormone, *Biol. Reprod. 21*, 1239-1245.

Lui, I.K.M., Bernoco, M. and Feldman, M. (1989) Contraception in mares heteroimmunized with pig zonae pellucidae, *J. Reprod. Fertil. 85*, 19-29.

Mariana, J.C., Monniaux, D., Caraty, A., Pisselet, C., Fontaine, J. and Solari, A. (1998) Immunization of sheep against GnRH early life : effects on gonadotropins, follicular growth and responsiveness of granulosa cells to FSH and IGF-I in two breeds of sheep with different prolificacy (Romanov and Ile-de-France), *Domest. Anim. Endocrinol. 15*, 195-207.

Martin, T.E., Henricks, D.M., Hill, J.R. and Rawlings, N.C. (1978) Active immunization of the cow against oestradiol-17-beta, *J. Reprod. Fert. 53*, 173-178.

McCauley, G.M., Kolek, M., Suster, D., Oliver, W.T., Harrell R.J. and Dunshea, F.R. (2000) An immunocastration vaccine (Improvac) and porcine somatotropin (Reporcin) have synergistic effects upon growth performance in boars, *J. Anim. Sci. 78 suppl 1*, 138.

McNeilly, A.S., Jonassen, J.A. and Fraser, H.M. (1986) Suppression of follicular development after chronic LHRH immunoneutralization in the ewe, *J. Reprod. Fertil. 76*, 481-490.

Naz, R.K. (1990) Sperm surface antigens involved in mammalian fertilization : their role in contraceptive vaccine development for humans, *Curr. Opin. Immunol. 2*, 748-751.

Norrie, C. (1997) Vaccin contraceptif pour les souris australienne, *Biofutur 169*, 10.

Oonk, H.B., Turkstra, J.A., Schaaper, W.M.M., Erkens, J.H.F., Schuitemaker de Weerd, M.H.., Verheijden, J.H.M., Meloen, R.H. and Van Nes, A. (1998) New GnRH-like peptide construct to optimize efficient immunocastration of male pigs by immunoneutralization of GnRH, *Vaccine 16*, 1074-1082.

Price, C.A., Morris, B.A. and Webb, R. (1987) Reproductive and endocrine effects of active immunization against a testosterone conjugate in the heifer, *J. Reprod. Fertil. 81*, 149-160.

Renaville, R., Massart, S., Prandi, A., Syndic, M., Fazzini, U., Nicolay, E., Burny, A. and Portetelle, D. (1996) Aspects on the use of anabolic steroids in animal production, in : *"Scientific conference on growth promotion in meat production"*, EU editors, pp 63-86.

Renaville, R., Sneyers, M., Rochart, S., Thewis, A., Portetelle, D. and Burny, A. (1988) Effects of anabolic treatment before puberty with trenbolone-oestradiol alone on growth rate, testicular development and hormonal status in young bulls, *Theriogenology 29*, 461-476.

Rider, V., Heap, R.B., Wang, M.Y. and Feinstein, A. (1987) Anti-progesterone monoclonal antibody affects early cleavage and implantation in the mouse by mechanisms that are in fluenced by genotype, *J. Reprod. Fertil. 79*, 33-43.

Roberts, A.J. and Reeves, J.J. (1989) Reproductive and endocrine changes in ewes actively immunized against luteinizing, *J. Reprod. Immunol. 16*, 187-197.

Robertson, I.S., Fraser, H.M., Innes, G.M and Jones, A.S. (1982) Effect of immunological castration on sexual and production characteristics in male cattle, *Vet. Rec. 111*, 529-531.

Robertson, I.S., Wilson, J.C. and Fraser, H.M. (1979) Immunological castration in male cattle, *Vet. Rec. 105*, 556-557.

Robertson, I.S., Wilson, J.C. and Rowland, A.C. (1981) Further studies on immunological castration in male cattle, *Vet. Rec. 108*, 381-382.

Safir, J.M., Loy, R.G. and Fitzgerald, B.P. (1987) Inhibition of ovulation in the mare by active immunization against LHRH, *J. Reprod. Fert. (Suppl.) 35*, 229-237.

Schanbacher, B.D. (1984) Manipulation of endogenous and exogenous hormones for red meat production, *J. Anim. Sci. 59*, 1621-1630.

Schanbacher, B.D. (1982) Responses of ram lambs to active immunization against testosterone and luteinizing hormone-releasing hormone, *Amer. J. Physiol. 242*, 201-205.

Sheridan, P.J., Austin, F.H., Bourke, S. and Roche, J.F. (1990) The effect of anabolic agents on growth rate and reproductive organs of pigs, *Liv. Prod. Sci. 26*, 263-275.

Suresh, R., Medharmurthy, R. and Moudgal, N.R. (1995) A role of nocturnal serum testosterone surge in regulating spermatogenesis in the adult non-human primate, *Endocrine 3*, 487-492.

Talwar, G.P., Singh, O., Pal, R., Chatterjee,N., Upadhyay, S.N., Kaushic, C. Garg, S., Kaur, R., Singh, M., Chandrasekhar, S., et al (1993) A birth vaccine is on the horizon for family planning, *Ann. Med. 25*, 207-12.

Thompson, D.L. (2000) Immunization against GnRH in male species (comparative aspects), *Anim. Reprod. Sci. 2*, 459-469.

Traywick, G.B. and Esbenshade, K.L. (1988) Pulsative administration of gonadotropin-releasing hormone agonist to gilts actively immunized against gonadotropin-releasing hormone, *J. Anim. Sci. 66*, 2209-2215.

Tshewang, U., Dowsett, K.F., Knott, L.M. and Trigg, T.E. (1997) Preliminary study of ovarian activity in fillies treated with a GnRH vaccine, *Aust. Vet. J. 75*, 663-667.

Tyndale-Biscoe, C.H. (1991) Fertility control in wildlife, *Reprod. Fertil. Develop. 3*, 339-343.

Walker, M.P., Thompson, D.L., Godke, R.A. and Honey, P.G. (1984) Active immunization of prepubertal bulls against testosterone : seminal and testicular characteristics after puberty, *Theriogenology 22*, 269-278.

Wang, M.W., Heap, R.B. and Taussig, M.J. (1989) Blocking of pregnancy in mice by immunization with anti-idiotype directed against monoclonal anti-progesterone antibody, *Proc. Natl. Acad. Sci. USA 86*, 7098-7102.

Wang, M.Y., Rider, V., Heap R.B. and Feinstein, A. (1984) Action of anti-progesterone monoclonal antibody in blocking pregnancy after postcoital administration in mice, *J. Endocrinol. 101*, 95-100.

Wassarman, P.M. (1990) Profile of mammalian sperm receptor, *Development 108*, 1-17.

PLACENTAL PROTEINS IN RUMINANTS:

Biochemical, physiological and zootechnical aspects

SOUSA N.M., FIGUEIREDO J.R. AND BECKERS J.F.*
*Faculty of Veterinary Medicine, Federal University of Santa Maria, Brazil and *Physiology of Reproduction, Faculty of Veterinary Medicine, Liège, Belgium.*

Abstract

During the last decade, investigations were carried out by several research groups in order to characterize proteins or glycoproteins synthesized in the ruminant placenta. Recently, as results of this research, a large family of pregnancy-associated glycoproteins (PAGs) was discovered. Using molecular biology techniques, they were found to be members of the superfamily of the aspartic proteinases which also contains pepsinogen, chymosin, renin, beta-secretase, cathepsin D and E etc. Synthesized in the mono and/or binucleate cells of the trophoblast, some forms of PAG seem to lack of proteinase activity. It is likely they are synthesized together with molecules involved in the tissue remodeling of the placenta. Their release in large quantities into the maternal blood circulation results in measurable plasma concentrations.

Thanks to international collaborative studies, we have shown that PAG levels are a good indicator of feto-placental well-being and that sharp decreases in PAG levels occur just before pregnancy failure in cows and in goats. In some countries, the PAG assay is available for veterinarians in the regional laboratories responsible for animal health including immunodiagnosis for brucellosis, IBR, BVD, CAEV, VISNA-MEDI etc.

1. Introduction

Placenta is a polyvalent organic system with a vital biological role for the perpetuation of the eutherian mammals. Fetal development is related to that of the placenta from the anatomical, genetic and metabolic points of view. As far as metabolism is concerned, endocrine function is particularly important. The endocrine function of the mammalian

R. Renaville and A. Burny (eds.), Biotechnology in Animal Husbandry, 179–208.

placenta includes various hormones: progesterone, oestrogens, chorionic gonadotropins, placental lactogens, also designed "chorionic somatomammotropins", prolactins, growth hormones and a series of growth factors, proteins and glycoproteins interfering with the establishment of pregnancy, corpus luteum maintenance, immunotolerance of the conceptus by the mother, intermediate maternal metabolism, fetal growth and mammary growth.
The study of the endocrine function of the ruminant placenta will be discussed below. Emphasis will be given to pregnancy-associated (specific) proteins, but information related to other polypeptide hormones will be freely utilized in order to develop our comparative understanding of the structure, function and release in maternal circulation of the placental proteins.

2. The Placenta

The placenta plays an essential role in both establishment and maintenance of pregnancy as well as in fetal development, working as a respiratory, nutritional, epurative, endocrine and immunological organ. The placenta is not only a barrier, as it was thought before, but it has a real active and selective exchanger role which is developed in harmony with other systems.

The topography and structure of the placenta vary according to the species and depend on the behavior of the egg which, at the time of its attachment, either remains in the uterine lumen or implants interstitially in the uterine mucosa.
Various classifications have been proposed though the most widely used, despite some imperfections, is the classification based on the histological structure, or, more precisely, the number of histological layers separating the maternal and fetal blood. Four types of placenta may be distinguished histologically: the epitheliochorial placenta (Equidae, Suidae), the synepitheliochorial placenta (ruminants), the endotheliochorial placenta (carnivores) and the haemochorial placenta (human, primates and rodents).
The uterine connective tissue is modified in both endothelial and haemochorial placentas and tearing of the uterine tissues accompanied by bleeding occurs during parturition. The uterine regions shed in this way are said to be deciduous and the species with these types of placenta are called deciduates.

The mature ruminant placenta is neither entirely syndesmochorial with no uterine epithelium, nor epitheliochorial with two apposed cell layers whose the only anatomical interaction observed is the presence of interdigitated microvilli. The uterine epithelium persists but is modified to a variable degree into a hybrid fetomaternal syncytium formed by the migration and fusion of the fetal chorionic binucleate cells with those of the uterine epithelium (Wooding, 1992).

2.1. BINUCLEATE CELLS

The most characteristic element of the ruminant placenta is the presence of binucleate cells or diplokaryocytes in the trophectoderm (Wimsatt, 1951). First described by Assheton in 1906, these cells have been the subject of numerous investigations over the last 5 decades (Drieux and Thiery, 1951; Amoroso, 1952; Wooding and Wathes, 1980; Klisch *et al.*, 1999).

Binucleate cells derive from chorionic mononucleate cells by karyokinesis without subsequent cytokinesis. The youngest cells are located deep in the trophectoderm and maturation takes place progressively as they rise to the maternal surface (Wooding, 1983).

Young binucleate cells have a small volume of dense ribosome-filled cytoplasm, which, after cytoplasmic re-organization to a spherical or oval cell, has no contact with either the basement membrane or the atypical trophectodermal tight junction. As the binucleate cells increase in size it develops an extensive array of rough endoplasmic reticulum and a large Golgi body. The latter produces a considerable number of characteristic granules that contain numerous protein and glycoprotein constituents (Wooding, 1992).

The first binucleate cells appear just before implantation (Wango *et al.*, 1990a). They can be recognized from day 14 after fertilization in the ovine trophectoderm (Wooding, 1984) and from day 16-17 in the bovine.

The binucleate cells represent 15-20% of the total population within the mature placenta (Wooding, 1983; Wango *et al.*, 1990b). Of these, about one in seven are migrating towards the uterine epithelium at any time of gestation (Wooding *et al.*, 1986). The migration of binucleate cells plays an important role in the remodeling of placental structure and leads to a reaction with the uterine cells: the formation of a hybrid fetomaternal tissue in the ewe and goat (the syncytium plaques) and the maternal giant cells in the cow (the trinucleate cells) (Wooding, 1984; Wooding and Beckers, 1987).

Trinucleate cells are short-lived structures, which, after exocytosis, are resorbed by the trophectoderm (Wooding and Wathes, 1980). Contrary to the ewe, in which binucleate cell migration is essential to support the enormous growth in area of the fetomaternal interface layer as the placentome develops (Wooding *et al.*, 1981), in cow placentomes the binucleate cell migration has no barrier or structural role (Wooding and Wathes, 1980).

Special cells of the ruminant placenta were suspected as responsible for an important endocrine function during gestation as early as 1940. In this decade, it was observed for the fist time a reduction in the pituitary gonadotropic activity in the pregnant cow and suggested that the corpus luteum was replaced by additional luteotropic substances produced elsewhere. Using microsections of cotyledons and staining for carbohydrates like periodic acid-Schiff (PAS), Weeth and Herman (1952) and Björkman (1954) described the presence of numerous trophoblastic cells containing glycoproteins. About 10 years later, Foote and Kaushik (1963) demonstrated the presence of a LH-like activity in the cotyledons. They used the bioassay of Parlow (1961) in order to identify this LH-like activity. These studies were pioneers in this field: they opened the way for further identification and characterization of placental hormones and proteins in the ruminant species.

Binucleate cells are directly involved in the production of progesterone (Wango *et al.*, 1991; Wango *et al.*, 1992; Wooding *et al.*, 1996), prostaglandins (Reimers *et al.*,

1985b), placental lactogen hormones (Verstegen *et al.*, 1985; Wooding *et al.*, 1992) and pregnancy-associated (specific) glycoproteins (Reimers *et al.*, 1985a; Gogolin-Ewens *et al.*, 1986; Morgan *et al.*, 1989; Zoli *et al.*, 1992a; Atkinson *et al.*, 1993). These secretory products are stored in dense granules which occupy more than 50% of the cytoplasm (Lee *et al.*, 1986) and deliver their content directly in the maternal system after binucleate cell migration (Wooding, 1984).

3. Placental Proteins

Proteins secreted by the placenta, when detected in the peripheral circulation of the mother, can be useful indicators of both pregnancy and feto-trophoblast well-being.
Despite many attempts to investigate the physiological functions of the placental proteins, the exact biofunction of most of them (e.g. pregnancy-associated glycoprotein) is still not known.

In ruminant species, interferon tau allows the maintenance of the corpus luteum, acting as an antiluteolytic factor (Martal *et al.*, 1979). In primates this role in corpus luteum maintenance is played by the chorionic gonadotropin (Aschheim, 1927). This glycoprotein acts as a luteotropin-like hormone having also an important role in progesterone synthesis stimulation. The interferon tau, in addition to its antiluteolytic function also has an important role in the first steps of immunotolerance of the conceptus by the mother (Fillion *et al.*, 1991; Martal *et al.*, 1997).

Later in pregnancy placenta produces a placental lactogen hormone, responsible for the stimulation of the mammogenic activity (Forsyth, 1986). This hormone is also involved, at least partially, in the fetal growth (Chene *et al.*, 1988) and in the long term changes of the maternal metabolism (Freemark *et al.*, 1992).
The main groups of placental proteins and their relevant aspects are discussed below.

3.1. THE PREGNANCY-ASSOCIATED GLYCOPROTEINS (PAGS)

Characterized in the last 20 years (Butler *et al.*, 1982; Beckers *et al.*, 1988a,b; Zoli *et al.*, 1991, Xie *et al.*, 1991), the pregnancy-associated glycoproteins (PAGs) constitute a large family of glycoproteins specifically expressed in the outer epithelial cell layer (chorion/trophectoderm) of the placenta of ungulate species (Guruprasad *et al.*, 1996; Xie *et al.*, 1997b). They are members of the aspartic proteinase family having high sequence homology with pepsinogens (Green *et al.*, 1998a).

By using biochemical procedures, some molecules of the PAG family were isolated from cotyledons of cow (Zoli *et al.*, 1991), ewe (Zoli *et al.*, 1995; Xie *et al.*, 1997a) and goat (Garbayo *et al.*, 1998). These molecules were used to immunize rabbits and the antisera obtained allowed the development of homologous and heterologous radioimmunoassay systems (Zoli *et al.*, 1992b; Ranilla *et al.*, 1994; González *et al.*, 1999).

In veterinary practice, the measurement of the PAG concentrations in maternal blood is useful for both pregnancy confirmation and follow-up of the trophoblastic function. The first aspect can help breeders in the management of reproduction, while the second concerns more specifically clinicians and researchers in their investigation to establish differential diagnosis of pathologies affecting pregnancy (Zarrouk *et al.*, 1999a,b).

Screening of placental libraries with nucleic acid probes has identified additional cDNAs that are very abundant and code for polypeptides related but structurally (amino acid sequence) distinct from PAGs isolated by biochemical procedures. Some of these molecules have amino acid sequences that are consistent with their being potentially functional proteinases (Roberts *et al.*, 1995). It remains to be determined whether the PAGs expressed in the ungulate placenta are catalytically active and, even if they are, whether their physiological function is that of proteinases.

3.1.1. Historical Overview of PAGs and Related Proteins Isolated by Biochemical Procedures

The PAGs, also known under a variety of other names including pregnancy-specific protein B (Butler *et al.*, 1982; Lynch *et al.*, 1992) and pregnancy-specific protein 60 (Mialon *et al.*, 1993), were first described as placental antigens of cattle placenta that were also present in the blood serum of the mother soon after implantation.

The methodology employed for bovine PAG purification by Zoli *et al.* (1991) was similar to that previously described by Butler *et al.* (1982). Briefly, an antiserum was first generated against placental extracts from which antibodies against common maternal antigens had been removed (or neutralized) by immunoadsorption. This reagent was then used to follow the immunoreactive fractions throughout the isolation procedure (a series of extractions of soluble proteins, acid and ammonium sulfate precipitations, gel filtration and ion exchange chromatographies).

The cleared antisera obtained by Sasser and coworkers (Butler *et al.*, 1982) allowed the partial purification of two antigens; pregnancy-specific protein A (PSP-A) and pregnancy-specific protein B (PSP-B). PSP-A was identified as α-fetoprotein, which is not strictly limited to pregnancy; PSP-B represented a novel antigen. Subsequent biochemical analysis indicated that PSP-B probably represented a mixture of related proteins since multiple molecular weight forms (47 000-90 000) and isoeletric variants (pI 3.7-4.4) were reported (Butler *et al.*, 1982).

Alternatively to the results obtained by Butler *et al.* (1982), the use of antisera raised against specific antigens of fetal cotyledons by Zoli *et al.* (1991) resulted in the purification of one major placental protein: the bovine pregnancy-associated glycoprotein 1 (boPAG-1).

3.1.1.1. The Bovine PAG Family. After the isolation of Zoli *et al.* (1991) and the first characterization of Xie *et al.* (1991), different PAG molecules were identified justifying an arbitrary numbered nomenclature adopted till now.

3.1.1.1.1. boPAG-1. The bovine PAG (bPAG) purified by Zoli *et al.* (1991), later designed boPAG-1 (Xie *et al.*, 1994; Xie *et al.*, 1995), was shown to be an acidic glycoprotein with 67 000 molecular weight. Four isoforms (I, II, II and IV) with different isoeletric points (4.4, 4.6, 5.2 and 5.4) were detected in this initial preparation.

The pI were closely related to the amount of sialic acid of each isoform (2.12%; 0.83%, 0.64% and 0.29%, respectively). Moreover, pI, sialic acid content and immunoreactivity were closely related, with the most basic form being the most immunoreactive (Zoli *et al.*, 1991). Immunocytochemical investigations allowed the localization of boPAG-1 predominantly in the cytoplasm of large binucleate cells present in the fetal cotyledonary tissue (Zoli *et al.*, 1992a). The presence of antigens immunologically similar to boPAG-1 has also been demonstrated in testicular and ovarian extracts (Zoli *et al.*, 1990b,c), justifying the adjective "associated" and not "specific" given to this glycoprotein.

The most surprising feature of boPAG-1, revealed by Roberts and coworkers soon after its purification (Xie *et al.*, 1991), was the fact that this protein belongs to a family of proteolytic enzymes known as aspartic proteinases, with more than 50% amino acid sequence identity to pepsin, cathepsin D and cathepsin E.

The boPAG-1 has the bilobed structure typical of all known eukaryotic aspartic proteinases and possesses a cleft between the two lobes capable of accommodating peptides up to 7 amino acid long (Xie *et al.*, 1997b). However, despite its structural similarity with other active aspartic proteinases, this molecule is considered as catalytically inactive because of key mutations close to the active site (Figure 1) (Xie *et al.*, 1991).

Aspartic Proteinas	Specie	Proteolytic Activity ?	NH$_2$-terminus	COOH-terminus
Pepsinogen A	Monkey	Yes	Phe **Asp** Thr Gly Ser Ser	Val **Asp** Thr Gly Thr Ser
Progastricsin	Human	Yes	Phe **Asp** Thr Gly Ser Ser	Val **Asp** Thr Gly Thr Ser
Pepsinogen F	Rabbit	Yes	[Leu] **Asp** Thr Gly Ser [Ala]	Val **Asp** Thr Gly Thr Ser
Chymosin	Bovine	Yes	Phe **Asp** Thr Gly Ser Ser	[Leu] **Asp** Thr Gly Thr Ser
Renin	Rat	Yes	Phe **Asp** Thr Gly Ser [Ala]	Val **Asp** Thr Gly Thr Ser
Cathepsin D	Human	Yes	Phe **Asp** Thr Gly Ser Ser	Val **Asp** Thr Gly Thr Ser
Cathepsin E	Human	Yes	Phe **Asp** Thr Gly Ser Ser	Val **Asp** Thr Gly Thr Ser
Beta-secretase	Human	Yes	[Val] **Asp** Thr Gly Ser Ser	Val **Asp** [Ser] Gly Thr [Thr]
boPAG-1	Bovine	Probably no	Phe **Asp** Thr [Ala] Ser Ser	Val **Asp** Thr Gly Thr Ser
boPAG-2	Bovine	Unknown	Phe **Asp** Thr Gly Ser [Ala]	[Leu] **Asp** Thr Gly Thr Ser
ovPAG-1	Ovine	Probably no	Phe **Asp** Thr Gly Ser Ser	Val [**Gly**] Thr Gly Thr Ser
poPAG-1	Porcine	Probably no	Phe **Asp** Thr [Leu] Ser Ser	[Leu] **Asp** [Ser] Gly [Ser] [Ala]
poPAG-2	Porcine	Unknown	Phe **Asp** Thr Gly Ser Ser	Val **Asp** Thr Gly Thr Ser
eqPAG-1	Equine	Unknown	Phe **Asp** Thr Gly Ser [Ala]	Val **Asp** Thr Gly Thr Ser

Figure 1. Alignment at the active site residues found in a variety of aspartic proteinases. The open boxes denote amino acid substitutions. Data for aspartic proteinase sequences are obtained in the GenBank database.

Catalytic activity of aspartic proteinases is dependent upon two conserved aspartic acid residues (Asp) that are in close contiguity in the center of the substrate-binding cleft, even if they are localized in opposite lobes of the molecule. Each of these residues is preceded by a hydrophobic amino acid (usually phenylalanine - Phe) and followed by invariant threonine (Thr) and glycine (Gly) (Davies, 1990). A twofold symmetry, with a second conserved region centering around each aspartic acid is also required to polarize the bound water molecule which hydrolyses the scissile bond of the substrate (James and Sielecki, 1986).

As demonstrated by Xie *et al.* (1991), the apparent lack of proteolytic activity of boPAG-1 is due to the presence of an alanine (Ala-76) in place of a normally invariant glycine (Gly-76) residue in the Phe-Asp-Thr-Gly-Ser-Ser sequence (N-terminal lobe). This mutation has as consequence a subtle change in the placement of the catalytic water molecule from its normal symmetric position (Davies, 1990), rendering this protein catalytically inactive.

3.1.1.1.1.1 PAG and PSPB radioimmunoassays. In practice, antisera raised in rabbits against pregnancy-associated (specific) proteins have allowed the development of specific radioimmunoassays for bovine PAG and PSP-B (Sasser *et al.*, 1986; Humblot *et al.*, 1988a,b; Sasser *et al.*, 1989; Zoli *et al.*, 1992b). Higher PAG concentrations are observed in maternal than in fetal serum, suggesting that this glycoprotein is delivered preferentially in the maternal system (Zoli *et al.*, 1992b).

PAG can be detected in maternal circulation at around the time when the trophoblast forms definitive attachment to the uterine walls. In early and mid gestation, concentrations increase slowly and gradually (Figure 2). Around parturition concentrations increase rapidly to reach peak values of 1 to 5 μg/ml only few days before delivery (Figure 3) (Zoli *et al.* al., 1992b; Patel *et al.*, 1997). Concentrations decrease steadily in the postpartum period reaching undetectable levels only by day 100 postpartum (Zoli *et al.*, 1992b). Detectable levels of boPAG-1 have also been found in about 20% of unbred heifers and non-pregnant cows and 15% of bull sera (Zoli *et al.*, 1992b). The relatively long time needed for boPAG-1 to be cleared from maternal circulation can be explained by the very high concentrations present in maternal blood at parturition and by a long half-life of this glycoprotein, estimated to be 7.4 to 9 days (Kiracofe *et al.*, 1993; Ali *et al.*, 1997).
PAG detection in serum or plasma samples is currently used as a serological method for pregnancy diagnosis in cattle from days 28 (Szenci *et al.*, 1998a,b) to 30 (Sasser *et al.*, 1986; Humblot *et al.*, 1988a) after breeding.

Investigations realized in peripartum clearly demonstrated the positive influence of both maternal environment and fetal genotype (sex and race) on peripheral blood concentrations of boPAG-1. In fact, PAG concentrations were found to be more elevated in maternal circulation of the intra-species crossbreeds than in inter-species ones (Guilbault *et al.*, 1991). Moreover, Ectors and coworkers (1996) showed that in nuclear transfer programs, even if the majority of calves are of normal size, some of them can present morphological abnormalities like edema of the umbilical cord with placental hypertrophy, responsible for higher concentrations of PAG in maternal blood. In the same way, abnormally high PAG concentrations in maternal circulation can indicate the presence of a abnormal amount of trophoblast tissues, as observed in hydatiform molar pregnancy (Figures 4A and 4B) (Ectors *et al.*, 1996b).

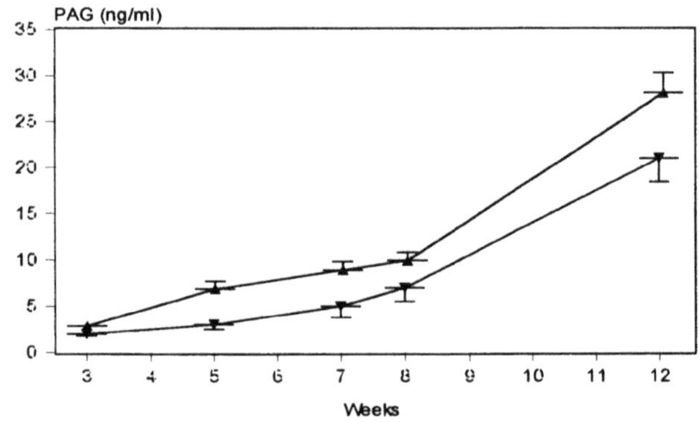

Figure 2. *PAG profiles (mean ± S.E.M) during early pregnancy in heifers embryo transferred with either single (-▼-▼-) or multiple (-▲-▲-) gestations. Adapted from Ectors* et al. *(1996a).*

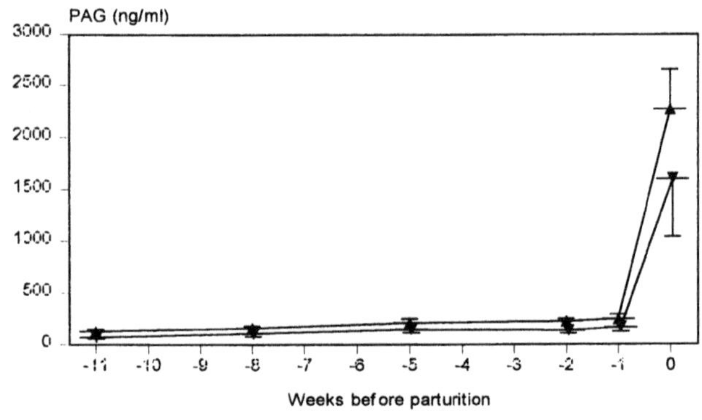

Figure 3. *PAG profiles (mean ± S.E.M) in heifers embryo transferred with either single (-▼-▼-) or multiple (-▲-▲-) gestations during the last 11 weeks before parturition. Adapted from Ectors* et al. *(1996a).*

After IVF or cloning, the PAG follow-up in plasma samples collected weekly was able to monitor embryonic and fetal deaths (Ectors *et al.*, 1996a). However, due to the large variations of PAG concentrations in maternal blood, only a marked decrease (or disappearance) in plasma concentrations of this protein can be a no controversial predictive sign of embryonic or fetal death (Szenci *et al.*, 1999).

3.1.1.1.2. boPAG-2. In 1994, Xie *et al.* identified the poorly characterized bovine chorionic gonadotropin isolated by Beckers *et al.* (1988a) as a new member of the aspartic proteinase family: the boPAG-2 (Beckers *et al.*, 1994).

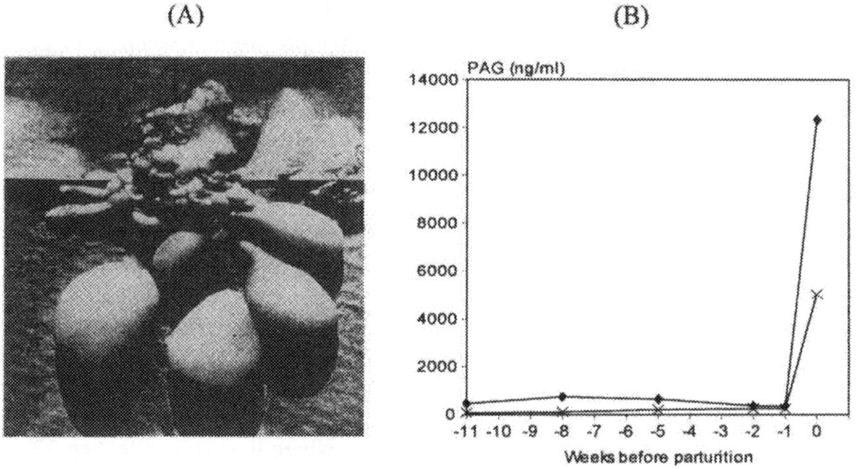

Figure 4. *A) Hydatiform molar pregnancy; B) Abnormally high PAG concentrations observed in hydatiform molar pregnancy (-▲-▲-) versus normal gestations (-×-×-). Adapted from Ectors* et al., *1996(a).*

The classification of the LH-like factor of the bovine placenta as a proteinase was surprising but not the first observation on similarities between placental hormones and proteinases. In fact, previous studies developed by Willey and Leindenberger (1989) had already demonstrated a high analogy between primary sequences of human chorionic gonadotropin (hCG) and the serine proteinase chymotrypsin. In that study it was also observed that proteinaceous proteinase inhibitors like soy bean trypsin (SBI) are capable to neutralize the agonistic activity of hCG and to reduce the binding of hCG to its receptor and to its specific antisera.

The boPAG-2 was characterized by Xie *et al.* (1994) as a polypeptide of 372 amino acids long, structurally related to boPAG-1, ovPAG-1 and pepsin (58%, 58% and 51% amino acid sequence identity, respectively). Unlike boPAG-1, boPAG-2 has a catalytic center with the amino acid consensus sequence of other active aspartic proteinases. However, it remains unclear whether boPAG-2 has a proteolytic activity.

Expression of boPAG-2 mRNA is detected as early as 17-19 days of pregnancy, coinciding with the beginning of implantation. Its mRNA is expressed in fetal placenta but not in other fetal organs, and is localized in both mononucleate and binucleate cells. Interestingly, boPAG-2 is synthesized by placental explants as a 70 000 molecular weight isoform that is processed to smaller molecules (Xie *et al.,* 1996).

3.1.1.2. The Ovine PAG Family. The identification of antigen(s) immunologically related to boPAG-1 (bPSP-B) in the peripheral circulation of pregnant ewes (Ruder *et al.*, 1988) has led to the isolation and partial purification of pregnancy-associated glycoproteins in ewes (Zoli *et al.*, 1990a; Zoli *et al.*, 1995; Willard *et al.*, 1995).

3.1.1.2.1 ovPAG-1. The ovine PAG, later designated ovPAG-1, was initially identified as a molecule close to the boPAG-1, containing 382 amino acids (Xie *et al.*, 1991). Parallel to the cDNA characterization, several attempts were developed in order to purify the ovine PAG starting from fetal cotyledons. In comparison to bovine PAG, the ovine PAG was more difficult to purify and only semi-purified preparations were made available (Zoli *et al.*, 1990a, Zoli *et al.*, 1995; Willard *et al.*, 1995). The ovPAG-1 showed multiple molecular weight isoforms (43 000-67 000) with different pI ranging from 4.06 to 4.65 (Willard *et al.*, 1995).

Molecular cloning studies have shown that the ovPAG-1, like boPAG-1, belongs to the aspartic proteinase family showing 73% amino acid sequence identity with this protein (Xie *et al.*, 1991).

As observed for boPAG-1, it seems that ovPAG-1 is a catalytically inactive form of aspartic proteinase secreted by binucleate cells of the trophoblast (Xie *et al.*, 1991). In this case, the lack of enzymatic activity results from the mutation of the catalytically essential aspartic acid within the C-terminal lobe (Asp-257) to one glycine (Gly-257) (Xie *et al.*, 1991). A comparable change in pepsin, where the active site Asp-32 of the first lobe is replaced by alanine (a Asp-32-Ala mutation) abolishes all activity, even though the molecule is still capable of binding pepstatin A (a potent inhibitor of aspartic proteinases) with high affinity (Lin *et al.*, 1989).

Time-course studies of placental explants recently showed that ovPAG-1 is first detectable as a precursor with high molecular weight (67 000-70 000), which is gradually processed to smaller products (Xie *et al.*, 1996). The basis of the variety of molecular weight forms remains to be elucidated, but the differences seem likely to be due to some unusual form of posttranslational modification introduced in the binucleate cells (Xie *et al.*, 1996; Green *et al.*, 1998a).

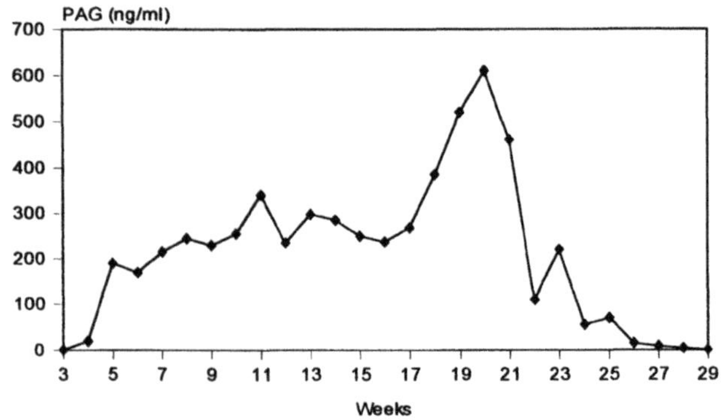

Figure 5. *PAG profiles during gestation and postpartum periods in Berrichon ewes (Adapted from Gajewski* et al., *1999).*

3.1.1.2.1.1. PAG and PSPB Radioimmunoassays. The development of heterologous radioimmunoassays for ovPAG-1 and oPSP-B has allowed its detection in maternal blood 3 weeks (Willard *et al.*, 1995) to 4 weeks (Ranilla *et al.*, 1994) after breeding. The PAG-RIA is based on the utilization of boPAG-1 as tracer and standard, and antisera raised in rabbits immunized against a preparation of ovine PAGs (Zoli *et al.*, 1995; Willard *et al.*, 1995).

PAG profiles in pregnant ewes (Figure 5) (Ranilla *et al.*, 1994; Ranilla *et al.*, 1997a; Gajewski *et al.*, 1999) appeared to be quite different those obtained in cattle (Zoli *et al.*, 1992b). In Churra and Merino ewes, for example, after a first period of high concentrations around day 60, the concentrations decrease until day 90 and increase again to remain elevated and stable until parturition (Ranilla *et al.*, 1994).
After parturition, the rate of decline in PAG concentrations is faster in ewes (4 weeks) than in cows (about 14 weeks), with no correlation being observed between this rate and the interval to first ovulation (Ranilla *et al.*, 1997b).

3.1.1.2.2. Other ovPAGs. An additional member of PAG family (ovPAG-2) was isolated from a cDNA library prepared from day 13 whole conceptuses (Nagel *et al.*, 1993; Guruprasad *et al.*, 1996). This protein is expressed in both mononucleate and binucleate cells of the placenta (Nagel *et al.*, 1993).

Several other different PAGs have been recently purified from ovine placental conditioned media by using a series of chromatography procedures (Xie *et al.*, 1997a). Amino-terminal sequencing of four of the purified products revealed that three of them were novel and that each of them had been processed by removal of the pro-peptide within what appeared to be a consensus sequence between the pro-peptide and the mature molecule. None of these purified molecules demonstrated any enzymatic activity in a standard proteinase assay with denatured haemoglobin (Xie *et al.*, 1997a).

Another subset of PAGs was identified in extracts of sheep placenta by Atkinson *et al.* (1993) who used the monoclonal antibody against SBU-3 developed by Gogolin-Ewens *et al.* (1986). The three proteins affinity-purified with the SBU-3 antibody had molecular weights of 57 000, 62 000 and 69 000, showing partial amino acid sequence homology to the ovPAG-1, boPAG-1 and rabbit pepsinogen F (Atkinson *et al.*, 1993).

3.1.1.3. The Caprine PAG Family. Three different PAGs were isolated and partially characterized from goat placenta (Garbayo *et al.*, 1998). These proteins differed in amino acid sequence and apparent molecular weight (55 000, 59 000 and 62 000) showing several isoforms with different pI: $caPAG_{55}$ (pI: 5.3, 5.1, 4.9), $caPAG_{59}$ (pI: 6.2, 5.9, 5.6) and $caPAG_{62}$ (pI: 5.1, 4.8).

Caprine PAGs showed high sequence homology to each other (60% to 73% residues identical) and a high sequence identity (from 30% to 81% between the first 27 amino acids sequenced) with proteins of the aspartic proteinase family like boPAG-1, ovPAG-1, boPAG-2, SBU-3 and rabbit pepsinogen F (Garbayo *et al.*, 1998).

3.1.1.3.1. PAG and PSPB Radioimmunoassays. The use of two semi-purified preparations (one containing the $caPAG_{55}$ and $caPAG_{59}$ forms, and the other containing the $caPAG_{55}$ and $caPAG_{62}$) allowed the development of an accurate system for pregnancy diagnosis in goats (González *et al.*, 1999). PAG concentrations are

significantly higher in pregnant than in non-pregnant females as early as 21 days after artificial insemination (González *et al.*, 1999).
During gestation, PAG concentrations reach maximal levels in week 8, decrease between weeks 12 and 14 and remain relatively constant until parturition (Figure 6) (Folch *et al.*, 1993; González *et al.*, 2000a). After parturition, concentrations decrease rapidly reaching lower levels at the 4th week postpartum (Sousa *et al.*, 1999).

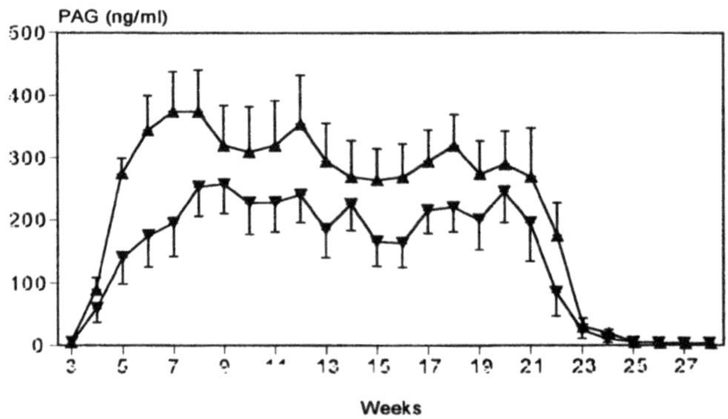

Figure 6. *PAG profiles (mean ± S.E.M) during gestation and postpartum periods in Blanca-Celtiberica goats with either single (-▼-▼-) or multiple (-▲-▲-) gestations. Adapted from Folch* et al., *1993.*

As observed in cattle (Guilbault *et al.*, 1991; Patel *et al.*, 1997), both number and genotype of fetus can influence PAG concentrations over gestation. From days 21-24 and throughout gestation, twin bearing goats have higher PAG (or PSP-B) concentrations than goats bearing one fetus (Humblot *et al.*, 1990; Sousa *et al.*, 1999; González *et al.*, 2000a,b). PAG levels in maternal circulation of inter-specific pregnancies (spanish ibex embryos transferred to domestic goats) are also higher (about ten times) when compared to that found in normal intra-specific gestations (Fernández-Arias *et al.*, 1999).
Sequential measurement of PAG in goats also allows for the determination of the onset of the disturbance of trophoblastic activity associated with the death of a fetus (Zarrouk *et al.*, 1999a,b). As demonstrated by Zarrouk *et al.* (1999a), in goats carrying a single fetus, PAG levels fall under the positive pregnancy threshold when the trophoblast died, while in goats carrying 2 or 3 fetuses, analysis of the profiles clearly shows marked drops in concentration that could indicate placental distress at different times. Therefore, systematic application of this test in herds with a high rate of pregnancy failure could help to pinpoint phenomena which might be implicated in triggering these events.

3.1.2. Molecular Studies of New PAG Related Molecules

By cloning expressed genes from ovine and bovine placental cDNA libraries, by Southern genomic blotting, by screening genomic libraries, and by using PCR to amplify portions of PAG genes from genomic DNA, it was recently estimated that cattle, sheep and most probably all ruminant *Artiodactyla* possess many, possibly 100 or more, PAG genes, many of them being placentally expressed (Xie *et al.*, 1997b; Green *et al.*, 1998a; Garbayo *et al.*, 1999; Green *et al.*, in press). New members of the PAG gene family have also been found to be expressed in the porcine (Baumbach *et al.*, 1988; Baumbach *et al.*, 1990a; Szafranska *et al.*, 1995; Doré *et al.*, 1996), equine (Green, *et al.*, 1998b; Green *et al.*, 1999), carnivora (Gan *et al.*, 1997) and zebra (Green *et al.*, 1994; Green *et al.*, 1999) placentas, showing that PAGs are not restricted to ruminant species.

The equine PAG (eqPAG-1) is expressed in the placenta of the horse and zebra and secreted from cultured placental tissues as both a processed (mature) and unprocessed (zymogen) form (Green *et al.*, 1999). The processed form is an active proteinase and may, therefore, be the horse homolog of pepsinogen F (Xie *et al.*, 1997b).

In porcine species, two different PAG cDNAs have been found: the poPAG-1 and poPAG-2. They share 64% homology sequence to each other, and about 50% amino acid sequence identity with the PAGs of ruminants. Interestingly, poPAG-1 has amino acid substitutions within its catalytic center that together were likely to render it enzymatically inactive, whereas poPAG-1 retained sequences identical to pepsin in these regions (Szafranska *et al.*, 1995). They seem occupy an intermediate position between the enzymatically functional aspartic proteinases and the PAGs from cattle and sheep (Xie *et al.*, 1997b).

3.1.3. Identification of Pregnancy-Associated (Specific) Proteins in Wild Ruminants

Serologically similar antigens to PAG or PSP-B (Table 1) have been found in maternal circulation of wild ruminants like mule deer (Wood de *et al.*, 1986), white-tailed deer (Wood *et al.*, 1986; Osborn *et al.*, 1996), mountain goats (Houston *et al.*, 1986), red deer (Haigh *et al.*, 1988), musk-oxen (Rowell *et al.*, 1989), bison (Haigh *et al.*, 1991), moose (Haigh *et al.*, 1993), sika deer (Willard *et al.*, 1994a; Willard *et al.*, 1996), elk (Willard *et al.*, 1994b), fallow deer (Willard *et al.*, 1994c; Willard *et al.*, 1998; Willard *et al.*, 1999) and reindeer (Russel *et al.*, 1998; Ropstat *et al.*, 1999).

The recent isolation and characterization of new forms of pregnancy-associated (specific) proteins from elk and moose placenta (Huang *et al.*, 1999) allowed the development of a more specific radioimmunoassay to quantify this protein in maternal serum of wild species (Huang *et al.*, 2000).

3.2. THE PLACENTAL LACTOGENS

Placental lactogens (PL), also known as chorionic somatomammotropins (CS), are proteinaceous hormones of placental origin that share structural and functional homology to growth hormone (GH) and prolactin (PRL).

The function of PL varies according to the species. In general it has been suggested that PL influences mammogenesis and lactogenesis (Forsyth, 1986), ovarian (Glaser *et al.*, 1984; Telleria *et al.*, 1998) and placental steroidogenesis (Deayton *et al.*, 1993), fetal growth (Chene *et al.*, 1988) and that it alters the maternal metabolism to accommodate the growth and development of the fetus (Freemark *et al.*, 1992).

3.2.1. Human Placental Lactogen (hPL) or Human Chorionic somatomammotropins (hCS) (see Martal and Cédart, 1993 for review)

The human placental lactogen (hPL), later called human chorionic somatomammotropin (hCS), is a non glycosylated single chain protein that contains 191 amino acids and two intra disulfide bonds. Its molecular weight is about 22 300, but polymeric forms with higher molecular weights have also been described. hPL is secreted by the syncytiotrophoblast cells of the placenta, however the mechanisms by which this hormone is released in the maternal circulation have not been totally elucidated.

Quantitatively, hPL is one of the most important soluble proteins of the human placenta, corresponding to approximately 10% of the total proteins. Its production rate is very high varying between 0.3 and 1 g/day at term. Human PL is detected in the placenta 5 to 10 days after implantation and in peripheral blood 3 to 4 weeks later. The level increases in the maternal serum as pregnancy progresses and the rate of secretion is related to the placental weight. As the half-life of hPL is very short (10-30 min), its assay gives an estimation of the placental activity at the time of collection.

Human PL is a regulator of lipids, carbohydrates and proteins metabolism. It increases the general metabolism of the adipose tissue provoking either lipolysis or lipogenesis depending on the circumstances. In pregnant women, it acts as an insulin antagonist and may be responsible for the diabetic acidosis observed during pregnancy.

3.2.2. Bovine Placental Lactogen (bPL) or Bovine Chorionic Somatomammotropin (bCS)

Studies on placental lactogen hormones preceded those of PAG in domestic ruminants. In bovine species, for example, the lactogenic activity of cotyledons was first demonstrated by Buttle and Forsyth in 1976 by use of a 5 day co-culture of mouse mammary tissue and cow cotyledonary tissue at different stages of pregnancy. These authors obtained a lactogenic substance with about 300 ng of equivalent bovine PRL (bPRL). In parallel studies, starting from bovine placenta, several research groups (Bolander and Fellows, 1976; Beckers *et al.*, 1980) identified and purified a protein with a somatomammotropic activity.

The purified bPL exists in multiple isoforms (Mr 32 000 to 34 000) with at least five isoeletric variants ranging from 4 to 6 (Murthy *et al.*, 1982). Bovine PL shares 48 to 50% sequence homology with PRL and 26 to 28% with GH. However, by use of radioimmunoassay (Beckers, 1983) it was demonstrated that there is no cross reaction between bPL, PRL or GH.

Bovine PL is produced by chorionic binucleate cells (Duello *et al.*, 1985; Milosavljenic *et al.*, 1989; Kappes *et al*; 1992) which seem to deliver their product to the maternal circulation by migrating to and fusing with endometrial epithelium. It co-localizes in the same cells with other proteins like SBU-3 antigen and boPAG-1 (Wooding, 1992). In contrast to other placental lactogens, bPL is a glycoprotein containing both N-linked and O-linked oligosaccharides (Shimomura and Bremel,

1988; Byatt *et al.*, 1990). This glycoproteic composition characteristic of placental proteins like bPL, boPAG-1 and SBU-3 can result of a highly developed posttranslational mechanism expressed in bovine giant trophoblastic cells (Gogolin-Ewens *et al.*, 1986).

Bovine PL is detectable in maternal circulation by the 4th month of gestation, but its concentration remains lower than 1-2 ng/ml through gestation. In the peripheral circulation of the fetus, bPL concentrations are higher (about 5 to 25 ng/ml) and shows a decline with advancing gestational age (Beckers *et al.*, 1982; Byatt *et al.*, 1987). These ratios of fetal and maternal bPL concentrations are quite different from those observed for boPAG-1, despite the co-localization of both placental proteins in the same cells.

In cattle, the mammary growth seems to be, at least partially, controlled by hormones of placental origin like bPL. Both calf birth weight and placental mass are positively correlated with milk production during subsequent lactation (Erb *et al.*, 1980; Collier *et al.*, 1982).

Table 1. *Pregnancy-associated glycoproteins and related molecules belonging to the aspartic proteinases family.*

Family	Species	Detection in maternal blood or milk*	Proteic Biochemical Characterization	cDNA clonage
Bovidae	Bovine (*Bos taurus*)	Sasser *et al.*, 1986,1989 Humblot *et al.*, 1988a,b Zoli *et al.*, 1992	Butler *et al.* 1982 Zoli *et al.*, 1991	Xie *et al.*, 1991,1994
	Ovine (*Ovis aries*)	Ranilla *et al.*, 1994,1997a Gajewski *et al.*, 1999	Atkinson *et al.*, 1993 Zoli *et al.*, 1995 Willard *et al.*, 1995 Xie *et al.*, 1997b	Xie *et al.*, 1991,1997
	Caprine (*Capra hircus*)	Humblot *et al.*, 1990 Folch *et al.*, 1993 Sousa *et al.*, 1999 González *et al.*, 2000a,b*	Garbayo *et al.*, 1998	Garbayo *et al.*, 1999
	Spanish ibex (*Capra pyrenaica)*	Fernández-Arias *et al.*, 1999	-	-
	Montain goat (*Oreamnos americanus*)	Houston *et al.*, 1986	-	-
	Musk oxen (*Ovibus moschatus*)	Rowell *et al.*, 1989	-	-
	Bison (*Bison bison*)	Haigh *et al.*, 1991	-	-
Cervidae	Moose (*Alces alces*)	Haigh *et al.*, 1993	Huang *et al.*, 1999, 2000	-
	Elk (*Cervus elephus*)	Haigh *et al.*, 1988 Willard *et al.*, 1994b	Huang *et al.*, 1999, 2000	-

Table 1. *Pregnancy-associated glycoproteins and related molecules belonging to the aspartic proteinases family.*

Family	Species	Detection in maternal blood or milk*	Proteic Biochemical Characterization	cDNA clonage
	Mule deer (*Odocoileus hemionus*)	Wood *et al.*, 1986	-	-
	White-tailed deer (*Odocoileus virginianus*)	Wood *et al.*, 1986 Osborn *et al.*, 1996	-	-
	Sika deer (*Cervus nippon*)	Willard *et al.*, 1994a	-	-
	Fallow deer (*Dama dama*)	Willard *et al.*, 1994c	-	-
	Rein deer (*Rangifer tarandus*)	Russel *et al.*, 1998 Ropstat *et al.*, 1999	-	-
Equidae	Horse (*Equus caballus*)	-	-	Green *et al.*, 1994, 1998b
	Zebra (*Equus zebra*)	-	-	Gan *et al.*, 1997 Green *et al.*, 1998
Felidae	Cat (*Felis domestica*)	-	-	Gan *et al.*, 1997
Suidae	Pig (*Sus scrofa*)	-	Baumbach *et al.*, 1988 Doré *et al.*, 1996	Szafranska *et al.*, 1995

Interestingly, endometrium, corpus luteum as well as maternal and fetal liver contain mRNA encoding the bGH and bPRL receptors (bGHR and bPRLR) (Scott *et al.*, 1992). Specific binding sites for bPL, with little affinity for bGH and bPRL, have also been identified in endometrium (Galosy *et al.*, 1991; Kessler *et al.*, 1991).

Recent evidence (Scott *et al.*, 1992) indicates that bPL competes with equal affinity with bPRL for a recombinant bPRLR (rbPRLR). Diversely, bPL seems to bind the bGHR in a 1:1 stoichiometry rather than 1:2 stoichiometry reported for bGH (Staten *et al.*, 1993).

It is possible that, in some adult tissues (e.g. liver and corpus luteum), bPL may exert its actions through either the bGH or bPRL receptors. This action was supposed to be exerted either by a heterodimer of bPRLR and bGHR monomers, or through a heterodimer of either one monomer of the bPRLR of bGHR and a unique receptor monomer (Anthony *et al.*, 1995).

3.2.3. Ovine Placental Lactogen (oPL) or Ovine Chorionic Somatomammotropin (oCS)

Ovine placental lactogen (oPL) was studied and purified by Handwerger *et al.* (1974), Martal and Djiane (1975) and Chan *et al.* (1976). However, due to the fact that the amino terminal residue of oPL was blocked, the full sequence of ovine PL was not elucidated until quite recently (Colosi *et al.*, 1989; Warren *et al.* 1990). Although the primary structure of oPL and bPL are similar in 67%, oPL has a molecular weight of 22 000 and no potential N-linked glycosylation sites (Chan *et al.*, 1976). The degree of amino acid sequence homology between of oPL and PRL is higher than that to GH (49% and 28%, respectively). So, it is not surprising to observe that oPL can demonstrate a higher lactogenic than somatogenic activity.

Binding sites for oPL have been identified in fetal ovine liver approximately at day 70 of gestation and increase in number through pregnancy to reach a peak 3 to 7 days before parturition (Freemark *et al.*, 1986). These sites are also found in maternal liver during gestation as well as in neonatal liver for more than 1 week postpartum. Freemark and coworkers (1990) demonstrated a decreased binding of oPL to fetal liver during maternal fasting, and later (Freemark *et al.*, 1992) they suggested that oPL receptor may be either a variant or a shortened form of somatotropin receptor.

In trophoblastic tissue oPL is detectable at day 16-17 of gestation (Reddy and Watkins, 1978). In maternal blood, oPL can be detected at day 40 to 50 of gestation and peaks during the last trimester of pregnancy (Chan *et al.*, 1978). These values are positively correlated to fetal number, milk yield and placental mass. Ovine PL fetal concentrations decline from mid gestation until 1 to 2 weeks before parturition. Although fetal sera concentration and total fetal weight were not correlated (Kappes *et al.*, 1992), the entry rate of oPL into fetal vasculature increases with advancing gestation age (Schoknecht *et al.*, 1992).

Biological activities of oPL have been observed in ovarian steroidogenesis, maternal intermediary metabolism, mammary and fetal growth. Anthony *et al.* (1995) suggest that oPL acts through a structurally distinct receptor within fetus, influencing the fetal production of one or more insulin growth factors (IGFs) and playing an important role in fetal growth regulation.

The factors that regulate secretion of oPL are still poorly understood but it is clear that oPL is not regulated by the same way in the fetal and maternal compartments, and nutritional factors seem to be involved (Byatt *et al.*, 1992). A better knowledge about PL hormones probably will help to understand the metabolic way of dysfunctions occurring in late pregnancy, in postnatal development and milk production.

3.2.4. Caprine Placental Lactogen (cPL) or Caprine Chorionic Somatomammotropin (cCS)

Caprine placental lactogen (cPL) was the first placental lactogen described in ruminants (Buttle *et al.*, 1972). Several studies described the presence of prolactin-like and growth hormone-like activities in the blood of pregnant goats and several research groups (Becka *et al.*, 1977; Nugent and Hayden, 1987) tried to purify this protein. Early nineties, cPL has been purified to homogeneity (Currie *et al.*, 1990).

Caprine PL has a molecular weight of 22 000 and an isoeletric point of 8.35. Despite its purification, the primary amino acid sequence of cPL was not determined. Based on the close phylogenic relationship between caprine and ovine species, it is likely that oPL and cPL have a high percentage of amino acid sequence similarity to each other. Also by similarity, it seems that cPL, like oPL, is a non-glycosylated form of placental lactogen (Byatt *et al.*, 1992).

Caprine PL is detected from day 44 of pregnancy in jugular plasma (Currie *et al.*, 1990) and can be used as a late pregnancy diagnosis from the day 60 (Sardjana *et al.*, 1988). The level of cPL increases during pregnancy. Maximal concentrations are observed from mid to late pregnancy, followed by a gradual decline through the last week of gestation (Card *et al.*, 1988). Caprine PL concentrations are undetectable as early as 18 h postpartum (Currie *et al.*, 1990). As observed for PAG, cPL concentrations are higher in multiple than in single gestations (Currie *et al.*, 1990) and are directly related with the total weight of placentomes and with the total fetal weight.

Placental lactogen seems to have a significant role in the control of normal mammary development and function in goats. Milk yield is related with the weekly mean of PL between week 11 and term (Hayden *et al.*, 1979). The increase in secretion of PL coincides with the onset of rapid lobule alveolar development of the mammary gland.

3.3. PROTEINS ASSOCIATED WITH THE EMBRYONIC SIGNAL

In some species, even before implantation, the embryo produces some "signals" which are recognized by the mother and induce an inhibition of luteolysis and the transformation of a cyclic corpus luteum into a pregnancy corpus luteum. Among these signals, it can be mentioned the chorionic gonadotropin (CG), a luteotropic factor produced by the human and equine placentas, and the interferon tau (IFNτ), the antiluteolytic factor responsible for the rescue of the corpus luteum in ruminant species, and probably, at least partially in porcine and equine species.

3.3.1. The Chorionic Gonadotropins (CG) (see Leymarie and Martal (1993) and Derivaux et al. (1988) for review)

Human and equine species have long been known to display a placental gonadotropic activity. As early as 1927, Aschheim reported the presence of a gonadotropic factor in the urine of pregnant woman and during the following year they based their famous method for the early diagnosis of pregnancy on this finding (Aschheim and Zondek, 1928a,b). Due to its properties, this gonadotropic factor was later called human chorionic gonadotropin (hCG). Three years later, Cole and Hart (1930) showed the existence of a gonadotropic activity in the serum of pregnant mares. This pregnant mare serum gonadotropin (PMSG), after a better characterization, was called equine chorionic gonadotropin (eCG).

3.3.1.1. Human Chorionic Gonadotropin (hCG). The human chorionic gonadotropin (hCG) is a glycoprotein with a molecular weight of 38 400 and isoeletric points varying from 3.8 to 5.1. In early gestation, the synthesis of hCG increased in parallel with the trophoblastic activity.

hCG is secreted by the trophoblast as soon as the blastocyst leaves its zona pellucida. It is first found in the maternal blood from day 8-12 after fertilization, to achieve maximal concentrations during the first trimester of pregnancy. Afterwards, the levels decrease, but are still detectable till parturition.

Structurally, the hCG is composed by two subunits: alpha and beta. The alpha subunit comprises a peptide chain of 92 amino acids with 2 oligosaccharide chains attached at residues 52 and 78 (Asn-52 and Asn-78). This alpha subunit is similar to the alpha subunits of human, ovine and porcine LH. The beta hCG subunit is very similar to the beta subunit of human LH. It comprises 145 amino acids, of those, 115 identical to the beta subunit of LH, with an additional fragment (30 amino acids) in the carboxy terminal side. The beta subunit contains several additional carbohydrate chains linked to the serines (Ser-118 or Ser-119 and Ser-131) and to the asparagines (Asn-13 and Asn-30). Due to its high content in sialic acid, the half life of hCG (about 8 hours) is much longer then that of LH, estimated to be only 12 to 50 min. The biological activity of hCG is rapidly abolished by hydrolysis with trypsin or chymotrypsin.

Human CG produces an effect on luteal cells of LH type which involves the same receptors as the hypophyseal gonadotropin. However, its action is more sustained since the hCG receptor complex is internalized 50 times more slowly than the LH receptor complex.

3.3.1.2. Equine Chorionic Gonadotropin (eCG). The equine chorionic gonadotropin (eCG), also known as PMSG, is specific to the mare. It represents the main luteotropic factor responsible for progesterone ovarian secretion until the placenta becomes able to carry out this progestagens secretion.

The eCG is synthesized by the endometrial cups which are composed by typical trophoblastic cells with one or two nuclei. These cells release the eCG only during the first trimester of pregnancy (between 45 and 130 days of gestation). Initial eCG secretion coincides with the formation of the secondary corpus luteum in the ovaries.

Equine CG is a glycoprotein with an apparent molecular weight between 45 000 and 64 000. Like hCG, eCG is composed by subunits alpha and beta. Its plasmatic half-life (about 6 days) is due to its high sialic acid content (10 to 13.5%) and to the length of its carbohydrate chains. Contrary to the hCG, eCG was detected in urine of pregnant females.

The activity of this hormone presents analogies with that of the hypophyseal gonadotropin FSH. It can be therefore used in hormonal control programs when a FSH-like activity is required (see chapter on superovulation).

3.3.2. Interferon tau (IFNτ)

In domestic ruminants, ovarian production of progesterone is required for about 55 days in ewes, for at least 165-180 days in cattle (Estergreen *et al.*, 1967) and till the end of pregnancy in goats. In these species, rescue of the corpus luteum occurs before implantation (around day 15 to 22 after fertilization) and appears to be initiated by the release of a proteinaceous factor present in the pre-implantation conceptus. This factor, initially called trophoblastin (Martal *et al.*, 1979) or ovine trophoblast protein-1 (Godkin *et al.*, 1984), was recently officially designated as a distinct subclass of the interferon alpha (IFNα) family: the IFN tau (IFNτ) (Charpigny *et al.*, 1988).

The first studies that demonstrated the presence of an antiluteolytic factor produced by the conceptus were realized by Moor and Rowson (1966), who transferred 14-16 days old ovine embryos to ewes during the estrus cycle, before the 12th day, and observed the maintenance of the corpus luteum and the interruption of the cycle. The following year, the same authors homogenized 14-16 day ovine embryos and injected the homogenate into the uterus before the 12th day of the cycle. The same results were obtained: maintenance of the corpus luteum and interruption of the estrous cycle. This phenomenon was not observed if the homogenates were obtained from 21-23 day embryos.

Northey and French (1980) and Humblot and Dalla Porta (1984) carried out similar experiments in the cow. They transferred 15-17 days old embryos or their homogenates to recipients during the first 16 days of the estrous cycle obtaining the maintenance of the corpus luteum.

3.3.2.1. Bovine Interferon Tau (boIFNτ). The boIFNτ is a glycoprotein with approximately 172 amino acids, produced by mononucleate cells of the trophectoderm (Morgan *et al.*, 1993). It is the major secretory product produced by day 16-25 cow conceptus. When analyzed by two-dimensional gel electrophoresis, the boIFNτ fell into two major molecular weight classes (22 000 and 24 000) with isoeletric points between 6.5 and 6.7 (Helmer *et al.*, 1987). The 24 000 molecular weight form is complex in nature whereas the 22 000 form is a single high-mannose type glycoprotein, indicating that these products undergo differential posttranslational glycosylations (Helmer *et al.*, 1988). The entire cDNA sequence of the boIFNτ shared 79% identity with bovine interferon alpha II (IFNα II) (Imakawa *et al.*, 1989).

IFNτ is produced by the trophectoderm during the phase of trophoblast elongation. It alters uterine release of $PGF_{2\alpha}$ which results in rescue of the corpus luteum and continued release of progesterone. The mechanism of action of IFNτ includes inhibition of oestradiol receptors, consequent reduction in oxytocin receptors, activation of a cyclooxygenase inhibitor, and a shift in the prostaglandins to favour PGE2 over $PGF_{2\alpha}$ which also induces several endometrial proteins that may be critical for surviving of the developing embryo (Hansen *et al.*, 1999).

3.3.2.2. Ovine Interferon Tau (ovIFNτ). In ewes, ovIFNτ is produced transiently and when introduced into the non-pregnant female uterus, extends corpus luteum life only for a short period of a few days in most ewes, although a functional corpus luteum can persist much longer in others (Martal *et al.*, 1979; Godkin *et al.*, 1982).

Martal *et al.* (1979) were the first to suggest that the antiluteolytic factor produced by the ovine conceptus was a termolabile protein of trophoblastic origin. This protein contains 172 amino acids (Imakawa *et al.*, 1987; Roberts *et al.*, 1991) and presents a molecular weight of 20 000 (Godkin *et al.*, 1982). Its mRNA is expressed during a relatively short period (11 to 22 days) (Charlier *et al.*, 1989) that corresponds closely to the time at which the embryo acts to extend luteal lifespan. The highest level of mRNA expression was observed on the 22 days of pregnancy (Hansen *et al.*, 1988). There is a greater structural identity between ovine and bovine IFNτ (85%) than between this protein and the closest homologous IFN omega (70%) (Charlier *et al.*, 1991).

Like boIFNτ, the ovIFNτ binds receptors on the uterine endometrium (Bazer, 1989) and seems to suppress transcription of the estrogen and oxytocin receptor genes to block pulsatile release of $PGF_{2\alpha}$ (Godkin *et al.*, 1997).

3.3.2.3. Caprine Interferon Tau (caIFNτ). IFNτ molecules have also been identified in caprine species (Gnatek *et al.*, 1989). Two classes of proteins (17 000 and 22 000-24 000) (Guillomot *et al.*, 1998), each with different isoeletric points, were identified from goat conceptus culture medium (Baumbach *et al.*, 1990b). The goat IFNτ is present in the trophoblastic cells as early as day 14 and until day 17. However, by day 18, as implantation proceeded, goat IFNτ is no longer detected suggesting that, in this species, other factors need to be present by day 18 to take over a role in the maintenance of luteal function (Guillomot *et al.*, 1998).

3.3.3. Early Pregnancy Factor (EPF)

Early-pregnancy factor was first discovered in the early stages of gestation by use of the rosette inhibition test (Morton *et al.*, 1979). Despite several attemps, the use of this assay is not yet reliable as a current pregnancy diagnosis test in animal species (Takagi *et al.*, 1998).

Recently, it was observed that approximately 70% of the amino acid sequence of EPF are identical to that of rat chaperonin 10, a member of the highly conserved heat-shock family. Investigations with the latter confirmed that chaperonin 10 is the moiety in pregnancy serum which initiates response in the EPF bioassay (Cavanagh and Morton, 1994). As well as being a monitor of the presence of a viable embryo (Shu-Xin and Zhen-Qun, 1993), it is necessary for embryonic survival. In this capacity it acts as both an immunosuppresant and growth factor (Morton *et al.*, 1992; Morton, 1998).

Acknowledgements

This review is part of a study supported by the following grants. The investigations realized in Belgium were funded by FNRS and IRSIA grants to J.-F. Beckers. N.M. Sousa is supported by a grant from CAPES. We also thank the NIH Bethesda USA for gift of hormones.

References

Ali, M., Gebbie, F.E., Sinclair, K.D., Hutchinson, J.S.M., Beckers, J.F. and Broadbent, P.J. (1997) Half-life of serum bovine pregnancy-associated glycoprotein (bPAG), postpartum anoestrus and performance of beef cows, *J. Reprod. Fert. Abstract Series 20*, 25.

Amoroso, E.C. (1952) Placentation, in Marshall (ed.), *Physiology of Reproduction*, A.S. Parkes Ed., London, pp. 127-311.

Anthony, R.V., Pratt, S.L., Liang, R. and Holland, M.D. (1995) Placental-fetal hormonal interactions: impact on fetal growth, *J. Anim. Sci. 73*, 1861-1871.

Aschheim, S. (1927). Weitere Untersuchungen ueber Hormone and Schwangerschaft. Das Vorkommen des Hormone im Harn des Schwangeren. *Arch. Gynaekol.* 132, 179-183.

Aschheim, S. and Zondek, B. (1928a) Schwangerschaftsdiagnose aus dem Harn (durch Hormonnachweis), *Klin. Wschr.* 7, 8-9.

Aschheim, S. and Zondeck, B. (1928b) Die Schwangershaftsdiagnose au dem Harn durch nachweis des hypophysenvorder lappenhormones, *Klin. Wschr.* 7, 1404-1411.

Assheton, R. (1906) The morphology of the ungulate placenta. Particularly the development of that organ in sheep, and notes upon the placenta of the elephant and hyrax, *Phil. Trans. R. Soc. 198*, 143-220.

Atkinson, Y.H., Gogolin-Ewens, K.J., Hounsel, E.F., Davies, M.J., Brandon, M.R. and Seamark, R.F. (1993) Characterization of placentation-specific binucleate cell glycoproteins possessing a novel carbohydrate, *J. Biol. Chem. 268*, 26679-26685.

Baumbach, G.A., Climer, A.H., Bartley, N.G., Kattesh, H.G. and Godkin, J.D. (1988) Purification, characterization and immunocytochemical localization of the major basic protein of the pig conceptus, *Biol Reprod 39*, 1171-1182.

Baumbach, G.A., Bartley, N.G., Kattesh, H.G. and Godkin, J.D. (1990a) Immunolocalization of BP, a conceptus secretory protein in trophectoderm during pig trophoblast elongation, *Cell Biol. Intl. Rep. 14*, 815-821.

Baumbach, G.A., Duby, R.T. and Godkin, J.D. (1990b) N-glycosylated and unglycosylated forms of caprine trophoblast protein-1 are secreted by preimplantation goat conceptuses, *Biochem Biophys Res Commun. 172*, 16-21.

Bazer, F.W. (1989) Establishment of pregnancy in sheep and pigs, *Reprod. Fertil. Dev. 1*, 237-242.

Becka, S., Bilek, J., Slaba, J. and Mikulas, I. (1977) Some properties of the goat placental lactogen, *Experientia 33*, 771-772.

Beckers, J.F. (1983) L'hormone placentaire somatomammotrope bovine, Thèse d'agrégation, Université de Liège.

Beckers, J.F., Fromont-Lienard, Ch., Van der Zwalmen, P., Wouters-Ballman, P. and Ectors, F. (1980) Isolement d'une hormone placentaire bovine présentant une activité analogue à la prolactine et à l'hormone de croissance, *Ann. Méd. Vét. 124*, 584-601.

Beckers, J.F., De Coster, R., Wouters-Ballman, P., Fromont-Liénard, Ch., Van der Zwalmen, P. and Ectors, F. (1982) Dosage radioimmunologique de l'hormone placentaire somatotrope et mammotrope bovine. *Ann. Méd. Vét. 126*, 9-21.

Beckers, J.F., Dewulf, M., Verstegen, J., Wouters-Ballman, P. and Ectors, F. (1988a) Isolation of a bovine chorionic gonadotrophin (bCG), *Theriogenology 29*, 218 (abstract).

Beckers, J.F., Wouters-Ballman, P. and Ectors, F. (1988b) Isolation and radioimmunoassay of a bovine pregnancy specific protein, *Theriogenology 29*, 219 (abstract).

Beckers, J.F., Roberts, R.M., Zoli, A.P., Ectors, F. and Derivaux, J. (1994) Molecules of the family of aspartic proteinases in the placenta of ruminants: hormones or proteins? *Bull. Mem. Acad. R. Med. Belg. 149*, 355-367.

Björkman, N. (1954) Morphological and histochemical studies on the bovine placenta, *Acta Anat. 22 (Suppl. 2)*, 1-31

Bolander, F.F.Jr. and Fellows, R.E. (1976) Purification and characterization of bovine placental lactogen. *J. Biol. Chem. 251*, 2703-2708.

Butler, J.E., Hamilton, W.C., Sasser, R.G., Ruder, C.A., Hass, G.M. and Williams, R.J. (1982) Detection and partial characterization of two bovine pregnancy-specific proteins, *Biol. Reprod. 26*, 925-933.

Buttle, H.L. and Forsyth, I.A. (1976) Placental lactogen in the cow, *J. Endocrin. 68*, 141-146.

Buttle, H.L., Forsyth, I.A. and Knaggs, G.S. (1972) Plasma prolactin measured by radioimmunoassay and bioassay in pregnant and lactating goats and the occurrence of a placental lactogen, *J. Endocr. 53*, 483-491.

Byatt, J.C., Wallace, C.R., Bremel, R.D., Collier, R.J. and Bolt, D.J. (1987) The concentration of bovine placental lactogen and the incidence of different forms in fetal cotyledons and in fetal serum. *Anim. Endocrinol. 4*, 231-241.

Byatt, J.C., Welply, J.K., Leimgruber, R.M. and Collier R.J. (1990) Characterization of glycosylated bovine placental lactogen and the effect of enzymatic deglycosylation on receptor binding and biological activity, *Endocrinology 127*, 1041-1049.

Byatt, J.C., Warren, W.C., Eppard, P.J., Staten, N.R., Krivi, G.G. and Collier, R.J. (1992) Ruminant placental lactogens: structure and biology, *J. Anim. Sci. 70*, 2911-2923.

Card, C.E., Butler, W.R., Currie, W.B. and Michel, F. (1988) Caprine placental lactogen (PL): fetomaternal levels, short term changes and ketosis, *Biol. Reprod. 38*, 138 (abstract).

Cavanagh, A.C. and Morton, H. (1994) The purification of early-pregnancy factor to homogeneity from human platelets and identification as chaperonin 10, *Eur. J. Biochem. 222*, 551-60.

Chan, J.S.D., Robertson, H.A. and Friesen H.G. (1976) The purification and characterization of ovine placental lactogen, *Endocrinology 98*, 65-76.

Chan, J.S.D., Robertson, H. and Friesen, H.G. (1978) Maternal and fetal concentrations of ovine placental lactogen measured by radioimmunoassay, *Endocrinology 102*, 1606-1613.

Charlier, M., Hue, D., Martal, J. and Gaye, P. (1989) Cloning and expression of cDNA encoding ovine trophoblastin: its identity with a class-II alpha interferon, *Gene 77*, 341-348.

Charlier, M., Hue, D., Boisnard, M., Martal, J. and Gaye, P. (1991) Cloning and structural analysis of two distinct families of ovine interferon-alpha genes encoding functional class II and trophoblast (oTP) alpha-interferons, *Mol. Cell Endocrinol. 76*, 161-171.

Charpigny, G., Reinaud, P., Huet, J.C., Guillomot, M., Charlier, M., Pernollet, J.C. and Martal, J. (1988) High homology between a trophoblastic protein (trophoblastin) isolated from ovine embryo and alpha-interferons, *FEBS Lett. 228*, 12-16.

Chene, N., Martal, J. and Charrier, J. (1988) Ovine chorionic somatomammotropin and fetal growth, *Reprod. Nutr. Dev. 28*, 1707-1729.

Cole, H.H. and Hart, G.H. (1930) The potency of blood serum of mares in progressive stages of pregnancy in effecting the sexual maturity of the immature rat, *Am. J. Physiology 93*, 57-68.

Collier, R.J., Doelger, S.G., Head, H., Thatch, W.W. and Wilcox, C.J. (1982) Effects of heat stress during pregnancy on maternal hormone concentrations, calf birth weight and postpartum milk yield, *J. Anim. Sci. 54*, 309-319.

Colosi, P., Thordarson, G., Hellmiss, R., Singh, K., Forsyth, I.A., Gluckman, P. and Wood, W.I. (1989) Cloning and expression of ovine placental lactogen. *Mol. Endocrinology 3*, 1462-1469.

Currie, W.B., Card, C.E., Michel, F.J. and Ignotz, G. (1990) Purification, partial characterization, and development of a specific radioimmunoassay for goat placental lactogen, *J. Reprod. Fert. 90*, 25-36.

Davies, D.R. (1990) The structure and function of the aspartic proteinases, *Annu. Rev. Biophys. Chem. 19*, 189-215.

Deayton, J.M., Young, I.R. and Thorburn, G.D. (1993) Early hypophysectomy of sheep fetuses: effects on growth, placental steroidogenesis and prostaglandin production, *J. Reprod. Fertil. 97*, 513-520.

Derivaux, J., Ectors, F. and Beckers, J.F. (1988) The ruminant placenta: structure and endocrine function, Monography, University of Liège, pp. 1-54.

Doré, J.J.E, Kattesh, H.G. and Godkin, J.D. (1996) Isolation and identification of porcine embryonic basic protein as a fragment of pregnancy-associated glycoprotein-2, *Int. J. Biochem. Cell Biol. 28*, 1249-1255.

Drieux, H. and Thiery, M. (1951) La placentation chez les mammifères domestiques. III. Placenta des bovidés, *Rec. Méd. Vét. 127*, 5-25.

Duello, T.M., Byatt, J.C., Bremel; R.D. (1985) Immunohistochemical localization of placental lactogen in binucleate cells of bovine placentome, *Endocrinology 119*, 1351-1355.

Ectors, F.J., Drion, P.V., Delval, A., Smith, L.C., Sulon, J., Zaaijer, D., Szenci, O., Remy, B., Beckers, J.F. and Ectors, F. (1996a) Interests of pregnancy follow-up in cows after embryo transfer: special focusing on IVP and NT, *12ème Colloque Association Européenne de Transfert Embryonnaire* (Lyon), pp. 95-103.

Ectors, F.J., Schmidt, M., Smith, L.C., Sulon, J., Delval, A., Remy, B., Avery, B. and Beckers, J F. (1996b) bPAG profiles in recipient heifers after transfer of IVF and nuclear transfer embryos, *Theriogenology 45*, 283 (abstract).

Erb, R.E., Chew, B.P., Malven, P.V., D'Amico, M.F., Zamet, C.N. and Colenbrander, V.F. (1980) Variables associated with peripartum traits in dairy cows. VII. Hormones, calf traits peripartum traits and subsequent milk yield. *J. Anim. Sci. 51*, 143-152.

Estergreen, V.L.Jr., Frost, O.L., Gomes, W.R., Erb, R.E. and Bullard, J.F. (1967) Effect of ovariectomy on pregnancy maintenance and parturition in dairy cows, *J. Dairy Sci. 50*, 1293-1295.

Fernández-Arias, A., Alabart, J.L., Folch, J. and Beckers, J.F. (1999) Interspecies pregnancy of Spanish ibex (*Capra pyrenaica*) fetus in domestic goat (*Capra hircus*) recipients induces abnormally high plasmatic levels of pregnancy-associated glycoprotein, *Theriogenology 51*, 1419-1430.

Fillion, C., Chaouat, G., Reinaud, P., Charpigny, J.C. and Martal, J. (1991) Immunoregulatory effects of ovine trophoblastin protein (oTP): all five isoforms suppress PHA-induced lymphocyte proliferation, *J. Reprod. Immunol. 19*, 237-49

Folch, J., Benitez-Ortiz, W., Alabart, J.L. and Beckers, J.F. (1993) Determinacion de la concentracion plasmatica de PAG (pregnancy-associated glycoprotein) en cabras blanca celtibèrica y su utilizacion como diagnostico de gestacion, *ITEA 12*, 364-366.

Foote, W.C. and Kaushik, D.K. (1963) LH activity in bovine reproductive tissue, *J. Anim. Sci. 22*, 857-858.

Forsyth, I.A. (1986) Variation among species in the endocrine control of mammary growth and function: the roles of prolactin, growth hormone, and placental lactogen, *J. Dairy Sci. 69*, 886-903.

Freemark, M., Comer, M. and Handwerger, S. (1986) Placental lactogen and GH receptors in sheep liver: striking differences in ontogeny and function, *Am. J. Physiol. 251*, 328-333

Freemark, M., Comer, M., Mularoni, T., D'Ecorle, A.J., Grandis, A. and Kodack, L. (1990) Placental lactogen receptors in maternal sheep liver: effects of fasting and refeeding, *Am. J. Physiol. 258*, 338-346.

Freemark, M., Keen, A., Fowlers, J., Mularoni, T., Comer, M., Grandis, A. and Kodack, L. (1992) The placental lactogen recptor in maternal and fetal sheep liver: regulation by glucose and role in the pathogenesis of fasting during pregnancy, *Endocrinology 130*, 1063-1070.

Gajewski, Z., Beckers, J.F., Sousa, N.M., Thun, R., Sulon, J. and Faundez, R. (1999) Oznaczanie stezenia PAG u owiec badania retrospektywne, *Adv. Cell Biol. 26 (Suppl. 12)*, 89-96.

Galosy, S.S., Gertler, A., Elberg, G. and Laird, D.M. (1991) Distinct placental lactogen and prolactin (lactogen) receptors in bovine endometrium, *Mol. Cell. Endocrinol. 78*, 229-236.

Gan, X., Xie, S., Green, J.A. and Roberts, R.M. (1997) Identification of transcripts for pregnancy associated glycoprotein (PAG) in carnivora and perissodactyla, *Biol. Reprod. 56 (Suppl)*, 431 (abstract).

Garbayo, J.M., Remy, B., Alabart, J.L., Folch, J., Wattiez, R., Falmagne, P. and Beckers, J.F. (1998) Isolation and partial characterization of a pregnancy-associated glycoprotein family from the goat placenta. *Biol. Reprod. 58*, 109-115.

Garbayo, J.M., Green, J.A., Beckers, J.F. and Roberts, M.R. (1999) Cloning and expression of pregnancy-associated glycoproteins (PAGs) from the caprine placenta, *Biol. Reprod. 60 (Suppl)*, 493 (abstract).

Glaser, L.A., Kelly, P.A. and Gibori, G. (1984) Differential action and secretion of rat placental lactogens, *Endocrinology 115*, 969-976.

Gnatek, G.C., Smith, L.D., Duby, R.T. and Godkin, J.D. (1989) Maternal recognition of pregnancy in the goat: effects of conceptus removal on interestrus intervals and characterization of conceptus protein production during early pregnancy, *Biol. Reprod. 41*, 655-663.

Godkin, J.D., Bazer, F.W., Moffat, J., Sessions, F. and Roberts, M.R. (1982) Purification and properties of a major low molecular weight protein released by the trophoblast of sheep blastocyst at day 13-21, *J. Reprod. Fert. 65*, 141-150.

Godkin, J.D., Bazer, F.W. and Roberts, M.R. (1984) Ovine trophoblast protein 1, an early secreted blastocyst protein, binds specifically to uterine endometrium and affects protein synthesis, *Endocrinology 114*, 120-130.

Godkin, J.D., Smith, S.E., Johnson, R.D. and Dore, J.J. (1997) The role of trophoblast interferons in the maintenance of early pregnancy in ruminants, *Am. J. Reprod. Immunol. 37*, 137-143.

Gogolin-Ewens, K.J., Lee, C.S., Mercer, W.R., Moseby, A.M. and Brandon, M.R. (1986) Characterization of a sheep trophoblast-derived antigen first appearing at implantation, *Placenta 7*, 243-255.

Gonzalez, F., Sulon, J., Garbayo, J.M., Batista, M., Cabrera, F., Calero, P., Gracia, A. and Beckers, J.F. (1999) Early pregnancy diagnosis in goats by determination of pregnancy-associated glycoprotein concentrations in plasma samples, *Theriogenology 52*, 717-725.

González, F., Sulon, J., Garbayo, J.M., Batista, M., Cabrera, F., Calero, P.O., Gracia, A. and Beckers, J.F. (2000a) Secretory profiles of pregnancy-associated glycoproteins at different stages of pregnancy in the goat, *Reprod. Domest. Anim. 35*, 65-68.

González, F., Sulon, J., Calero P.O. Batista M., Quesada, E., Garbayo, J.M., Gracia, A. and Beckers, J.F. (2000b) Pregnancy-associated glycoproteins (PAGs) detection in milk samples as an aid for pregnancy diagnosis in goats. *7th International Conference on Goats* (Tours), France.

Green, J.A., Xie, S., Newman, A., Szafranska, B., Roberts, R.M., Baker, C.B. and Mcdowell, K. (1994) Pregnancy associated glycoproteins of the horse. *Biol. Reprod. 50 (Suppl. 1)*, 152 (abstract).

Green, J.A., Xie, S. and Roberts, R.M. (1998a) Pepsin-related molecules excreted by trophoblast. *Reviews of Reproduction 3*, 62-69.

Green, J.A., Xie, S., Gan, X. and Roberts, R.M. (1998b) An aspartic proteinase expressed in the equine placenta: In James, M.N.G (ed.), *Aspartic proteinases*, Plenum press, New York, pp. 163-168.

Green J.A., Xie, S., Szafranska, B., Gan, X., Newman, A.G., McDowell, K. and Roberts, R.M. (1999) Identification of a new aspartic proteinase expressed by the outer chorionic cell layer of the equine placenta, *Biol. Reprod. 60*, 1069-1077.

Green, J.A., Xie, S., Quan, X., Bao, B., Gan, X., Mathialagan, N. and Roberts, R.M. (2000) Pregnancy-associated glycoproteins exhibit spatially and temporally distinct expression patterns during pregnancy. Biol/ Reprod. 62: 1624-1631.

Guilbault, L.A., Beckers, J.F., Lapierre, S., Zoli, A.P., Benitez-Ortiz, W. and Roy, G.L. (1991) Peripartum concentrations of placental protein hormones (hPL and bPAG) in holstein and hereford recipients carrying pubered holstein fetuses, *Theriogenology 35*, 208 (abstract).

Guillomot, M., Reinaud, P., La Bonnardiere, C. and Charpigny, G. (1998) Characterization of conceptus-produced goat interferon tau and analysis of its temporal and cellular distribution during early pregnancy, *J. Reprod. Fertil. 112*, 149-156.

Guruprasad, K., Blundell, T.L., Xie, S., Green, J.A., Szafranska, B., Nagel, R.J., McDowell, K., Baker, C.B. and Roberts, R.M. (1996) Comparative modeling and analysis of amino acid substitutions suggests that the family of pregnancy-associated glycoproteins includes both active and inactive aspartic proteinases, *Protein Engineer. 9*, 849-856.

Haigh, J.C., Cranfield, M. and Sasser, R.G. (1988) Estrus synchronization and pregnancy diagnosis in red deer, *J. Zoo. Anim. Med. 19*, 202-207.

Haigh, J.C., Gates, C., Ruder, A. and Sasser, R.G. (1991) Diagnosis of pregnancy in wood bison using a bovine assay for pregnancy-specific protein B, *Theriogenology 36*, 749-754.

Haigh, J.C., Dalton, W.J., Ruder, C.A. and Sasser, R.G. (1993) Diagnosis of pregnancy in moose using a bovine assay for PSPB, *Theriogenology 40*, 905-911.

Handwerger, S., Maurer, W., Barrett, J., Hurley, W. and Fellows, R.E. (1974) Evidence for homology between ovine and human placental lactogens, *Endocr. Res. Commun. 1*, 403-413.

Hansen, T.R., Imakawa, K., Polites, H.G., Marotti, K.R., Anthony, R.V. and Roberts, R.M. (1988) Interferon RNA of embryonic origin is expressed transiently during early pregnancy in the ewe, *J. Biol. Chem. 263*, 12801-12804.

Hansen, T.R., Austin, K.J., Perry, D.J., Pru, J.K., Teixeira, M.G. and Johnson, G.A. (1999) Mechanism of action of interferon-tau in the uterus during early pregnancy, *J. Reprod. Fertil. 54*, 329-339.

Hayden, T.J., Thomas, C.R and Forsyth, I.A. (1979) Effect of number of young born (litter size) on milk yield of goats: role for placental lactogen, *J. Dairy Sci. 62*, 53-57.

Helmer, S.D., Gross, T.S., Hansen, P.J. and Thatcher, W.W. (1988) Bovine conceptus secretory proteins (bCSP) and bovine trophoblast protein-1 (bTP-1), a component of bCSP, alter endometrial prostaglandin (PG) secretion and induce an intracellular inhibitor of PG synthesis in vitro, *Biol. Reprod. 38 (Suppl)*, 153 (abstract).

Helmer, S.D., Hansen, P.J. and Thatcher, W.W. (1988) Differential glycosylation of the components of the bovine trophoblast protein-1 complex, *Mol. Cell Endocrinol. 58*, 103-107.

Houston, B.D., Robbins, C.T., Ruder, C.A. and Sasser, R.G. (1986) Pregnancy detection in mountain goats by assay for PSPB, *J. Wild. Manage. 50*, 740-742.

Huang, F., Cockrell, D.C., Stephenson T.R., Noyes, J.H. and Sasser, R.G. (1999) Isolation, purification, and characterization of pregnancy-specific protein B from elk and moose placenta, *Biol. Reprod. 61*, 1056-1061.

Huang, F., Cockrell, D.C., Stephenson T.R., Noyes, J.H. and Sasser, R.G. (2000) A serum pregnancy test with a specific radioimmunoassay for moose and elk pregnancy-specific protein B, *J. Wild. Manag. 64*, 492-499.

Humblot, P. and Dalla Porta, M.A. (1984) Effect of conceptus removal and intrauterine administration of conceptus tissue on luteal function in the cow, *Reprod. Nutr. Dev. 24*, 529-541.

Humblot, P., Camous, S., Martal, J., Charlery, J., Jeanguyot, N., Thibier, M. and Sasser, R.G. (1988a) Diagnosis of pregnancy by radioimmunoassay of a pregnancy-specific protein in the plasma of dairy cows, *Theriogenology 30*, 257-268.

Humblot, P., Camous, S., Martal, J., Charlery, J., Jeanguyot, N., Thibier, M. and Sasser, R.G. (1988b) Pregnancy specific protein B, progesterone concentrations and embryonic mortality during early pregnancy in dairy cows. *J. Reprod. Fertil. 83*, 215-223.

Humblot, P., Montigny, G., Jeanguyot, N., Tetedoie, F., Payen, B., Thibier, M. and Sasser R. G. (1990) Pregnancy-specific protein B and progesterone concentrations in french alpine goats throughout gestation, *J. Reprod. Fert. 89*, 205-212.

Imakawa, K., Anthony, R.V., Kazemi, M., Marotti, K.R., Polites, H.G. and Roberts, R.M. (1987) Interferon-like sequence of ovine trophoblast protein secreted by embryonic trophectoderm, *Nature 330*, 377-379.
Imakawa, K., Hansen, T.R., Malathy, P.V., Anthony, R.V., Polites, H.G., Marotti, K.R. and Roberts, R.M. (1989) Molecular cloning and characterization of complementary deoxyribonucleic acids corresponding to bovine trophoblast protein-1: a comparison with ovine trophoblast protein-1 and bovine interferon-alpha II, *Mol. Endocrinol. 3*, 127-39.
James, M.N. and Sielecki, A.R. (1986) Molecular structure of an aspartic proteinase zymogen, porcine pepsinogen, at 1.8 A resolution, *Nature 319*, 33-38.
Kappes, S.M., Warren, W.C., Pratt, S.L., Liang, R. and Anthony, R.V. (1992) Quantification and cellular localization of ovine placental lactogen messenger ribonucleic acid expression during mid- and late gestation, *Endocrinology 131*, 2829-2838.
Kessler, M.A., Duello, T.M. and Schuler, L.A. (1991) Expression of prolactin-related hormones in the early bovine conceptus, and potential for paracrine effect on the endometrium, *Endocrinology 129*, 1885-1895.
Kirakofe, G.H., Wright, J.M., Schalles, R.R., Ruder, C.A., Paris, S. and Sasser, R.G. (1993) Pregnancy-specific protein B in serum of postpartum beef cows, *J. Anim. Sci. 71*, 2199-2205.
Klisch, K., Hecht, W., Pfarrer, C., Schuler, G., Hoffmann, B. and Leiser, R. (1999) DNA content and ploidy level of bovine placentomal trophoblast giant cells. *Placenta 20*, 451-458.
Lee, C.S., Wooding, F.B. and Brandon, M.R. (1986) Immunogold co-localization of ovine placental lactogen and the antigen recognized by the SBU-3 monoclonal antibody in sheep placental granules, *J. Reprod. Fertil. 78*, 653-662.
Leymarie, P. and Martal, J. (1993) The corpus luteum from cycle to gestation, in Thibault, M.C. (ed.), *Reproduction in mammals and man*, Ed. Lavasseur, Paris, pp. 423-433.
Lin, X.L., Wong, R.N. and Tang, J. (1989) Synthesis, purification, and active site mutagenesis of recombinant porcine pepsinogen, *J. Biol. Chem. 264*, 4482-4489
Lynch, K.A., Alexander, R.M. and Sasser, R.G. (1992) The cloning and expression of the bovine pregnancy specific protein B (bPSPB) gene, *Biol. Reprod. 46 (Suppl. 1)*, 73 (abstract).
Martal, J. and Cédart, L. (1993) Endocrine functions of the placenta, in Thibault, M.C. (ed.), *Reproduction in mammals and man*, Ed. Lavasseur, Paris, pp. 435-459.
Martal, J. and Djiane, J. (1975) Purification of a lactogenic hormone in sheep placenta, *Bioch. Biophys. Res. Commun. 65*, 770-778.
Martal, J., Lacroix, M.C., Lourdes, C., Saunier, M. and Winternberger-Torres, S. (1979) Trophoblastin, an antiluteolytic protein present in early pregnancy in sheep, *J. Reprod. Fert. 56*, 63-73.
Martal, J., Chene, N., Camous, S., Huynh, L., Lantier, F., Hermier, P., L'Haridon, R., Charpigny, G., Charlier, M. and Chaouat, G. (1997) Recent developments and potentialities for reducing embryo mortality in ruminants: the role of IFN-tau and other cytokines in early pregnancy, *Reprod. Fertil. Dev. 9*, 355-380.
Mialon, M.M., Camous, S., Renand, G., Martal, J. and Menissier, F. (1993) Peripheral concentrations of a 60-kDa pregnancy serum protein during gestation and after calving and in relationship to embryonic mortality in cattle, *Reprod. Nutr. Dev. 33*, 269-282.
Milosavljenic, L., Duello, T.M. and Schuler, L.A. (1989) In situ localization of two prolactin-related messenger ribonucleic acids to binucleate cells of bovine placentomes, *Endocrinology 125*, 883-889.
Moor, R.M. and Rowson, L.E. (1966) Local uterine mechanisms affecting luteal function in the sheep, *J. Reprod. Fertil. 11*, 307-310.
Morgan, G., Wooding, F.B., Beckers, J.F. and Friesen, H.G. (1989) An immunological cryo-ultrastructural study of a sequential appearance of proteins in placental binucleate cells in early pregnancy in the cow, *J. Reprod. Fertil. 86*, 745-752.
Morgan, G., Wooding, F.B. and Godkin, J.D. (1993) Localization of bovine trophoblast protein-1 in the cow blastocyst during implantation: an immunological cryoultrastructural study, *Placenta 14*, 641-649.
Morton, H. (1998) Early pregnancy factor: an extracellular chaperonin 10 homolog, *Immunol. Cell Biol. 76*, 483-496.
Morton, H., Clunie, G.J. and Shaw, F.D. (1979) A test for early pregnancy in sheep, *Res. Vet. Sci. 26*, 261-262.
Morton, H., Cavanagh, A.C., Athanasas-Platsis, S., Quinn, K.A. and Rolfe, B.E. (1992) Early pregnancy factor has immunosuppressive and growth factor properties, *Reprod. Fertil. Dev. 4*, 411-422.

Murthy, G.S., Schelllenberg, C. and Friesen, H.G. (1982) Purification and characterization of bovine placental lactogen, *Endocrinology 111*, 2117-2124.
Nagel, R.J., Xie, S. and Roberts, R.M. (1993) Aspartic proteinases as markers of trophoblast differentiation in sheep, *Biol. Reprod. 48 (Suppl.)*, 139 (abstract).
Northey, D.L. and French, L.R. (1980) Effect of embryo removal and intrauterine infusion of embryonic homogenates on the lifespan of the bovine corpus luteum, *J. Anim. Sci. 50*, 298-302.
Nugent, P.F. and Hayden, T.J. (1987) The purification and preliminary characterization of caprine placental lactogen, *J. Endocrinol. 115 (Suppl.)*, 5 (abstract).
Osborn, D.A., Beckers, J.F., Sulon, J., Gasset, J.W., Muller, L.I., Murphy, B.P., Miller, K.V. and Marchinton, R.L. (1996) Use of glycoprotein assays for pregnancy diagnosis in white-tailed deer, *J. Wildl. Manage. 60*, 388-393.
Parlow, A.F. (1961) Bio-assay of pituitary luteinizing hormone by depletion of ovarian ascorbic acid. In: Albert A. (eds.), *Human pituitary gonadotropins*, Ed. Charles Thomas, pp. 300-310.
Patel, O.V., Sulon, J., Beckers, J.F., Takahashi, T., Hirako, M., Sasaki, N. and Domeki, I. (1997) Plasma bovine pregnancy-associated glycoprotein concentrations throughout gestation in relationship to fetal number in the cow, *Eur. J. Endocrinol. 137*, 423-428.
Ranilla, M.J., Sulon, J., Carro, M.D., Mantcc6n, A.R. and Beckers, J.E. (1994) Plasmatic profiles of pregnancy-associated glycoprotein and progesterone levels during gestation in churra and merino sheep. *Theriogenology 42*, 537-545.
Ranilla, M.J., Sulon, J., Mantec6n, A.R., Beckers, J.E. and Carro, M.D. (1997a) Plasma pregnancy-associated glycoprotein and progesterone concentrations in pregnant assaf ewes carrying single and twin lambs. *Small Rum. Res. 24*, 125-131.
Ranilla, M.J., Gebbie, F.E., Mitchell, L.M., King, M.E., Skinner, J.G., Beckers, J.F. and Robinson, J.J. (1997b) Pregnancy-associated glycoprotein (PAG) in postpartum ewes. *Brit. Soc. Anim. Sci. 1*, 185.
Reddy, S. and Walkins, W.B. (1978) Immunofluorescent localization of ovine placental lactogen, *J. Reprod. Fert. 52*, 173-174.
Reimers, T.J., Sasser, R.G. and Ruder, C.A. (1985a) Production of pregnancy-specific protein by bovine binucleate trophoblastic cells. *Biol. Reprod. 32 (Suppl. 1)*, 55 (abstract).
Reimers, T.J., Ullmann, M.B. and Hansel, W. (1985b) Progesterone and prostanoid production by bovine binucleate trophoblastic cells, *Biol. Reprod. 33*, 1227-1236.
Roberts, R.M., Klemann, S.W., Leaman, D.W., Bixby, J.A., Cross, J.C., Farin, C.E., Imakawa, K. and Hansen, T.R. (1991) The polypeptides and genes for ovine and bovine trophoblast protein-1, *J. Reprod. Fert. 43 (Suppl.)*, 3-12.
Roberts, R.M., Xie, S., Nagel, R.J., Low, B., Green, J.A. and Beckers, J.F. (1995) Glycoproteins of the aspartyl proteinase gene family secreted by the developing placenta, in Takahashi, K. (ed.), *Aspartic proteinases: structure, function, biology, and biomedical implications*, Plenum Press, Tokyo, pp. 231-240.
Ropstad, E., Johansen, O., King, C., Dahl, E., Albon, S.D., Langvatn, R.L., Irvine, R.J., Halvorsen, O. and Sasser, G. (1999) Comparison of plasma progesterone, transrectal ultrasound and pregnancy specific proteins (PSPB) used for pregnancy diagnosis in reindeer, *Acta Vet. Scand. 40*, 151-162.
Rowell, J.E., Flood, P.F., Ruder, C.A. and Sasser, R.G. (1989) Pregnancy-associated protein in the plasma of captive musk-oxen, *J. Wildl. Manag. 53*, 899-901.
Ruder, C.A., Stellflug, J.N., Dahmen, J.J. and Sasser, R.G. (1988) Detection of pregnancy in sheep by radioimmunoassay of sera for pregnancy-specific protein B, *Theriogenology 29*, 905-912.
Russel, D.E., Gerhart, K.L., White, R.G. and Wetering, D.V.D. (1998) Detection of early pregnancy in caribou: evidence for embryonic mortality, *J. Wildl. Manag. 62*, 1066-1075.
Sardajana, I.K.W., Tainturier, D. and Djiane, J. (1988) Etude de l'hormone chorionique somatomammotrophique dans le plasma et le lactoserum au cours de la gestation et du postpartum chez la chèvre (application au diagnosic tardif de gestation), *Rev. Méd. Vét. 139*, 1045-1052.
Sasser, R.G., Ruder, C.A., Ivani, K.A., Butler, J.E. and Hamilton, W.C. (1986) Detection of pregnancy by radioimmunoassay of a novel pregnancy-specific protein in serum of cows and a profile of serum concentrations during gestation, *Biol. Reprod. 35*, 936-942.
Sasser, R.G., Crock, J. and Ruder, C.A. (1989) Characteristics of pregnancy-specific protein B in cattle, *J. Reprod. Fert. 37 (Suppl.)*, 109-113.
Schoknecht, P.A., Currie, M.A. and Bell, A.W. (1992) Kinetics of placental lactogen in mid- and late gestation ovine fetus, J. *Endocrin. 133*, 95-100.

Scott, P., Kessler, M.A. and Schuler, L.A. (1992) Molecular cloning of the bovine prolactin receptor and distribution of prolactin and growth hormone receptor transcription fetal and utero-placental tissues, *Mol. Cell. Endocrinol. 89*, 47-58.

Shimomura, K. and Bremel, R.D. (1988) Characterization of bovine placental lactogen as a glycoprotein with N-Linked and O-linked carbohydrate side chains, *Mol. Endocrinol. 2*, 845-853.

Shu-Xin, H. and Zhen-Qun, Z. (1993) A study of early pregnancy factor activity in the sera of patients with unexplained spontaneous abortion, *Am. J. Reprod. Immunol. 29*, 77-81.

Sousa, N.M., Garbayo, J.M., Figueiredo, J.R., Sulon, J., Gonçalves, P.B.D. and Beckers, J.F. (1999) Pregnancy-associated glycoprotein and progesterone profiles during pregnancy and postpartum in native goats from the north-east of Brazil, *Small Rum Res 32*, 137-147.

Staten, N.R., Byatt, J.C. and Krivi, G.G. (1993) Ligand-specific dimerization of the extracellular domain of the bovine growth hormone recptor, *J. Biol. Chel. 268*, 18467-18473.

Szafranska, B., Xie, S., Green, J.A. and Roberts, R.M. (1995) Porcine pregnancy-associated glycoproteins: new member of the aspartic proteinase gene family expressed in trophectoderm, *Biol. Reprod. 53*, 21-28.

Szenci, O., Taverne, M.A., Beckers, J.F., Sulon, J., Varga, J., Börzsönyi, L., Hanzen, Ch. and Schekk, G. (1998a) Evaluation of false ultrasonographic diagnoses in cows by measuring plasma levels of bovine pregnancy-associated glycoprotein 1, *Vet. Rec. 142*, 304-306.

Szenci, O., Beckers, J.F., Humblot, P., Sulon, J., Sasser, R.G., Taverne, M.A., Varga, J., Baltusen, R. and Schekk, G. (1998b) Comparison of ultrasonography, bovine pregnancy-specific protein B, and bovine pregnancy-associated glycoprotein 1 tests for pregnancy detection in dairy cows, *Theriogenology 50*, 77-88.

Szenci, O., Humblot, P., Beckers, J.F., Sasser, R.G., Sulon, J., Baltusen, R., Varga, J., Bajcsy C.A. and Taverne, M.A., (1999) Plasma profiles of progesterone and conceptus proteins in cows with spontaneous embryonic/foetal mortality as diagnosed by ultrasonography, *Vet. J. 159*, 287-290.

Takagi, M., Nishimura, K., Oguri, N., Ohnuma, K., Ito, K., Takahashi, J., Yasuda, Y., Miyazawa, K. and Sato, K. (1998) Measurement of early pregnancy factor activity for monitoring the viability of the equine embryo, *Theriogenology 50*, 255-262.

Telleria, C.M., Zhong, L., Deb, S., Srivastava, R.K., Park, K.S., Sugino, N., Park-Sarge, O.K. and Gibori, G. (1998) Differential expression of the estrogen receptors alpha and beta in the rat corpus luteum of pregnancy: regulation by prolactin and placental lactogens, *Endocrinology 139*, 2432-2442.

Verstegen, J., Fellmann, D. and Beckers, J.F. (1985) Immunodetection of bovine chorionic somatomammotrophin (bCS), *Acta Endocrinol. (Copenh) 109*, 403-410.

Wango, E.O., Wooding, F.B. and Heap, R.B. (1990a) The role of trophoblastic binucleate cells in implantation in the goat: a morphological study, J. Anat. 171, 241-57.

Wango, E.O., Wooding, F.B. and Heap, R.B. (1990b) The role of trophoblast binucleate cells in implantation in the goat: a quantitative study, *Placenta 11*, 381-394.

Wango, E.O., Heap, R.B. and Wooding, F.B. (1991) Progesterone and 5 beta-pregnanediol production by isolated fetal placental binucleate cells from sheep and goats, *J. Endocrinol. 129*, 283-289.

Wango, E.O., Heap, R.B. and Wooding, F.B. (1992) Regulation of steroid synthesis and metabolism in isolated binucleate cells of the placenta in sheep and goats, *J. Reprod. Fertil. 94*, 203-211.

Warren, W.C., Keisler, D.H. and Anthony R.V. (1990) Synthesis and secretion of ovine placental lactogen and its biochemical properties, *Dom. An. Endocrin. 7*, 331-342.

Weeth, H.J. and Herman, H.A. (1952) A histological and histochemical study of the bovine oviducts, uterus and placenta, *Res. Bull. Univ. Miss. Agric. Exp. 501*.

Willard, S.T., Sasser, R.G., Bringans, M., Hughes, D.M., Welsh, T.H., Jaques, J.T. and Randel, R.D. (1994a) Methods for pregnancy detection in sika deer (*Cervus nippon*), *J. Anim. Sci. 72 (Suppl. 1)*, 1309 (abstract).

Willard, S.T., Sasser, R.G., Gillespie, J.C., Jaques, J.T., Welsh, T.H.Jr. and Randel, R.D. (1994b) Methods for pregnancy determination and the effects of body condition on pregnancy status in rocky mountain elk (*Cervus elephus nelsoni*), *Theriogenology 42*, 1095-1102.

Willard, S.T., Sasser, R.G., Gillespie, J.C., Jaques, J.T., Welsh, T.H. and Randel, R.D. (1994c) Early pregnancy detection and embryonic mortality in fallow deer (*Dama dama*), *J. Anim. Sci. 72 (Suppl. 2)*, 67 (abstract).

Willard, J.M., White, D.R., Wesson, C.A.R., Stellflug, J. and Sasser, R.G. (1995) Detection of fetal twins in sheep using a radioimmunoassay for pregnancy-specific protein B, *J Anim Sci 73*, 960-966.

Willard, S.T., Hughes, D.M.Jr., Bringans, M., Sasser, R.G., White, D.R., Jaques, J.T., Godfrey, R.W., Welsh, T.H.Jr. and Randel, R.D. (1996) Artificial insemination, hybridization and pregnancy detection in sika deer (*Cervus nippon*), *Theriogenology 46*, 779-789.

Willard, S.T., Sasser, R.G., Jaques, J.T., White, D.R., Neuendorff, D.A. and Randel, R.D. (1998) Early pregnancy detection and the hormonal characterization of embryonic-fetal mortality in fallow deer (Dama dama). *Theriogenology 49*, 861-869.

Willard, S.T., Petty, S.J., Sasser, R.G., White, D.R., and Randel, R.D. (1999) Pregnancy detection and the effects of age, body weight, and previous reproductive performance on pregnancy status and weaning rates of farmed fallow deer (*Dama dama*), *J. Anim. Sci. 77*, 32-38.

Willey, K.P. and Leidenberger, F. (1989) Functionally distinct agonist and receptor-binding regions in human chorionic gonadotropin, *J. Biol. Chem. 33*, 19716-19729.

Wimsatt, W.A. (1951) Observations on the morphogenesis, cytochemistry and significance of the binucleate giant cells of the placenta of ruminants, *Am. J. Anat. 89*, 233-240.

Wood, A.K., Short, R.E., Darling, A.E., Dusek, G.L., Sasser, R.G. and Ruder, C.A. (1986) Serum assays for detecting pregnancy in mule and white-tailed deer, *J. Wildl. Manag. 50*, 684-687.

Wooding, F.B. (1983) Frequency and localization of binucleate cells in the placentomes of ruminants, Placenta 4, 527-539.

Wooding, F.B. (1984) Role of binucleate cells in fetomaternal cell fusion at implantation in the sheep, *Am. J. Anat. 170*, 233-250.

Wooding, F.B. (1992) Current topic: the synepitheliochorial placenta of ruminants: binucleate cell fusions and hormone production, *Placenta 13*, 101-113.

Wooding, F.B. and Beckers, J.F. (1987) Trinucleate cells and the ultrastructural localisation of bovine placental lactogen, *Cell Tissue Res. 247*, 667-673.

Wooding, F.B. and Wathes, D.C. (1980) Binucleate cell migration in the bovine placentome, *J. Reprod. Fertil. 59*, 425-430.

Wooding, F.B., Flint, A.P., Heap, R.B. and Hobbs, T. (1981) Autoradiographic evidence for migration and fusion of cells in the sheep placenta: resolution of a problem in placental classification, *Cell Biol. Int. Rep. 5*, 821-827.

Wooding, F.B., Flint, A.P., Heap, R.B., Morgan, G., Buttle, H.L. and Young, I.R. (1986) Control of binucleate cell migration in the placenta of sheep and goats, *J. Reprod. Fertil. 76*, 499-512.

Wooding, F.B., Morgan, G., Forsyth, I.A., Butcher, G., Hutchings, A., Billingsley, S.A. and Gluckman, P.D. (1992) Light and electron microscopic studies of cellular localization of oPL with monoclonal and polyclonal antibodies, *J. Histochem. Cytochem. 40*, 1001-1009.

Wooding, F.B., Morgan, G., Monaghan, S., Hamon, M. and Heap, R.B. (1996) Functional specialization in the ruminant placenta: evidence for two populations of fetal binucleate cells of different selective synthetic capacity, *Placenta 17*, 75-86.

Xie, S., Low, R.C., Nagel, R.J., Kramer, K.K., Anthony, R.V., Zoli, A.P., Beckers, J.F. and Roberts, R.M. (1991) Identification of the major pregnancy-specific antigens of cattle and sheep as inactive members of the aspartic proteinase family, *Proc. Natl. Acad. Sci. USA 88*, 10247-10251.

Xie, S., Low, B.G., Nagel, R.J., Beckers, J.F. and Roberts, R.M. (1994) A novel glycoprotein of the aspartic proteinase gene family expressed in bovine placental trophectoderm, *Biol. Reprod. 51*, 1145-1153.

Xie, S., Green, J.A., Beckers, J.F. and Roberts, R.M. (1995) The gene encoding bovine pregnancy-associated glycoprotein-l, an inactivc member of the aspartic proteinase family, *Gene 159*, 193-197.

Xie, S., Nagel, R.J., Green, J.A, Beckers, J.F. and Roberts, R.M. (1996) Trophoblast-specific processing and phosphorylation of pregnancy-associated glycoprotein-1 in day 15 to 25 sheep placenta, *Biol. Reprod. 54*, 122-129.

Xie, S., Green, J.A, Bao, B., Beckers, J.F., Valdez, K.E., Hakami, L. and Roberts, R.M. (1997a) Multiple pregnancy-associated glycoproteins are secreted by day 100 ovine placental tissue, *Biol. Reprod. 57*, 1384-1393.

Xie, S., Green, J.A, Bao, B., Bixby, J.B., Szafranska, B., DeMartini, J.C., Hecht, S. and Roberts, R.M. (1997b) The diversity and evolutionary relationships of the pregnancy-associated glycoproteins, an aspartic proteinase subfamily consisting of many trophoblast-expressed genes, *Proc. Natl. Acad. Sci. USA 94*, 12809-12816.

Zarrouk, A., Engeland, I., Sulon, J. and Beckers, J.F. (1999a) Determination of pregnancy-associated glycoprotein concentrations in goats (*Capra hircus*) with unsuccessful pregnancies: a retrospective study, Theriogenology 51, 1321-1231.

Zarrouk, A., Engeland, I.V., Sulon, J. and Beckers, J.F. (1999b) Pregnancy-associated glycoprotein levels in pregnant goats inoculated with *Toxoplasma gondii* or *Listeria monocytogenes*: a retrospective study, *Theriogenology 52*, 1095-1104.

Zoli, A.P., Beckers, J.F. and Ectors, F. (1990a) Isolation of a ovine pregnancy specific protein, *Theriogenology 33*, 366 (abstract).

Zoli, A.P., Beckers, J.F. and Ectors, F. (1990b) Evidence of a protein immunologically related to pregnant specific proteins in ruminant gonads, *3th International Ruminant Reproduction Symposium* (Nice), 304.

Zoli, A.P., Ectors, F. and Beckers, J.F. (1990c) Ruminant gonads as accessory sources of pregnant specific protein ?, *72nd Annual Meeting of the Endocrine Society* (Atlanta), 118.

Zoli, A.P., Beckers, J.F., Wouters-Ballman, P., Closset, J., Falmagne, P. and Ectors, F. (1991) Purification and characterization of a bovine pregnancy-associated glycoprotein, *Biol. Reprod. 45*, 1-10.

Zoli, A.P., Demez, P., Beckers, J.F., Reznik, M. and Beckers, A. (1992a) Light and electron microscopic immunolocalization of bovine pregnancy-associated glycoprotein in the bovine placentome, *Biol. Reprod. 46*, 623-629.

Zoli, A.P., Guilbault, L.A., Delahaut, P., Benitez-Ortiz, W. and Beckers, J.F. (1992b) Radioimmunoassay of a bovine pregnancy-associated glycoprotein in serum: its application in pregnancy diagnosis, *Biol. Reprod. 46*, 83-92.

Zoli, A.P., Beckers, J.F., Ectors, F. (1995) Isolement et caractérisation partielle d'une glycoprotéine associée à la gestation chez la brebis, *Ann. Méd. Vét. 139*, 177-184.

INDUCTION OF SUPEROVULATION IN DOMESTIC RUMINANTS

GONZÁLEZ F., CALERO P. AND BECKERS J.F.*
*Reproduction and Obstetrics, Faculty of Veterinary, Las Palmas de Gran Canaria, Spain and *Physiology of Reproduction, Faculty of Veterinary Medicine, Liège, Belgium*

Abstract

The embryo transfer (ET) is a technique of increasing both commercial production and genetic potential in livestock. The ET provides a rapid rate of improvement of the genetic quality offering access to the highest-quality genetics at a lower cost than purchasing a live animal. Techniques inducing superovulation are used in conjunction with ET to expedite the propagation of animals with genetic merit for desirable traits. The considerable variation in superovulatory response as well as the fact that the percentage of the superovulated donors do not produce transferable embryos, are the major limiting factors to the routine application of ET techniques. This review discusses the most commonly used protocols for induction of superovulation in domestic ruminants and focuses on the current understanding of factors affecting the superovulatory response and on the new approaches to reduce the variability in the superovulatory response.

1. Introduction

The main purpose of embryo transfer (ET) in domestic ruminants is to spread the genetic quality of livestock production for desirable traits. Although the basic procedures employed in ET are now well established, there is considerable scope for improvement of ET technology in various areas.

Superovulation plays an important role in the ET programs. It aims at inducing a high number of ovulations and a high yield of embryos of good quality. One of the more problematic aspects of the ET procedure is the variable response by the donor to superovulatory treatment and the percentage of embryos available for transfer from each donor. Despite much attention, little progress has been made during the last years,

R. Renaville and A. Burny (eds.), Biotechnology in Animal Husbandry, 209–223.

and the injection of exogenous gonadotrophins is still the only practical means of producing multiple ovulations from donor (Armstrong, 1993).

2. Superovulation in Cattle

The principle of superovulation in cattle is basically simple: to induce more ovulations than normal rate by giving a gonadotrophin stimulus (at critical moments of follicular development), followed by control of luteolysis, synchronous ovulation, high fertilisation and early embryonic development rates. The majority of donor cows will give the best superovulatory response if superovulation treatment is established between days 8 and 14 of the cycle. Therefore, the donor cow must always be examined prior to treatment to detect any abnormality and to establish the presence of a normal, functional corpus luteum (Mapletoft, 1986).

Although used widely in standardised protocols, superovulation is still not a well-controlled technique. The superovulation treatments usually yield an average of 6 transferable embryos, but this technique continues to be associated with variable and unpredictable responses in ovulation rates and recovery of transferable embryos (Armstrong, 1993). Normally, no transferable embryos are recovered from about 20 percent of donors and only 1 to 3 transferable embryos are obtained from another 20 percent. An ideal response of 5 to 12 good quality embryos is obtained from about one-third of the donors (Seidel & Seidel, 1991; Greve *et al.,* 1995; Callesen & Greve, 1996).

The preparations to induce superovulation include: equine chorionic gonadotrophin (eCG) derived from the serum of pregnant mares (usually called pregnant mare's serum gonadotrophin, PMSG); extracts of domestic animal pituitaries, particularly those of the pig, of various degrees of purity and FSH to LH ratios; recombinant FSH; and gonadotrophins of pituitary origin extracted from human post-menopausal urine (human menopausal gonadotrophin, hMG) (McGowan *et al.*, 1985).

2.1. PMSG

The pregnant mare's serum gonadotrophin (PMSG) is a glycoprotein that produces both FSH and LH biological effects. Due to its high content of carbohydrate side chains and sialic acid, PMSG has a very long half-life of about 5 days (Schams *et al.,* 1978). Traditionally, a single dose of 1500 to 3000 IU of PMSG during the mid-luteal phase of the estrous cycle has been used to superovulate cows. A luteolytic dose of prostaglandin $F_{2\alpha}$ or an analogue is administered 2 to 3 days later. The donor is expected to show heat signs 2 days after prostaglandin injection. The result is an interval of about 4 days between starting PMSG treatment and the onset of oestrus.

The advantage of using PMSG for superovulation is its availability in large quantities for a low cost and that it can be used in countries where the import of certain porcine products is banned. PMSG can be also administered, as a single dose compared with the multiple injections normally required when using pituitary preparations (Alfuraiji *et al.,* 1993).

Both the long half-life of PMSG and its LH activity might produce adverse effects on superovulatory response, particularly on the quality of embryos. The long half-life of PMSG often induces extrafollicular growth during the first follicular wave post-ovulation, with a resultant increased estradiol secretion that persists during the post-ovulatory period. This may have a deleterious effect on early embryonic development. To minimise the adverse effect of PMSG on fertilisation and embryonic development, the injection of an antiserum against PMSG has been proposed (Bouters *et al.,* 1983). The PMSG-antiPMSG treatment resulted in better superovulatory responses, including higher ovulation rate, decreased number of unruptured follicles and decreased number of cysts. Recent data suggest that the time of administration of the PMSG antibodies in relation to the time of occurrence of the LH surge is of crucial importance (Vos *et al.,* 1994). Neutralisation of PMSG at any time before the LH surge, worthy interferes with the normal stimulation of follicles. The administration of the anti-PMSG serum shortly (6 to 8 hours) after the preovulatory LH surge, decreases the excessive follicular development subsequent to PMSG injection, and therefore results in an increase in the ovulation rate and the number of normal embryos recovered (Dieleman *et al.,* 1993).

2.2. FSH

There is evidence supporting that a higher and more consistent response can be obtained with FSH preparations than with PMSG. The most widely used procedures consists of administrating two daily injections of FSH at approximately 12-hours intervals over a 4-day period, more frequently in decreasing doses. The half-life of FSH is believed to be approximately 5 hours (Demoustier *et al.,* 1988), which justifies the multiple injection protocols. Over the last years, the most common protocol for superovulation with FSH was 5, 5, 4, 4, 3, 3, 2 and 2 mg with prostaglandin $F_{2\alpha}$ or an analogue. The latter is given simultaneously to the 5th or 6th injection of FSH to induce the lysis of the donor's corpus luteum (Mapletoft, 1986).

One of the main factors influencing the effectiveness of superovulation is the quality of the gonadotrophin preparation used. Especially the proportion of FSH/LH plays an important role, since it controls the growth of the ovarian follicles. The activity ratio of LH to FSH preparation used affects the response, with the higher LH content of the gonadotrophin preparation generally resulting in a lower superovulatory response (Kelly *et al.,* 1997). According to Murphy *et al.,* (1984), the LH must be present in preparations used for superovulation with a FHS:LH ratio of approximately 5:1.

The recombinant bovine FSH (rbFSH) has been shown to posses high biological activity. Wilson *et al.,* (1993) concluded that the embryo production with rbFSH appears to be comparable with data in the literature for other superovulatory compounds, but embryo quality seems to be increased using rbFSH when compared with data from superovulation with pituitary FSH.

Although reducing the number of injections of FSH results in decreased superovulatory responses, superstimulation with a single subcutaneous injection of FSH has been reported with encouraging results.Probably the FSH dissolved in polyvinylpyrrolidone is capable of achieving a similar profile to that obtained with conventional multiple-injection procedure (Yamamoto *et al.,* 1993; Takedomi *et al,* 1995; Satoh *et al.,* 1996).

2.3. OTHER HORMONES

The availability of recombinant bovine somatotrophin (rbST) led to its use in combination with FSH treatment, being a co-gonadotrophin. Herrler *et al.,* (1994) found that the co-treatment with rbST enhances the superovulatory response and embryo yield, when Gray *et al.* (1993) did not confirm these data.

The hMG is a protein purified from urine collected from menopausal women; at presents both FSH and LH activities. Sugano *et al.* (1995) showed that 600 IU of hMG administered twice daily over a 3-day period produced a recovery rate of high quality embryos higher than that obtained following treatment with 20 mg of FSH. More recently, it was described that superovulation can be adequately induced using only one injection of 450 to 600 IU of HMG in polyvinylpyrrolidone (Sugano and Shinogi, 1999)

Treatment with a horse anterior pituitary extract (HAP) was found to be an acceptable alternative to FSH, but there is no reduction in the variability of the superovulatory response (Staigmiller *et al.,* 1992).

2.4. FACTORS AFFECTING SUPEROVULATORY RESPONSE

2.4.1. Follicular Status

Evidence exists that the large variation in superovulatory response may be due in part to the differences in the developmental stage of follicles present in the ovary at the beginning of treatment (Monniaux *et al.,* 1983; Bungartz and Niemann, 1994). Consequently, considerable attention has been given over the last several years to controlling follicular development so that it can be appropriately synchronised with ovarian superstimulation protocol (Bo *et al.,* 1993). According to Lussier *et al.* (1995), the presence of a dominant follicle when gonadotrophin treatment is initiated reduces the superovulatory response. Therefore, the time at which the superovulatory treatment provides the best results, is at days 9 to 13 of the estrous cycle, when the dominant follicle has reached its maximum size and the next dominant follicle has appeared (Gordon, 1996).

Likewise, the presence of a large follicle does not indicate functional dominance and it may or not influence the ovarian response to superstimulation. Furthermore, any effect of a dominant follicle may be influenced by its relation with other factors yet undefined (Staigmiller *et al.,* 1995). Anyway, it seems clear that the synchronization of the follicular wave development may be useful in improving the effectiveness of superovulatory treatments. This synchronisation can be reached by GnRH injection (Kohram *et al.,* 1996) or by combination of progestagen and estradiol treatment (Bo *et al.*, 1995; Duffy *et al.,* 1995).

2.4.2. Age and Breed of Donors

According to Hasler (1992), age was not considered to be a determinant factor in the superovulatory response nor in the viability of the embryos produced in cattle. However, Lerner *et al.* (1986) indicated that the maximal superovulatory response occurs at 5-6 years, followed by a decrease in superovulatory response related to a reduction in the number of follicles sensitive to gonadotrophin stimulus in older cattle. The cause of minor superovulatory response in aged cows could also be related to different reproductive disorders (Desaulniers *et al.,* 1995). In relation to differences among breeds, animals with a higher natural ovulation rate normally show a greater superovulatory response (Synder, 1986). According to Breuel *et al.* (1991), there are cattle breeds such as Simmental that might be more sensitive to gonadotrophin stimulation than other breeds such as Angus, Charolais and Hereford.

2.4.3. Extrinsic Factors

In the matter of seasonal effects on the superovulatory response, evidence has been conflicting. While several authors found that season had a significant effect on ovulation rate and embryo yield (Hasler *et al.,* 1983; Agarwal *et al.,* 1993), others did not find a consistent effect of season on ovarian response (Massey and Oden, 1984; Shea *et al.,* 1984). The existence of many factors involved in the superovulatory response and the variation of the weather pattern among years and regions, makes difficult an accurate evaluation of the effect of season on the superovulatory response. However, it is well known that the stress of cows during a superovulation programme results in a significant decrease in ovulation rate and early embryo survival (Dobson and Smith, 1995; Freytag *et al.,* 1995). Consequently, the thermal stress associated with elevated ambient temperature has been associated with the recovery of a higher number of abnormal and retarded embryos as compared to the number recovered from animals maintained in a thermoneutral environment.

Inadequate nutrition impairs reproductive performance in cattle, and its adverse effects may be due to hormonal disturbance (Schillo, 1992). Normally, selected donors will have an adequate level of feeding, but the possible nutritional influences on superovulation are poorly understood. Several reports did not find a significant relationship between feed regime and the number of embryos recovered after superovulatory treatment (Delacharlerie *et al.,* 1995; McEvoy *et al.,* 1996). Similarly, a negative effect of excessive subcutaneous fat depots has been noted in donors (Bielansky and Yadav, 1990), and an excessive feeding has been associated with a reduction in the yield and quality of embryos (Blanchard *et al.,* 1990). A recent study indicates that excessive concentrate intake reduced the superovulatory response and had a negative influence on the quality of embryos following superovulation, probably through changes in the number of available follicles or the sensitivity of follicles to FSH (Yaakub *et al.,* 1999).

3. Superovulation in Sheep and Goats

Although ET has been used more frequently in sheep than in goats, the procedures that have been employed are essentially similar. The induction of superovulation follows much the same lines as those employed in cattle. The donor female goats receive a gonadotrophic preparation either near the end of the luteal phase of the cycle or around two days before the end of a progestagen treatment which is employed to control oestrus.

The two most widely used gonadotrophin preparations in current use for superovulation are pregnant mare serum gonadotrophin (PMSG) and follicle stimulating hormone (FSH), but other hormones such as horse anterior pituitary (HAP) extracts and human menopausal gonadotrophin (HMG) have been also used for superovulation treatments.

3.1. PMSG

The PMSG molecule is able to induce follicle growth, oestrogen production, ovulation, luteinization and progesterone synthesis *(*Bindon and Piper, 1982*)*. Greater effectiveness of PMSG in terms of FSH activity is due to its longer biological half-life, which leads to require only one application. PMSG is administered as a single subcutaneous or intramuscular injection given 1 or 2 days prior the last synchronisation treatment at a dose of 750 to 2000 IU *(*Armstrong *et al.,* 1983*)*.

Treatments based on PMSG administration provide a high superovulatory response but the number of good quality embryos obtained following superovulation is too low to be acceptable for commercial embryo transfer (Pendleton *et al.,* 1992; Kiessling *et al.,* 1986). Several factors could explain the lack in the embryo's quality. Results obtained by Moor *et al.* (1985) in sheep and Kumar *et al.* (1990) in goats confirm that PMSG causes excessive follicular steroid secretion (predominantly estrogens) which can alter the endogenous endocrine environment, thereby disrupting sperm and gamete transport, oocyte maturation and early preimplantation embryo development. Moreover, the ovarian stimulation with PMSG in sheep ad goats decreases when females are repeatedly treated, and this is related to the presence of anti-PMSG antibodies (Roy *et al.,* 1999). Despite the now well-recognised failures in using PMSG in superovulation, reports continue on its use in the goat, probably for reasons related to cost and availability.

The adverse effects of PMSG (prolonged action, which may result in a second postovulatory wave of follicle growth and high levels of oestrogen production by follicles), can be minimised if oestrogen concentrations are reduced by neutralising PMSG after initial follicle stimulation. This can be achieved as in cattle by injection of anti-PMSG, to improve ovarian response and the yield of transferable embryos (Jabbour and Evans, 1991; Martemucci *et al.,* 1995).

3.2. FSH

The FSH is the gonadotrophin most frequently used, providing the best results in small ruminants. Both in sheep and in goat, FSH treatment results in high ovulation rate and better quality embryos per donor than treatment with PMSG.
Currently, FSH is administered at 12-hours intervals in decreased or constant doses for 3-4 days starting two days before sponge withdrawal or implant removal, and around the time at which prostaglandin $F_{2\alpha}$ is administered. The dose of total FSH administered varies from 16 to 20 mg Armour depending on type of FSH preparation and genetic features of the sheep and goats donors (Baril *et al.*, 1993).

Commercial pituitary FSH preparations became available for use in ET programmes. In the last years it was demonstrated that porcine FSH could exhibit a wide range in the ratio of FSH to LH activity and that a high LH content adversely affected the ovarian response, fertilisation rate and embryo quality in sheep and cattle (Chupin *et al.,* 1987). Recently, highly purified ovine and porcine FSH were made available. However, it appears to be clear justifications for administering an ovulating hormone as part of a superovulation regimen as this increases the ovulation rate and the number of embryos. Several experiments carried out with sheep during the non-breeding season suggested that a decreasing ratio FSH to LH during the last injections seems to improve superovulatory response (D'Alessandro *et al.,* 1996; Cognie, 1999). In goats, Nowshari *et al.* (1995) suggested that supplementation of FSH with approximately 40 percent of LH seems to be close to the optimum in ovulation rate and number of transferable embryos.

It was attempted to reduce multiple FSH injections to a single administration of the gonadotrophin being dissolved in polyvinylpyrrolidone (PVP), a long acting vehicle, to save time and labour with the FSH (Dattena *et al.,* 1994) or by a single normal injection of FSH (Meinecke-Tillmann *et al.,* 1993; Peebles and Kidd, 1994). A similar superovulatory response was found with these alternatives when compared to that with multiple injections of FSH.

Many comparisons have been made between PMSG and FSH in the superovulatory response in sheep and goat. Several authors observed that mean ovulation rate and number of transferable embryo were slightly higher in FSH than PMSG treated females. Furthermore, there is a higher percentage of large follicles which failed to ovulate and incidence of premature luteal regression in the PMSG-treated than in FSH-treated goats (Mahmood *et al.,* 1991; Pendlenton *et al.,* 1992; Rosnina *et al.,* 1992).

3.3. FSH PLUS PMSG

An alternative approach to induce superovulation is the administration of a single dose of FSH together with a low dose of PMSG. Thus, the use of PMSG associated with FSH can allow reduction in dose and number of injection of FSH (Maxwell and Wilson, 1989). The administration of a moderate dose of PMSG (400-800 UI) did markedly not increase the oestrogen production, but produced a prolonged gonadotrophic ovarian stimulation (Baril *et al.,* 1993).

In sheep, it was reported that a higher proportion of donors showed a higher superovulatory response when PMSG was used in conjunction with porcine FSH (Jabbour and Evans, 1991; Ryan *et al.* 1991). In goats, the protocols based on a single injection of FSH combined with PMSG may give results similar to protocols based on multiple doses of FSH (Batt *et al.,* 1993). On the contrary, others did not find benefit in a combined gonadotrophin treatment. Cseh and Seregi (1993) compared the efficiency of PMSG alone and PMSG plus FSH and did not found significant differences, either in ovulation rate or in transferable quality embryos. Similarly, Boland *et al.* (1993) did not record advantage in using a combined PMSG plus FSH treatment when compared with FSH alone.

Finally, Peebles and Kidd (1994) induced the superovulatory response in Cashmere goats by a single injection of FSH with or without PMSG: not all the goats showed an ovarian stimulatory effect, the authors found no difference between these treatments.

3.4. OTHER HORMONES

Horse anterior pituitary extract (HAP) has been used for ovarian stimulation in sheep and goats, giving similar results to those obtained with conventional protocols. In Angora goats HAP was as effective as PMSG in inducing superovulation, when administered as three subcutaneous injections on consecutive days starting one day before ending the synchronising treatment. In goats, the rates of recovery of embryos and of fertilisation following PMSG treatment were lower than with HAP (Moore and Eppleston, 1979).

Human menopausal gonadotrophin (hMG) was demonstrated to provide a superovulatory response as well as a rate of transferable embryos comparable to FSH (Schiewe *et al.,* 1990; Baldasarre *et al.,* 1992). Even, it has been reported that hMG induced a significantly higher ovulation rate as well as better embryo quality than did FSH (Stefani *et al.,* 1991).

In order to improve the synchronisation of the ovulation of the follicles stimulated by treatment with FSH or PMSG, several workers administered GnRH after progestagen sponge removal. Walker *et al.* (1989) noted in ewes that use of GnRH in embryo collection programs appears to be justified and is likely to improve embryo yields owing to improved rates of fertilisation. The timing of GnRH 24 hours after progestagen treatment was the preferred time in ewes treated with PMSG while 36 hours was preferred in ewes treated with FSH. Krisher *et al.* (1994) found a significantly higher ovulation rate in goats treated with GnRH administered 24 hours after implant removal than in untreated goats. Akinlosotu and Wilder (1993) observed in non cyclic goats treated with the FSH and GnRH treatment regimen, an increased follicular development, ovulation rate and higher blood progesterone levels. Moreover, GnRH treatment improves embryo quality. Parallel, some authors have established evidence of increased embryo production in superovulated ewes pre-treated for two weeks with a GnRH agonist (50 μg per day, Buserelin). This pretreatment reduces the ovarian activity before the initiation of gonadotrophin treatment (Briois *et al.,* 1992).

3.5. FACTORS AFFECTING SUPEROVULATORY RESPONSE

3.5.1. Premature Regression of CL

One of the limitations of these superovulatory treatments is the early regression of corpora lutea in 10-35 percent of the treated females, about 6-8 days after estrus. The main causes of this premature luteal regression still remains unknown (Chemineau *et al.*, 1999). This phenomenon affects a varying number of females, from which no embryos or non-viable embryos showing clear degenerative features are collected. Embryo recovery rates are often reduced, and this is thought to be associated with abnormal tubal transport of embryos (Pintado *et al.*, 1996).

It has been shown how endogenous secretion of prostaglandins is involved in premature regression of CL, but other factors such as high progesterone and estradiol levels at the onset of estrus might also contribute (Ishwar and Memon, 1996). Low body condition score, could be one of the reasons of such luteal regression (Chemineau *et al.*, 1999).

In sheep, it has been suggested that exogenous prostaglandin administration during superovulation treatments could also be a cause of this disorder (Schiewe *et al.*, 1990). However, Pintado *et al.* (1996) concluded that in the superovulated Murciano goats, the administration of prostaglandin $F_{2\alpha}$ did not increase the number of animals showing premature luteal regression.

3.5.2. Age and Breed of Donor

Some workers have examined the influence of age in donor sheep and goats. Dingwall *et al.* (1993) reported that embryo quality and survival were lower in the younger sheep. Wolf and Mylne (1994) found no differences in ovulations and embryo recovery for aged Texel ewes and yearling sheep, but embryo quality and survival tended to be lower in the younger donors. Mahmood *et al.* (1991) demonstrated that the superovulatory response (number of corpora lutea) and the embryo recovery rate was better in the higher age groups (4 to 6 years) than younger goats (1.5 to 3 years).

It is known that some breeds are more responsive to gonadotrophins than others. Differences among sheep breed are reported, with very higher rate superovulatory response in Romanov-Prealpes, Merino and Ile de France ewes than in Prealpes and Romney Marsh ewes (Armstrong and Evans, 1983; Torres and Cognie, 1984; Cognie *et al.*, 1986). In 1987, Nuti *et al.* superovulating Alpine and Nubian donor goats showed that the percentage of females responding to superovulatory treatment was higher in Nubian than in Alpine goats. On the contrary, Kiessling *et al.* (1986) found no breed differences for FSH and PMSG superovulated Saanen, Alpine, Nubian, and La Mancha goats.

3.5.3. Season and Nutrition

López *et al.* (1990), working in Spain with Manchega ewes recorded that a dose of 16 mg FSH in decreasing doses yielded a mean of 4.2 viable embryos, regardless of time of year when the ewes were treated.

In a study with five breeds of sheep that were either superovulated in the breeding season or during the anoestrous, ovulation rate was found to be significantly higher in the non-breeding than in the breeding season (Greaney *et al.,* 1991).

Rosnina *et al.* (1992) working in Malaysia reported that a tropical breed could be superovulated throughout the year with a decreased response during the dry months. However, this seasonality seems to be more related with feeding level and body condition; those factors might have contributed to the better superovulatory response during the rainy months of the year.

In goats, Baril and Vallet (1990) used Alpine goats synchronised with vaginal sponges of FGA and superovulated with porcine FSH during and out of the breeding season. They found a similar ovulation rate. However, in the 20 % of the goats treated in anoestrus all the CL were regressing. Senn and Richardson (1992) observed that a significantly higher number of viable embryos could be obtained from Nubian does when treated for estrus synchronisation and superovulation early in the breeding season than late in the season.

A study carried out by McEvoy *et al.* (1995) showed that the flushing of the superovulated ewe before mating did not increase ovulation rate; on the contrary, by adversely affecting preovulatory progesterone levels, high plane feeding potentially incurs serious failures in terms of embryos development and survival. Despite controversial effects were reported, it appeared that the body condition score has a significant effect on ovulatory response and embryo yield (Boland *et al.,* 1993).

Jabbour *et al.* (1991) demonstrated a beneficial effect of supplementary lupin grain feeding in reducing the incidence of premature luteal regression in superovulated Merino ewes. Therefore, a poor feeding regime during superovulation can lead to the premature regression of corpora lutea.

4. Conclusions

In embryo transfer, success depends on knowledge, practice and constant attention to detail. By the way to optimize and improve the general reproductive physiology and management, researchers may bring one hope to reduce the incidence of donors that do not produce transferable embryos (Greve *et al.,* 1995). Future research to reduce the variability in the superovulatory response should focus on the effects of gonadotrophin treatments on oocyte maturation and follicular development as well as the use of alternative protocols for superovulation (Cognie, 1999). Since controlling the follicular growth could minimize the variability of the ovulatory response, new methods to synchronize ovarian follicular development may offer the possibility to develop superovulatory programs at random stages of the estrous cycle and under optimal conditions for embryo production (Guilbault *et al.*, 1996). Methods for oocyte collection in unstimulated donors by ultrasound-guided aspiration, in vitro maturation of oocytes, and *in vitro* fertilisation (IVF) with subsequent laboratory embryo production, have been proposed. If these procedures prove widely applicable for embryo yield under field conditions, they could have a major impact on the future practice of ET, including the decrease of the dependence on superovulation (Armstrong,

1993). During the next decade, the practionners and researchers will be able to compare the different techniques (multiple ovulations and ovum pick up combined with in vitro fertilisation) in animal production and genetic studies.

Acknowledgements

This review is issued in part from investigations funded by Belgium FNRS and Ministry of Agriculture. The NIH Bethesda USA is acknowledged for gift of gonadotropin

References

Agarwal, S.K., Taneja, V.K., Yadav, M.C. and Shankar, U. (1993) Effect of season on superovulation response, recovery rate and quality of embryos in crossbred cattle, *Indian J. Anim. Sci. 63*, 505-510.

Akinlosotu, B.A. and Wilder, C.D. (1993) Fertility and blood progesterone levels following LHRH-induced superovulation in FSH treated anestrous goats, *Theriogenology 40*, 895-904.

Alfuraiji, M.M., Atkinson, T., Broabdent, P.J. and Hutchinson, J.S.M. (1993) Superovulation in cattle using PMSG followed by PMSG-monoclonal antibodies, *Anim. Reprod. Sci. 33*, 99-109.

Armstrong, D.T. (1993) Recent advances in superovulation of cattle, *Theriogenology 39*, 7-24.

Armstrong, D.T. and Evans, G. (1983) Factors influencing success of embryo transfer in sheep and goats, *Theriogenology 19*, 31-42.

Armstrong, D.T. Pfitzner, A.P., Warnes, G.M. and Seamark, R.F. (1983) Superovulation treatment and embryo transfer in Angora goats, *J. Repr. Fertil. 67*, 403-410.

Baldassarre, A., Aste, F., Argerich, M.C. and de Matos, D.G. (1992) Comparative study between human menopausic gonadotrophin (HMG) and hipofisisary follicle stimulating hormone (FSH-P) for superovulation in sheep, in *Proc. 12th Int. Cong. Anim. Reprod. 1*, 182-183.

Baril, G. and Vallet, J.C. (1990) Time of ovulation in dairy goats induced to superovulated with porcine follicle stimulating hormone and out of the breeding season, *Theriogenology 34*, 303-311.

Baril, G., Brebion, P. and Chesne, P. (1993) *Manuel de Formation Practique pour la Transplantation Embryonnaire chez la Brebis et la Chevre*, Food and Agriculture Organization of the United Nations, Rome.

Batt, P.A., Kilben, J.D. and Cameron, A.W.N. (1993) Use of a single or multiple injections of FSH in embryo collection programmes in goats, *Reprod. Fertil. Develop. 5*, 49-56.

Bielansky, A. and Yadav, B.R. (1990) A note on fertilization and embryo production in superovulated cattle with various levels of subcutaneous fat tissue, *Anim. Prod. 51*, 426-430.

Bindon, B.M. and Piper, L.R. (1982) Physiological basis of ovarian response to PMSG in sheep and cattle, in *Embryo Transfer in Cattle, Sheep and Goats*, Australian Society Reproductive Biology, Canberra, pp.1-5

Blanchard, T., Ferguson, J., Love, L., Takeda, T., Henderson, B., Hasler, J. and Chalupa, W. (1990) Effect of dietary crude-protein type on fertilization and embryo quality in dairy cattle, *Am. J. Vet. Res. 51*, 905-908.

Bo, G.A., Adams, G.P., Nasser, L.F., Pierson, R.A. and Mapletoft, R.J. (1993) Effect of estradiol valerate on ovarian follicles, emergence of follicular waves and circulating gonadotropins in heifers, *Theriogenology 40*, 225-239.

Bo, G.A., Adams, G.P., Pierson, R.A. and Mapletoft, R.J. (1995) Exogenous control of follicular wave emergence in cattle, *Theriogenology 43*, 31-40.

Boland, M.P., O'Doherty, J.V. and Crosby, T.F. (1993) Superovulation in sheep influenced by FSH and body condition, in *Proc. 9th Meet. Eur. Embryo Trans. Assoc.*, p. 158.

Bouters, R., Moyaert, I., Coryn, M. and Vandepplassche, M. (1983) The use of a PMSG antiserum in superovulated cattle: endocrinological changes and effects on timing of ovulation, *Zuchthygiene 18*, 172-177.

Breuel, K.F., Baker, R.D., Butcher, R.L., Townsend, E.C., Inskeep, E.K., Dailey, R.A. and Lerner, S.P. (1991) Effects of breed, age of donor and dosage of follicle stimulating hormone on the superovulatory response of beef cows, *Theriogenology 36*, 241-255.

Briois, M., Belloc, J.P., Brebion, P., Guerin, Y. and Cognie, Y. (1992) Increased embryo production in superovulated elite Lacaune ewes pretreated with a GnRH agonist, in *Proc. 8th Meet. Eur. Embryo Trans. Assoc.*, p. 132.

Bungartz, L. and Niemann, H. (1994) Assessment of the presence of a dominant follicle and selection of dairy cows suitable for superovulation by a single ultrasound examination, *J. Reprod. Fertil. 101*, 583-591.

Callesen, H. and Greve, T. (1996) Superovulation of cattle, lessons learned and to be learned, in *Proc.12th Meet. Eur. Embryo Trans. Assoc.*, pp. 85-87.

Chemineau, P., Baril, G., Leboeuf, B., Maurel, M.C., Roy, F., Pellicer, M., Malpaux, B. and Cognie, Y. (1999) Implications of recent advances in reproductive physiology for reproductive management of goats, *J. Reprod. Fertil. Suppl. 54*, 129-142.

Chupin, D., Cognie, Y., Combarnous, Y., Procureur, R. and Saumande, J. (1987) Effect of purified LH and FSH on ovulation in the cow and ewe, in J.F. Roche and D. O'Callaghan (eds.), *Follicular Growth and Ovulation Rate in Farm Animals*, Martinus Nijhoff, pp. 63-71.

Cognie, Y. (1999) State of the art in sheep-goat embryo transfer, *Theriogenology 51*, 105-116.

Cognie, Y., Chupin, D. and Saumande, J. (1986) The effect of modifiying FSH/LH ratio during the superovulatory treatment in ewes, *Theriogenology 25*, 148.

Cseh, S. and Seregi, J. (1993) Practical experiences with sheep embryo transfer, *Theriogenology 39*, 207.

D'Alessandro, A., Martemucci, G., Colonna, M.A., Cafueri, C. and Toteda, F. (1996) Some effects of adding p-LH in defined amounts to purified p-FSH to modify FSH/LH ratios during the superovulatory treatment of anestrous ewes, *Anim. Reprod. Sci. 47*, 91-98.

Dattena, M., Vespignani, S., Branca, A., Gallus, M., Ledda, S., Naitana, S. and Cappai, P. (1994) Superovulatory response and quality of embryos recovered from anestrus ewes after a single injection of porcine FSH dissolved in polyvinylpyrrolidone, *Theriogenology 42*, 235-239.

Delacharlerie, P.F., Manciaux, L., Charreaux, F. and Marie, M. (1995) Relationships between diet and response to superovulation in dairy cows, in *Proc.11th Meet. Eur. Embryo Trans. Assoc.*, p. 160.

Demoustier, M.M., Beckers, J.F., Van der Zwalmen, P., Closset, J., Gillard, J.L. and Ectors, F. (1988) Determination of porcine plasma follitropin levels during superovulation treatment in cows, *Theriogenology 30*, 379-386.

Desaulniers, D.M., Lussier, J.G., Goff, A.K., Bousquet, D. and Guilbault, L.A. (1995) Follicular development and reproductive endocrinology during a synchronized estrous cycle in heifers and mature cows displaying contrasting superovulatory responses, *Dom. Anim. Endocrinol. 12*, 117-131.

Dieleman, S.J., Bevers, M.M., Vos, P.L.A.M. and de Loos, F.A.M. (1993) PMSG/anti-PMSG in cattle: a simple and efficient superovulatory treatment?, *Theriogenology 39*, 25-41.

Dingwall, W.S., McKelvey, W.A.C., Mylne, J. and Simms, G. (1993) An evaluation of MOET in Suffolk sheep, in *Proc.British Soc. Anim. Prod.*, 103.

Dobson, H. and Smith, R.F. (1995) Stress and reproduction in farm animals, *J. Reprod. Fertil. Suppl. 49*, 451-46.

Duffy, P., Duby, R.T., Overstrom, E.W., Hill, J.T. and Boland, M.P. (1995) Effect of administration of oestradiol benzoate and progesterone on superovulation in heifers, in *Proc. 11th Meet. Eur. Embryo Trans. Assoc.*, p. 168.

Freytag, A., Lange, H., Gehrmeyer, D. and Jongeling, C. (1995) Effect of temperature, atmospheric pressure and relative humidity on the response of superovulation in cattle, in *Proc. 11th Meet. Eur. Embryo Trans. Assoc*, p. 170.

Gordon, I. (1996) *Controlled Reproduction in Cattle and Buffaloes*, CAB International, Oxon.

Gray, B.W., Stringfellow, D., Riddell, M., Riddell, K., Davenport, G. and Wright, J. (1993) The effect of treatment with bovine somatotropin (BST) on the superovulatory response of cattle, *Theriogenology 39*, 227 (abstract).

Greaney, K.B., Mcdonald, M.F., Vivanco, H.A. and Tervit, H.R. (1991) Out of season embryo trasfer in five breeds of imported sheep, in *Proc. New Zealand Soc. Anim. Prod. 51*, 129-121.

Greve, T., Callesen, H., Hyttel, P., Høier, R. and Assey, R. (1995) The effects of exogenous gonadotropins on oocyte and embryo quality in cattle, *Theriogenology 43*, 41-50.

Guibault, L.A., Kohram, H., Twagiramungu, H., Bousquet, D. and Dufour, J.J. (1996) Control of follicular dominance and synchronization of follicular waves to improve superovulation in cattle, in *Proc. 12th Meet. Eur. Embryo Trans. Assoc*, pp. 61-71.

Hasler, J.F. (1992) Current status and potential of embryo transfer and reproductive technology in dairy cattle, *J. Dairy Sci. 75*, 2857-2879.

Hasler, J.F., McCauley, A.D., Schermerhorn, E.C. and Foote, R.H. (1983) Superovulatory responses of Holstein cows, *Theriogenology 19*, 83-99.

Herrler, A., Einspanier, R., Schams, D. and Niemann, H. (1994) Effect of a recombinant bovine somatotropin (rBST) on follicular IGF-I contents and the ovarian response following superovulatory treatment in dairy cows: a preliminary study, *Theriogenology 41*, 601-611.

Ishwar, A.K. and Memon, M.A. (1996) Embryo transfer in sheep and goats: a review, *Small Rum. Res. 19*, 35-43.

Jabbour, H.N. and Evans, G. (1991) Ovarian and endocrine responses to Merino ewes following treatment with PMSG and GnRH or PMSG antiserum, *Anim. Reprod. Sci. 24*, 259-270.

Jabbour, H.N., Ryan, J.P., Evans, G. and Maxwell, W.M.C. (1991) Effects of season, GnRH administration and lupin supplementation on the ovarian and endocrine responses of Merino ewes treated with PMSG and FSH-P to induce superovulation, *Reprod. Fertil. Dev. 3*, 699-707.

Kelly, P., Duffy, P., Roche, J.F. and Boland, M.P. (1997) Superovulation in cattle: effect of FSH type and method of administration on follicular growth, ovulatory response and endocrine patterns, *Anim. Reprod. Sci. 46*, 1-14.

Kiessling, A.A., Hughes, W.H. and Blankevoort, M.R. (1986) Superovulation and embryo transfer in the dairy goat, *J. Am. Vet. Med. Assoc. 188*, 829-832.

Kohram, H., Twagiramungu, H., Bousquet, D., Durocher, J., Brassard, P., Dufour, J.J. and Guilbault, L.A. (1996) Superovulation at random stages of the estrous cycle in heifers, *Theriogenology 45*, 331 (abstract).

Krisher, R.L., Gwazdauskas, F.C., Page, R.L., Russell, C.G., Canseco, R.S., Sparks, A.E.T., Velander, W.H., Johnson, J.L. and Pearson, R.E. (1994) Ovulation rate, zygote recovery and follicular populations in FSH superovulated goats treated with PGF2α and/or GnRH, *Theriogenology 41*, 491-498.

Kumar, J., Osborn, J.C., Cameron, A.W.N., Batt, P.A. and Trouson, A.O. (1990) Premature condensation of chromatin induced in goat (Capra hircus) oocytes after gonadotrophin treatment, *Reprod. Fertil. Dev. 2*, 661-670.

Lerner, S.P., Thayne, W.V., Baker, R.D., Henschen, T., Meredith, S., Inskeep, E.K., Daily, R.A., Lewis, P.E. and Butcher, R.L. (1986) Age, dose of FSH and other factors affecting superovulation in Holstein cows, *J. Anim. Sci. 63*, 176-183.

Lopez, A., Cognie, Y., Cocero, M.J., De la Fuente, J. and Poulin, N. (1990) Effect of season and duration of FSH treatment on embryo production in sheep, *Theriogenology 34, 175-180.*

Lussier, J.G., Lamothe, P. and Pacholek, X. (1995) Effects of follicular dominance and different gonadotrophin preparations on the superovulatory response in cows, *Theriogenology 43*, 270.

Mahmood, S., Koul, G.I. and Biswas, J.C. (1991) Comparative efficacy of FSH-P and PMSG on superovulation in Pashmina goats, *Theriogenology 35*, 1191-1196.

Mapletoft, R.J. (1986) Bovine embryo transfer, in D.A. Morrow (ed.), *Current Therapy in Theriogenology*, W. B. Saunders, Philadelphia, pp. 54-58.

Martemucci, G., D'Alessandro, A., Toteda, F. and Facciolongo, A.M. (1995) Embryo production and endocrine response in ewes superovulated with PMSG, with or without monoclonal anti-PMSG administered at different times, *Theriogenology 44*, 691-703.

Massey, J.M. and Oden, A.J. (1984) No seasonal effect on embryo donor performance in the southwest region of the USA, *Theriogenology 21*, 196-217.

Maxwell, W.M.C. and Wilson, H.R. (1989) Superovulation and embryo recovery in ewes treated with a single injection of PMSG and FSH-p, in *Proc. 21th Ann. Conf. Austr. Soc. Reprod. Biol.*, p. 50.

McEvoy, T.G., Broadbent, P.J., Gebbie, F.E., Dolman, D.F., Watt, R.G. and Higgins, L.C. (1996) Progesterone profiles and superovulatory responses of Simmental heifers in relation to preovulatory energy intake and progesterone priming treatment, *Theriogenology 45*, 330.

McEvoy, T.G., Robinson, J.J., Aitken, R.P., Findlay, P.A., Palmer, R.M. and Robertson, I.S. (1995) Dietary induced supression of preovulatory progesterone concentrations in superovulated ewes impairs the subsequent in vivo and in vitro development of their ova, *Animal Reproduction Science 39*, 89-107.

McGowan, M.R., Braithwaite, M., Jochle, W. and Mapletoft, R.J. (1985) Superovulation of beef heifers with pergonal (HMG): a dose response trial, *Theriogenology 24*, 173-184.

Meinecke-Tillman, S., Lewalski, H. and Meineke, B. (1993) Induction of superovulation in Merinolandschaf ewes after single or multiple FSH injections, *Reproduction in Domestic Animals 28*, 433-440.

Monniaux, D., Chupin, D. and Saumande, J. (1983) Superovulatory response of cattle, *Theriogenology 19*, 55-81.

Moor R.M., Osbourne, J.C. and Crosby, I.M. (1985) Gonadotrophin-induced abnormalities in sheep oocytes after superovulation, *J. Reprod. Fertil. 74*, 167-172.

Moore, N.W. and Eppleston, J. (1979) Embryo transfer in the Angora goats, *Austr. J. Agr. Res.* 30, 973-981.

Murphy, D.B., Mapletoft, R.J., Manns, J. and Humphrey, W.D. (1984) Variability in gonadotropin preparations as a factor in the superovulatory response, *Theriogenology 21*, 117-125.

Nowshari, M.A., Beckers, J.F. and Holtz, W. (1995) Superovulation of goats with purified FSHp supplemented with defined amounts of LHp, *Theriogenology 43*, 797-802.

Nuti, L.C., Minhas, B.S., Baker, W.C., Capehart, J.S. and Marrack, P. (1987) Superovulation and recovery of zygotes from Nubian and Alpine dairy goats, *Theriogenology 28*, 481-488.

Peebles, I.R. and Kidd, J.F. (1994) Single treatment superovulation in the Cashmere goat using porcine follicle stimulating hormone, *Theriogenology 41*, 210.

Pendleton, R.J., Youngs, C.R., Rorie, R.W., Pool, S.H., Memon, M.A. and Godke, R.A. (1992) Follicle stimulating hormone versus pregnant mare serum gonadotropin for superovulation of dairy goats, *Small Rum. Res. 8*, 217-224.

Pintado, B., Gutierrez, A., Perez, B. and Garde, J. (1996) Influence of the synchronization treatment on the superovulatory response of Murciana goats, *Small Rum. Res. 23*, 135-141.

Rosnina, Y., Jainudeen, M.R. and Nihayah, M. (1992) Superovulation and egg recovery in goats in the tropics, *Vet. Rec. 130*, 97-99.

Roy, F., Combes, B., Vaiman, D., Cribiu, E.P., Pobel, T., Deletang, F., Combarnous, Y., Guillou, F. and Maurel, M.C. (1999) Humoral immune response to equine chorionic gonadotropin in ewes: association with major histocompatibility complex and interference with subsequent fertility, *Biol. Reprod. 61*, 209-218.

Ryan, J.P., Hunton, J.R. and Maxwell, W.M.C. (1991) Increased production of sheep embryos following superovulation of Merino ewes with a combination of pregnant mare serum gonadotrophin and follicle stimulating hormone, *Reprod. Fertil. Dev. 3*, 551-560.

Satoh, H., Numabe, T., Takada, T., Oikawa, T., Kifune, A., Watanabe, G. and Taya, K. (1996) Superovulation in Japanese beef cows using polyvinylpyrrolidone (PVP) as the vehicle for porcine FSH (pFSH), *Theriogenology 45*, 332.

Schams, D., Menzer, Ch., Schallenberger, E., Hoffman, B. and Hahn, R. (1978) Some studies on pregnant mare serum gonadotrophin (PMSG) and on endocrine responses after application for superovulation in cattle, in J.M. Sreenan (ed.), *Control of Reproduction in the Cow*, Martinus Nijhoff, The Hague, pp. 122-143.

Schiewe, M.C., Howard, J.G., Goodrowe, K.L., Stuart, L.D. and Wildt, D.E. (1990) Human menopausal gonadotrophin induces ovulation in sheep but embryo recovery after prostaglandin F2α synchronization is compromised by premature luteal regression, *Theriogenology 34*, 469-486.

Schillo, K.K. (1992) Effects of dietary energy on control of luteinizing hormone secretion in cattle and sheep, *J. Anim. Sci. 70*, 1271-1282.

Seidel, G.E. and Seidel, S.M. (1991) *Training Manual for Embryo Transfer in Cattle*, Food and Agriculture Organization of the United Nations, Rome.

Senn, B.J. and Richardson, M.E. (1992) Seasonal effects on caprine response to synchronization of estrus and superovulatory treatment, *Theriogenology 37*, 579-585.

Shea, B.F., Janzen, R.E. and McDermand, D.P. (1984) Seasonal variation in response to stimulation and related embryo transfer procedures in Alberta over a nine year period, *Theriogenology 21*, 186-195.

Staigmiller, R.B., Bellows, R.A., Anderson, G.B., Seidel, G.E., Foote, W.D., Menino, A.R. and Wright, R.W.Jr. (1992) Superovulation of cattle with equine pituitary extract and porcine FSH, *Theriogenology 37*, 1091-1099.

Staigmiller, R.B., MacNeil, M.D., Bellows, R.A., Short, R.E. and Phelps, D.A. (1995) The effect of estrus synchronization scheme, injection protocol and large ovarian follicle on response to superovulation in beef heifers, *Theriogenology 43*, 823-834.

Stefani, J.S., Palma, M.D.C. and Rodrigues, J.L. (1991) Ovine superovulation response to HMG, in *Proc. 7th Meet. Eur. Embryo Trans. Assoc.*, p. 210.

Sugano, M., Takano, T., Sato, N., Ohsaki, J. and Watanabe, S. (1995) Effects of human menopausal gonadotrophin (hMG) for superovulation in Japanese Black cattle, *J. Vet. Med. Sci. 57*, 1113-1115.

Sugano, M. and Shinogi, T. (1999) Superovulation induction in Japanese Black cattle by a single intramuscular injection of hMG or FSH dissolved in polyvinylpyrrolidone, *Anim. Reprod. Sci. 55*, 175-181.

Synder, D.A. (1986) Superovulation of cows and heifers selected for twinning, *Theriogenology 25*, 200 (abstract).

Takedomi, T., Aoyagi, Y., Konishi, M., Kishi, H., Taya, K., Watanabe, G. and Sasamoto, S. (1995) Superovulation of Holstein heifers by a single subcutaneous injection of FSH dissolved in polyvinylpyrrolidone, *Theriogenology 43*, 1259-1268.

Tervit, H.R. (1997) Factors affecting the success of goat embryo trasfer, in *Proc. 4th AAAP Anim. Sci. Congress*, pp. 262-266.

Torres, S. and Cognie, Y. (1984) Superovulation and egg transfer in the ewe, *Reprod. Nutr. Dev. 24*, 623-631.

Vos, P.L.A.M., Van der Schans, A., de Wit, A.C.C., Bevers, M.M., Willemse, A.H. and Dieleman, S.J. (1994) Effects of neutralization of pregnant mare's serum gonadotrophin (PMSG) shortly before the preovulatory LH surge in PMSG-superovulated heifers on follicular function and development, *J. Reprod. Fertil. 100*, 387-393.

Walker, S.K., Smith, D.H., Frensham, A., Ashman, R.J. and Seamark, R.F. (1989) The use of synthetic gonadotrophin hormone treatment in the collection of sheep embryos, *Theriogenology 31*, 741-752.

Wilson, J.M., Jones, A.L., Moore, K., Looney, C.R. and Bondioli, K.R. (1993) Superovulation of cattle with a recombinant-DNA bovine follicle stimulating hormone, *Anim. Reprod. Sci. 33*, 71-82.

Wolf, B.T. and Mylne, M.J.A. (1994) Influence of age of donor ewe on MOET in Texel sheep, in *Proc. British Soc. Anim. Prod.*, 86.

Yaakub, H., Callaghan, D.O. and Boland, M.P. (1999) Effect of type and quantity of concentrates on superovulation and embryo yield in beef heifers, *Theriogenology 51*, 1259-1266.

Yamamoto, M., Suzuky, T., Ooe, M. and Takagi, M. (1993) Superovulation in Beef cows and heifers with a single injection of FSH dissolved in polyvinylpyrrolidone, *J. Reprod. Dev. 39*, 353-356

SEX PRESELECTION IN MAMMALS

RENAVILLE R., HAEZEBROECK V., PARMENTIER I., PIRARD M., FONTAINE S. AND PORTETELLE D.
Faculté universitaire des Sciences agronomiques, Unité de Biologie animale et microbienne, 13 Avenue Maréchal juin, 5030 Gembloux, Belgique

Abstract

Since a long time, sex preselection has been a goal of the dairy and meat industry to increase the rate of response to selection, to reduce the cost of progeny testing for elite males and to produce desired specialized and genetically superior offsprings. Over the past 10 years, the technology has been further developed and proven effective for sex preselection based on the fact that thousands of animals have been born. New biotechnology methods authorize sexing of preimplantion embryos using Y-chromosome-specific primers and PCR amplification. More recently, the successful separation into predominant X and Y sperm populations by the high-speed flow cytometric method opened a new window in the control of mammalian reproduction. If it has yet to optimize some aspects of sorted spermatozoa (capacitation, the acrosome reaction) compared with unsexed sperm, this technology can be used by the livestock industry to achieve sex selected progeny to fit a particular market application.

1.Introduction

Two general mechanisms of sex determination have been identified among gonochoristic vertebrates: environmental sex determination where offspring become male or female in response to an environmental factor(s) during development (for example, some fishes and reptiles); and genetic sex determination where sex is determined by genotype at conception (as in birds and mammals).

Over the past century, scientists have learned to manipulate that cardinal characteristics of life, reproduction, with powerful techniques like artificial insemination, contraception, embryo transfer, cryopreservation and cloning by nuclear transfer. Interest in this area, probably starting with semen storage technology was

R. Renaville and A. Burny (eds.), Biotechnology in Animal Husbandry, 225–233.

revolutionized approximately 50 years ago by the discovery that glycerol could act as a cryoprotectant. This important observation enabled spermatozoa to be frozen, stored for prolonged periods, and then used successfully for artificial insemination.

However, in animal husbandry, pre-selection of sex prior to conception will dramatically impact a farmer's productivity and income because in each of the chosen target industries there is a strong preference for one sex over the other. For example, the dairy industry must have females to produce milk whereas the beef industry prefers males for their higher feed efficiency and muscle mass. For pork production, females are preferred for their higher quality and lower cost of production. There are other advantages that result in faster improvement of the genetic quality of breeding herds which, in turn, results in greater productivity and better quality.

Sex preselection is one of the most sought after biotechnology of all times. The hunt has been on for several decades to find the scientific breakthrough that will allow one to use spermatozoa which will produce offspring of the desired sex (commonly called sexed semen). Several reviews have been written on this subject (Kiddy and Hafs, 1971; Garner *et al.*, 1983; Bradley, 1989; Amann, 1999). Methods for sexing preimplantation embryos range from karyotyping to identification of Y-chromosome-specific DNA using the polymerase chain reaction. However, using this approach, sex is not selected, unless one uses the inefficient route of selective abortion.

Through the past years, the Beltsville Sperm Sexing Technology (Johnson, 1997) has consistenly shown that offspring of the predicted sex can be produced when it is applied to rabbits (Johnson *et al.*, 1989), swine (Rath *et al.*, 1999), cattle (Cran *et al.*, 1995) or human (Sills *et al.*, 1998). The results have shown an average of 90% of the predicted sex. Flow cytometric sperm sorting for sex preselection is now a tool that can be used by the livestock industry to achieve sex selected progeny to fit a particular market application. In this mini review, some of these aspects will be developed.

2. Factors Affecting Sex Ratio

Considerable folklore particularly in humans has arisen regarding preconception methods to manipulate animal sex ratio including suggestions that the ingestion of sweet foods will increase the chances of having a female, whereas the ingestion of sour foods will produce males. For exemple, in pigs, Toriumi *et al.* (1993) reported on the effect of a Ca/P-enriched diet (2.15% Ca and 1.28% P) on the sex ratio; the male:female ratio for sows receiving a standard and enriched diet was 0.89 and 1.32, respectively. However, in theory, since the testes produce an equal number of X- and Y-bearing sperm, there should be an exactly equal number of males and females born in mammals. In fact, this is not observed due to a variety of environmental or physiological factors (such as litter size, maternal age, nutrition, timing of insemination, stress, date of birth, and maternal parity of the previous breeding season) influencing the sex ratio of mammals in order to help assure survival of the species.

Sex steroid hormones were found to have potent effects on sex determination and sexual differentiation in various fishes and, to a lesser extent birds (Crews, 1994). However, there was no evidence that these chemicals could influence gonadal differentiation in mammals, although they were recognized to mediate differentiation of accessory and secondary sex characteristics. Indeed, conventional wisdom holds that steroid hormones play no role in sex determination in mammals, and it is only following gonadal differentiation that steroid hormones produced by the ovaries or testes sculpt the characters that distinguish males from females. However, in birds and fishes, a naturally occurring sex reversal such as seen in experimentally induced sex reversal by administration of steroids can be observed (Bogart, 1987).

Distorted sex ratio, attributable to a temperature manipulation during embryonic development, was observed in different species including fish (Conover and Heins, 1987; Craig *et al.*, 1996) and chicken (Romanov *et al.*, 1994). For example, Conover and Heins (1987) show that in a fish with temperature-dependent sex determination, populations at different latitudes compensate for differences in thermal environment

For a number of years, the time of insemination or mating during the estrus was believed to influence the sex ratio of offspring. Possible mechanisms of altering the sex ratio include facilitating or inhibiting the transport of either X- or Y-chromosome bearing sperm through the reproductive tract, preferential selection of sperm at fertilization, or sex-specific death of embryos after fertilization. In deer and sheep, evidence exists that early insemination results in more females and late insemination in more males (Ericsson and Ericsson, 1999; Rorie, 1999). In bovine, conflicting reports on the effect of time of insemination on sex ratio make the premise less clear and evidence that treatments used for synchronization of oestrus or ovulation may influence the sex ratio (Rorie, 1999). Dominko and First (1997) noted that, in bovine, maturational state of oocytes at the time of insemination affected the male:female ratio (0.71 for immediately after and 2.52 for 8 hours after polar body extrusion, respectively ($P<0.05$)). These results have been confirmed recently by Gutierez *et al.* (1999). In an experiment to evaluate the reproductive performance of heifers after estrus synchronization with a combination of progesterone, estradiol benzoate, and PGF_{2alpha} and fixed-time artificial insemination (50 h to 54 h), Xu and Burton (1999) noted that synchronized heifers gave birth to more female calves (53.8%) than did control heifers (45.7%). The time of artificial insemination also appeared to affect gender of calf: cows bred at 0 and 32 h having a higher percentage of female offspring (Pursley *et al.*, 1998). However, Rorie *et al.* (1999) indicate that inseminating beef cattle at approximately 20 or 10 h before an expected ovulation does not alter the gender ratio of the resultant calves. This observation was also confirmed by Lonergan *et al.* (1999) who reported no effect of the time of insemination in embryo transfer on sex ratio and by Soede *et al.* (2000) who noted that sex ratio was not influenced by litter size ($P>0.05$), and its distribution was normally dispersed (i.e., as expected under a binomial distribution) in different intervals between insemination and ovulation ($P>0.05$)

3. Sexing of Preimplantation Embryos

Available techniques for the collection and direct transplantation of embryos in different species are simple and efficient and could be used for the expansion of new lines, for increasing selection pressure in nucleus herds and for extracting healthy stock from a diseased source. The efficiency of breeding schemes could be improved by the sexing of embryos and the possibility to identify gene markers associated with animal performances (figure 1)

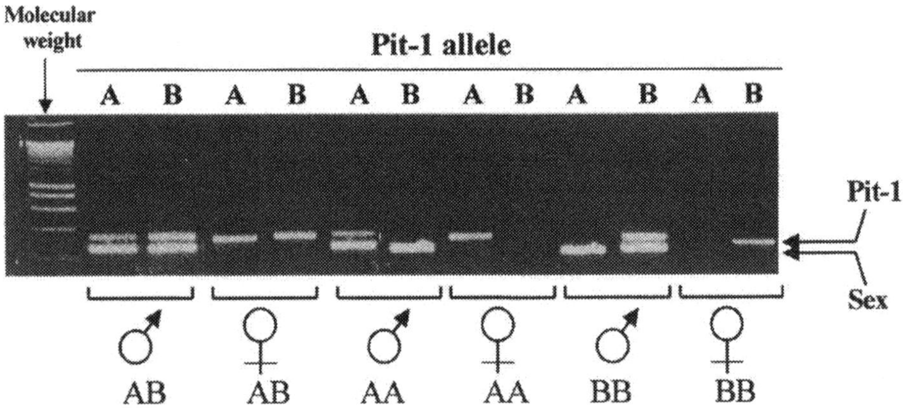

Figure 1. Sexing and identification of Pit-1 gene marker by a DNA unique amplification test (see Parmentier et al. *(2000) for more informations on Pit-1 gene marker).*

The sex determination of embryos routinely used on a commercial scale is to biopsy embryo (a small number of cells are removed from a seven day old embryo) and to test for the presence of the Y-chromosome using a DNA amplification process and primers which are designed to allow amplification of sex specific gene i.e. like Y-chromosome SRY (Rao and Totey, 1992; Le Bourhis *et al.*, 1998; Kawarasaki *et al.*, 2000), amelogenin (Chen *et al.*, 1999) or UcdO43 random amplified polymorphic DNA marker (Gutierrez-Adan *et al.*, 1997).

This procedure is now successfully applicable in various species including bovine (Cotinot *et al.*, 1991), buffalo (Rao *et al.*, 1993; Rao and Totey, 1993, 1999), sheep and goat (Rao and Totey, 1992) and porcine (Kawarasaki *et al.*, 2000). An accuracy of about 95% can be reached (Bredbacka, 1998; Seidel, 1999). The pregnancy rate from fresh sexed embryos runs approximately 5 % less than with non-sexed fresh embryos (50-70%) while a pregnancy rate of 50% can be achieved with frozen biopsied embryos (Bredbacka, 1998).

However, this procedure is an invasive procedure that requires opening the protective covering of the embryo and removing a few embryonic cells to obtain the DNA. The embryo may not grow well after some of its cells have been removed, and breakage in the covering may lead to infection. Moreover, the PCR sex approach traditionally utilizes electrophoresis which involves the risk of DNA contamination of subsequent assays and misdiagnosis (Bredbacka, 1998).

4. Immunological Sexing

An attractive alternative to DNA embryo sexing would be to identify an immunological marker confined to one sperm type and therefore, significant scientific effort has been devoted in examining antibodies that appear to recognize approximately 50% of spermatozoa in an ejaculate.

Many methods have been devised on the basis of supposed differences in various sperm characteristics (Windsor *et al.*, 1993; Hendriksen, 1999). Also, numerous schemes have been promoted on the identification of plasma-membrane protein specific for either X and Y sperm. In 1989, Bradley reported that male-specific antigen is located on both the postacrosomal region of the head and the midpiece of the flagellum. Using high resolution two-dimensional SDS-PAGE, Howes *et al.* (1997) isolated sex specific proteins (SSPs) and more particularly X-specific proteins which unfortunatelly are not associated with the surface membrane and unlikely profitable for sexing.

H-Y antigen is the first plasma membrane or cell surface sex-specific protein to which a specific organogenesis function has been assigned. Studies designed to answer the question wheteher or not H-Y antigen is preferentially expressed on Y chromosome bearing sperm have resulted in conflicting results. This is probably due to the absence of reliable methods for estimating the percentage of X and Y chromosome bearing sperm in fractions, enriched or depleted for H-Y antigen positive sperm. However, using seven different monoclonal antibodies against H-Y, Hendriksen *et al.* (1993) tested the ability of these anti-H-Y antibodies to bind to fractions enriched for X and Y chromosome bearing sperm by using a flow cytometer cell sorter. They concluded that their experiments do not yield evidence that H-Y antigen is preferentially expressed in Y chromosome-bearing sperm. In various experiments, Blecher *et al.* (1999) obtained around 90% of male embryos with sperm sexing using antibodies against SSPs.

Finally, using a similar immunological approach, Veerhuis *et al.* (1994) calculated the accuracy of sexing bovine embryos ranged from 58% to 71% which is insufficient for its application in farm practise.

The inability to detect sex-specific antigens at sperm surface suggests that further work on this immunological approach to semen sexing is unlikely to be profitable.

5. High-Speed Flow Cytometric Sorting of X and Y Sperm

Intact, viable X and Y chromosome-bearing sperm populations of various species according to DNA content are separated with a flow cytometer/cell sorter (Morrell *et al.*, 1988; Hammano *et al.,* 1999). Indeed, the most critical differences in sorting procedures relative to species are the differences in DNA between X and Y sperm within a species. Mammalian sperm are inherently different in that the X sperm carries from 2.8 to 7.5% more DNA than the Y sperm (rabbit (3.0%), pig (3.6%), bull (3.8%), horse (3.7%) and sheep (4.2 %)). The narrower the DNA difference, the more care is required in setting the sort windows.

In flow cytometry and cell sorting technology, the process utilizes the fluorochrome Hoechst 33342 to bind to the DNA. The relative DNA is measured by passing the living sperm through a laser beam and collecting the light energy from the individual sperm. Data is acquired and used to select the particular sperm for deflection into collection tubes. The proportions of sorted X and Y sperm in each tube can be validated by reanalyzing an aliquot for DNA content (figure 2). This value is then used to predict the outcome of fertilization and subsequent gestation.

Numerous improvements have been made in the sexing technology since it was first reported in 80s (Rens *et al.*, 1998; Stap *et al.*, 1998). Three aspects of instrumentation are especially critical to sort sperm efficiently for use in various semen-delivery situations: the first of these is the need for a forward-fluorescence detector to replace the normative-light-scatter detector present on most cell sorters; a second instrumental factor is the nozzle through which one can control the presentation of individual sperm to the laser beam ; the third instrumental factor that has boosted sorting efficiency is the adaptation of the orienting nozzle to high-speed cell sorters. A production rate of 6 million per h of each X- and Y- sperm fractions is achieved for routine conditions with the expectation of 90% or greater of one sex or the other being born (Johnson & Welch, 1999). Increasing the speed of the sexing process to make the application of the technology available to a larger segment of the livestock industry is paramount, even with insemination technology designed for small numbers of spermatozoa. In addition, the recent advent of ultrasound guided insemination in cattle may provide and opportunity to use this technology for much lower numbers of sperm per insemination than previously thought possible

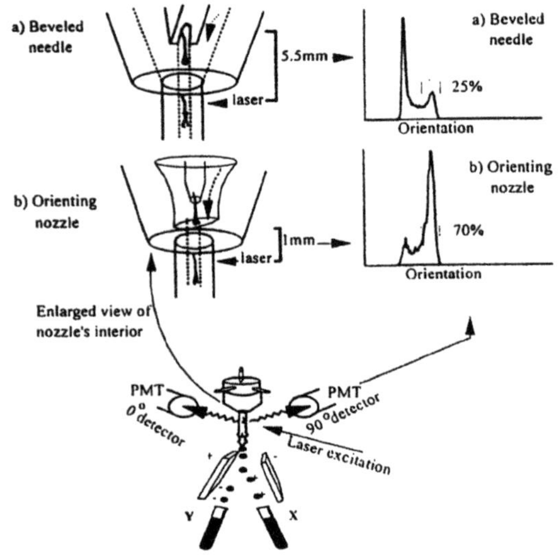

Figure 2. Schematic diagram of basic cell sorter modified for sperm.

Because of the inability to obtain large numbers of sorted sperm in a short amount of time, the technologies use for regular artificial insemination would not be practical in most domestic species. This sexing technology however is very applicable where IVF, intrauterine or intratubal insemination are convenient means for producing offspring.

Artificial insemination or in vitro fertilisation using isolated populations of X- and Y- chromosome bearing spermatozoa is now applied in rabbits (Johnson *et al.*, 1989), swine (Rath *et al.*, 1999), cattle (Cran *et al.*, 1995) or human (Sills *et al.*, 1998). In these cases, sex ratios are shifted from the normal 50:50 to 85 to 90% of one sex or the other. Fertility rates will likely be 12 to 25% lower that with conventional semen however, all progeny produced have exhibited completely normal morphological appearance and normal reproductive function. Likewise, cattle, swine and rabbit offspring have been reproduced through the second generation with normal morphology and reproductive function (Johnson, 1997).

Efficiency of sperm sexing could be also improved by optimization of artificial insemination techniques to be used with the small numbers of flow cytometrically separated X or Y sperm populations. Finally, no one is sure of the price, but it will most likely cost an extra $ 50 to $ 60 dollars per straw over and above the retail price of the semen.

6. Conclusion

Economics dictate that livestock producers are under increasing pressure to produce a given number of progeny of the desired sex. Two techniques are presently available with some advantages and disadvantages.

Using the DNA of embryos to determine their sex has been successfully accomplished, but it's an invasive procedure that requires opening the protective covering of the embryo and removing a few embryonic cells to obtain the DNA. The embryo may not grow well after some of its cells have been removed, and breakage in the covering may lead to infection.

Currently, the only successful method for separating X and Y chromosomes-bearing spermatozoa is fluorescence activated cell sorting. Although effective, this technique is of limited usefulness to the animal breeding industry as it can not produce the large volumes of sexed spermatozoa needed for artificial insemination.

References

Amann, R.P. (1999) Issues affecting commercialization of sexed sperm, *Theriogenology 52*, 1441-1457.

Blecher, S.R., Howie, R., Li, S., Detmar, J. and Blahert, L.M. 1999. A new approach to immunological sexing of sperm, *Theriogenology* 52, 1309-1321.

Bogart, M.H. (1987) Sex determination: a hypothesis based on steroid ratios, *J. Theor. Biol. 128*, 349-357.

Bradley, M.P. (1989) Immunological sexing of mammalian semen: current status and future options, *J. Dairy Sci. 72*, 3372-3380.

Bredbacka, P. (1998) Recent developments in embryo sexing and its field application, *Reprod. Nutr. Dev. 38*, 605-613.

Chen, C.M., Hu, C.L., Wang, C.H., Hung, C.M., Wu, H.K., Choo, K.B. and Cheng, W.T. (1999) Gender determination in single bovine blastomeres by polymerase chain reaction amplification of sex-specific polymorphic fragments in the amelogenin gene, *Mol. Reprod. Dev. 54*, 209-214.

Conover, D.O. and Heins, S.W. (1987) Adaptive variation in environmental and genetic sex determination in a fish, *Nature 326*, 496-498.

Cotinot, C., Kirszenbaum, M., Leonard, M., Gianquinto, L. and Vaiman, M. (1991) Isolation of bovine Y-derived sequence: potential use in embryo sexing, *Genomics 10*, 646-653.

Craig, J.K., Foote, C.J. and Wood, C.C. (1996) Evidence for temperature-dependent sex determination in sockeye salmon (Oncorhynchus nerka), *Canadian J. Fisheries Aquatic Sci. 53*, 141-147.

Cran, D.G., Johnson, L.A. and Polge, C. (1995) Sex preselection in cattle: A field trial, *Vet. Rec. 136*, 495-496.

Crews, D. (1994) Temperature, steroids and sex determination, *J. Endocrinol. 142*, 1-8.

Dominko, T. and First, N.L. (1997) Relationship between the maturational state of oocytes at the time of insemination and sex ratio of subsequent early bovine embryos, *Theriogenology* 47, 1041-1050.

Ericsson, R.J. and Ericsson, S.A. (1999) Sex ratios, in: *Encyclopedia of Reproduction*, vol 4, Academic press, pp 431-443.

Garner, D.L., Gledhill, B.L., Pinkel, D., Lake, S., Stephenson, D., Van Dilla, M.A. and Johnson, L.A. (1983) Quantification of the X and Y chromosome-bearing spermatozoa of domestic animals by flow cytometry, *Biol. Reprod. 28*, 312-321.

Gutierrez, A.A., Perez-Garnelo, A., Granados, J., Garde, J.J., Perz-Guzman, M., Pintado, B. and de la Fuente, J. (1999) Relationship between sex ratio and time of insemination according to both time of ovulation and maturational state of oocyte, *Zygote 7*, 37-43.

Gutierrez-Adan, A., Cushwa, W.T., Anderson, G.B. and Medrano, J.F. (1997) Ovine-specific Y-chromosome RAPD-SCAR marker for embryo sexing, *Anim. Genet. 28*, 135-138.

Hamano, K., Li, X., Qian, X.Q., Funauchi, K., Furudate, M. and Minato, Y. (1999) Gender preselection in cattle with intracytoplasmically-injected, flow cytometrically-sorted sperm heads, *Biol. Reprod. 60*, 1194-1197.

Hendriksen, P.J. (1999) Do X and Y spermatozoa differ in proteins?, *Theriogenology 52*, 1295-1307.

Hendriksen, P.J., Tieman, M., Van der Lende, T. and Johnson, L.A. (1993) Binding of anti-H-Y monoclonal antibodies to separated X and Y chromosome-bearing porcine and bovine sperm, *Mol. Reprod. Dev. 35*, 189-196.

Howes, E.A., Miller, N.G., Dolby, C., Hutchings, A., Butcher, G.W. and Jones, R. (1997) A search for sex-specific antigens on bovine spermatozoa using immunological and biochemical techniques to compare the protein profiles of X and Y chromosome-bearing sperm populations separated by fluorescence-activated cell sorting, *J. Reprod. Fertil. 110*, 195-204.

Johnson, L.A. (1997) Advances in gender preselection in swine, *J Reprod Fertil Suppl 52*, 255-266.

Johnson, L.A., Flook, J.P. and Hawk, H.W. (1989) Sex preselection in rabbits: live births from X and Y sperm separated by DNA and cell sorting, *Biol. Reprod. 41*, 199-203.

Johnson, L.A. and Welch, G.R. (1999) Sex preselection: high-speed flow cytometric sorting of X and Y sperm for maximum efficiency, *Theriogenology 52*, 1323-1341.

Kawarasaki, T., Matsumoto, K., Chikyu, M., Itagaki, Y. and Horiuchi, A. (2000) Sexing of porcine embryo by in situ hybridization using chromosome Y and 1-specific DNA probes, *Theriogenology 53*, 1501-1509.

Kiddy, C.A. and Hafs, H.D. (1971) Sex ratio at birth. Prospects for control. *American Society of Animal Science*, 104 pages.

Le Bourhis, D., Chesne, P., Nibart, M., Marchal, J., Humblot, P., Renard, J.P. and Heyman, Y. (1998) Nuclear transfer from sexed parent embryos in cattle: efficient birth of offsprings, *J. Reprod. Fertil. 113*, 343-349.

Lonergan, P., Khatir, H., Piumi, F., Rieger, D., Humlot, P. and Boland, M.P. (1999) Effect of time interval from insemination to first cleavage on the developmental characteristics, sex ratio and pregnancy rate after transfer of bovine embryos, *J. Reprod. Fertil. 117*, 159-167.

Morrell, J.M., Keeler, K.D., Noakes, D.E., Mackenzie, N.M. and Dresser, D.W. (1988) Sexing of sperm by flow cytometry, *Vet. Rec. 122*, 322-324.

Parmentier, I., Portetelle, D., Bertozzi, C., Haezebroeck, V., Pirard, M. and Renaville, R. (2000) Marker Genes in farm animals, in: Renaville R. and Burny A. (eds), *Biotechnology in animal Husbandry*, Kluwer academic publishing, pp 37-54.

Pursley, J.R., Silcox, R.W. and Wiltbank, M.C. (1998) Effect of time of artificial insemination on pregnancy rates, calving rates, pregnancy loss, and gender ratio after synchronization of ovulation in lactating dairy cows, *J. Dairy Sci. 81*, 2139-2144.

Rao, K.B. and Totey, S.M. (1992) Sex determination in sheep and goats using bovine Y-chromosome primers via polymerase chain reaction: potential for embryo sexing, *Indian J. Exp. Biol. 30*, 775-777.

Rao, K.B. and Totey, S.M. (1999) Cloning and sequencing of buffalo male-specific repetitive DNA sexing of in-vitro developed buffalo embryos using multiple nested polymerase chain reaction, *Theriogenology 51*, 785-797.

Rao, K.B., Pawshe, C.H. and Totey, S.M. (1993) Sex determination of in vitro developed buffalo (Bubalus bubalis) DNA amplification, *Mol. Reprod. Dev. 36*, 291-296.

Rath, D., Long, C.R., Dobrinsky, J.R., Welch, G.R., Schreier, L.L. and Johnson, L.A. (1999) *In vitro* production of sexed embryos for gender preselection: high-speed sorting of X-chromosome-bearing sperm to produce pigs after embryo transfer, *J. Anim. Sci. 77*, 3346-3352.

Rens, W., Welch, G.R. and Johnson, L.A. (1998) A novel nozzle for more efficient sperm orientation to improve sorting efficiency of X and Y chromosome-bearing sperm, *Cytometry 33*, 476-481.

Romanov, M.M., Marushin, V.G. and Sakhatsky, N.I. (1994) Studies on chick embryo sex determination and manipulation in the countries of the former USSR. A review. 2. Shifting of sex ratio, *Procc. 9th Eur. Poult. Conf.*, Glasgow, UK, 7-12 August, pp 324-326.

Rorie, R.W. (1999) Effect of timing of artificial insemination on sex ratio, *Theriogenology 52*, 1273-1280.

Rorie, R.W., Lester, T.D., Lindsey, B.R. and McNew, R.W. (1999) Effect of timing of artificial insemination on gender ratio in beef cattle, *Theriogenology 52,* 1035-1041.

Seidel, G.E. Jr. (1999) Sexing mammalian spermatozoa and embryos--state of the art, *J. Reprod. Fertil. Suppl. 54*, 477-487.

Sills, E.S., Kirman, I., Colombero, L.T., Hariprashad, J., Rosenwaks, Z. and Palermo, G.D. (1998) H-Y antigen expression patterns in human X- and Y-chromosome-bearing spermatozoa, *Am. J. Reprod. Immunol. 40*, 43-47.

Soede, N.M., Nissen, A.K. and Kemp, B. (2000) Timing of insemination relative to ovulation in pigs: effects on sex ratio of offspring, *Theriogenology 53,* 1003-1011.

Stap, J., Hoebe, R.A., Merton, J.S., Haring, R.M., Bakker, P.J. and Aten, J.A. (1998) Improving the resolution of cryopreserved X- and Y-sperm during DNA flow cytometric analysis with the addition of Percoll to quench the fluorescence of dead sperm, *J. Anim. Sci. 76*, 1896-1902.

Toriumi, H., Ohba, S., Kuwabara, Y., Takagi, K., Tsumagari, S. and Takeishi, M. (1993) Effects of high calcium and phosphorus administration on secondary sex ratio in pigs, *J. Reprod. Develp. 39*, 73-77.

Veerhuis, R., Hendriksen, P.J., Hengst, A.M., Kruijy, L., Tieman, M. and Booman, P. (1994) The production of anti-H-Y monoclonal antibodies: their potential use in a sex test for bovine embryos, *Vet. Immunol. Immunopathol. 42*, 317-330.

Windsor, D.P., Evans, G. and White, I.G. (1993) Sex predetermination by separattion of X and Y chromosome-bearing sperm: A review, *Reprod. Fertil. Dev. 5,* 155-171.

Xu, Z.Z. and Burton, L.J. (1999) Reproductive performance of dairy heifers after estrus synchronization and fixed-time artificial insemination, ***J. Dairy Sci. 82***, 910-917.

CLONING AND TRANSGENESIS IN CATTLE: POTENTIAL APPLICATIONS

Heyman Y.
INRA- Unité de Biologie du Développement et Biotechnologies
78352 Jouy en Josas Cedex, France

Abstract

This article reviews the current state of generating genetically identical animals (clones) in the bovine species through nuclear transfer. In this process, the nucleus of a donor cell is transferred to a recipient oocyte in which the genetic material has been removed (cytoplast). First success have been achieved using totipotent embryonic cells as a source of nuclei for cloning. The different steps of the technique for embryo cloning are described , such as the enucleation of recipient oocytes, processing of the donor cells, reconstitution by micromanipulation and fusion and in vitro culture up to the blastocyst stage. The importance of the nucleo-cytoplasmic interactions is underlined and the conditions for chromatin remodelling greatly affect the development potential of the reconstituted embryo. The actual efficiency of embryo cloning is limited by the low number of nuclei available in a donor morula stage embryo. Recent research on the use of differentiated cells for nuclear transfer has completely reactivated the impact of cloning . In bovine, somatic cloning has been proven to be feasible from several types of cells taken on a live valuable cow or bull.This opens new possibilities for application in breeding schemes . Calculations indicate that the use of somatic cloning instead of embryo cloning from elite cows could improve by 20% the annual genetic gain in dairy cattle. For endangered breeds, somatic cloning could be a useful tool to maintain or even multiply the existing phenotypes. Furthermore sets of cloned calves can be used as animal models for experiments on pathology studies, reproduction or nutrition. However the main application of somatic cloning in cattle will be its utilisation for improved transgenesis efficiency . Gene targetting in cultured somatic cells through homologous recombination becomes possible in domestic species. Using such genetically modified somatic cells as a source of nuclei for cloning will enable to produce a first generation of identical offspring which are all transgenic without mosaicism. Potential applications in biomedecine to produce a variety of pharmaceuticals in milk, or in animal production to generate cattle in which undesirable genes would be knocked out, are discussed.

R. Renaville and A. Burny (eds.), Biotechnology in Animal Husbandry, 235–259.

1. Introduction

A clone can be defined as a set of two or more individuals with identical genetic make up derived by asexual reproduction from a single common parent. Such animals would be very useful in animal selection for the determination of the genetic component of zootechnical traits (Smith, 1989) or in research to reduce the number of experimental animals (Kubota *et al.,* 2000). A first possibility of producing identical animals was achieved by splitting early embryos into two parts, but this approach is limited to producing 2 or exceptionally 4 identical individuals. Another strategy is provided by nuclear transplantation. This concept was first proposed by Speeman in 1938 in amphibians, he assumed that all the nuclei within an early embryo are genetically identical and each of them could be grafted to an enucleated recipient oocyte. This would result in the possibility of a large number of identical single cell embryos. In this process, the nucleus of a donor cell (karyoplast) is transferred to a recipient cell (cytoplast) from which the genetic material has been removed. Pioneer work has demonstrated in amphibians (Gurdon *et al., 1975*) that adult frogs can be obtained by the introduction of nucleus from cells of tadpoles into the cytoplasm of an oocyte.
Successful cloned mammals were obtained by nuclear transfer of donor nuclei from early stages embryos into enucleated mature oocytes. The first farm animal to be cloned by nuclear transfer was the sheep in 1986 by Willadsen, followed by cattle (Robl *et al.,* 1987), pig (Prather *et al.,* 1989), rabbit (Heyman *et al.,* 1990) and goat (Yong *et al.,* 1991). Since a decade, embryo cloning has been developed by many groups throughout the world especially in the bovine species (see review Yang (1991)).

Recent research on the use of cultured embryonic cells and then differentiated cells as a possible source of nuclei for cloning has completely reactivated the impact of cloning. The report in sheep on the birth of three viable offspring from nuclei of embryo-derived epithelial cells in 1996 (Campbell *et al.,* 1996) has generated new insights into cloning and was followed one year later by the announcement of Dolly's birth (Wilmut *et al.,* 1997). This was really the start of the new area of somatic cloning in domestic mammals.This article relates the current status of the cloning technology in the bovine species with special emphasis on some of its potential applications for animal production and transgenesis.

2. Nuclear Transfer from Embryonic Cells

2.1. TECHNIQUE OF BOVINE NUCLEAR TRANSFER

2.1.1. Preparation of Recipient Cytoplasms

The mature oocyte at the metaphase II stage has been proven to provide a favourable cytoplasmic environment for nuclear reprogramming in cattle compared to zygotic cytoplasm (Barnes *et al.,* 1993; Bondioli *et al.,* 1990). However, large numbers of bovine mature oocytes were difficult to obtain *in vivo.* Since the progress achieved in the *in vitro* maturation (IVM) of germinal vesicle stage oocytes (Leibfried-Rutledge *et al.*, 1987), one of the most prominent changes in bovine embryo cloning, has been the possibility of using *in vitro* matured oocytes to prepare large numbers of recipient cytoplasms (Barnes *et al.*, 1993; Yang *et al.*, 1993). The standard procedures now commonly used in IVM allow use of batches of metaphase II oocytes from slaugtherhouse ovaries at relatively low cost. Therefore, cattle ovaries are collected at the slaughterhouse and transported to the laboratory in sterile buffer. Antral follicles 2 to 8 mm in diameter are aspirated and cumulus-oocytes complexes (COCs) at the germinal vesicle stage are isolated. Intact COCs with several dense cumulus layer are selected and cultured *in vitro* for 22 to 24h in a maturation medium (M 199) supplemented with fetal calf serum, FSH, LH and estradiol-17ß. Normally more than 80% of the cultured oocytes achieve their maturation *in vitro* and reach the metaphase II stage as attested by the presence of the first polar body. Immature oocytes can also be recovered *in vivo* repeatedly on the same animal using ultrasonic guided aspiration or 0vum Pick Up technique (OPU)(Pieterse *et al.*, 1989).

To prepare the recipient cytoplasm, the genetic material from the mature oocyte i.e at this stage the chromosomes of the metaphase plate and the first polar body have to be removed. This can be done by micromanipulation but two technical difficulties have to be overcome with bovine oocytes. First of all, the introduction of an enucleation micropipette would normally induce the rupture of the plasma membrane and lysis of the cell but using pretreatment of the oocyte by a drug which blocks polymerization of microfilaments can circumvent this problem. In the presence of cytochalasin, the plasmic membrane of the oocyte becomes smooth enough to allow aspiration of the chromosomes and collapsing without its rupture.

The second difficulty concerning oocytes from cattle and other ruminants is the darkness of the oocyte cytoplasm due to its high lipid concentration. This makes the visualisation of the metaphase plate through the microscope very difficult. The control of the enucleation can be achieved by labelling the chromosomes with a specific marker of DNA (Hoechst dye 33342) and visualized by epifluorescence under UV light. However, excessive stain concentrations and/or prolonged exposure to UV light during observation by microscope may inhibit further development. Techniques using a very low concentration of Hoechst dye (0.5 mg/ml) together with a very sensitive low level light camera have been used so that it is possible to directly control the efficiency of the enucleation process by visualizing the genetic material of the oocyte in the enucleation pipette (Chesné *et al.*, 1993). Recently, Dominko *et al.* (1999) used other vital fluorochromes aimed to label the tubulin spindle or DNA, and which do not require UV light irradiation.

After pretreatment with cytochalasin, each oocyte freed from its cumulus cells, is placed on an inverted microscope equipped with micromanipulators and maintained by a holding pipette. With a sharp enucleation pipette, the zona pellucida is perforated and the first polar body plus a small adjacent volume of cytoplasm containing the metaphase II plate are aspirated. The enucleated oocyte can then be used as the recipient cytoplasm for nuclear transfer in this stage or after an activation treatment.

2.1.2. Processing of Donor Cells

For embryo cloning, donor cells are generally isolated from a pre-or compacted morula produced in vivo or in vitro. At this stage, the embryo contains between 30 and 60 totipotent cells which are not yet differentiated but have already junctions between polarised cells. Just before the nuclear transfer procedure, the zona pellucida is removed by pronase treatment and the morula is incubated in a calcium free medium for 15 min to ease separation of the cells. If the dispersion of the embryonic cells is not sufficient, it is possible to separate them mechanically with a fine bore polished pipette.

For somatic cloning, the donor cells are generally cultured in vitro over several passages during a few weeks and are of small size compared to embryonic blastomeres. Before nuclear transfer procedure, the monolayer of confluent cells has to be trypsinised in order to isolate them. This step is of importance and has to be carefully handled in order to maintain intact the membrane of the somatic cell for further cell fusion to the recipient cytoplasm.

2.1.3. Reconstitution

Donor blastomeres or somatic cells are aspirated singly into the micropipette and transferred under the zona pellucida of each enucleated oocyte, in the perivitellin space. When transfer of cells is completed, couplets oocyte-cell have to be fused to allow the blastomere nucleus to enter the recipient oocyte cytoplasm. This is generally achieved by electrofusion techniques (Zimmerman and Vienken, 1982). Therefore, couplets are transferred into the cell fusion solution. Generally, the fusion medium is a non-conductive solution of mannitol 0.3 M containing calcium and magnesium. Furthermore, sucrose or glucose can be substituted for mannitol. The fusion chamber consists of two stainless steel wire electrodes mounted 0.2 to 1 mm apart on a glass microscope slide and connected to an electric stimulator. Alignment of the cells (oocyte and donor cell) such that the membranes to be fused are parallele to the electrodes is necessary to achieve high rates of fusion. Alignment can be done either manually with a pipette or using an AC field. Electro-fusion itself is achieved by DC pulses. Concerning field strenght and pulse duration, 1kV/cm during 50µs twice give good fusion rate when using embryonic blastomeres.

An alternative to cell fusion could be direct microinjection of the donor nucleus into the recipient oocyte cytoplasm. The donor cell is aspirated in a narrow injection pipette in order to provoke the rupture of its plasmic membrane. With the same pipette, the oocyte cytoplasm is sucked into the pipette and the nucleus is then injected into the oocyte cytoplasm. Reconstitution by microinjection has successfully been achieved in cattle (Collas and Barnes, 1994) and in the mouse (Wakayama *et al.*, 1998). The main condition is that the size of karyoplast should be very small, which is the case for somatic cells rather than embryonic cells.

The main requirements for efficient nuclear transfer technique are of course specific equipment such as a good microforge to prepare the microtools, an inverted microscope equipped with epifluorescence, two micromanipulators: one for securing oocytes with the holding pipette and one for aspiration and cell transfer, and two aspiration syringes:

one for the holding pipette and the second for enucleation. A reliable electro-stimulator system is also necessary to achieve fusion.

The quality of the microtools, i e : cleanliness, appropriate size, diameter, sharpness... of the micropipettes which are commonly made from borosilicate glass capillary, is an important factor for nuclear transfer. Success in nuclear transfer experiments is of course largely dependent upon the people involved in the program. The micromanipulation itself requires well trained technicians with solid experience in embryo handling. They should be very precise and able to focus their attention for several consecutive hours behind the microscope. It is also important to have a good coordination within the team so that every step is ready on time and synchronous (preparation of recipient oocytes, enucleation , fusion, in vitro culture) as the final result greatly depends upon the cumulative efficiency of each of the techniques involved.

2.1.4. In vitro Culture of Reconstituted Eggs

The progress in in vitro culture methods for IVF embryos during the past years has also benefitted the development of nuclear transfer embryos which are successfully developed in vitro up to the blastocyst stage in different culture systems such as SOF or CR1 medium (Thompson, 2000) or coculture system with Vero cells. Bovine nuclear transfer blastocysts developed entirely *in vitro* have been assessed in our laboratory by nuclei counting and morphological evaluation using electron microscopy; they did not differ from control IVF embryos in terms of kinetics of development, cell number or degree of differentiation (Heyman *et al.,* 1995).

2.2. NUCLEO-CYTOPLASMIC INTERACTIONS

One of the main progress in embryo cloning by nuclear transfer concerns the recipient cytoplasm with the use of in vitro matured oocytes in cattle. The influence of the recipient oocyte cell cycle stage on DNA synthesis, nuclear envelope breakdown and chromatin behavior has been intensively studied (Barnes *et al.,* 1993; Chan, 1999) indicating that DNA synthesis is regulated differently if the recipient cytoplasm is in MII or S phase at the time of fusion.

The introduction of blastomere nuclei into enucleated young metaphase II oocytes results in poor development, probably due to damage induced by premature chromosome condensation (PCC) caused by high levels of maturation promoting factor activity (MPF) in such cytoplasts (Szollosis *et al.,* 1988; Collas and Robl, 1991). Several authors have shown that preactivation of the recipient cytoplast before fusion with the blastomere resulted in the absence of chromatin condensation and a higher rate of in vitro development in cattle (Kono *et al.,* 1994; Stice *et al.,* 1994).

So, the decay in maturation promoting factor activity (MPF) in recipient cytoplasm was found to be favourable for nuclear remodelling of most of the embryonic donor nuclei which are in a high proportion interphasic in a given parent embryo. It is now clearly established that when the donor blastomere is fused after MPF decline in the recipient oocyte, the nuclear membrane is maintained and premature chromosome condensation does not occur and thus DNA synthesis proceeds without interruption

(Barnes *et al.*, 1993). Under those conditions, nuclear swelling occurs and this remodelling is the result of protein exchanges between the nucleus and the recipient cytoplasm (Prather *et al.*, 1990). The proportion of normal chromosome constitution of one or two cell nuclear transfer embryos derived from preactivated oocyte cytoplasms is much higher than that derived from early MII oocytes (Barnes *et al.*, 1993) and this could explain the higher rates of development when S-phase nuclei (which is the case for 80% of the nuclei in a morula stage donor embryo) are transferred to early S-phase oocyte cytoplasms.

For bovine oocytes several preactivation treatments have been developed in order to provide cytoplasms with low level of MPF at time of fusion. We have defined conditions in which the preparation of the oocyte cytoplasm includes enucleation of IVM oocytes, in vitro aging and exposure to reduced low temperature (10°C) before fusion (Chesné *et al.*, 1993). Under these conditions, the development of nuclear transfer embryos to blastocysts did not differ from that obtained with control IVM-IVF embryos (Heyman *et al.*, 1994). Such oocyte cytoplasms have been characterized at the biochemical level in terms of phosphorylation patterns and H1 kinase activities (Gall *et al.*, 1996). These recipient cytoplasms have initiated an interphase like stage as revealed by the disappearance of an 80 kD phosphoprotein and the shift of a 42 and a 34 kD phosphoprotein to a 38 and a 32 kD phosphoprotein respectively. Such patterns are similar to that of activated oocytes, furthermore in these recipient cytoplasms the H1 kinase activity has dropped to basal levels while MAPK (Mitogen Activated Protein Kinase) activity remained elevated. This indicates an interphase like stage of the recipient cytoplasm with low MPF activity before the time of fusion with the foreign nucleus; this is confirmed by the existence of an interphasic microtubule network which is clearly observed by immunochemistry on material fixed at the time of fusion (Adenot *et al.*, 1997)). The disappearance of MPF activity has been described to occur in bovine *in vitro* matured oocytes about 3h after activation and preceeds pronuclear formation (Campbell *et al.*, 1993). Oocyte pre-activation may also be induced by electric stimulus (Kono *et al.*, 1989), or chemical treatments such as ethanol/cycloheximide (Presicce and Yang, 1994) or ionophore followed by 6-DMAP incubation (Loi *et al.*, 1998). The use of a preactivated recipient cytoplasm at the time of nucleus introduction resulted in improved development of the nuclear transfer embryos into blastocysts *in vitro* [Kono *et al.*, 1994; Stice *et al.*, 1994; Du *et al.*, 1995).

2.3 EFFICIENCY OF EMBRYO CLONING IN CATTLE

2.3.1. In vitro Development

Using fresh donor morulae as the source of nuclei for cloning results in blastocysts rate which are similar to that obtained after In vitro fertilisation (IVF). In a previous study (Heyman *et al.*, 1994, 1995), we have cultured nuclear transfer embryos under the same conditions as control IVF embryos derived from the same batches of *in vitro* matured oocytes, 30.2% of the reconstituted eggs developed into blastocysts (150/497) which was not different from control IVF blastocyst rate 33.8% (222/657). This suggests that the recipient cytoplasm plays a major role in cloning efficiency and that any progress in the control of *in vitro* maturation would benefit to the in vitro development of cloned embryos. Different culture systems and media are being used for the *in vitro* production of blastocysts after nuclear transfer, from sophisticated coculture methods to more simple defined media such as SOF. Nuclei from late compacted morulae support the development of nuclear transfer embryos as well as that of younger morulae or early cleavage stages (Zakhartchenko *et al.,* 1996). However, the percentage of blastocysts per donor embryo is significantly higher when using nuclei from late compacted morulae (15.7 ± 4.1) rather than nuclei from early morulae (9.8 ± 5.5). Furthermore at the time of nuclear transfer, the sex of the donor embryo can efficiently be assessed on a single blastomere in order to produce clones of blastocysts of the desired sex (Le Bourhis *et al.,* 1998). No effect of sex of the donor nuclei on in vitro development was observed.

2.3.2 In vivo Development

In vivo development of NT blastocysts after transfer to recipient uteri is still limited and somewhat lower than that reported for IVF embryos [34]. Increased embryonic or fetal mortality have been reported (Bondioli *et al.,* 1990; Heyman *et al.,* 1995). A survey of at least 200 transfers of cloned embryos in the experimental farm of our institute indicates a decrease of the confirmed pregnancy rate from 49.5% by day 35 of pregnancy to a calving rate of 31.4%. Then 80% of the calves born survived over 2 weeks after calving (Heyman *et al.,* 1997). It is also noteworthy that some neonatal anomalies such as "big calf syndrome" have been observed (Gary *et al.,* 1994; Behboodi *et al.,* 1995) with different incidence according to the different laboratories and culture conditions (Kruip and den Daas, 1997).

2.3.3. Overall Efficiency

The number of calves born from nuclear transfer world-wide during the past years was estimated in the year 1995 to be between 1000 and 2000 in 1995 (Seidel, 1985) but has increased since that time. Almost all have been produced at private companies in North America, however, the size of the clones obtained is very limited and only exceptionally exceeds 5 identical animals from one donor embryo. The biggest clone that was obtained from embryo cloning in cattle was 11 calves using serial cloning (cited by Bondioli (1993)). The poor overall efficiency of the procedure is mainly due to low pregnancy rates and to the variability encountered in the ability of nuclei from different donor embryos to reprogramming after nuclear transplantation. Meanwhile the blastocysts rate obtained *in vitro* is satisfactory and similar to that of IVF, the potential for development of such nuclear transfer blastocysts into live calves is only about 20-30% (Westhusin *et al.,* 1991; Heyman *et al.,* 1994; Stice *et al.,* 1994) which is nearly half of that obtained with *in vitro* produced embryos through IVF (Hasler *et al.,* 1995). From compiled data of the literature, it can be estimated that the overall efficiency of embryo cloning in cattle is about 10% . That means we can expect the birth of 10 live calves from 100 blastomere nuclei fused to recipient cytoplasms.

2.3.4. Limiting Factors

Using embryo cloning, the potential number of offspring is restricted by the limited number of donor nuclei available in an early stage parent embryo. One way to increase the number of blastocysts available from one parent embryo is to recycle first generation nuclear transfer embryos as the nuclear donors for a second and third generation nuclear transfer (recloning). Experiments by Stice and Keefer (1993) reported as many as 54 genetically identical bovine embryos derived from one parent donor morula through recloning. Birth of calves have been reported for second and third generation NT embryos [Bondioli, 1993; Ectors *et al.,* 1995; Le Bourhis *et al.,* 1996) but pregnancy rates seem to be dramatically reduced for third generation NT embryos (Stice and Keefer, 1993) and to our knowledge, no term calves have yet been reported for bovine embryos cloned past the fourth generation. Takano *et al.* in 1996 reported that NT blastocysts of fith generation were able to initiate pregnancy when transferred to recipients but resulted in abortion and it is suggested that the incidence of chromosomal abnormalities may have increased with repeated nuclear transfer and extension of the culture period.

Recloning experiments however indicate a high variability between different parent embryos to support development by cloning, a maximum of 43 cloned blastocysts was produced from one morula while other donor morula yielded only 5 blastocysts after repeated nuclear transfer (Tanako *et al.,* 1996). Experiments in Australia (Lewis *et al.,* 1998; Peura *et al.,* 1998) combined multigenerational nuclear transfer and freezing by vitrification according to the open pulled straws (OPS) method (Kruip and den Daas, 1997) and reported the production of over fifty identical embryo clones in 3 rounds of nuclear transfer. However, limited number of calves have been born after transfer of vitrified multigenerational cloned embryos to conclude on its efficiency.

Due to the the great variability in full term development according to the parent embryo, it remains quite impossible to predict the number of offspring that could be produced from one specific donor morula, therefore the application of cloning in genetic programmes will greatly depend upon the improvement of efficiency of this technology. With the recent possibilities offerred by somatic cloning, the limitation due to the low number of nuclei in a morulae will no more exist as differentiated tissues will provide large number of donor cells.

3. Somatic Cloning in Cattle

3.1. STRATEGIES FOR SOMATIC NUCLEAR TRANSFER AND CELL CYCLE

Somatic cells, used as the source of nuclei, can be situated in four different stages of the cell cycle indicated as G_1, S, G_2 and M. The recipient cytoplasm of the enucleated oocyte can be used either in the MII-phase (before activation of oocyte) or in the S-phase of the cell cycle (after activation of oocyte) as described for embryo cloning. At the time of nuclear transfer, the donor nucleus can be in a different stage of the cell cycle than that of the the recipient cytoplasm, with possible consequences on the ploidy of the reconstructed embryos (Campbell *et al.*, 1996). The transfer of nuclei in S or G_2 phase to non-activated cytoplasts, where a high concentration of maturation promoting factor is present, leads to nuclear envelope breakdown (NEBD) and premature chromosome condensation (PCC), resulting in damaged chromatin or tetraploidy respectively. At the opposite, in the strategy of fusion in an activated oocyte (interphase), all phases of the cell cycle of the donor cell are theoretically compatible with a successful nuclear transfer because neither nuclear envelope breakdown nor PCC occurs before the first division of the reconstituted embryos (see review by Renard (1998)).

3.1.1. Synchronisation of the Donor Cells

To improve the matching of the donor cell cycle with that of the recipient cytoplast, the somatic cells can be synchronized during in vitro culture. Donor cells can be induced to enter the G_0 phase or quiescence, by serum starvation during a short period of time before fusion with a MII cytoplast. Induction of quiescence in somatic donor cells is considered as a pre-requisite for successfull somatic nuclear transfer in sheep (Wilmut *et al.*, 1997). In cattle, Zakhartchenko *et al.* (1999b) could significantly improve the cloning efficiency by using the serum starvation of the donor cell and numerous other studies claim the use of G_0 cells for the embryo reconstruction (Katot *et al.*, 1998; Baguisi *et al.*, 1999; Wells *et al.*, 1999). At the opposite, in other studies, no difference was found in the rate of blastocysts obtained either with serum starved cells or non-serum starved cells (Hill *et al.*, 2000; Kubota *et al.*, 2000; Vignon *et al.*, 2000) with adult cells (Table 1). For unclear reasons, it seems that the serum starvation increased the potential of development to the blastocyst stage only in the case of fetal cells (Zakhartchenko *et al.*, 1999b; Hill *et al.*, 2000), whereas no benefit was recorded in the case of adult cells (Zakhartchenko *et al.*, 1999b; Hill *et al.*, 2000; Vignon *et al.*, 2000). However, it must be mentionned that, in the majority of the cultured somatic cells used as donor for the nuclear transfer, less than half of the population is really in an actively dividing state and 60 to 80% are in the G_0+G_1 phases of the cell cycle. According to the rate of blastocysts generally obtained in cloning with somatic cells, it remains possible that it is just the nuclei which are issued from those cells that give the reconstructed embryos with the potential to further develop.

New insights are given by recent experiments of mouse cloning with embryonic stem (ES) cells. Nuclei from ES cells taken from actively dividing culture were efficiently able to direct developement to term after nuclear transfer by microinjection in enucleated oocytes (Wakayama *et al.*, 1999). Meanwhile it has frequently been emphasized that the success of cloning is dependent on nucleus donors being in the G_0-phase of the cell cycle, it seems reasonable to conclude that this hypothesis is unlikely to be universally true. All the groups using different methods of somatic cloning in cattle have reported the full term development of at least some nuclear transfer embryos (Table 2) and the question of whether the small proportion of cells developing into live animals were really in G0 remains open.

Table 1. In vitro *development of bovine somatic nuclear transfer embryos derived from either non-starved or starved bovine fibroblasts.*

Cell culture type	Origin of cells	Recipient cytoplast at fusion	No. of fused embryos	Blastocysts (%)	Reference
Non-starved fibroblasts	Fetal	Interphase	116	10 (8.6)	Vignon *et al.*, 1999
		MII	67	8 (12)	Hill *et al.*, 2000
		MII	174	35 (20)	Zakhartchenko *et al.*, 1999b
	Adult	Interphase	391	12 (3.1)	Vignon *et al.*, 2000
		MII	102	28(28)	Hill *et al.*, 2000
		MII	102	28 (28)	Kubota *et al.*, 2000
		MII	152	49 (32)	Vignon *et al.*, 2000
Starved fibroblasts	Fetal	Interphase	275	23 (8.4)	Vignon *et al.*, 1999
		MII	73	31(43)	Hill *et al.*, 2000
		MII	205	80 (39)	Zakhartchenko *et al.*, 1999b
	Adult	Interphase	343	12 (3.5)	Vignon *et al.*, 1999
		MII	88	25(28)	Hill *et al.*, 2000
		MII	114	24 (21)	Kubota *et al.*, 2000
		MII	124	44 (35)	Vignon *et al.*, 2000

3.1.2 Different Somatic Cells Used

In the bovine, successful cloning of adult animals has been achieved by using various type of cells issued from postnatal tissues such as skin, muscle, ovary, or even blood (Table 3). Due to the relatively limited number of live offspring obtained to date, the relationship between cloning success or failure and the origin of the donor cells cannot yet be established. The results suggest that in most of the tissues, at least some cells can be multiplied in vitro culture and can be fully reprogrammed after cloning. Since most of the tissues, even in old animals, contain progenitor cells which are able to divide in culture and which are not in a fully differentiated state, it is not yet possible to certify that a viable offspring obtained after nuclear transfer has been derived from a terminally differentiated cell rather than a stem cell or a partially differentiated cell. Thus the complete molecular reprogramming of mammalian differentiated cell nuclei is still a challenge. On the other hand, the somatic cells population might demonstrate a high potential for nuclear reprogramming if those progenitors (stem cells) which are present in immature tissues could be isolated and cultivated as ES-like cells. A number of tissues would then be good candidates to provide these stem cells: bone marrow, liver, lung, skin, intestine, but they will not be necessarily easy to collect from the donor animal. It should be kept in mind that the accessibility and non-invasiveness of the tissue biopsy is an important condition for further application of somatic cloning from adult animals.

Table 2. *Comparative data on the full term developmental potential of embryos derived from non-starved (proliferating) or starved (quiescent) bovine fibroblasts.*

Cell culture type	Source of cells	Blastocysts transferred	Fetus on day 30 (%)	Offspring (% of blastocysts)	Reference
Non-starved fibroblasts	Fetal	16	3 (18.7)	2 (12.5)	Vignon *et al.*, 1999
		2	0		Hill *et al.*, 2000
		7	2 (28.6)	2 (28.6)	Zakhartchenko *et al.*, 1999b
	Adult	18	1 (5.5)	1 (5.5)	Vignon *et al.*, 1999
		12	2 (16.7)	0	Hill *et al.*, 2000
		15	0		Kubota *et al.*, 2000
		148	34 (22.3)	7 (4.7)	Vignon *et al.*, 2000
Starved fibroblasts	Fetal	43	5 (11.6)	2 (4.6)	Vignon *et al.*, 1999
		30	5 (16.7)	0	Hill *et al.*, 2000
		16	9	0	Zakhartchenko *et al.*, 1999b
	Adult	11	1 (9.0)	0	Vignon *et al.*, 2000
		14	4 (28.6)	1 (7.1)	Hill *et al.*, 2000
		15	3 (20.0)	0	Kubota *et al.*, 2000
		24	9 (37.5)	1 (4.2)	Vignon *et al.*, 2000

3.1.3. Storage of Cell Lines

A variety of differentiated cell types have now been used for nuclear transfer experiments, most of them have been cultured in vitro over several passages and then frozen before use as a source of nuclei. Meanwhile the overall efficiency of adult somatic cloning is still low, when there is the need to conserve domestic endangered breeds, it is not possible to wait a decade for the cloning technology to be developed, without implying a decade of further animal genetic resources erosion. In this context it may be urgent to sample and store cells from somatic tissues for further use as a source of nuclei for cloning.

The most practical tissue to collect and store is probably skin tissue. It can be obtained easily and rapidly from a simple skin biopsy which is non invasive and can be done with very minimal health risk to the donor animal. There is now increasing experience available concerning the collection, culture and cryo-preservation of the cells. It is especially the case for fibroblast cells which are grown in vitro from skin biopsies taken on a live cow (Vignon *et al.*, 2000). Generally we proceed as follows: after shaving a 5 by 5 cm patch of skin at the ear level and disinfection, a minimal biopsy of about 5 mm is cut and immediately plunged in a sterile isotonic saline and kept cool (+4°C). The sample is transported to the laboratory for the cell multiplication step. A primary culture is developed for about 6-8 days and then subcultures are performed by passage every 3 days. At subconfluence of fibroblasts, the cells are trypsinized for dissociation and pelleted in order to get a concentration of 10.10^6 cells/ml. The cell suspension is equilibrated in serum containing 10% dimethylsulfoxide as cryoprotectant and aliquoted in cryovials or straws before freezing and storage. Long term storage has to be cryopreservation in liquid nitrogen (-196°C) as no other technology is currently available to provide the storage temperature that ensures total cessation of biological activity at the molecular level, i.e. lower than -150°C. Of course the cryopreservation process must be totally reversible and the cells must be able to grow again in vitro after thawing. A better freezing efficiency is obtained when the cells are homogeneous and not over-diluted. In such conditions frozen somatic cell suspensions can be stored for years without any loss of potential which is very important for projects aimed at the conservation of genetic resources.

Table 3. *Developmental potential of nuclear transfer embryos reconstituted with different sources of adult cells in cattle.*

Type of cells (tissue)	No. of reconstructed embryos (%)	No of blastocysts		Calves born (%)	Reference
		Total (%)	Transferred		
Cumulus (ovary)	47 (47)	18 (49)	6	5 (83)	Kato *et al.*, 1996
Granulosa (ovary)	552 (77)	152 (27)	100	10 (10)	Wells *et al.*, 1999
Epithelial (oviduct)	94 (63)	20 (23)	4	3 (75)	Kato *et al.*, 1996
Epithelial (mammary gland)	140 (63)	25 (18)	4	1 (25)	Zakhartchenko *et al.*, 1999a
Fibroblasts (skin)	840 (70)	159(19)	133	9(6.8)	Heyman *et al.*, 2000 Vignon *et al.*, 2000
Leukocytes (blood)	698 (79)	124 (18)	50	1 (2.0)	Galli *et al.*, 1999

3.2. ACTUAL EFFICIENCY OF BOVINE SOMATIC CLONING

3.2.1. In vitro Development

Increasing rate of development are reported after nuclear transfer of bovine somatic cells. This is mainly due to progress in oocyte preparation, improved *in vitro* culture methods as well for somatic cells or for nuclear transfer embryos. In vitro development rates of reconstituted embryos into blastocysts range from 510% to up to 50 % according to the cell line and the tprotocole used (Tables 1, 3). In somatic cloning, the possible influence of the age of the donor animal on efficiency is still a matter of debate. Compilation of recent data from the litterature does not reveal an influence of age at least for *in vitro* development potential. If we consider fibroblasts originating from skin biopsies as the source of cells, the proportion of NTE developed into blastocysts in vitro is in the same range whatever the cells are of fetal or adult origin (Table 1) and successfull cloning has been achieved from somatic cells taken on old animals such as a 21 years old Brahman bull (Hill *et al.,* 2000) and a 17 years old Japanese Black beef bull (Kubota *et al.,* 2000).

3.2.2. Prenatal Development

Data obtained in the bovine species indicate that, whatever the method used for the reconstitution of the embryos, only about 10% of the blastocysts transplanted in recipient heifers result in a full term development. This is four times less than the rate obtained in the same species with embryos issued from IVF technique (Heyman *et al.,* 1999). However after transfer of blastocysts derived from somatic cloning the initiated pregnancy rate as assessed by ultrasonic scanning as early as one month after transfer, is acceptable and close to 50% but then many embryonic and fetal loss occur at different periods of gestation (Heyman *et al.,* 1999). High rate of abortion have been reported (Cibelli *et al.,* 1998; Vignon *et al.,* 1998). These late abortions are frequently related to the presence of hydroallantois, or to an abnormal development of the placenta, (Heyman *et al.,* 1999; Hill *et al.,* 1999). The origin of the high rate of abortion and frequent perinatal death in somatic clones is not clearly identified, it may relate to one or a combination of such parameters as nuclear reprogramming, cultured donor cells, in vitro culture systems or nuclear transfer manipulation itself. An incomplete nuclear reprogramming of donor cell nuclei after transfer is often suspected because if some imprinted genes which are involved in the regulation of placentation escape complete reprogramming, then they can be aberrantly expressed during placental development, leading to incomplete communication between the placenta and the embryo. Some of the genes responsible for fetal growth and development are imprinted genes which are involved in the regulation of growth factors, and recent findings indicated that the concentration of factor IGF-I was lower in the blood of cloned calves as compared to aged-matched control calves (Chavatte-Palmer *et al.,* 2000). Such observations suggest that epigenetic events can affect the correct reprogramming of gene activity in somatic nuclear transfer embryo.

3.2.3. Viability of Offspring

The total number of calves born from somatic cloning to date over the world is still limited and can be estimated to be around 200 animals mainly in North America, Japan, Europe, Australia and New Zealand. These animals were born since 1998, so that the oldest ones are just becoming adults ones. Therefore, there is very few information available on the long term viability of cattle produced through somatic nuclear transfer.

Data from sheep and especially from the pioneer cloned animal Dolly indicate that this female developed normally and was proven to be fertile. In cattle, a limited number of male cloned offspring have now reached puberty and these young bulls are able to produce semen (Heyman, unpublished data). However, some pathologies can affect part of the somatic cloned calves even after birth. The Large Offspring Syndrome (LOS) already reported after the transfer of *in vitro* cultured embryos in sheep and cattle, also occurs also after somatic cloning. Pathologies may affectIt appears also that only one function of the organism such ascan be apparently affected : the respiratory function, the circulatory system (Hill *et al.*, 1999) or even the immune system as attested by the occurrence of one case of thymic hypoplasia (Renard *et al.*, 1999). Possible long term effects are difficult to study on a species such as cattle who has a long generation interval, but recent success of somatic cloning in the mouse model (Wakayama *et al.*, 1998) will provide a useful tool to study eventual consequences of somatic cloning over several generations of animals.

4. Applications of Bovine Cloning

4.1. POTENTIAL USE IN BREEDING SCHEMES

In cattle, most of the genetic improvement is achieved through Artificial Insemination (AI) from selected bulls and embryo transfer from selected bull dams to generate the next generation of AI bulls. However the diffusion potential from highly valuable bulls can be impaired by accidental death of the progenitor. Somatic cloning allows to secure this risk by generating a genocopy of the bull once it has been proved to be superior. Recent work in Italy has shown that a live offspring could be obtained from an aged AI bull (Galli *et al.*, 1999) using lymphoid blood cells as a source of nuclei.

Female cloning could also be a very useful tool to improve annual genetic gain in conventional dairy breeding schemes that are already using intensive embryo transfer (Colleau, 1992). Cloning of female embryos recovered from elite cows mated to a selected bull would give birth to genetically identical heifers among which the future donors for the genetic programme could be chosen. If we consider milk production trait (heritability $H^2 = 0.25$), calculations indicate that the precision for genetic evaluation of a clone of 5 females is as high as that of a bull evaluated from the lactations of at least 25 daughters. Furthermore, the precision of evaluation is increased by the fact that the performances of one given genotype are evaluated through several copies of this genotype that are placed in different environmental conditions. Calculations by Colleau *et al.* (1998) indicate that the use of somatic cloning instead of embryo cloning from elite cows could improve by 20% the annual genetic gain provided that sets of 3 to 5 cloned heifers could be obtained from each selected donor. In this situation, adult somatic cloning would result in a higher annual genetic progress because the first lactation of the bull dam donor of somatic cells is already known (Table 4).

Cloning could also be used for diffusion of some elite genotypes that are well characterized and have interesting genes such as those involved in longevity or disease resistance. A good example of such application is provided by the work of Wells *et al.* in New Zealand (1999). They established a primary cell line from mural granulosa cells collected by aspirating the ovarian follicles from a Friesan dairy cow of high genetic merit, using ultrasound-guided Ovum Pick Upvaginal probe (OPU) and demonstrated that it was possible to obtain at least 10 cloned calves genetically identical to the adult elite donor cow.

Table 4. *Possible impact of cloning on annual genetic gain in dairy cattle.*

Origin of donor cells	Age of female donors	Nb of calves per clone	Annual genetic gain*
Embryos	4 years	3	+5%
		5	+10%
Fetal fibroblasts	4 years	3	-3%
		5	+1%
Adult fibroblasts	2 years	3	+21%
		5	+27%

* comparison with standard breeding scheme involving embryo transfer from 2 years old females. Adapted from Colleau *et al.* (1998).

In beef breeds, potential applications for cloning do exist (Smith, 1989), breeding schemes have to consider criteria of meat quality. Such measurements could be made directly on the carcass of clones of a selected bull and not on his offspring. This would also save time in the selection programme by allowing to multiply very valuable phenotypes. That is the case in Japan where there is a shortage of top quality meat with specific infiltration of fat in the muscle. As it is known that in the same breed, the carcass quality is a highly variable trait, adult cloning from the best phenotypes of this breed could contribute to standardize top quality meat. Therefore, different groups are actively developing nuclear transfer in cattle in order to fit this objective.

4.2. CLONING FOR ENDANGERED BREEDS

If used on a very large scale, somatic cloning could lead to a potential risk of decreasing genetic diversity. This could happen if thousands of copies of the same animal were diffused in livestock which is far from being is not the case. On the opposite, somatic cloning could be a valuable tool for maintaining and developing endangered breeds of cows or other species. Provided semen from different bulls is stored, a very limited number of females can be multiplied by cloning and then further inseminated by the stored semen in order to re-create different families and re-establish a minimum of genetic diversity. In New Zealand, somatic cloning was used to preserve the last surviving cow of the Enderby island cattle breed and at least 4 calves were generated from this single surviving female (Wells *et al.*, 1998). This application of cloning has to be considered right now as it is already possible to store the genetic material from endangered breeds after a very small skin biopsy of the live animal, multiplication of the cells *in vitro* and freezing of the cells in liquid nitrogen for a further use when nuclear transfer efficiency has improved.

4.3. BOVINE CLONES AS ANIMAL MODELS

The generation of geFnetically identical offspring from the same parent animal can provide a valuable tool for agronomical research. In cattle, sets of cloned calves can be used as animal models for experiments on pathology studies, nutrition behaviour or reproduction evaluation. In France, at INRA, we have developed such an approach to minimize Tthe number of cattle that are required for an experimental project can be minimized (Lambertson, 1994). For example, feeding behavior as assessed by quantity of food intake, number of daily meals or rumination time, indicate that the coefficient of variation for these criteria is significantly reduced between cloned animals compared to control non cloned animals (Baumont *et al.*, 1995). For reproduction studies, the developmental oocyte competence can be studied through repeated ovum pick up and IVF on a limited number of cloned heifers.

At the laboratory level, *in vitro* development of nuclear transfer embryos derived from differentiated cells fused to enucleated oocytes provides a valuable tool to investigate nuclear reprogramming. The interactions between the oocyte cytoplasm components and the nucleus can be studied in relation to their respective cell cycle stage, in order to define the optimal conditions for development up to the blastocyst stage. The most important point is to avoid any DNA damage and to maintain a correct ploidy in the reconstituded embryo by adequate matching of the donor and recipient cell cycles. Furthermore the recent work on interspecific nuclear transfer using cattle oocytes as recipients for a variety of donor nuclei from other species (Dominko *et al.*, 1999) opens the way to better understand the species-specificities in early development. Trans-species nuclear transfer would have potential clinical application for human medicine by allowing the generation of human embryonic stem cells from a somatic cell of a patient. The use of bovine oocytes would bypass some of the ethical issues associated with the use of human IVF embryos to generate human therapeutic cell lines.

4.3.1. Are Clones Identical ?

Genetically identical animals obtained by nuclear transfer have the same nuclear genome but may present some phenotypical differences due to epigenetic phenomena. The most visible of these differences concern the coat color. For example a clone of Hholstein breed calves may show black and white spotting differences. The black color is due to the presence of a pigment (melanin) contained in melanocytes. These melanocytes are derived from stem cells, melanoblasts located in the neural crest. The migration of melanocytes towards hair follicles is determined genetically but also depends on fetal growth (Seidel, 1985). In cattle, cloned embryos derived from the same cell source are usually transferred to the uterus of different recipient females which provide different uterine environment for fetal growth that can partly explain different migration of melanocytes.

Furthermore, during the nuclear transfer procedure itself, the recipient cytoplasms are prepared from oocytes of different genetic origin. These recipient cytoplasm contain mitochondrial DNA and maternal transcripts. Meanwhile the mitochondrial DNA sequence is very limited compared to the bovine nuclear genome, it is more susceptible to mutation and its role is important for the energy metabolism of the cell. During nuclear transfer, differential mitochondrial genotype segregation can occur within the same clone (Smith *et al.*, 2000) but in cattle, the factors affected by this cytoplasmic inheritence remain to be determined.

5. Transgenesis and cloning

A transgenic animal is one carrying in its genome foreign recombinant DNA molecules that were introduced by manipulation in the laboratory instead of by conventional genetic selection and animal breeding. The transgene is a recombinant DNA molecule that includes at least 2 parts : a regulatory element or promoter which confers tissue specificity, modulation of expression, and a structural part which consists of DNA sequence that encodes the genetic information necessary to synthetise the desired protein.

5.1. GENERATION OF TRANSGENIC ANIMALS THROUGH SOMATIC CLONING

5.1.1. Limits of Transgenes is Through Micro Injection

During the past decade, limited numbers of transgenic livestock animals have been produced in species such as pigs, sheep or cattle (for review, see Wall (1996)). These transgenic animals have been obtain ed by pronuclear injection of a foreign DNA construct. In cattle, this technology of microinjection was restricted to very few laboratories with enough technical and economical potential for processing thousands of bovine zygotes before getting a transgenic offspring. In fact, transgenesis in livestok through pronuclear injection suffers several serious limitations. The most important limitation is that the DNA can only be added but not deleted or modified in situ. Microinjection of the construct into the pronucleus results in random integration of the foreign DNA which can lead to erratic expression due to effects at site of incorporation.

The second point is that microinjection is of very low efficiency, most of the animals derived from pronuclear injectoion do not express the transgene or are mosaic. Finally, the proportion of injected zygotes that result in transgenic offspring able to transmit into its germline is very low, in the range of 1 out of 1600 according to Wall (1996). For all these reasons, generation of transgenic cattle through microinjection remains inefficient, very expensive and far from ecomomic objectives.

5.1.2. Gene Targetting in Somatic Cells

In contrast, with random insertion that characterizes pronuclear injection, the technique of homologous recombination offers several advantages as it allows the precise introduction of a transgene into selected areas of the genome. Such technology not only permits the introduction of a new gene at the right place but also is a way to correct a specific mutation, replace or inactivate a specific gene. Unfortunately the frequency of homologous recombination events per cell modified is very low and can be as low as 1 per 10 million. In this situation, the carrier cell line must be cultured for a long time and still be able to generate a live offspring after modification. This is the case for embryonic stem (ES) cells derived from totipotent cells of early embryos. Such cells are capable of unlimited proliferation in vitro but have only been obtained in the mouse species (Evans and Kaufman, 1981). Targetted mutagenesis of murine ES cell is widely used and hundreds of knockout mice strains have been generated according to the 1999 database. In domestic species and particularly in cattle, no real ES cell lines have been established despite intensive research focussed on this aim (Strelchenko, 1996) and only ES-like cells could be obtained but their totipotency is not maintained over culture passages.

With the advent of somatic cloning in domestic species, it is possible to circumvent the need for bovine true ES cells. Ccultured somatic cells have been used to introduce genetic changes in the germline in sheep (Campbell *et al.*, 1996; Schnieke *et al.*, 1997), and also in cattle (Cibelli *et al.*, 1998). Clearly, both fetal and adult somatic cells can be used for cloning after a prolonged culture period and genetic modification during the in vitro culture of fibroblasts. It appears possible to consider somatic cells as carriers for genetic manipulation through homologous recombination. A first report indicates that homologous recombination has been successfully achieved in sheep by McCreath *et al.* (2000). These authors describe a reproducible gene targetting in fetal fibroblasts to place a therapeutic transgene encoding for human α-antitrypsin, at the ovine α 1 (I) procollagen (COL1A1) locus. These modified cells have then been used as donor cells for nuclear transfer and have generated cloned sheep expressing human α-a 1 antitrypsin in their milk. This opens the way for successful utilisation of homologous recombination on somatic cells of domestic mammals in associtation with cloning technology especially for the bovine species. Gene mapping and the utilisation of genetic markers will allow the identification of important new candidate genes for transgenesis in large animals.

5.1.3. Examples of Genetic Modifications in Cattle

5.1.3.1 Milk Modifications. The most studied transgene systeme in livestock species is the modification of proteins produced in the mammary gland. The annual protein production from dairy cows is over 3 billion of kg of protein in the US. (USDA, Agriculture Fact Book, 1994). Such a protein source could provide a variety of pharmaceuticals for human medecine through « gene pharming », or milk with enhanced health value (neutraceuticals) or milk that can be easily transformed by the industry into cheese (caseins).

The bovine mammary gland as a bioreactor for therapeutic agents has been extensively reviewed (Houdebine, 1995; Wall *et al.*, 1997; Ziomek, 1998) but direct impact of biopharming in the dairy industry is low as a limited number of herds of transgenic dairy cows would produce enough to cover the world demand. The situation could be very different for milk modification aimed at improvement of nutritional and manufacturing properties.These objectives could benefit the dairy industry in the future. Among the possible changes in milk composition through transgenesis, the main applications are:

- suppression of β lactoglobulin in cow milk. This protein only present in the milk of ruminants is thethe main allergen of bovine milk for human. and Iits reduction would clearly improve human consumption by limiting allergy problems;
- increase of ing casein content. Caseins are the major proteins in milk, they entrap fat and warter during cheese manufacture and have an important role in the cheese yield. Experiments on the mouse model have already shown that the presence of extra κ casein in transgenic mice resulted in significant increased curd strength and a smaller micelle diameter (Gutierrez-Adan *et al.*, 1996);
- alteration of fat content. Fat reduction in milk through genetic selection is difficult because of high correlation with proteins. Modifications such as decrease of the acetyl CoA expression could reduce de novo fat synthesis (Bremel *et al.*, 1989). It is also possible to increase the expression of stearoyl CoA desaturase and thus reduce the presence of saturated fatty acids in milk, improving its dietetical quality for direct human consumption;
- reduction of lactose. This sugar in bovine milk is responsible for digestive disorders in a high proportion of the human adult population. The production of milk with reduced lactose would not only benefit human consumption but also result in increased efficiency in cheese manufacturing (Hettinga, 1989). Transgenic mice that were deficient for α lactalbumin (this protein is a component of lactose synthetase complex) have been produced through homologous recombination (Stinnakre *et al.*, 1994). As a consequence of this genetic manipulation, mice produced milk with little or no lactose. Such an approach using targetted transgenes is appears to be feasible in the cow.

Other qualitative changes of cow milk are aimed to resemble the composition of human milk. Cow milk normally contains only very little lactoferrin and lysosyme compared to human milk. Such proteins which have a bacteriostatic effect can be added to cow milk via transgenesis. In the same way, specific monoclonal antibodies and vaccinating antigens that are active orally could be present in cow milk and fit into the definition of neutraceuticals.

5.1.3.2. Knock Out of the Prion Gene. Observations that mice expressing mutant forms of PrP develop spontaneous scrapie-like disease, suggest that a normal PrP protein is a necessary link for the development of Prion-like diseases (Bueler *et al.*, 1993). Removal of this prion Protein gene from cattle would result in animals resistant tp spongiform encephalopathy (BSE). Such a goal could be achieved on bovine somatic cells through homologous recombination to inactivate the gene and through cloning to generate knockout PrP cattle.

Table 5. Comparison of transgenic cattle production efficiency using cloning or injection method. Adapted from Chan (1999).

Methods	Cloning (Nuclear transfer)	Injection (into pronucleus of Zygote)
Nb oocytes or zygotes for gene transfer	276	25 023
Nb developed to blastocysts	33 (12%)	1282 (5%)
Embryos transferred to recipient animals	28	978
Nb of calves born alive (% of embryos transferred)	5 (18%)	134 (14%)
Nb transgenic (% of calves born)	5 (100 %)	9 (7%)
Transgenes	pCMV/□-GEO	Human α-lactalbumin
Overall efficiency:	55	2780
Nb of oocytes used /transgenic calf	5.6	108
Nb of recipients used /transgenic calf		

5.2. SOMATIC CLONING FOR IMPROVED TRANSGENESIS EFFICIENCY

Most of the somatic cells can be cultured in vitro over many passages and result in live offspring after nuclear transfer. Calves have been produced using adult fibroblasts cultured over 15 passages before nuclear transfer (50). Such a culture period allows already to insert a new gene into the nuclei of the somatic cells by transfection, and select the modified cell line. Recently, gene targetting by homologous recombination has been proven to be feasible in sheep somatic cells (57), similarly to what has been achieved in murine ES cells. Homologous recombination will allow a gene of interest to be inserted at a specific chromosome site of the bovine genome in order to facilitate high expression, in the mammary gland for example, and produce therapeutic or nutritional proteins directly in milk. It would also be possible to replace bovine serum albumin with the human form in order to have a cost effective production of human serum albumin in the milk of cows. Using such modified somatic cells as a source of nuclei for cloning will enable to produce a first generation of identical offspring which are all transgenic without mosaicism. In the longer term, the combination of this technology with the information on the bovine genome mapping may allow the targetting of DNA segments coding for specific production traits such as those coding for disease resistance, meat quality or growth. The marriage of nuclear transfer and transgenesis will allow to save time and money in generating transgenic cattle. The minimum interval for producing a generation of transgenic bulls able to transmit their characters in the germline could be about 2 years when using modified cells in nuclear transfer compared to at least 4 years by the way of microinjection and chimeras. Furthermore, the nuclear transfer approach to generate transgenic calves can be 50 times more efficient than microinjection in terms of number of cattle oocytes necessary to get one transgenic calf Table 5).

Provided targetted transgenesis is achieved on the somatic cell line, then we can expect that 100% of the calves born from nuclear transfer using this cell line are effectively transgenic without any mosaicism.That is not the case after pronuclear injection were mosaicism rates over 80% are reported (Chan *et al.,* 1999).

6. Conclusions

Generating clones through nuclear transfer has met with a considerable evolution during the few past years with the possibilities offered by somatic cloning. At the same time, progress with the tools of molecular biology and homologous recombination opens a new era in transgenic biotechnology. In livestock, the combination of cloning and genetic modification of somatic cells opens the way for inducible knockin and knockout and other targetted integrations. However, meanwhile recent progress in cattle or sheep, this technology is still in its active research phase and far from the standard methodology used in the mouse model. Somatic cloning is at its very beginning and needs a huge amount of basic research for better understanding of the mechanisms of correct nucleus reprogramming and improvement of efficiency. Besides the technical aspects, the possible use of transgenic cattle for biomedical purpose or for food production will require very high standards for safety and genetic security. Introduction of genes into livestock meets also the barrier of societal concerns and public acceptance of the use of transgenic animals. It is also clear the all the possible applications of these biotechnologies must consider animal welfare and ethical concerns.

References

Adenot, P., Szollosi, M., Chesne, P., Chastant, S. and Renard, J.P. (1997) *In vivo* aging of oocytes influences the behavior of nuclei transferred to enucleated rabbit oocytes, *Mol. Reprod. Dev.* 46, 325-329.

Baguisi, A., Behboodi, E., Melican, D.T., Pollock, J.S., Destrempes, M.M., Cammuso, C., Williams, J.L., Nims, S.D., Porter, C., Midura, P., Palacios, M.J., Ayres, S.L., Denniston, R.S., Hayes, M.L., Ziomek, C.A., Meade, H.M., Godke, R.A., Gavin, W.G., Overström, E.W. and Echelard, Y. (1999) Production of goats by somatic cell nuclear transfer, *Nature Biotech.* 17, 456-461.

Barnes, F., Endebrock, M., Looney, C., Powell, R., Westhusin, M. and Bondioli, K. (1993) Embryo cloning in cattle: the use of *in vitro* matured oocytes. *J. Reprod. Fert.* 97, 317-320.

Barnes, F.L., Collas, P., Powell, R., King, W.A., Westhusin, M. and Shepherd, D. (1993) Influence of recipient oocyte cell cycle stage on DNA synthesis, Nuclear envelope breakdown, chromosome constitution, and development in nuclear transplant bovine embryos, *Mol. Reprod. Dev.* 36, 33-41.

Baumont, R., Detour, A., Jamot, J., Cellarier, E., Jaillier, R., Geay, Y. and Renard, J.P. (1995) Individual variations of voluntary intake, feeding behaviour and development of the digestive tract in holstein cloned growing bulls, *Proc. IVth Congress on ruminant nutrition*, **Clermont-Ferrand.**

Behboodi, E., Anderson, G., Bondurant, R., Cargill, S., Kreusher, B., Medrano, J. and Murray, J. (1995) Birth of large calves that developed from in vitro-derived bovine embryos, *Theriogenology* 44, 227-232.

Bondioli, K.R., Westhusin, M.E. and Looney C.R. (1990) Production of identical bovine offspring by nuclear transfer, *Theriogenology* 33, 165-174 .

Bondioli, K.R. (1993) *Proc. Symp. Recent Develop. Embryo Biotech. Farm Anim.* Baton-Rouge, pp. 21-23

Bremel, R.D., Yom, H.C. and Bleck, G.T. (1989) Alteration of milk composition using molecular genetics, *J. Dairy Sci.* 72, 2826-2833.

Bueler, H., Aguzzi, A., Sailer, A., Greiner, R., Autenried, P., Aguet, M. and Weissman, C. (1993) Mice devoid of PrP are resistant to scrapie, *Cell* 73, 1339-1347.

Campbell, K., Ritchie, W. and Wilmut, I. (1993) Disappearence of maturation promoting factor activity and the formation of pronuclei in electrically activated *in vitro* matured bovine oocytes, *Theriogenology* 39, 199.

Campbell, K.H.S., Loi, P., Otaegui, P. and Wilmut, I. (1996) Cell cycle coordination in embryo cloning by nuclear transfer, *Rev. Reprod* 1, 40-46.

Campbell, K., McWhir, J., Ritchie, W. and Wilmut I. (1996) Sheep cloned by nuclear transfer from a cultured cell line, *Nature* 380, 64-66.

Chan, A. (1999) Transgenic animals: current and alternative strategies, *Cloning* 1, 25-46.

Chan, A., Kukoil, G., Skalka, A.M. and Bremel, R.D. (1999) Timing of DNA integration, transgenic mosaicism, and pronuclear injection, *Mol. Reprod. Dev.* 52, 406-413.

Chavatte-Palmer, P., Heyman, H., Monget, P., Richard, C., Kann, G., LeBourhis, D., Vignon, X. and Renard, J.P. (2000) Plasma IGF-1, IGF-BP, and GH concentrations in cloned neonatal calves from somatic and embryonic cells : preliminary results, *Theriogenology* 53, 213.

Chesné, P., Heyman, Y., Peynot, N. and Renard, J.P. (1993) Nuclear transfer in cattle: birth of cloned cattle and estimation of blastomere totipotency in morulae used as source of nuclei, *C. R. Acad. Sci.* 316, 487-491.

Cibelli, J.B., Stice, S.L., Golueke, P.J., Kane, J.J., Jerry, J., Blackwell, C., Ponce de Leon, F.A.P. and Robl, J.M. (1998) Cloned transgenic calves produced from nonquiescent fetal fibroblast, *Science* 280, 1256-1258.

Collas, Ph. and Robl, J.M. (1991) Development of rabbit nuclear transfer embryos from morula and blastocyst stage donor nuclei, *Theriogenology* 35, 190.

Collas, P. and Barnes, F.L. (1994) Nuclear transplantation by microinjection of inner cell mass and granulosa cell nuclei, *Mol. Reprod. Dev.* 38, 264-267.

Colleau, J.J. (1992) Combining use of embryo sexing and cloning within mixed MOETs for selection on dairy cattle, *Genet. Sel. Evol.* 24, 345-361.

Colleau, J.J., Heyman, Y. and Renard, J.P. (1998) Les biotechnologies de la reproduction chez les bovins et leurs applications reelles ou potentielles en selection, *INRA Prod. Anim.* 11, 41-56.

Dominko, T., Ramalho-Santos, G., Chan A., *et al.* (1999) Optimization strategies for production of mammalian embryos by nuclear transfer, *Cloning* 1, 143-152.

Dominko, T., Mitalipova, M., Haley, B., Beyhan, Z., Memili, E., McKusick, B. and First, N.L. (1999). Bovine oocyte cytoplasm supports development of embryos produced by nuclear transfer of somatic cell nuclei froma various mammalian species, *Biol. Reprod.* 60, 1496-1502.

Du, F., Jiang, S. and Yang, X. (1995) Beneficial effect of oocyte activation prior to and during nuclear transfer in cattle using in vitro matured oocytes 24 h of age, *Reprod. Nutr. Dev.* 35, 703-712.

Echelard, Y. (1996) Recombinant protein production in transgenic animals, *Curr. Opinion Biotech.* 7, 536-540.

Ectors, F.J., Delval, A., Smith, L., Touati, K., Remy, B., Beckers, J.F. and Ectors, F. (1995) Viability of cloned bovine embryos after one or two cycles of nuclear transfer and in vitro culture, *Theriogenology* 44, 925-933.

Evans, M.J. and Kaufman, M.H. (1981) Establishment in culture of pluripotential cells from mouse embryos, *Nature* 292, 154-156.

Gall, L., Dedieu, T., Chesné, P., Ruffini, S., Sevellec, C., Peynot, N., Renard, J.P. and Heyman, Y. (1996) Bovine embryo cloning : Characterization of the recipient cytoplasts by phosphorylation patterns and kinase activities, *Develop. Growth Differ.* 38, 517-525.

Galli, C., Duchi, R., Moor, R.M. and Lazzari, G. (1999) Mammalian leukocytes contain all the genetic information necessary for the development of a new individual, *Cloning* 1, 161-170.

Garry, F., Adam, R. and Odde, K. (1994) Birth weight and survivability of neonatal calves produced by nuclear transfer cloning, *J. Anim. Sci.*, 72 (suppl. 1), 374.

Gurdon, J.B., Laskey, R.A. and Reeves, O.R. (1975) The development capacity of nuclei transplanted from keratinized skin cells from adult frogs, *J. Embryol. Exp. Morphol.* 34, 93-112.

Gutierrez-Adan, A., Maga, E.A., Meade, H.M., Shoemaker, C.F., Medrano, J.F., Anderson, G.B. and Murray, J.D. (1996) Alterations of physical characteristics of milk from transgenic mice producing bovine k-casein, *J. Dairy Sci.* 79, 791-799.

Hasler, J.F., Henderson, W.B., Hurtgen, P.J., Jin, Z.Q., McCauley, A.D., Mower, S.A., Neely, B., Shuey, L.S., Stokes, J.E. and Trimmer, S.A. (1995) Production, freezing and transfer of bovine IVF embryos and subsequent calving results, *Theriogenology* 43, 141-152.

Hettinga, D.H. (1989) Why alter milk composition? *J.Dairy Sci.* 72, 2790-2800.

Heyman, Y., Chesne, P. and Renard, J.P. (1990) Reprogrammation complete de noyaux embryonnaires congeles apres transfert nucleaire chez le lapin, *C. R. Acad. Sci.Paris.* (ser III) 311, 321-326.

Heyman, Y., Chesne, P., Lebourhis, D., Peynot, N. and Renard, J.P. (1994) Developmental ability of bovine embryos after nuclear transfer based on the nuclear source: *in vivo* versus *in vitro*, *Theriogenology* 42, 695-702.

Heyman, Y., Camous, S., Chesne, P., Marchal, J. and Renard, J.P. (1995) Gestational profiles following transfer of cloned bovine blastocysts developed *in vitro*, *Theriogenology* 43, 234 .

Heyman, Y., Degrolard, J., Adenot, P., Chesne, P., Flechon, B., Renard, J.P. and Flechon, J. (1995) Cellular evaluation of bovine nuclear transfer embryos developed *in vitro*, *Reprod. Nutr. Dev.* 35, 713-723.

Heyman, Y., Marchal, J., Chesné, P., Le Bourhis, D., Deniau, F. and Renard, J.P. (1997) Outcome of pregnancies after single or twin transfer of cloned embryos in cattle, *Proc. 13th AETE Meet.*, Lyon, p156.

Heyman, Y., Chavatte-Palmer, P., Le Bourhis, D., Deniau, F., Laigre, P, Vignon, X. and Renard, J.P. (1999) Evolution of pregnancies after transfer of cloned bovine blastocysts derived from fetal or adult somatic cells, *Proc. 15th AETE Meet.*, Fondation Mérieux - Lyon, France, p. 166

Heyman, Y., Chavatte-Palmer, P., Le Bourhis, D., Camous, S., Vignon, X. and Renard, J.P. (2000) Pregnancy profiles and outcome after transfer of bovine blastocysts derived from embryonic and somatic cloning, submitted.

Hill, J.R., Roussel, A.J., Cibelli, J.B., Edwards, J.F., Hooper, N.L., Miller, M.W., Thompson, J.A., Looney, C.R., Westhusin, M.E., Robl, J.M. and Stice, S.L. (1999) Clinical and pathological features of cloned transgenic calves and fetuses (13case studies), *Theriogenology* 51, 1451-1466.

Hill, J.R., Quinton, A.W., Long, C.R., Looney, C.R., Thompson, J.A. and Westhusin, M.E. (2000) Development rates of male bovine nuclear transfer embryos derived from adult and fetal cells, *Biol. Reprod.* 62, 1135-1140.

Houdebine, L.M. (1995) The production of pharmaceutical proteins from the milk of transgenic animals, *Reprod. Nutr. Dev.* 35, 609-617.

Kato, Y., Tani, T., Sotomaru, Y., Kurokawa, K., Kato, J., Doguchi, H., Yasue, H. and Tsunoda, Y. (1998) Eight calves cloned from somatic cells of a single adult, *Science* 282, 2095-2098.

Kono, T., Iwasaki, S. and Nakahara, T. (1989) Parthenogenic activation by electric stimulus of bovine oocytes matured in vitro, *Theriogenology* 32, 569-576.

Kono, T., Sotomaru, Y., Aono, F., Takahasi, T., Ogiwara, I., Sekizawa, F., Arai, T. and Nakahara, T. (1994) Effect of ooplast activation on the development of oocytes following nucleus transfer in cattle, *Theriogenology* 41, 1463-1471.

Kruip, T.A.M. and den Daas, J.H.G. (1997) *In vitro* produced cloned embryos : effects on pregnancy, parturition and offspring, *Theriogenology* 278, 2130-2133.

Kubota, C., Yamakuchi, H., Todoroki, J., Mizoshita, K., Tabara, N., Barber, M. and Yang, X. (2000) Six cloned calves produced from adult fibroblast cells after long-term culture, *Proc. Natl. Acad. Sci.* 97, 990-995.

Lambertson, W.R. (1994) Production of identical swine: uses for families with reduced phenotypic variations, *Theriogenology* 41, 453-460.

Le Bourhis, D., Deniau, F., Marchal, J., Renard, J.P. and Heyman, Y. (1996) Bovine nuclear transfer in vitro and in vivo development of second generation clones, ***Proc. 12th AETE Meet.***, p 160.

Le Bourhis, D., Chesne, P., Nibart, M., Marchal, P., Humblot, P., Renard, J.P. and Heyman, Y. (1998) Nuclear transfer from sexed parent embryos in cattle: efficiency and birth of offspring, *J. Reprod. Fertil.* 113, 343-348.

Leibfried-Rutledge, M.L., Critser, E.S., Eyestone, W.H., Nothey, D.L. and First, N.L. (1987) Development potential of bovine oocytes matured *in vitro* or *in vivo*, *Biol. Reprod.* 36, 376-383.

Lewis, I.M., Peura, T.T., Lane, M.W., Vajta, G. and Trounson, A.O. (1998) Pregnancies and live birth achieved from vitrified multigenerational cattle nuclear transfer embryos, *Theriogenology* 49, 323.

Loi, P., Ledda, S., Fulka, J., Cappai, P. and Moor, R.M. (1998) Development of parthenogenetic and cloned ovine embryos effect of activation protocols, *Biol. Reprod.* 58,1177-1187.

McCreath, K.J., Howcroft, J., Campbell, K.H.S., Colman, A., Schnieke, A.E. and Kind, A.J. (2000) Production of gene targetted sheep by nuclear transfer, *Nature* 405, 1066-1069.

Peura, T.T., Lane, M.W., Lewis, I. and Trounson, A.O. (1998) The use of vitrified multigenerational cloned cattle embryos as donors in nuclear transfer, *Theriogenology* 49, 326.

Pieterse, M.C., Kappen, K.A., Kruip, Th.A.M. and Taverne, M.A. (1988) Aspiration of bovine oocytes during transvaginal ultrasound scanning of the ovaries, *Theriogenology* 30, 751-762.

Prather, RS., Sims, M. and First, N.L. (1989) Nuclear transplantation in early pig embryos, *Biol. Reprod.* 41, 414-418.

Prather, R.S., Sims, M. and First, N.L. (1990) Nuclear transplantation in the pig embryo: nuclear swelling, *J. Exp. Zool.* 255, 355-358.

Presicce, G.A. and Yang, X. (1994) Parthenogenetic development of bovine oocytes matured in vitro for 24 h and activated by ethanol and cycloheximid, *Mol. Reprod. Dev.* 38, 380-385.

Renard, J.P. (1998) Chromatin remodelling and nuclear reprogramming at the onset of embryonic development in mammals, *Reprod. Fertil. Dev.* 10, 573-580.

Renard, J.P., Chastant, S., Chesne, P., Richard, C., Marchal, J., Cordonnier, N., Chavatte, P. and Vignon, X. (1999) Lymphoid hypoplasia and somatic cloning, *The Lancet* 353, 1489-1491.

Robl, J., Prather, R., Barnes, F., Eyestone, W., Northey, D. and First, NL. (1987) Nuclear transplantation in bovine embryos. *J. Anim. Sci.* 64, 642-647.

Schnieke, A.E., Kind, A.J., Ritchie, W.A., Mycock, K., Scott, A.R., Ritchie, M., Wilmut, I., Colman, A. and Campbell, K.H. (1997) Human factor IX transgenic sheep produced by transfer of nuclei from transfected fetal fibroblasts, *Science* 278, 2130-2133.

Seidel, G.E. (1985) Are identical twins produced from micromanipulation always identical? *Proc. Ann. Conf. AI and embryo transfer*, Denver, 50-53.

Seidel, G.E. (1995) Sexing, bisection and cloning embryos :perspectives and applications to animal breeding. *Proc. Symp. Reprod. Anim. Breed.*, Milan, 147-154.

Smith, C. (1989) Cloning and genetic improvement of beef cattle. *Anim. Prod.* 49, 163-169.

Smith, L., Bordignon, V., Garcia, J.M. and Meirelles, F.V. (2000) Mitochondrial genotype segregation and effects during mammalian development: application to biotechnology, *Theriogenology* 53, 35-46.

Speeman, H. (1938) Embryonic development and induction, Hafner publishing Co., New York, 210-211.

Stice, S. and Keefer, C.L. (1993) Multiple generational bovine embryo cloning, *Biol. Reprod.* 48, 715-719.

Stice, S., Keefer, C. and Matthews, L. (1994). Bovine nuclear transfer embryos: oocyte activation prior to blastomere fusion, *Mol. Reprod. Dev.* 38, 61-68.

Stinnakre, M.G., Vilotte, J.L., Soulier, S. and Mercier, J.D. (1994) Creation and phenotypic analysis of a lactalbumin-deficient mice, *Proc. Natl. Acad. Sci. USA* 91, 6544-6548.

Strelchenko, N. (1996) Bovine pluripotent stem cells, *Theriogenology* 45, 131-140.

Szollosi, D., Czolowska, R., Szollosi, M. and Tarkowski, A.K. (1988) Remodelling of mouse thymocyte nuclei depend on the time of their transfer into activated homologous oocytes. *J. Cell. Sci.* 91, 603-613.

Tanako, H., Kozai, C., Shimazu, S., Kato, Y. and Tsunoda, Y. (1996) Cloning by multiple nuclear transfer, *Theriogenology* 47, 1365-1373.

Thompson, J.G. (2000) *In vitro* culture and embryo metabolism of cattle and sheep embryos, a decade of achievements, *Anim. Reprod. Sci.* 60, 263-275.

Vignon, X., Chesné, P., Le Bourhis, D., Fléchon, J.E., Heyman, Y. and Renard, J.P. (1998) Developmental potential of bovine embryos reconstructed from enucleated matured oocytes fused with cultured somatic cells, *C. R. Acad. Sci. Paris, Life Sci.* 321, 735-745.

Vignon, X., Le Bourhis, D., Chesné, P., Marchal, J., Heyman, Y. and Renard, J.P. (1999) Development of bovine nuclear transfer embryos reconstituted with quiescent and proliferative skin fibroblasts, *Theriogenology* 51, 216.

Vignon, X., Le Bourhis, D., Adenot, P., Marchal, J., Lavergne, Y. and Laloi, E. (2000) Production of bovine embryos by nuclear transfer of adult skin fibroblasts: the effect of serum starvation, *Theriogenology* 53, 245.

Wakayama, T., Perry, A.C.F., Zuccotti, M., Johnson, K.R. and Yanagimachi, R. (1998) Full-term development of mice from enucleated oocytes injected with cumulus cells nuclei, *Nature* 394, 369-374.

Wakayama, T., Rodriguez, I., Perry, A.C.F., Yanagimachi, R. and Mombaerts, P. (1999) Mice cloned from embryonic stem cells. *Proc. Natl. Acad. Sci. USA* 96, 14984-14989.

Wall, R.J. (1996) Transgenic livestock: Progress and prospects for the future, *Theriogenology* 45, 57-68.
Wall, R.J., Kerr, D.E. and Bondioli, K.R. (1997) Transgenic dairy cattle: genetic engineering on a large scale, *J. Dairy Sci.* 80, 2213-2224.
Wells, D.N., Misica, P.M., Tervit, H.R. and Vivanco, W.H. (1998) Adult somatic cell nuclear transfer is used to preserve the last surviving cow of the Enderby Island cattle breed, *Reprod. Fertil. Dev.* 10, 369-378.
Wells, D.N., Misica, P.M. and Tervit, H.R. (1999) Production of cloned calves following nuclear transfer with cultured adult mural granulosa cells, *Biol. Reprod.* 60, 996-1005.
Westhusin, M.E., Pryor, J.H. and Bondioli, K., (1991) Nuclear transfer in the bovine embryo: a comparison of 5-day, 6-day, frozen-thawed, and nuclear transfer donor embryos, *Mol. Reprod. Dev.* 28, 119-123.
Willadsen, S.M. (1986) Nuclear transplantation in sheep embryos, *Nature* 320, 63-65.
Wilmut, I., Schnieke, A.E., McWhir, J., Kind, A.J. and Campbell, K.H.S. (1997) Viable offspring derived from fetal and adult mammalian cells, *Nature* 385, 810-813.
Yang, X. (1991) Embryo cloning by nuclear transfer in cattle and rabbit, *Embryo Transfer Newsletter* 9, 10-22.
Yang, X., Jiang, S., Farrell, P., Foote, RH. and McGrath, A. (1993) Nuclear transfer in cattle : effect of nuclear donor cells, cytoplast age, coculture and embryo transfer, *Mol. Reprod. Dev.* 35, 29-36.
Yong, Z., Jianchen, W., Jufen, Q. and Zhiming, H. (1991) Nuclear transplantation in goats, *Theriogenology* 35, 299.
Zakhartchenko, V., Reichenbach, H., Reidl, J., Palma, G., Wolf, E. and Brem, G. (1996) Nuclear transfer in cattle using in vivo derived vs in vitro produced donor embryos: effect of developmental stage, *Mol. Reprod Dev.* 44, 493-498.
Zakhartchenko, V., Alberion, R., Stojkovic, M., Prelle, K., Schernthaner, W., Stojkovic, P., Wenigerkind, H., Wanke, R., Düchler, M., Steinborn, R., Mueller, M., Brem, G. and Wolf, E. (1999a) Adult cloning in cattle: Potential of nuclei from a permanent cell line and from primary cultures, *Mol. Reprod. Dev.* 54, 264-272.
Zakhartchenko, V., Durcova-Hills, G., Stojkovic, M., Schernthaner, W., Prelle, K., Steinborn, R., Mueller, M., Brem G. and Wolf, E. (1999b) Effects of serum starvation and re-cloning on the efficiency of nuclear transfer using bovine fetal fibroblasts, *J. Reprod. Fertil.* 115, 325-331.
Zimmerman, U. and Vienken, J. (1982) Electric field induced cell-to-cell fusion, *J. Membrane Biol.* 67, 165-182.
Ziomek, C. (1998) Commercialization of proteins produced in the mammary gland, *Theriogenology* 49,139-144.

TRANSGENIC FISH: PRODUCTION, TESTING, AND RISK ASSESSMENT

MUIR W.M. AND HOSTETLER H.A.
Purdue University, Department of Animal Sciences, W. Lafayette IN 47907-1151, USA

Abstract

Domestic fish production through transgenic techniques offers many potential economics advantages for commercial aquaculture production, including introduction of new or novel traits and increased response to selection for faster growth. The traditional method of producing transgenic fish is still microinjection, however, advancements have been made using pseudotyped retroviral vectors and electroporation. Some success has also been shown using particle bombardment, but other methods such as sperm-mediated gene transfer, liposome-mediated DNA uptake, MPG peptide vector-oligonucleotide complex-mediated transfer, and nuclear transfer have had limited success in aquaculture. Several factors influence success of integration and expression, these include: the initial amount of DNA used in the injection (or electroporation), form of the DNA used (linear versus supercoiled), site of integration, copy number, and orientation of multiple transgene copies and matrix attachment regions (MARs). The species of origin and insertion of the transgene has also been shown to be important. Most promoters used in aquaculture are either constitutive or inducible, and except for being able to turn on or off the expression with inducible promoters, the level of expression remains uncontrolled. Progress has been made toward tissue specific promotion. The escape or introduction of transgenic fish into natural communities is a major ecological concern. Risk assessment can be addressed using the Net Fitness Approach (NFA) whereby critical fitness parameters are estimated on transgenic fish relative to wild type and incorporated into a model to predict risk. The net fitness parameters include viability to sexual maturity, age at sexual maturity, mating success, fecundity, fertility, and longevity. In this way, it is possible for regulators to develop a set of unambiguous tests to assess risk. Although it is impossible to measure these parameters under a totally natural setting, one can conclude if these parameters show no risk under a relatively favorable laboratory setting, they are even less likely to be a risk in nature's more rigorous conditions. On the other hand, if the model predicts risk, it is not possible to conclude that this will

R. Renaville and A. Burny (eds.), Biotechnology in Animal Husbandry, 261–281.

translate into a real risk in nature, but adequate protection measures should be taken to prevent a *de novo* test of the hypothesis.

1. Introduction

Generation of transgenic fish for enhanced production has many advantages over conventional breeding. The major advantage is the introduction of novel genes for traits which may be lacking in the species. Another advantage is the ability to increase the rate of genetic gain for desirable traits. The goal of traditional selective breeding is to increase frequencies of alleles with small effects that cumulatively have a large effect. These so called "polygenes" have been of critical importance in domesticating all species reared for human consumption. However, improvement by selective breeding occurs in a series of small steps. These small steps can take tens, or even hundreds of generations to accomplish what is possible with transgene technology in a few generations. As such, transgenic technology has attracted commercial interest in aquaculture, particularly for growth hormone, because fast-growing fish have several economic advantages (Lakra and Das, 1998; Dunham, 1999). Developing countries are also interested in this technology as a way of increasing food production (Dunham, 1999).

The escape or introduction of transgenic fish into natural communities, however, is a major ecological concern. Exotic species sometimes adversely affect biotic communities in many ways including eliminating populations of other species (Drake and Mooney, 1986; Lodge, 1993; Bright, 1996). For species that can thrive in the wild, release of transgenic organisms into natural environments could pose additional ecological risks, because these transgenic individuals retain most of the characteristics of their wild type counterparts while possessing some novel advantage. As a consequence, transgenic organisms might threaten the survival of wild type conspecifics as well as other species in a community (Devlin and Donaldson, 1992; Hallerman and Kapuscinski, 1990; Kapuscinski and Hallerman, 1990; Muir and Howard, 2000; Tiedje *et al.*, 1989).

The objective of this chapter is to review methods of producing transgenic fish, issues related to construct design, insertion and expression, and finally examine methods for risk assessment of transgenic fish.

2. Methods of Production

Different methods of gene transfer have been attempted with varying success. Although relatively successful in mice, methods such as liposome-mediated DNA uptake (Schaefer-Ridder *et al.*, 1982; Bachiller *et al.*, 1991), MPG peptide vector-oligonucleotide complex-mediated transfer (Morris *et al.*, 1997), and nuclear transfer as in bovine (Prather *et al.*. 1987; Sims and First, 1993; Krisher *et al.*, 1995) have had limited success in aquaculture. Khoo *et al.* (1992) showed low-level integration and transmission of the reporter plasmid pUSVCAT through the use of sperm-mediated gene transfer in zebrafish, although there were no signs of expression. Some success has also been shown using particle bombardment in rainbow trout, loach, and zebrafish (Zelenin *et al.*, 1997). The primary methods used to produce transgenic fish are pseudotyped retroviral vectors, microinjection, and electroporation

2.1. PSEUDOTYPED RETROVIRAL VECTORS

There has been some success with pseudotyped retroviral vectors. Retroviral pseudotypes are made by encapsulating the genome of one virus with the envelope protein of a second virus (Burns *et al.*, 1993). Pseudotyped retroviral vectors have the transgene of interest integrated into their genome such that when the retroviruses infect cells of interest, the transgene becomes integrated into the genome.

Lin *et al.* (1994) developed a pseudotyped retroviral vector LZRNL(G), in which the Moloney murine leukemia virus long terminal repeat drives the expression of the hepatitis B surface antigen and the Rous sarcoma virus long terminal repeat drives the expression of the neomycin resistance gene, and were able to achieve integration and germ-line transmission in zebrafish. Microinjection has also been used to aid infection of the embryos. Linney *et al.* (1999) attempted microinjection into four different regions near the base of 500- to 1000-cell stage embryos with a retroviral vector which contained an enhanced green fluorescent protein (eGFP). While all embryos injected showed GFP expression; only 10% demonstrated germ-line transmission.

2.2. MICROINJECTION

To date, the most common technique used for DNA transfer is microinjection, where a construct is injected into the cytoplasm of embryos at the one or two cell stage using ultra fine glass needles (Winkler and Schartl, 1997). Many of the microinjection techniques used have been derived from techniques used in mammalian microinjection. However, significant differences in egg physiology of fish require that various details of the technique be modified. The majority of a teleost fish egg is made up of yolk and is protected by a thick outer membrane, the chorion or zona radiata, which is impermeable, except the micropyle through which sperm enters to fertilize the egg (Fletcher and Davies, 1991). Upon fertilization, the contact between the fertilizing sperm and ooplasm causes the egg to be activated (Shears *et al.*, 1992). This activation in turn causes many effects: meiosis is completed, the perivitelline space and fluid are formed, and there is migration of the peripheral cytoplasm (which will form the blastodisc) to the animal pole. The developing egg is now free to rotate within the chorion. In most eggs, the blastodisc remains on top and there is no longer any link between its location and that of the micropyle (Fletcher and Davies, 1991).

In many species, it is at this time that the chorion becomes a hard protective shell (Ginsburg, 1968) and poses a problem for microinjection, although in some species the chorion remains relatively soft following fertilization, such as in the common carp, channel catfish, and loach (Fletcher and Davies, 1991). In these species the chorion can be easily penetrated with glass needles. In species where the chorion is impenetrable, it is possible to circumvent the problem by injecting through the micropyle (Fletcher *et al.*, 1992). The fertilizing sperm nucleus remains in the ooplasm until the union of the male and female pronuclei. Because this union appears to be activated by water, microinjection through the micropyle in fertilized, but non-activated, eggs puts the DNA in close proximity to the pronuclei (Fletcher and Davies, 1991). In some strains of coho and chinook salmon, incubating fertilized eggs in fish saline solution enhances localization of the micropyle (Devlin, 1997).

Attempts have been made to keep the chorion soft with glutathione (Devlin, 1997), or drilling small holes into the hardened chorion (Chourrout *et al.*, 1986). In medaka one solution has been to stagger the mating times such that small quantities of eggs are collected at a time and can be injected before hardening occurs (Inoue *et al.*, 1990). Rahman and Maclean (1992) kept tilapia eggs at 21°C to slow development until they were ready to be injected. Another method of solving the penetration problem is to dechorinate the eggs. Incubation of medaka eggs in 3-cyclohexylaminopropanesulfonic acid (CAPS) buffer (pH 10.0) containing 2 mM glutathione has been used successfully to dechorinate the eggs (Iwamatsu, 1983). Ozato *et al.* (1986) mechanically dechorinated medaka eggs in Dulbecco's phosphate-buffered saline with fine forceps. Zebrafish eggs have been mechanically dechorinated (Stuart *et al.*, 1990; Culp *et al.*, 1991), as well as chemically dechorinated, in embryo medium and modified Hanks saline (Bayer and Campos-Ortega, 1992).

Because of the opaqueness of the chorion and inability to discern the nucleus, it is not possible to directly inject DNA into the nucleus. Therefore, microinjection is usually into the cytoplasm. The thin layer of ooplasm just under the blastodisc is the main target of cytoplasmic injections in fish eggs (Ozato *et al.*, 1989; Fletcher and Davies, 1991). Because the DNA cannot be precisely injected into the nucleus, some time is required for the DNA to migrate to the nucleus before it can be incorporated and integrated. Prior to incorporation, cell development and division continues and is thought to result in transgene mosaicism (Wall and Burdon, 1997). Mosaicism results when not all cells have at least one complete copy of the transgene, and thus only certain tissues test positive for the transgene.

2.3. ELECTROPORATION

Although microinjection is an efficient and successful method, it is time consuming, requires specialized skills and training, and requires that each egg be handled individually. Electroporation is an alternative that alleviates some of these problems and makes gene transfer more efficient. Techniques for electroporation have not been standardized and considerable variation exists among methods. The general procedure was derived from cell culture studies in which an electric current was used to stimulate the uptake of DNA by permeabilizing the cell membrane (Neumann *et al.*, 1982; Shigekawa and Dower, 1988).

Inoue *et al.* (1990) were the first to create transgenic fish using electroporation. They used multiple square electronic pulses, in which the pulse remained at a constant voltage for a set duration, to introduce a construct containing rainbow trout growth hormone cDNA fused to the mouse metallothionein I promoter into medaka fish. Although the success rate was low (4% of hatching fish), they were able to demonstrate germ-line transmission of the transgene. Inoue and Yamashita (1997) found several factors that affected the success of electroporation in fish embryos, including voltage, resistance, and capacitance. Sheela *et al.* (1998) reported that optimal conditions for hatchability in zebrafish were 0.07 KV, 25 μF, 1-3 pulses (one second apart), and infinite resistance to give a survival of 53-81%, as compared to 85% for the control group. However, optimal conditions in the Indian catfish were reported as 0.100 KV, 100 μF, infinite resistance, and 2 pulses to give a survival of 56% as compared to 70% for the control group (Sheela *et al.*, 1999). The state of the chorion can also affect the efficiency of transfer, but efficiency also depends upon the species type whether removal of the chorion is required. Müller *et al.* (1993) successfully transferred a transgene into dechorionated African catfish eggs, but were unable to duplicate their results with chorionated eggs. In contrast, medaka and zebrafish embryos were successfully made transgenic using this technique without removing the chorion (Inoue *et al.*, 1990; Buono and Linser, 1991; Murakami *et al.*, 1994).

Electroporation in conjunction with pseudotyped retroviral vectors has also been successful. Lu *et al.* (1996, 1997) showed that retroviral vectors could be used along with electroporation techniques to obtain germ-line transmission in dwarf surfclams and medaka fish. Electroporation of 5000-50,000 surfclam eggs (1 hr post-fertilization) in 250 μl of either sea water, Dulbecco's minimal essential medium (DMEM) with high glucose, and 10% fetal calf serum or sea water and virus medium with approximately 1 x 104 colony-forming units (cfu) of the LZRNL-(VSV-G) vector was carried out at 700V using the Baekon model 2000 electroporator. With this technique, 13.6% of the pooled gamete samples were positive for the transgene (Lu *et al.*, 1996). Dechorinated and non-dechorinated fertilized medaka eggs were electroporated in groups of 50 in 250 μl of modified Yamamoto saline with approximately 4 x 104 cfu of the MoMLV retroviral vector and 4 μg/ml of polybrene. Electroporation voltage in the range of 2.5 – 5.0 KV resulted in 29% and 61% positive transgenics for dechorinated and non-dechorinated embryos respectively (Lu *et al.*, 1997).

More recently, pulses of DC-shifted oscillating electric fields in the radio-frequency range have been examined to induce cell poration. This method was first proposed by Chang (1989) based on dynamics of the lipid bilayer which is responsible for the opening and closing of the outer cell membrane. Chang and Reese (1990) observed that electropores (discrete pore-like structures within the cell membrane) expanded after electroporation, and after a few seconds of electric pulse, these electropores would have openings large enough for supercoiled DNA to enter. By applying an oscillating pulse, it was hypothesized that these electropores remain open long enough for DNA to enter, while not so long that the outer membrane is destroyed. The Gene Pulser II RF module (Bio-Rad, Hercules, CA) was designed based on this hypothesis. Recent experiments in our laboratory using this electroporator have been successful, with a 20% efficiency rate of gene transfer in hatching medaka fish and require ten times less DNA than the earlier experiments involving microinjection of medaka fish.

3. Factors Determining Success of Integration

Besides the technique used, there are many factors that appear to determine success of transgene integration. For example, the initial amount of DNA used in the injection (or electroporation) effects survival as well as integration of the transgene. Because there is a relatively low probability of the injected DNA reaching the nucleus, many researches have injected large quantities of DNA; however, there is a maximum tolerance level for each species (for review see Fletcher and Davies, 1991).

Another factor that influences success is the form of the DNA used. Use of linear versus supercoiled DNA has been debated. Studies have shown success using both linearized and supercoiled DNA. There have been several studies that found that vector-free linear DNA produced more positive fish than supercoiled DNA (Chourrout *et al.*, 1986; Penman *et al.*, 1990). There are also studies that show that the form of DNA does not affect the efficiency of transgene transfer (Chong and Vielkind, 1989). Stuart *et al.* (1990, 1988) demonstrated a higher percentage of transgenics from supercoiled DNA than from linearized DNA. However, since there are more factors involved in these studies than just the DNA form, the issue remains a controversy.

Variation in expression may also be due to site of integration, copy number, and orientation of multiple transgene copies (Clark, 1997), and even matrix attachment regions (MARs). Studies show that foreign DNA fragments rapidly form large concatemers (tandem arrays), which replicate after microinjection regardless of the actual sequence, (Houdebine and Chourrout, 1991). In fish, the concatemers tend to form head to tail arrangements when supercoiled DNA is used and random arrangements when linear DNA is used (for review, Vielkind, 1992). This result may be due to the storage of required proteins (such as DNA polymerases and ligases) in the cytoplasm of the egg prior to fertilization (Iyengar *et al.*, 1996). These concatemers are then replicated during early development of the embryo. The initial amount of DNA injected, as well as the form of the DNA, may determine the length and number of these concatemers formed. In turn, the length and number of concatemers formed may influence the copy number and site of integration.

4. Constructs

A construct is composed of a promoter, the gene itself, and a termination sequence. The promoter sequence functions to regulate gene expression and may be comprised of only the minimal sequence needed to start transcription and may also contain enhancer or repressor elements which control the amount of transcription. The termination sequence serves as a stopping point for the end of transcription. The construct may also contain leader sequences to protect against methylation and site specific effects. Constructs are generally of two types: reporter and functional. Those containing reporter genes are designed to examine efficiency of alternative promoters, methods of producing transgenic fish, or determine tissue of expression. Those containing functional genes are designed to have potential economic impact.

4.1. REPORTER GENES

There are many assays to determine gene expression in transgenic animals. Reporter genes are used to help quantify expression and to rapidly determine success of gene transfer techniques. Reporter genes by definition need to have quick and efficient assays, either by use of immunohistochemistry, chemoluminescence, or autoradiography. The most common reporter genes are chloramphenicol acetyl transferase (CAT), β-galactosidase (β-*gal*), luciferase, and green fluorescent protein (GFP). Each reporter has advantages and disadvantages, some require that the fish be sacrificed to perform the test while others can be detected in live animals.

This CAT reporter gene has been used to show gene expression in zebrafish (Stuart *et al.*, 1990), medaka (Kinoshita *et al.*, 1996), and rainbow trout (Iyengar and Maclean, 1995). There are several techniques used to observe the expression of CAT; however, all techniques commonly used require a radiolabelled substrate. One such method involves autoradiography following separation of products by thin layer chromatography (Stuart *et al.*, 1990). Quantification of the amount of expression can be accomplished by dot blot or scintillation counting (Gong *et al.*, 1991). Additionally, antibodies specific for the CAT protein can be used to determine the location of the expression through immunohistochemistry.

Tests for expression of β-*gal* can be accomplished by simple staining of specific tissues or the entire organism. Another technique for β-*gal* detection involves the use of a commercially available antibody for immunochemical localization. This reporter has proven useful in zebrafish (Culp *et al.*, 1991; Bayer and Campos-Ortega, 1992; Lin *et al.*, 1994). However, non-specific staining of the gut in control embryos has been a problem (Iyengar *et al.*, 1996; H. H. unpublished). To address this problem, Nam *et al.* (1998) devised a reliable method for visualizing the β-*gal* activity without background interference.

Although not widely used, the luciferase assay has some advantages. Specifically the assay does not require post-transcriptional modifications for assay, there is no staining or probing required, and background interference for viewing is negligible (for review see Iyengar *et al.*, 1996). Patil *et al.* (1994) used the luciferase gene in order to study the fate of transgenes in zebrafish. Unfortunately, the luciferase assay is not simple. Gibbs *et al.* (1994) required the use of a sensitive scintillation counter assay to show expression of the luciferase gene.

More recently, *Aequorea victoria* green fluorescent protein (GFP) assays have become popular. There also exist several different GFP mutations which allow for spectra excitation and emission at different wave lengths. One of the major benefits of GFP assays over others is the ability to detect expression in living organisms. Most GFP assays require only a fluorescent microscope fitted with a filter of a particular wavelength; this wavelength is dependent upon which mutant GFP is being used (Muldoon *et al.*, 1997). Such studies have been reported in medaka (Takagi *et al.*, 1994; Hamada *et al.*, 1998) and zebrafish (Amsterdam *et al.*, 1995) embryos. Nevertheless, we have observed problems with autofluorescnence in several fish species (including medaka and zebrafish) when observed with a fluorescent microscope.

4.2. FUNCTIONAL GENES

The primary objective of making transgenic fish is to modify the phenotype for traits of economic importance. Although most previous transgenic fish research was aimed toward either low temperature resistance, enhanced growth rate, or developmental gene expression (Ozato *et al.*, 1989), research is now extending to genes which impact other traits of economic importance, such as glucose transporters and hexokinases to increase carbohydrate utilization. As new genes are discovered and cloned in other species, the number of functional genes used to transform fish will increase dramatically. From a commercial perspective, the primary areas where transgenics are needed are for improved disease resistance, feed efficiency, rate of growth, stress resistance, and behavior modification.

The antifreeze protein (AFP) gene was the first functional gene used in fish transgenics. Antifreeze proteins, and glycoproteins, evolved in certain marine teleosts in response to cold seawater at high latitudes (Davies *et al.*, 1988). The high concentration of solutes in seawater, which is much greater than that of the teleost plasma, causes the freezing point of the sea to be lower than the freezing point of the plasma. In species that produce AFPs, the freezing temperature of the serum is lowered without changing the osmotic balance. A winter flounder antifreeze protein gene has been successfully transferred, expressed, and passed on to offspring in Atlantic salmon (Shears *et al.*, 1991). However, the amounts of protein were too low to provide useful freeze resistance (Fletcher *et al.*, 1992). Tests have also been conducted on the functional expression of promoter regions of AFP genes (Gong *et al.*, 1991).

More recently, studies have looked at the possibility of improving the L-ascorbic acid biosynthesis and carbohydrate utilization in salmonid fish. Due to the low rate of sugar transport across plasma membranes, and subsequent phosphorylation, sugars are poorly utilized in fish (Krasnov *et al.*, 1999b). Glucose transporters (GLUT) and hexokinases (HK) catalyze this process. Tse and Young (1990) reported that rainbow trout do not have a facilitative glucose transporter in their red blood cells. Walton and Cowey (1982) found that the HK activity in salmonid fish was the lowest of the glycolytic enzymes. In order to compensate for this problem, transgenic studies have involved the addition of both a glucose transporter and a hexokinase. Several studies have used such methods to attempt to aid the utilization of sugars in fish, but only minimal changes have been noted (Pitkänen *et al.*, 1999a; Krasnov *et al.*, 1999a, 1999b).

There have been numerous studies utilizing growth hormone genes in a variety of fish species. Most of these experiments were based on results obtained from studies of growth hormone transfer into mice (Palmiter *et al.*, 1983; Brinster *et al.*, 1985). In 1986, human growth hormone cDNA was injected into fertilized rainbow trout eggs, although stable integration was not proven (Chourrout *et al.*, 1986). Dunham *et al.* (1987) had limited success in transferring a metallothionein-human growth hormone fusion gene (MT-hGH) into channel catfish. Several others have been able to transfer the human growth hormone gene into tilapia (Brem *et al.*, 1988), loach (Xie *et al.*, 1993), and medaka (Muir, 1995; Muir *et al.*, submitted).

The amount of gene expression seems to vary depending on the fish species of interest and on the gene constructs used (Sin, 1997). This phenomenon has been shown with the insertion of GH genes from a variety of species. The low expression of human GH constructs in fish led to the use of "all fish" constructs. Du *et al.* (1992) developed an "all fish" chimeric growth hormone gene construct. In this construct, an ocean pout antifreeze protein gene (AFP) promoter was linked to a chinook salmon GH cDNA. With this construct, they were able to produce transgenic Atlantic salmon that were 2 to 6 fold larger than non-transgenic controls. Devlin *et al.* (1994a,b) used an "all-salmon" construct in which the sockeye salmon metallothionein B promoter was linked to the sockeye salmon growth hormone gene (OnMTGH1) to produce salmon that were, on average, 11-fold heavier than non-transgenic controls and one individual 37 times larger than controls. Devlin *et al.* (1994a,b, 1995a,b) and Devlin (1996) have produced transgenic salmonids: coho salmon (*O. kisutch*), rainbow trout (*O. mykiss*), cutthroat trout (*O. clarki*), and chinook salmon (*O. tshawytscha*) which are, on average, ten to fifteen times larger than non-transgenic controls. Using the same or similar construct, Arctic charr have been produced which are 14-fold larger than controls (Pitkanen *et al.*, 1999b). Tilapia have also been produced with a similar construct which are 3-fold larger than controls (Rahman and Maclean, 1999). Several other species have also been transformed with GH, but with growth enhancements less than 100% that of wild-type fish (reviewed by Chen *et al.*, 1995; Devlin, 1996, 1997; Fletcher and Davies, 1991; Hew *et al.*, 1995; Houdebine and Chourrout, 1991; Iyengar *et al.*, 1996; Maclean, 1998; Pandian and Marian, 1994; Pitkanen *et al.*, 1999; Sin, 1997; Sin *et al.*, 1997).

4.3. PROMOTERS

Promoters are the regulators of gene expression. The promoter determines the tissue and strength of expression. A promoter may be constitutive or inducible. Constitutive promoters are unregulated and always on. An inducible promoter, such as the metallothionein (Mt) promoter, can be turned on or off using heavy metals in the diet. A promoter can also be tissue specific or ubiquitous. A ubiquitous promoter expresses in all tissues. Promoters of viral origin, such as the human cytomegalovirus (CMV), tend to be both constitutive and ubiquitous; whereas many of the promoters isolated from fish for transgenic studies are tissue specific and/or inducible. A ubiquitous promoter would be advantageous for experiments testing for the efficiency of transgene integration and expression in founder fish. Whereas a tissue specific promoter may be better for attaining a desired effect of functional genes.

Since the goal of transgenics is to produce a gene product in a desired tissue at an appropriate level in order to obtain some desired effect (Chen *et al.*, 1995), it is important to understand how a given promoter or enhancer element will affect expression. Takagi *et al.* (1994) used a β-galactosidase reporter gene to test for expression of the medaka β-actin promoter in medaka embryos, whereas Hamada *et al.* (1998) used a mutant GFP to determine the efficiency of the medaka β-actin promoter in medaka. Expression was found to be localized to the epidermis of the head, trunk, fins, and yolk sac with the GFP construct. A zebrafish-derived GATA-1 promoter spliced to GFP showed expression in zebrafish blood cells (Long *et al.*, 1997). Several other promoters have been tested with the GFP gene. Higashijima *et al.* (1997) tested the efficiency of the zebrafish α-actin and β-actin promoters in zebrafish and found expression in whole muscle for α-actin and throughout the whole body for β-actin.

More recently, Ju *et al.* (1999) isolated the ck (cytokeratin), mck (muscle creatine kinase), and arp (acidic ribosomal phosphoprotein PO) promoter regions from genomic zebrafish DNA. These promoters were specifically chosen for their tissue specificity (or lack thereof); ck for skin specificity, mck for muscle specificity, and arp for ubiquitous expression. Mori and Devlin (1999) have examined areas of mRNA expression in transgenic salmon with the OnMT-GH1 transgene. They observed the highest level of expression in the pyloric caeca, with decreasing levels of expression in the kidney, spleen, and intestine respectively.

There have also been studies to determine which promoters express most efficiently. Chan and Devlin (1993) used cell culture techniques to characterize the functionality of three promoters derived from the sockeye salmon: metallothionein B (MT), histone type III (H3), and protamine (PT), as well as the CMV promoter. They found that the PT promoter was not active in fish cell lines, but the CMV, H3, and MT promoters were, with CMV being the strongest. Hanley *et al.* (1998) compared the Atlantic salmon H3 promoter to the viral RSV promoter in zebrafish and rainbow trout. They found the H3 promoter was not only stronger, but its expression was ubiquitous in zebrafish embryos. Pitkänen *et al.* (1999b) tested the efficiency of the CMV, piscine MT-B, and piscine H3 promoters for driving expression of the sockeye salmon GH against the Atlantic salmon GH (with its native promoter) in Arctic charr and rainbow trout. Although there was considerable variation in size, the group with the CMV promoter were overall larger.

5. Other Factors Determining Transgene Expression

Although a correlation has been observed between level of transient expression and copy number of an injected construct (Volckaert *et al.*, 1994), no correlation between stable expression and integrated copy number has been reported (Zhang *et al.*, 1990). This is an area in which more experiments are needed before a conclusion can be made.

Site of integration has often been suggested as a possible explanation for variable expression among transgenic founder animals (Stuart *et al.*, 1990; Devlin *et al*, 1995a; Iyengar *et al.*, 1996). Although not well characterized in fish, position effects have been studied in *Drosophila* and yeasts and are referred to as position-effect variegation (PEV) (Snibson and Adams, 1997). This phenomenon is thought to be due to the structure of the chromatin at the site of integration. If the transgene is inserted near an area of inactive chromatin, known as heterochromatin, the activity of the transgene seems to be repressed due to the spreading effects of the heterochromatin. Moreover, if a transgene is integrated into the genome into an area of active chromatin, known as euchromatin, activity may be higher than expected; especially if an enhancer element is in relatively close proximity.

A method to address problems associated with chromosomal position effects is to design transgene constructs with sequences of DNA that prevent heterochromatin from spreading and which are capable of acting as independent regulatory units for transcription (for review see Sippel *et al.*, 1997). These sequences are known as insulators, because they insulate the transgene from genomic effects, or as domain border regions (DBRs), because they function to form independent domains for gene expression. Caldovic *et al.* (1999) tested both a border element from *Drosophila* and a border element from chicken in creating transgenic zebrafish with position-independent expression. They found that the level of transgene expression was proportional to the number of integrated transgenes in both cases. Although detailed knowledge of these sequences is not available, it is generally accepted that they are universally conserved and A/T rich. One such family has been isolated and characterized from channel catfish (Liu *et al.*, 1998). Matrix attachment regions (MARs) have also been shown to function as insulators for transgene expression in *Drosophila* (Namciu *et al.*, 1998), rice, and tobacco (Vain *et al.*, 1999).

A second method of preventing chromosomal position effects involves pre-selecting the site of integration, rather than allowing random integration. If a reporter transgene is integrated into a genome and found to strongly express in the correct location, it has been suggested that a second transgene could be forced to integrate in close proximity to the first transgene by use of homologous recombination (Clark, 1997). Wallace *et al.* (2000) used a similar method using homologous recombination in embryonic stem (ES) cells for the introduction of single copy transgenes.

Transgene methylation may also be responsible for position-dependent levels of transgene expression (Donoghue *et al.*, 1992). As a defense mechanism, foreign sequences that become integrated into the genome are commonly *de novo* methylated, in which a methyl group is added to the C-5 position of cytosine (for review see Snibson and Adams, 1997). This process allows for differentiation of self DNA from foreign DNA. Unfortunately, this mechanism leads to suppression of the transgene. There have been many reports of transgene silencing in plants because of methylation (Matzke *et al.*, 1994; Matzke and Matzke, 1995; Park *et al.*, 1996; Matzke and Matzke, 1998). Salehi-Ashtiani *et al.* (1993) found the metallothionein promoter to be highly sensitive to methylation, and that methylation of the transgene in certain cell types (mice liver and kidney) was capable of repressing transcription. Another study in mice found the degree of methylation of the transgene to be inversely proportional to the level of its expression (Komatsu *et al.*, 2000). Most of these types of studies have been conducted in plants or mice, there is no information on effects of methylation in aquaculture.

6. Risk Assessment

Risk assessment has generally followed two approaches 1) a Net Fitness Approach (Muir and Howard, 2000) whereby critical fitness parameters are estimated on transgenic fish and incorporated into a model to predict risk and 2) observation of behavior in transgenic fish and inference to risk. The latter approach can also be simulated by injection or implantation of the transgene product direct into the fish. Alternatively, examination of wild populations using existing variation for the trait of interest is useful for projecting potential impact. This approach is not possible for transgenes which introduce new functionality, but for transgenes which influence such traits as growth, observation of natural size variation will provide important clues as to the potential impact of the transgene. However, inference from natural trait variation has limitations because the phenotype of transgenic animals may be beyond the natural size range. Further, the transgene may be poorly regulated or expressed in non-normal tissue resulting in unexpected viability impacts on the fish. The only transgene which has been studied in detail is that for GH and will be the primary focus of this section.

6.1. THE NET FITNESS APPROACH

A necessary prerequisite for any ecological risk associated with the introduction of a transgenic organism is that the initially rare transgene can spread in a natural population. Some scientists have assumed that such an increase in transgene frequency is unlikely because transgenic organisms typically have a viability disadvantage (e.g., Knibb, 1997; Dunham, 1999). Muir *et al.* (1994, 1995, 1996) and Muir and Howard (1999, 2000) provided the first comprehensive experimental studies to evaluate this assumption in transgenic fish. The objective of our research was to determine if the fate of a transgenic organism introduced into a natural population can be predicted by a computer model based on fitness parameters of the base population of transgenic fish, relative to the wild-type population, measured in a secure laboratory setting. This goal required three stages: estimation, prediction, and testing. Specifically, we determined what parameters need to be estimated to predict risk and developed a mathematical model which incorporates those parameters to predict risk.

Our research (Muir and Howard, 1999, 2000) clearly showed that, to assess ecological risk of a particular transgenic line, one needs to estimate six Net Fitness factors: viability to sexual maturity, fecundity, fertility, mating success, age at sexual maturity, and longevity. All biological factors which influence the fitness of a transgenic animal fall into one of these categories. Our results stressed the importance of measuring all factors because a disadvantage in one category can be offset by an advantage in another category. For example, reduced viability of transgenics could be offset by any one of the following size- or growth-related advantages of transgenics: 1) a reduced generation interval which would increase reproductive rate; 2) increased mating success of males; 3) increased egg production by females, and 4) reduced cannibalism on young offspring.

To determine the extent to which any one of these fitness factors could offset a viability disadvantage, quantitative values of the factor for the specific transgenic line under investigation needs to be put into our mathematical model for prediction (Muir and Howard, 1999, 2000). Our model brings together all six factors, using population genetics theory, to predict changes in transgene frequency and population size. Fortunately, it is not necessary to know why a transgenic line differs from its wild type counterparts in some fitness factor, only that a difference exists. The underlying cause of the difference may be of academic interest, but may require extensive study and funding to determine. For example, consider some viability difference: all that the model requires is the relative number of transgenics and wild types that reach sexual maturity; no information is needed on whether observed trends result from differences in disease resistance, foraging ability, escape from predation, improper gill size, or swimming ability, etc.

For biological parameterization of our model to examine specific risks, we (Muir and Howard, 2000) used Japanese medaka, *Oryzias latipes,* as a study organism. Medaka were convenient study organisms on which to obtain data on fitness components and propagate multiple generations within a relatively short period of time. Individuals readily bred in the lab, were easily cultured, and attained sexual maturity in about two months. We produced a stock of transgenic medaka by using cytoplasmic microinjection to insert a gene construct consisting of the human growth hormone gene (hGH) with a salmon promoter (sGH) into just fertilized eggs. Results of several experiments showed that: transgenic young had a viability disadvantage (survival was 69% of that of wild type); the transgenic line had a distinct early developmental advantage (peaking at 4 weeks of age with a 39% size advantage) which resulted in a 21% increase in escape from predation; and transgenic females produced 27% more eggs/spawn than wild type fish (Muir and Howard, 2000). For the SGH-hGH line our model predicted that, for a wide range of parameter values including those we documented experimentally, transgenes should spread in invaded populations despite high viability costs whenever the transgenes also have positive effects on fecundity or developmental rate (Muir *et al.*, 1996).

We also examined the risk to a natural population after a release of a few transgenic individuals when the transgene simultaneously increased transgenic male mating success and lowered transgenic offspring viability. Mating experiments using wild type medaka were performed to assess the degree to which large males obtained a mating advantage over small males. We found that large males obtained a four-fold mating advantage (Howard *et al.*, 1998). Large male mating advantages have also been reported in several salmonid species. Our deterministic equations predict that a transgene introduced into a natural population by a small number of transgenic fish will spread rapidly as a result of enhanced mating advantage, but the reduced viability of transgenic offspring will cause eventual local extinction of both the wild and transgenic populations, we call this the Trojan Gene Hypothesis (Muir and Howard, 1999). Furthermore, loss of a wild type population could also have cascading negative effects on the rest of the ecological community. The final stage of our research involves testing alternative predictions of our models using alternative founder lines and parameters estimated in the first stage.

6.2. RISK INFERENCE BASED ON NATURAL SIZE VARIATION

Based on the six Net Fitness parameters we identified (Muir and Howard, 2000), the following relationships were found using natural size variation: 1) Viability: large trout have reduced predator avoidance when foraging as compared to small trout (Johnsson, 1993), but, large trout are socially dominant to small (Johnsson, 1993) giving them a feeding advantage and they may have a higher early survival of young because large females can control better nest sites (de Gaudemar, 1998). 2) Mating Success: large males have a mating advantage over small males in several species (de Gaudemar, 1998; Fleming, 1996, 1998; Moran *et al.*, 1996; Howard *et al.*, 1998). 3) Age at sexual

maturity: Males with faster growth rates have precocious sexual maturation (de Gaudemar, 1998). 4) Longevity: Larger males and females have reduced longevity (Moran *et al.*, 1996). 5) Fecundity: larger females produce larger clutches and/or larger eggs (de Gaudemar, 1998; Fleming, 1996, 1998). 6) Fertility: Large anadromous males have high fertilization success because of more sperm/ejaculate; parr males may fertilize 11-40% of eggs and larger parr males have size advantage in fertilizing eggs (de Gaudemar, 1998; Fleming, 1996, 1998; Thomaz *et al.*, 1997).
Overall, larger fish have advantages and disadvantages for viability, but have greater mating success, earlier age at sexual maturity, shorter longevity, higher fecundity and fertility. Because there is no evidence that fish in their native habitat are becoming larger with time, a logical conclusion is that advantages and disadvantages of these traits are evenly balanced in nature. Thus, the benefit of being slightly larger than the current maximum size may not be offset by the costs of increased size.

Nevertheless, transgenes represent mega-mutations that may be substantially beyond the natural size range of the species and thus could change many aspects of its niche. Such a mega-mutation could involve a switch in the current adaptive peak of an organism to a different adaptive peak based on Wright's (1969, 1982) shifting balance theory. For example, GH could change an organism from being a prey of one species to being a predator of another.

6.3. RISK INFERENCE BASED ON OBSERVATIONS OF TRANSGENIC OR GH FISH

Based on the six Net Fitness parameters, the following relationships were found in transgenic fish (TR) for GH or fish which mimic transgenics by use of GH implant. Viability: early viability disadvantage for TR fish were observed by Muir and Howard (1999) and for predation of juveniles by Johnsson *et al.*, (1996, 1999), Abrahams and Sutterlin (1999) and Jönsson *et al.* (1996). In contrast, TR fry escape cannibalism sooner (Muir and Howard, 2000) and TR fish digest protein more efficiently than wild type (WT) (Fu *et al.*, 1998; Johnsson and Björnsson, 1994) indicating they could better compete when food is limited or poor in quality. TR are also more competitive than WT (Devlin *et al.*, 1999; Jönsson *et al.*, 1998). Conflicting results were observed for swimming speed. Farrell *et al.* (1997) observed that TR swimming speed was half that of 1-yr older WT of similar size while Abrahams and Sutterlin (1999) observed that TR alevins had a 3-fold speed advantage over WT. Age at Sexual Maturity: Saunders *et al.* (1998) and Devlin *et al.* (1994b, 1995a) observed that TR salmon attain smolt status earlier than WT. Similarly, Muir and Howard (2000) found that TR medaka females mature earlier than WT females. Mating Success Actual mating success has not been observed in any TR species but results of Rahman *et al.* (1998), Rahman and Maclean (1999), and Martinez *et al.* (1996) show that TR are much larger than WT at sexual maturity, suggesting a possible mating advantage. Fecundity: The only study to examine fecundity of TR fish was that of Muir and Howard (2000), they found that TR females had larger clutches than did WT females of the same size. To date there has been no examination of relative fertility or longevity of TR fish. Overall the data is inconclusive on viability, but TR fish appear to have a reduced age at sexual maturity, greater fecundity, and possible greater mating success.

7. Conclusion

Development of transgenic fish is still in the early stages. Methods of production are improving but control of transgene insertion site, copy number, regulation, and tissue specificity is still needed. Commercial interest has mainly been limited to enhancing growth. However, improvements in disease resistance, feed efficiency, nutrient utilization, stress resistance, and behavior modifications through transgenesis could be as important as increased growth. As the genomes of various species are sequenced and functions of those genes discovered, the array of possible types of transgenic fish is enormous.

As animal scientists, our goal should be to promote animal agriculture in an environmentally acceptable manner. Acceptance of this technology will require that we be able to convince environmentalists and the public that the food is wholesome and the fish do not pose a threat to native species or the ecosystem. Long-term persistence of a transgenic line seems unlikely but not impossible, and may represent the creation of a new species with a novel niche. The precedence for this result occurring in nature through mutations with large effects has been demonstrated; thus the possibility that such a mega mutation introduced in the laboratory having a similar impact cannot be ignored. The methods presented for estimation of Net Fitness parameters present a method whereby such risks can be determined in a secure environment before commercial production is initiated. Despite laboratory constraints that apply when studying a GH-transgenic fish in confinement, one can still test an important one-sided hypothesis: on one extreme, if a transgenic organism exhibits superior fitness in an optimum laboratory environment, in terms of growth, fecundity, competitive ability etc., that does not necessarily mean that the organism will be a hazard in the wild due to nature's less than optimal conditions and greater complexity. On the other hand, if the transgenic organism is shown to have a low fitness in an artificial optimum environment, the organism will not compete effectively or thrive in the wild where conditions are usually suboptimal.

Acknowledgements

This research was supported by grants 93-33120-9468 and 97-39210-4997 from the USDA National Biological Impact Assessment Program.

References

Abrahams, M. V. and Sutterlin, A. (1999) The foraging and antipredator behaviour of growth-enhanced transgenic Atlantic salmon, *Anim. Behav. 58*, 933-942.

Amsterdam, A., Lin, S. and Hopkins, N. (1995) The *Aequorea victoria* green fluorescent protein can be used as a reporter in live zebrafish embryos, *Dev. Biol. 171*, 123-129.

Bachiller, D., Schellander, K., Peli, J. and Rüther, U. (1991) Liposome-mediated DNA uptake by sperm cells, *Mol. Reprod. Dev. 30*, 194-200.

Bayer, T.A. and Campos-Ortega, J.A. (1992) A transgene containing *lacZ* is expressed in primary sensory neurons in zebrafish, *Dev. 115*, 421-426.

Brem, G., Brenig, B., Hörstgen-Schwark, G. and Winnacker, E.L.. (1988) Gene transfer in Tilapia (*Oreochromis niloticus*), *Aquaculture 68*, 209-219.

Bright, C. (1996) Understanding the threat of biological invasions. *in* Starke L. (ed.) *A World Watch Institute Report on progress toward a sustainable society* Norton W.W., New York, pp 95-113

Brinster, R.L., Chen, H.Y., Trumbauer, M.E., Yagle, M.K. and Palmiter, R.D. (1985) Factors affecting the efficiency of introducing foreign DNA into mice by microinjecting eggs, *Proc. Natl. Acad. Sci. USA 82*, 4438-4442.

Buono, R.J. and Linser, P.J. (1991) Transgenic zebrafish by electroporation, *Bio-Rad US/EG Bulletin 1354*, 1-3.

Burns, J.C., Friedmann, T., Driever, W., Burrascano, M. and Yee, J.K.. (1993) Vesicular stomatitis virus G glycoprotein pseudotyped retroviral vectors: concentration to very high titer and efficient gene transfer into mammalian and nonmammalian cells, *Proc. Natl. Acad. Sci. USA 90*, 8033-8037.

Caldovic, J., Agalliu, D. and Hackett, P.B. (1999) Position-independent expression of transgenes in zebrafish, *Transgenic Res. 8*, 321-334.

Chan, W.K. and Devlin, R.H. (1993) Polymerase chain reaction amplification and functional characterization of sockeye salmon histone H3, metallothionein-B, and protamine promoters, *Mol. Mar. Biol. Biotech. 2*, 308-318.

Chang, D.C. (1989) Cell poration and cell fusion using an oscillating electric field, *Biophys. J. 56*, 641-652.

Chang, D.C. and Reese, T.S. (1990) Changes in membrane structure induced by electroporation as revealed by rapid-freezing electron microscopy, *Biophys. J. 58*, 1-12.

Chen, T.T., Lu, J.K., Shamblott, M.J., Cheng, C.M., Lin, C.M., Burns, J.C., Reimschuessel, R., Chatakondi, N. and Dunham, R.A. (1995) Transgenic fish: ideal models for basic research and biotechnological applications, *Zool. Stud. 34*, 251-234.

Chong, S.S.C. and Vielkind, J.R. (1989) Expression and fate of CAT reporter gene microinjected into fertilized medaka (*Oryzias latipes*) eggs in the form of plasmid DNA, recombinant phage particles and its DNA, *Theor. Appl. Genet. 78*, 369-380.

Chourrout, D., Guyomard, R. and Houdebine, L.M. (1986) High efficiency gene transfer in rainbow trout (*Salmo gairdneri* RICH.) by microinjection into egg cytoplasm, *Aquaculture 51*, 143-150.

Clark, A. J. (1997) Transgene rescue, in Houdebine L.M. (ed., *Transgenic animals generation and use* Harwood academic publisher, France, pp 267-272.

Culp, P., Nüsslein-Volhard, C. and Hopkins, N. (1991) High-frequency germ-line transmission of plasmid DNA sequences injected into fertilized zebrafish eggs, *Proc. Natl. Acad. Sci. USA 88*, 7953-7957.

Davies, P.L., Hew, C.L. and Fletcher, G.L. (1988) Fish antifreeze proteins: physiology and evolutionary biology, *Can. J. Zool. 66*, 2611-2617.

de Gaudemar, B. (1998) Sexual selection and breeding patterns: insights from salmonids (salmonidae), *Acta Biotheoretica 46*, 235-251.

Devlin, R.H. (1997) Transgenic Salmonids, in *Transgenic animals generation and use*, Houdebine L.M. (ed., Harwood Academic Publishers, France, pp. 105-117.

Devlin, R.H. and Donaldson, E.M. (1992) Containment of genetically altered fish with emphasis on salmonids, in Hew, C.L. and Hew G.L. (eds.) *Transgenic Fish,* World. Sci. Pub. Co. Singapore, pp. 229-265.

Devlin, R.H, T.Y. Yesaki, C.A. Biagi, and E.M. Donaldson. (1994a) Extraordinary salmon growth. *Nature. 371*, 209-210.

Devlin, R.H., Yesaki, T.Y., Biagi, C., Donaldson, E.M. and Chan, W.K. (1994b) Production and breeding of transgenic salmon, *Proc. 5th World Cong. Genet. Appl. Livest. Prod.19*, 372-378.

Devlin, R.H., Yesaki, T.Y., Donaldson, E.M., Du, S.J. and Hew, C.L. (1995a) Production of germline transgenic Pacific salmonids with dramatically increased growth performance, *Can. J. Fish. Aquat. Sci. 52*, 1376-1384.

Devlin, R.H., Yesaki, T.Y., Donaldson, E.M. and Hew, C.L. (1995b) Transmission and phenotypic effects of an antifreeze/GH gene construct in coho salmon (Oncorhynchus kisutch), *Aquaculture 137*, 161-169.

Devlin, R.H., Johnsson, J.I., Smailus, D.E., Biagi, C.A., Jonsson, E. and Bjornsson, B.Th. (1999) Increased ability to compete for food by growth hormone-transgenic coho salmon Oncorhynchus kisutch (Walbaum), *Aquacult. Res. 30*, 479-482.
Donoghue, M.J., Patton, B.L., Sanes, J.R. and Merlie, J.P. (1992) An axial gradient of transgene methylation in murine skeletal muscle: genomic imprint of rostrocaudal position, *Development 116*, 1101-1112.
Drake, J.A. and Mooney, H.A. (1986) Ecology of Biological Invasions of North America and Hawaii, Springer-Verlag edit, New York.
Du, S. J., Gong, Z., Fletcher, G.L., Shears, M.A., King, M.J., Idler, D.R. and Hew, C.L. (1992) Growth enhancement in transgenic Atlantic salmon by the use of an "all fish" chimeric growth hormone gene construct, *BioTechnology 10*, 176-181.
Dunham, R.A. (1999) Utilization of transgenic fish in developing countries: potential benefits and risks, *J. World Aquacult. Soc. 30*, 1-11.
Dunham, R.A., Eash, J., Askins, J. and Townes, T.M. (1987) Transfer of the metallothionein-human growth hormone fusion gene into channel catfish, *Trans. Am. Fisheries Soc. 116*, 87-91.
Farrell, A.P., Bennett, W. and Devlin, R.H. (1997) Growth-enhanced transgenic salmon can be inferior swimmers, *Can. J. Zool. 75*, 335-337.
Fleming, I.A. (1998) Pattern and variability in the breeding system of Atlantic salmon (Salmo salar), with comparisons to other salmonids, *Can. J. Fisheries Aquatic Sci. Soc. 55 suppl1*, 59-76.
Fleming, I.A. (1996) Reproductive strategies of Atlantic salmon: ecology and evolution, *Rev. Fish Biol. Fisheries 6*, 379-416.
Fletcher, G.L. and Davies, P.L. (1991) Transgenic fish for aquaculture, *Genet. Engineer. 13*, 331-370.
Fletcher, G.L., Davies, P.L. and Hew, C.L. (1992) Genetic engineering of freeze resistant Atlantic salmon, in Hew, C.L. and Fletcher, G.L. (eds.) *Transgenic Fish*, World Scientific Publishing Co., Singapore, pp. 190-208.
Fu, C., Cui, Y., Hung, S.S.O. and Zhu, Z. (1998) Growth and feed utilization by F4 human growth hormone transgenic carp fed diets with different protein levels, *J. Fish Biol. 53*, 115-129.
Gibbs, P.D.L., Peek, A. and Thorgaard, G. (1994) An *in vivo* screen for the luciferase transgene in zebrafish, *Mol. Mar. Biol. Biotech.* 3, 307-316.
Ginsburg, A.S. (1968) Fertilization in fishes and the problem of polyspermy. Academy of Sci.s of the USSR, Institute of Developmental Biology. Translated from Russian by Z. Blake and edited by B. Golek. (Israel program for Scientific Translations, Jerusalem 1972, Keter Press Binding: Wiener Binding Ltd. Jerusalem. 366 pages.
Gong, Z., Hew, C.L. and Vielkind, J.R. (1991) Functional analysis and temporal expression of promoter regions from fish antifreeze protein genes in transgenic Japanese medaka embryos, *Mol. Mar. Biol. Biotech. 1*, 64-72.
Hallerman, E.M. and Kapuscinski, A. (1990) Transgenic fish and public policy: regulatory concerns, *Fisheries 15*, 12-20.
Hamada, K., Tamaki, K., Sasado, Y., Watai, Y., Kani, S., Wakamatsu, Y., Ozato, K., Kinoshita, M., Kohno, R., Takagi, S. and Kimura, M. (1998) Usefulness of the medaka β-actin promoter investigated using a mutant GFP reporter gene in transgenic medaka (*Oryzias latipes*), *Mol. Mar. Biol. Biotech. 7*, 173-180.
Hanley, S., Smith, T.J., Muller, F., Maclean, N., Uzbekova, S., Prunet, P. and Breton, B. (1998) Isolation and functional analysis of the histone H3 promoter from Atlantic salmon (*Salmo salar* L.), *Mol. Mar. Biol. Biotech. 7*, 165-172.
Higashijima, S.I., Okamoto, H., Ueno, N, Hotta, Y. and Eguchi, G. (1997) High frequency generation of transgenic zebrafish which reliably express GFP in whole muscles or the whole body by using promoters of zebrafish origin, *Develop. Biol. 192*, 289-299.
Houdebine, L.M. and Chourrout, D. 1991. Transgenesis in fish, *Experientia 47*, 891-897.
Howard, R.D., Martens, R.S., Innes, S.A., Drnevich, J.M. and Hale, J. (1998) Mate choice and mate competition influence male body size in Japanese medaka, *Anim. Behav. 55*, 1151-1163.
Inoue, K. and Yamashita, S. (1997) The techniques using electroporation to generate transgenic fish, in Houdebine, L.M. (ed.) *Transgenic animals generation and use*, Harwood academic publishers, France, pp. 129-132.
Inoue, K., Yamashita, S., Hata, J., Kabeno, S., Asada, S., Nagahisa, E. and Fujita, T. (1990) Electroporation as a new technique for producing transgenic fish, *Cell Diff. Develop. 29*, 123-128.

Iwamatsu, T. (1983) A new technique for dechorionation and observations on the development of the naked egg in *Oryzias latipes*, *J. Exper. Zool. 228*, 83-89.

Iyengar, A. and Maclean, N. (1995) Transgene concatemerization and expression in rainbow trout *(Oncorhynchus mykiss)*, *Mol. Mar. Biol. Biotech. 4*, 248-254.

Iyengar, A., Müller, F. and Maclean, M. (1996) Regulation and expression of transgenes in fish- a review, *Transgenic Res. 5*, 147-166.

Johnsson, J.I. (1993) Big and brave - size selection affects foraging under risk of predation in juvenile rainbow-trout, *Oncorhynchus mykiss*, *Anim. Behav. 45*, 1219-1225.

Johnsson, J.I., Petersson, E., Jönsson, E., Björnsson, B.T. and Jarvi, T. (1996) Domestication and growth hormone alter antipredator behaviour and growth patterns in juvenile brown trout (*Salmo trutta*), *Can. J. Fish. Aquatic Sci. 53*, 1546-1554.

Johnsson, J.I., Petersson, E., Jönsson, E., Jarvi, T. and Björnsson, B.T. (1999) Growth hormone-induced effects on mortality, energy status and growth: a field study on brown trout (*Salmo trutta*), *Funct. Ecol. 13*, 514-522.

Johnsson, J.I. and Bjornsson, B.T. (1994) Growth hormone increases growth rate, appetite and dominance in juvenile rainbow trout, *Oncorhynchus mykiss*, *Anim. Behav. 48*,177-186.

Jönsson, E., Johnsson, J.I. and Björnsson, B.T. (1998) Growth hormone increases aggressive behavior in juvenile rainbow trout, *Horm. Behav. 33*, 9-15.

Jönsson, E., Johnsson, J.I. and Björnsson, B.T. (1996) Growth hormone increases predation exposure of rainbow trout, *Proc. Royal Soc. London Series B 263*, 647-651.

Ju, B., Xu, Y., He, J., Liao, J., Yan, T., Hew, C.L., Lam, T.J. and Gong, Z. (1999) Faithful expression of green fluorescent protein (GFP) in transgenic zebrafish embryos under control of zebrafish gene promoters, *Develop. Genet. 25*, 158-167.

Kapuscinski, A.R. and Hallerman, E.M. (1990) Transgenic fish and public policy: anticipating environmental impacts of transgenic fish, *Fisheries 15*, 2-11.

Khoo, H.-W., Ang, L.H., Lim, H.B. and Wong, K.Y. (1992) Sperm cells as vectors for introducing foreign DNA into zebrafish, *Aquaculture 107*, 1-19.

Kinoshita, M., Toyohara, H., Sakaguchi, M., Inoue, K., Yamashita, S., Satake, M., Wakamatsu, Y. and Ozato, K. (1996) A stable line of transgenic medaka (*Oryzias latipes*) carrying the CAT gene, *Aquaculture 143*, 267-276.

Knibb, W. (1997) Risk from genetically engineered and modified marine fish, *Transgenic Res. 6*, 59-67.

Komatsu, S., Okazaki, Y., Tateno, M., Kawai, J., Konno, H., Kusakabe, M., Yoshiki, A., Muramatsu, M., Held, W.A. and Hayashizaki, Y. (2000) Methylation and downregulated expression of mac25/insulin-like growth factor binding protein-7 is associated with liver tumorigenesis in SV40T/t antigen transgenic mice, screened by restriction landmark genomic scanning for methylation (RLGS-M), *Bioch. Biophys. Res. Comm. 267*, 109-117.

Krasnov, A., Pitkänen, T.I. and Mölsä, H.. (1999a) Gene transfer for targeted modification of salmonid fish metabolism: genetic Analysis, *Biomol. Engineer. 15*, 115-119.

Krasnov, A., Pitkänen, T.I., Nishikimi, M. and Mölsä, H. (1999b) Expression of human glucose transporter type I and rat hexokinase type II complementary DNAs in rainbow trout embryos, effects on glucose metabolism, *Mar. Biotechnol. 1*, 25-32.

Krisher, R.L., Gibbons, J.R. and Gwazdauskas, F.C. (1995) Nuclear transfer in the bovine using microinjected donor embryos: assessment of development and deoxyribonucleic acid detection frequency, *J. Dairy Sci. 78*, 1282-1288.

Lakra, W.S. and Das, P. (1998) Genetic engineering in aquaculture, *Ind. J. Anim. Sci. 68*, 873-879.

Lin, S., Gaiano, N.,Culp, P., Burns, J.C., Friedmann, T., Yee, J.K. and Hopkins, N. (1994) Integration and germ-line transmission of a pseudotyped retroviral vector in zebrafish, *Science 265*, 666-669.

Linney, E., Hardison, N.L., Lonze, B.E., Lyons, L. and DiNapoli, L. (1999) Transgene expression in zebrafish: a comparison of retroviral-vector and DNA-microinjection approaches, *Develop. Biol. 213*, 207-216.

Liu, Z., Li, P. and Dunham, R.A. (1998) Characterization of an A/T-rich family of sequences from channel catfish (*Ictalurus punctatus*), *Mol. Mar. Biol. Biotech. 7*, 232-239.

Lodge, D.M. (1993) Biological invasions: lessons for ecology, *Trends Ecol. Evol. 8*, 133-137.

Long, Q., Meng, A., Wang, H., Jessen, J.R., Farrell, M.J. and Lin, S. (1997) GATA-1 expression pattern can be recapitulated in living transgenic zebrafish using GFP reporter gene, *Development 124*, 4105-4111.

Lu, J.K., Burns, J.C. and Chen, T.T. (1997) Pantropic retroviral vector integration, expression, and germline transmission in medaka (*Oryzias latipes*), *Mol. Mar. Biol. Biotech. 6*, 289-295.

Lu, J.K., Chen, T.T., Allen, S.K., Matsubara, T. and Burns, J.C. (1996) Production of transgenic dwarf surfclams, *Mulinia lateralis*, with pantropic retroviral vectors, *Proc. Natl. Acad. Sci. USA 93*, 3482-3486.

Martinez, R., Estrada, M.P., Berlanga, J., Guillen, I., Hernandez, O., Cabrera, E., Pimentel, R., Morales, R., Herrera, F., Morales, A., Pina, J.C., Abad, Z., Sanchez, V., Melamed, P., Leonart, R. and de la Fuente, J. (1996) Growth enhancement in transgenic tilapia by ectopic expression of tilapia growth hormone, *Mol. Mar. Biol. Biotech. 65*, 62-70.

Matzke, A., Neuhuber, F., Park, Y.D., Ambros, P.F. and Matzke, A.J.M. (1994) Homology-dependent gene silencing in transgenic plants- epistatic silencing loci contain multiple copies of methylated transgenes, *Mol. Gen. Genet. 244*, 219-229.

Matzke, M.A. and Matzke, A.J.M. (1995) Homology-dependent gene silencing in transgenic plants: what does it really tell us?, *Trends Genet. 11*, 1-3.

Matzke, M.A. and Matzke, A.J.M. (1998) Gene silencing in plants: relevance for genomic evolution and the acquisition of genomic methylation patterns, *Epigenetics 214*, 168-180.

Moran, P., Pendas, A.M., Beall, E. and Garcia-Vazquez, E. (1996) Genetic assessment of the reproductive success of Atlantic salmon precocious parr by means of VNTR loci, *Heredity 77*, 655-660.

Mori, T. and Devlin, R.H. (1999) Transgene and host growth hormone gene expression in pituitary and nonpituitary tissues of normal and growth hormone transgenic salmon, *Mol. Cell. Endocrinol. 149*, 129-139.

Morris, M.C., Vidal, P., Chaloin, L., Heitz, F., and Divita, G. (1997) A new peptide vector for efficient delivery of oligonucleotides into mammalian cells, *Nucl. Acid Res. 25*, 2730-2736.

Muir W.M., Hostetler, H.A., Rider, S.D.Jr, Schulte, S.J., and Howard, R.D. (2000) Effect of the Atlantic salmon growth hormone promoter (sGH) human growth hormone (hGH) gene construct on growth and survival in Japanese medaka (Oryzias latipes), submitted.

Muir, W.M. and Howard, R.D. (2000) Methods to Assess Ecological Risks of Transgenic Fish Releases, in Letourneau, D.K. and Burrows, B.E. (eds) *Genetically Engineered Organisms: Assessing Environmental and Human Health Effects*, CRC Press, in Press.

Muir, W.M. and Howard, R.D. (1999) Possible ecological risks of transgenic organism release when transgenes affect mating success: sexual selection and the Trojan gene hypothesis, *Proc. Natl. Acad. Sci. USA 96*, 13853-13856.

Muir, W.M., Howard, R.D., and Bidwell, C. (1994) Use of multigenerational studies to assess genetic stability, fitness, and competitive ability of transgenic Japanese medaka: I. Methodology, *Proc. Biotech. Risk Assess. Symp.*, pp. 170-193.

Muir, W.M., R. D. Howard, R. S. Martens, S. Schulte, and C. A. Bidwell. (1996) Use of multigenerational studies to assess genetic stability, fitness, and competitive ability of transgenic Japanese medaka, III. results and predictions, *Proc. 8th Inter. Conf. Risk Assess. Method.*

Muir, W.M., Martens, R., Howard, R.D. and Bidwell, C. (1995) Use of multigenerational studies to assess genetic stability, fitness, and competitive ability of transgenic Japanese medaka: II. Development of transgenic medaka and mating preferences, *Proc. 2nd Inter. Conf. Risk Assess. Method.*

Muldoon, R.R., Levy, J.P., Kain, S.R., Kitts, P.A. and Link, C.J. Jr. (1997) Tracking and quantitation of retroviral-mediated transfer using a completely humanized, red-shifted green fluorescent protein gene, *BioTech. 22*, 162-167.

Müller, F., Lele, Z., Váradi, L., Menczel, L. and Orbán, L. (1993) Efficient transient expression system based on square pulse electroporation and in vivo luciferase assay of fertilized fish eggs, *FEBS 324*, 27-32.

Murakami, Y., Motohashi, K., Yano, K., Ikebukruo, K., Yokoyama, K., Tamiya, E. and Karube, I. (1994) Micromachined electroporation system for transgenic fish, *J. Biotech. 34*, 35-42.

Nam, Y.K., Jeong, C.H. and Kim, D.S. (1998) A simple clearing method for improved visualization of *lacZ* expression in fish larvae, *Transgenic Res. 7*, 397-399.

Namciu, S.J., Blochlinger, K.B. and Fournier, R.E.K. (1998) Human matrix attachment regions insulate transgene expression from chromosomal position effect in *Drosophila melanogaster, Mol. Cell. Biol. 18*, 2382-2391.

Neumann, E., Schaefer-Ridder, M., Wang, Y. and Hofschneider, P.H. (1982) Gene transfer into mouse lyoma cells by electroporation in high electric fields, *EMBO J. 1*, 841-845.

Ozato, K., Kondoh, H., Inohara, H., Iwamatsu, T. Wakamatsu, Y. and Okada, T.S. (1986) Production of transgenic fish: introduction and expression of chicken δ-crystallin gene in medaka embryos, *Cell Diff. 19*, 237-244.

Ozato, K., Inoue, K. and Wakamatsu, Y. (1989) Transgenic fish: biological and technical problems, *Zool. Sci. 6*, 445-457.

Palmiter, R.D., Norstedt, G., Gelinas, R.E., Hammer, R.E. and Brinster, R.L. (1983) Metallothionein-hGH fusion genes stimulate growth of mice, *Science 222*, 809-814.

Park, Y.D., Papp, Y., Moscone, E.A., Iglesias, V.A., Vaucheret, H., Matzke, A.J.M. and Matzke, M.A. (1996) Gene silencing mediated by promoter homology occurs at the level of transcription and results in meiotically heritable alterations in methylation and gene activity, *Plant J. 9*, 183-194.

Patil, J.G., Wong, V. and Khoo, H.W. (1994) Assessment of pMTL construct for detection *in vivo* of luciferase expression and fate of the transgene in the zebrafish, *Brachydanio rerio*, *Zool. Sci. 11*, 63-68.

Penman, D.J., Beeching, A.J., Penn, S. and Maclean, N. (1990) Factors affecting survival and integration following microinjection of novel DNA into rainbow trout eggs, *Aquaculture 85*, 35-50.

Pitkänen, T.I., Krasnov, A., Teerijoki, H. and Mölsä, H. (1999b) Transfer of growth hormone (GH) transgenes into Arctic charr (*Salvelinus alpinus* L.) I. Growth response to various GH constructs. Genetic Analysis, *Biomol. Engin. 15*, 91-98.

Pitkänen, T.I., Krasnov, A., Reinisalo, M. and Mölsä, H. (1999a) Transfer and expression of glucose transporter and hexokinase genes in salmonid fish, *Aquaculture 173*, 319-332.

Prather, R.S., Barnes, F.L., Sims, M.M., Robl, J.M., Eyestone, M.H. and First, N.L. (1987) Nuclear transplantation in the bovine embryo: assessment of donor nuclei and recipient oocyte, *Biol. Reprod. 37*, 859-866.

Rahman, M.A. and Maclean, N. (1992) Production of transgenic tilapia (*Oreochromis niloticus*) by one-cell stage microinjection, *Aquaculture 105*, 219-232.

Rahman, M.A. and Maclean, N. (1999) Growth performance of transgenic tilapia containing an exogenous piscine growth hormone gene, *Aquaculture 173*, 333-346.

Rahman, M.A., Lyengar, A. and Maclean, N. (1997) Co-injection strategy improves integration efficiency of a growth hormone gene construct, resulting in lines of transgenic tilapia (*Oreochromis niloticus*) expressing an exogenous growth hormone gene, *Transgenic Res. 6*, 369-378.

Rahman, M.A., Mak, R., Ayad, H., Smith, A. and Maclean, N. (1998) Expression of a novel piscine growth hormone gene results in growth enhancement in transgenic tilapia (*Oreochromis niloticus*), *Transgenic Res 7*, 357-369.

Salehi-Ashtiani, K., Widrow, R.J., Marker, C.L. and Goldberg, E. (1993) Testis-specific expression of a metallothionein I-driven transgene correlates with undermethylation of the locus in testicular DNA, *Proc. Natl. Acad. Sci. USA 90*, 8886-8890.

Saunders, R.L., Fletcher, G.L. and Hew, C.L. (1998) Smolt development in growth hormone transgenic Atlantic salmon, *Aquaculture 168*, 177-193.

Schaefer-Ridder, M., Wang, Y. and Hofschneider, P.H. (1982) Liposomes as gene carriers: efficient transformation of mouse L cells by thymidine kinase gene, *Science 215*,166-168.

Shears, M.A., Fletcher, G.L., Hew, C.L., Gauthier, S., and Davies, P.L. (1991) Transfer, expression, and stable inheritance of antifreeze protein genes in Atlantic salmon (*Salmo salar*), *Mol. Mar. Biol. Biotech. 1*, 58-63.

Shears, M.A., King, M.J., Goddard, S.V. and Fletcher, G.L. (1992) Gene transfer in salmonids by injection through the micropyle, in Hew C.L. and Fletcher G.L. (eds), *Transgenic fish*. World Scientific Publishing Co., Singapore, pp. 44-60.

Sheela, S.G., Chen, J.D., Mathavan, S. and Pandian, T.J. (1998) Construction, electroporation transfer and expression of ZpβypGH and ZpβrtGH in zebrafish, *J. Biosci. 23*, 565-576.

Sheela, S.G., Pandian, T.J. and Mathavan, S. (1999) Electroporation transfer, stable integration, expression and transmission of pZpβypGH and pZpβrtGH in Indian catfish, *Heteropneustes* (Bloch), *Aquaculture Res. 30*, 233-248.

Shigekawa, K. and Dower, W.J. (1988) Electroporation of eukaryotes and prokaryotes: a general approach to the introduction of macromolecules into cells, *BioTech. 6*, 742-751.

Sims, M. and First, N.L. (1993) Production of calves by transfer of nuclei from cultured inner cell mass cells, *Proc. Natl. Acad. Sci. USA 90*, 6143-6147.

Sin, F.Y.T. (1997) Transgenic fish, *Rev. Fish Biol. Fish. 7*, 417-441.

Sippel, A.E., Saueressig, H., Huber, M.C., Faust, N. and Bonifer, C. (1997) Insulation of transgenes from chromosomal position effects, in Houdebine L.M. (ed.), *Transgenic animals generation and use*, Harwood academic publishers, France, pp. 257-265.

Snibson, K.J. and Adams, T.E. (1997) Transgenic methylation in transgenic animals, in Houdebine L.M. (ed.), *Transgenic animals generation and use*, Harwood academic publishers, France, pp 237-243.

Stuart, G.W., Vielkind, J.R., McMurray, J.V. and Westerfield, M. (1990) Stable lines of transgenic zebrafish exhibit reproducible patterns of transgene expression, *Develop. 109*, 577-584.

Stuart, G.W., McMurray, J.V. and Westerfield, M. (1988) Replication, integration and stable germ-line transmission of foreign sequences injected into early zebrafish embryos, *Develop. 103*, 403-412.

Takagi, S., Sasado, T., Tamiya, G., Ozato, K., Wakamatsu, Y., Takeshita, A. and Kimura, M. (1994) An efficient expression vector for transgenic medaka construction, *Mol. Mar. Biol. Biotech. 3*, 192-199.

Thomaz D, Beall E, and Burke, T. (1997) Alternative reproductive tactics in Atlantic salmon: Factors affecting mature parr success, *Proc. Royal Soc. London Series B-Biol. Sci. 264*, 219-226.

Tiedje, J.M., Colwell, R.K., Grossman, Y.L., Hodson, R.E., Lenski, R.E., Mack, R.N. and Regal, P.J. (1989) The planned introduction of genetically engineered organisms: ecological considerations and recommendations, *Ecology 70*, 298-315.

Tse, C.-M. and Young, J.D. (1990) Glucose transport in fish erythrocytes, variable cytochalasin-B-sensitive hexose transport activity in the common eel (*Anguilla japonica*) and transport deficiency in the paddyfish eel (*Monopterus albus*) and rainbow trout (*Salmo gairdneri*), *J. Exp. Biol. 148*, 367-383.

Vain, P., Worland, B., Kohli, A., Snape, J.W., Christou, P., Allen, G.C. and Thompson, M.F. (1999) Matrix attachment regions increase transgene expression levels and stability in transgenic rice plants and their progeny, *Plant J. 18*, 233-242.

Vielkind, J.R. (1992) Medaka and zebrafish, ideal as transient and stable transgenic systems, in Hew C.L. and Fletcher, G.L. ([eds.), *Transgenic fish*, World Scientific Publishing Co., Singapore, pp. 72-91

Volckaert, F.A., Hellemans, B.A., Galbusera, P., Sekkali, B., Belayew, A. and Ollevier, F. (1994) Replication, expression, and fate of foreign DNA during embryonic and larval development of the African catfish (*Clarias gariepinus*), *Mol. Mar. Biol. Biotech. 3*, 57-69.

Wall, R.J. and Burdon, T.G. (1997) The fate of microinjected genes in preimplantation embryos, in Houdebine L.M. (ed.), *Transgenic animals generation and use*, Harwood academic publishers, France, pp 215-217.

Wallace, H., Ansell, R., Clark, J. and McWhir, J. (2000) Pre-selection of integration sites imparts repeatable transgene expression, *Nucl. Acids Res. 28*, 1455-1464.

Walton, C.-M. and Cowey, C.B. (1982) Aspects of intermediary metabolism in salmonid fish, *Comp. Biochem. Physiol. B 73*, 59-79.

Winkler, C. and Schartl, M. (1997) Gene transfer in laboratory fish, model organisms for the analysis of gene function, in Houdebine L.M. (ed.), *Transgenic animals generation and use*, Harwood academic publishers, France, pp 387-395.

Wright, S. 1969. *Evolution and the Genetics of Populations. Vol. 2.* The Theories of Gene Frequencies, Univ. of Chicago Press.

Wright, S. 1982. The shifting balance theory of macroevolution, *Ann. Rev. Genet. 16*, 1-19.

Xie, Y., Liu, D., Zou, J., Li, G. and Zhu, Z. (1993) Gene transfer via electroporation in fish, *Aquaculture 111*, 207-213.

Zelenin, A.V., Alimov, A.A., Barmintzev, V.A., Beni, A.O., Zelenina, I.A., Krasnov, A.M. and Kolesnikov, V.A. (1997) The delivery of foreign genes into fertilized fish eggs using high microprojectiles. *FEBS 287*(1,2), 118-120.

Zhang, P., Hayat, M., Joyce, C., Gonzalez-Villaseñor, L.I., Lin, C.M., Dunham, R.A., Chen, T.T. and Powers, D.A. (1990) Gene transfer, expression and inheritance of pRSV-rainbow trout-GH cDNA in the common carp, *Cyprinus carpio* (Linnaeus), *Mol. Reprod. Dev. 25*, 3-13.

THE QUEST FOR TRANSGENIC POULTRY: BIRDS ARE NOT MICE WITH FEATHERS

PROUDMAN J.A., WENTWORTH A.L.* AND WENTWORTH B.C.*
*Germplasm and Gamete Physiology Laboratory, Agricultural Research Service, U.S. Department of Agriculture, Beltsville, Maryland 20705, U.S.A. and *Department of Animal Sciences, University of Wisconsin, Madison, Wisconsin 53706, U.S.A.*

Abstract

Although the production of transgenic mice for research is now routine, and the application of this technology to many livestock species has met with considerable success, the production of transgenic poultry has been severely hindered by biological obstacles that have prevented the direct application of mammalian technology to birds. Many of these obstacles have been overcome in recent years, resulting in success in the production of either chimeric or transgenic individuals by microinjection of naked DNA into the newly fertilized ovum, transfer of blastodermal cells or primordial germ cells from donor to recipient embryos, and development of replication-defective retroviral vectors for DNA insertion. Stable integration of foreign DNA into the poultry genome remains problematical, but recent research demonstrating the use of the *Drosophila* element *mariner* to integrate foreign DNA into the chicken chromosome with a high frequency of germline transmission seems promising. Much basic research remains in order to achieve the routine production of transgenic poultry and to regulate expression of transgenes in tissues and amounts that are appropriate for their intended use. The short generation time and large number of progeny possible with poultry species should ensure that interest in the production of transgenic poultry for research and commercial purposes will increase.

1. Introduction

The production of transgenic animals is now an essential tool in many aspects of biological and biomedical research, and the technology is increasingly being applied to the genetic advancement of livestock and the production of pharmaceutical products.

R. Renaville and A. Burny (eds.), Biotechnology in Animal Husbandry, 283–299.

The mouse has been the species of choice for transgenic research since the first reports 20 years ago of integration of naked DNA into the mouse chromosome by microinjection (Gordon *et al.*, 1980; Gordon and Ruddle, 1981). Subsequently, injection of naked DNA into the pronucleus or into the cytoplasm of a newly fertilized egg has resulted in its chromosomal integration in the rat (Zeng *et al.*, 1994; Charreau *et al.*, 1996), frog (Etkin *et al.*, 1984; Heasman *et al.*, 1991) , sea urchin (Flytzanis *et al.*, 1985), fish (Iyengar *et al.*, 1996), commercial livestock (Pursel *et al.*, 1989; Pursel and Rexroad, 1993; Wall *et al.*, 1996), and chicken (Love *et al.*, 1994). The isolation of mouse embryonic stem cells (Evans and Kaufman, 1981; Martin, 1981) provided a second, more precise method for making a transgenic animal by allowing gene targeting for precise genetic modifications. However, to date, this technology can only be applied in the mouse. An overwhelming majority of the transgenic animals produced for research are mice.

The more recent reports of the production of transgenic livestock by microinjection have demonstrated the economic potential of transgenic animals. Fleece weight of first-generation yearling sheep has been improved by stable expression of IGF-1 in wool follicles, although the difference did not persist in subsequent year's shearings or in second-generation lambs (Su *et al.*, 1998). Transgenic sheep carrying the growth hormone gene under the control of the metallothionein promoter grew faster without apparent deleterious health effects previously associated with excessive GH secretion (Ward and Brown, 1998). Human proteins and antibodies are now being produced in the milk of transgenic dairy animals for therapeutic use (Pollock *et al.*, 1999; Rudolph, 1999; Ziomek, 1998). Nuclear transfer of genetically modified cells has recently provided the technology to produce human proteins in transgenic dairy animals (Colman, 1999).

The use of similar techniques to produce transgenic poultry is hampered by major biological differences in the structure and development of mammalian and avian eggs. Many novel approaches are under investigation to access the poultry genome and introduce different types of DNA molecules in a heritable manner. Although the potential for gene transfer in poultry has been recognized for almost two decades (Shuman and Shofffner, 1982; Salter *et al.*, 1987; Lee and Shuman, 1990), few successes of germline transmission of foreign DNA have been reported.

The present paper will detail the obstacles to production of transgenic birds, review the progress achieved, and explore the strategies that are under investigation to access the poultry genome.

2. Unique Aspects of Reproduction and Embryonic Development in Poultry

2.1. BIOLOGICAL OBSTACLES TO MICROINJECTION

The technique of pronuclear microinjection that has proven successful in other species is not directly applicable to avian species. In the mouse, one-cell embryos are easily recovered from the oviducts following fertilization, and the haploid male and female pronuclei are readily accessible for microinjection. In contrast, the mature avian egg (ovum) consists of the diploid female germ cell enclosed in the germinal vesicle. The germinal vesicle is situated just under the perivitelline membrane on the periphery of an enormous amount of yolk (the food reserves for the developing amniotic embryo). Approximately one hour prior to ovulation the first meiotic division proceeds in the ovum and the first polar body is extruded. During the first 15 minutes following ovulation, the fertilization process begins with several spermatozoa penetrating the perivitelline layer of the egg. After the sperm nuclear membranes fuse with the perivitelline membrane, the male pronuclei (haploid gametes prior to fusion) are released into the egg, and an outer perivitelline membrane forms preventing entry of additional spermatozoa (Bakst and Howarth, 1977). Within the next two and one-half to three hours, the primary oocyte extrudes the second polar body and then all of the pronuclei enlarge. Finally, one male pronucleus moves adjacent to and fuses with the maternal pronucleus, completing syngamy and fertilization (Waddington *et al.*, 1998)(Figure 1). Despite the entry of several spermatozoa into the egg, usually only one male pronucleus fuses with the female pronucleus. The other remaining male supernumerary pronuclei divide once, the nuclei fragment and the cells disintegrate in the cytoplasm (Fofanova, 1964; Perry, 1987). The chicken egg spends approximately 24 to 27 hours in the oviduct to receive albumen, membranes, and shell. Cleavage division begins early in this process. By the time the completed egg is laid, the embryo consists of 40,000 to 60,000 cells.

Access to the newly fertilized ovum, therefore, necessitates sacrifice of the hen. Use of such an ovum as a target for gene transfer obligates either it's reinsertion into an oviduct, or external incubation in an *ex vivo* culture system that would support embryo development to hatching. The inability to identify the pronuclei that will fuse, the lack of access to the early embryo for microinjection, and the presence of a large yolk, have presented extraordinary obstacles to adapting this mammalian technology for use in poultry. Extensive research has overcome many of the obstacles to the successful application of microinjection techniques to poultry, but documented successes have been very limited.

2.2. EMBRYONIC STEM CELLS AND PRIMORDIAL GERM CELLS

Embryonic stem cells are totipotent cells that are capable of developing into any embryonic cell type in an appropriate environment. Such cells have yet to be identified in the avian embryo. Primordial germ cells (PGCs), the precursors to the male and female gametes of the adult animal, are cells that are first identified in the blastoderm of the freshly laid avian egg. They migrate to the gonads during embryonic development, but have not been shown to be totipotent. Blastodermal cells from the germinal disk of the freshly laid egg have been successfully used to modify recipient embryos. However, successes in the production of germline chimeras, and in the germline transmission of foreign DNA transfected by retrovirus into blastodermal cells, are thought to have resulted from the few PGCs present in the blastoderm. Aggressive research has markedly improved techniques for the isolation, culture and transfection of embryonic cells, particularly PGCs, and for their reintroduction into embryos that subsequently develop and hatch. The allure of genetic modification by homologous recombination demonstrated with mouse embryonic stem cell lines, coupled with occasional successes in production of transgenic poultry using manipulated embryonic cells, drives intensive efforts to identify and manipulate undifferentiated avian cells.

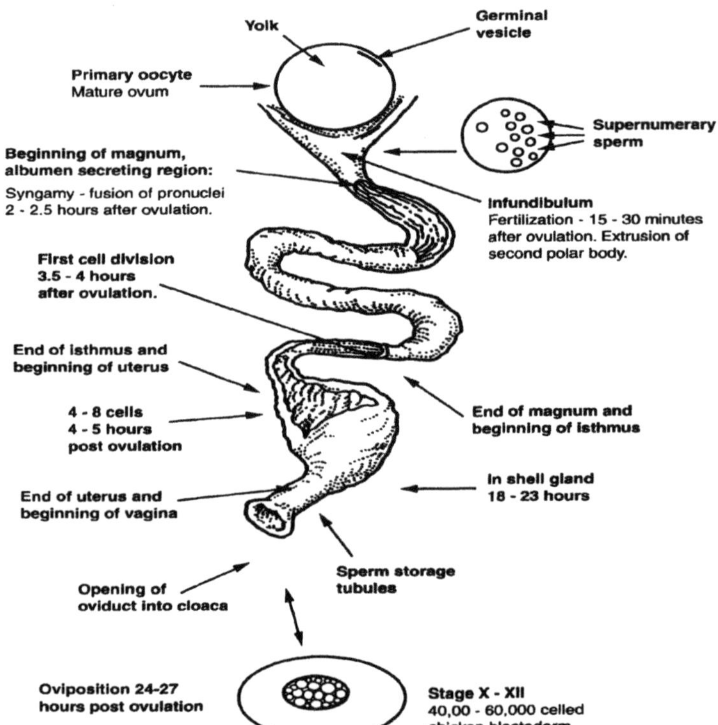

Figure 1. Diagramatic representation of the female reproductive tract of the chicken, demonstrating the location of the ovum at the time of fertilization, syngamy, and the start of embryonic cell division. (Adapted from Johnson, 1986).

3. Approaches to Accessing the Poultry Genome

3.1. INFECTION WITH RETROVIRUSES

Souza *et al.* (1984) were first to demonstrate the *in vivo* expression of a cloned chicken gene (growth hormone) coupled to a replication-competent viral vector in chicks that had been infected as embryos (F_0, or founder, generation). Germline transmission of foreign DNA was first reported by injecting replication-competent retroviruses into the blastoderm of freshly laid eggs (Salter *et al.*, 1986; Salter *et al.*, 1987). Although the natural insertion of viral DNA into the avian genome by retroviruses has often occurred with pathological results (Frisby *et al.*, 1979), the use of wild-type (pathogenic) or recombinant avian leukosis virus to intentionally produce viremic founder lines and pass viral DNA to progeny demonstrated the feasibility of using such viruses as vectors to introduce foreign genes. This work led to development of replication-defective retroviral vectors for the production of virus-free transgenic poultry. Bosselman *et al.* (1989) demonstrated germline transmission of replication-defective viral DNA to F_1 progeny, and this DNA was subsequently detected in F_2 progeny (Briskin *et al.*, 1991) and in fibroblasts of F_3 embryos (Thoraval *et al.*, 1995). However, concerns about the potential risk of generating replicating recombinant retroviruses, and limitations in the size of the transgene that can be inserted using retroviral vectors, have focused most research on other methods of gene insertion.

3.2. MICROINJECTION OF NAKED DNA INTO THE GERMINAL DISC OF THE NEWLY FERTILIZED OVUM

Despite the biological limitations to accessing the newly fertilized avian ovum and subsequently hatching offspring, exogenous DNA has been successfully introduced and transmitted to F_1 progeny using this technique (Love *et al.*, 1994). This approach necessitated the development of a complete *in vitro* culture system to maintain the chick embryo from fertilization through to hatching (Rowlett, 1987; Perry, 1988; Naito *et al.*, 1990; Ono *et al.*, 1994). The fertilized ovum was removed from the magnum 2.5 hours after oviposition of the previous egg. This is about two hours after ovulation, and is either prior to, or at about the time of, fusion of the male and female pronuclei. Cloned DNA was then injected into the center of the germinal disc, close to the putative site of the female pronucleus, and the egg was cultured *ex vivo* in surrogate eggshells until hatch. Initial studies injected either circular or linearized plasmids containing viral or chicken promoter-enhancer sequences that directed constitutive expression of reporter genes (Sang and Perry, 1989; Naito *et al.*, 1991; Perry, 1991). Results all indicated mosaic expression in early embryonic tissues that was gradually lost. Since no evidence for chromosomal integration of the injected DNA was observed, the general conclusion from these experiments was that the plasmids existed episomally and were gradually lost with successive cell divisions. Extension of these earlier experiments resulted in the production of one cockerel that showed transmission of the exogenous DNA into his progeny (Love *et al.*, 1994). Of the 128 ova injected, half of the resulting embryos contained plasmid DNA. Seven healthy chicks survived to sexual maturity, and one was identified as a somatic mosaic with approximately one copy of the transgene per 10 genomic equivalents. Semen from this male was used to inseminate laying hens and their progeny were screened by PCR for the presence of the transgene. The introduced DNA was transmitted to 14 out of a total of 412 (3.4%) F_1 progeny chicks tested.

Recently, Sherman *et al.* (1998) reported a unique adaptation of the microinjection method. The *Drosophila* element, *mariner,* was demonstrated to transpose into the chicken genome and was employed as a vector for transgenesis. Using this novel approach, these investigators constructed a plasmid carrying the active *mariner* element and injected it into 97 chick zygotes. Forty-four of the manipulated embryos survived for at least 12 days of development with 27% of these embryos containing *mariner* equivalent to one copy per genome. Three chicks survived to sexual maturity, and the one positive cockerel was crossed with stock hens. Of the 93 first-generation hatched offspring, 27 were identified by PCR as transgenic for *mariner*. No evidence for instability of *mariner* after transposition was detected. These data were confirmed by the generation of additional transgenic birds. The frequency of *mariner* transposition into the chicken genome was over 20% and the frequency of germline inheritance was just under 30%, a 10-fold higher transmission frequency than that obtained after introduction of standard gene constructs.

3.3. EMBRYONIC TRANSFER OF BLASTODERMAL CELLS

Blastodermal cells from the germinal disk of the freshly laid egg have been successfully used to modify recipient embryos. Early studies indicated that dispersed blastodermal cells from the freshly-laid egg could develop in a host blastoderm and differentiate into melanocytes (Marzullo, 1970; Reynaud, 1976). Petitte *et al.* (1990) were the first to demonstrate germline chimerism using donor blastodermal cells. A rooster that showed the pigmentation of the donor cell strain transmitted the donor genome to 0.3% of his progeny as revealed by both feather color and blood DNA analysis. These results demonstrated that primordial germ cells in the embryo blastoderm could develop in a recipient embryo, migrate to the host gonads, differentiate, and produce viable gametes.

More recently, Kagami *et al.* (1997) determined unequivocally that the most central area of the area pellucida was the primary area for the localization of primordial germ cells in the stage X chicken blastoderm. An average of 70 percent germline chimeras were produced from transfer of approximately 700 cells from the center of the area pellucida of one blastoderm to a recipient blastoderm from which the cells from the central-most portion of the area pellucida had been removed. Additionally several germline chimeras transmitted exclusively donor-derived gametes into the offspring. Cells transferred from either the area opaca or from the outer rim of the area pellucida were unable to contribute germline chimeras. Other studies have produced interspecies chimeras by transfer of blastodermal cells between quail and chickens (Naito *et al.*, 1992; Watanabe *et al.*, 1992) and between ducks and chickens (Li *et al.*, 1999).

One approach to improving the efficiency of germline chimerism has been to expose the recipient embryos to γ-irradiation, which presumably temporarily compromises the ability of the recipient's germ cells to divide (Carsience *et al.*, 1993; Etches *et al.*, 1997). In these experiments, 1105 irradiated embryos were injected with blastodermal cells, 38 hatched, and 11 of these subsequently produced progeny that were derived from donor cells. One of these individuals produced only donor-derived offspring, suggesting that only donor-derived PGCs populated the gonads during development.

The demonstration that embryonic transfer of blastodermal cells would produce germline chimeras led to efforts to insert foreign DNA into the poultry genome using transfected PGCs of the blastoderm. Liposome-mediated transfection of a *lacZ* reporter gene, which required culture of the blastodermal cells for two to four days, resulted in incorporation of the gene into extra- and intra-embryonic tissues of the recipient embryo, particularly after γ-irradiation of the recipient embryo (Carsience *et al.*, 1993). Rosenblum and Chen (1995) transfected blastodermal cells *in ovo* by injecting a DNA-liposome complex into the subgerminal cavity of the freshly laid egg and demonstrated tissue expression of the reporter gene for up to eight days, but stable incorporation was not observed. Interestingly, Inada *et al.* (1997) reported that such *in ovo* transfection did result in the subsequent appearance of transfected PGCs in embryonic blood at the time of PGC migration. Electroporation of blastodermal cells resulted in transient expression of the *lacZ* gene *in vitro*, but failed to result in expression in embryonic tissues (Etches *et al.*, 1997).

3.4. EMBRYONIC TRANSFER OF CULTURED BLASTODERMAL CELLS

Embryos from freshly-laid eggs have been cultured for 48 hours as explanted whole embryos, as dispersed cells from the central disc of the area pellucida, or as dispersed cells grown on a confluent layer of STO feeder cells (Etches *et al.*, 1996). When transferred to recipient embryos compromised by γ-irradiation, a total of nine out of 127 hatched recipient chicks (7%) were germline chimeras, while 28 of the 127 produced feather chimeras (22%), presumably from cells that were of neural crest origin. Of note in this experiment is the fact that six of the birds that produced germline chimeras did not show any feather chimerism. Although other somatic tissues were not tested, these data suggest that blastodermal cells maintained in culture for at least 2 days retain their predetermined potential to develop along their already determined pathway of differentiation, and to produce various somatic and some germline chimeras. The finding of germline chimeras without obvious somatic chimerism would indicate that these chimeras had probably incorporated mostly donor PGCs, along with cells from the hypoblast (extraembryonic tissue). The large consistent overall percentage difference between feather and germline chimeras in these experiments may be reflective of the low numbers of determined PGCs that are present in the blastoderm at this stage of development. These data would indicate that blastodermal cells are already committed to their pathway of differentiation.. In order to select a totipotent cell population from an avian blastoderm, comparable to the mouse embryonic stem cell, an earlier staged embryo may be required. The most successful murine embryonic stem cell line, line 129, was established from totipotent cells of the inner cell mass at the morula stage of development (i.e., from some 64 cells). To be used successfully for transgenesis, any cultured cell must retain its ability to enter the germline. Blastodermal cells maintained in culture for seven days were capable of producing both somatic and germline chimeras, although the percentage of germline chimeras was substantially reduced compared to freshly isolated blastodermal cells (Pain *et al.*, 1996; Etches *et al.*, 1997). Following several months of passage, these blastodermal cells did not produce germline chimeras, and gradually lost their ability to bind the EMA-1antibody, which identifies chicken PGCs in culture (Pain *et al.*, 1996). An immortalized blastodermal cell line similarly failed to express the EMA-1 epitope (Tsai, 1995). However, Tsai *et al.* (1999) have recently reported a blastodermal culture system that maintains cells for at least six months, and these authors have shown that

30% of these cultured cells expressed the EMA-1 epitope. Cells cultured for two months and loaded with a fluorescent dye prior to transplantation into recipient 18-hour incubated embryos were detected in gonadal tissue of embryos six days after injection (Tsai *et al.*, 1999). The cells, cultured two months or longer, were also transfected with a plasmid vector containing green fluorescent protein (pEGFP) and injected into recipient 18h incubated embryos in the gonadal crescent area. Blood was obtained from the four week old chickens and analyzed by flow cytometry. Donor $B^{21}B^{21}$ haplotype blood cells were detected in recipient B^2B^{15} blood. In some chicks the percent donor cells was as high as 25% to 33%. In one recipient of pEGFP, green fluorescent protein was detectable in the blood and feather pulp tissue. Taken together, these results suggest that the production of germline chimeras from long-term cultured blastodermal cells may be possible.

3.5. PRIMORDIAL GERM CELLS AS A ROUTE TO THE AVIAN GERMLINE

Since germline chimeras produced from blastodermal cells have likely arisen from the small population of PGCs present at this stage of development, the isolation of enriched populations of PGCs from later embryos may provide a method of increasing the efficiency of germline transmission. Unlike PGCs in mammals, avian PGCs migrate to the gonads primarily by means of the embryonic circulation (Gilbert, 1991). Stage 7 (Hamburger and Hamilton, 1951) chicken embryos (23-26 hours of incubation) possess from 125 to 250 PGCs in the germinal crescent as identified by periodic acid schiff (PAS) staining. As incubation proceeds, PGCs pass into the developing vascular system at Stage 12-18 (45-69 hr of incubation) and migrate to colonize the gonads, where they proliferate and differentiate into spermatogonia or oogonia (for review, Wentworth *et al.*, 1996).

The transfer of Stage 7 germinal crescent cells, containing PGCs, into the vascular system of Stage 15 (50-55 hr of incubation) recipient quail and chickens resulted in hatched chicks that produced offspring derived from the donor PGCs (Gonzales, 1989; Wentworth *et al.*, 1989). PGC-enriched germinal crescent cells were subsequently employed for retroviral transfection (Vick *et al.*, 1993) and ballistic insertion (Li *et al.*, 1995). Although manipulated embryos that hatched were observed to produce offspring of the donor germline, F_1 offspring lost their ability to transmit the donor DNA as they matured (Li *et al.*, 1995).

Simkiss *et al.* (1989) were first to show that PGCs in the blood could be successfully transferred between strains of chickens and become resident in the gonads of the recipient embryo. Although numbers of PGCs isolated per embryo were initially low (10 to 40 PGCs in 10 μl of blood; Al-Thani and Simkiss, 1991), current techniques have yielded estimates of 40 to 51 PGCs per μl of blood (Naito *et al.*, 1994; Kagami *et al.*, 1997), or a mean of 251.4 ± 140.6 PGCs per embryo after Ficoll density gradient centrifugation (Naito *et al.*, 1999).

The initial experiments of Nakamura *et al.* (1991; 1992) unequivocally showed that the optimum stage for transferring PGCs from the vascular system of the donor to that of the recipient embryo was at stages 13 to 15. When the PGCs were transferred into increasingly older embryos the percentage of donor cells residing in the gonads decreased so that in recipient Stage 20 embryos, over 90% were found ectopically in the head, surrounding the neural tube, trunk and limbs. Chimeric chicks have been hatched from vascular transfer of PGCs at Stage 13-15 and subsequently shown to produce progeny derived from the injected cells. Tajima *et al.* (1993) hatched 17 chicks from reciprocal transfers of White Leghorn and Rhode Island Red PGCs and eight produced progeny from the donor cells at rates ranging from 0.6% to 12%. Naito *et al.* (1994a) conducted similar experiments, but removed blood from recipient embryos prior to PGC transfer in an effort to increase the frequency of donor-derived offspring by increasing the proportion of donor-to-recipient PGCs in the recipient embryo. Nineteen of the 20 progeny that were test-mated produced donor-derived offspring at a rate ranging from 2 to 96%. Withdrawal of blood from recipient embryos prior to PGC injection appeared to increase the rate at which male chimeras produced donor-derived offspring by about 10%. Lastly, Vick *et al.* (1993) have provided the only report of production of transgenic offspring following transfection of blood-derived PGCs. Two males, derived from embryos receiving retrovirally-transduced PGCs, produced semen that contained vector DNA. One of these males produced transgenic progeny.

A fourth source of avian PGCs is the embryonic gonad. This source is particularly inviting because the number of PGCs is higher than in the blastoderm, the germinal crescent or the blood (Simon, 1960; Meyer, 1964; Singh and Meyer, 1967; Fritts-Williams and Meyer, 1972; Swartz and Domm, 1972; Simkiss *et al.*, 1989; Muniesa and Dominguez, 1990; Simkiss, 1990; Ginsburg, 1994). Embryonic gonads provide an easily accessible site for germ cells. They can be cultured, and will expand in culture (Wentworth *et al.*, 1992a; Wentworth *et al.*, 1992b; Wentworth *et al.*, 1996), providing a possible means to insert foreign DNA into these cells for subsequent injection into recipient embryos. A major concern in their use for transgenesis was whether gonadal PGCs retained the capacity to migrate to the gonads of the recipient embryos, since this phase of their normal development ends at Stage 19.

To test their capacity for migration, PGCs from the embryonic gonads of one pure strain of Japanese quail, an incomplete-recessive pure white bird that laid white eggs, were transferred into the embryo of a second pure strain, the incomplete dominant brown speckled wild type, which laid wild-type brown mottled eggs (Wentworth and Wentworth, unpublished). The PGCs were contained in a freshly-dispersed (Heath, 1978) single-cell suspension from Stage 27 white quail embryo gonads (about 400 gonadal cells) and were injected into the peripheral vitelline vessels of a Stage 15 wild type quail embryo. Of the 74 wild-type embryos injected, 22 hatched to produce the F_0 population. Upon maturity, 18 F_0 birds were individually mated to recessive white quail and over 800 progeny were hatched from these matings. Five white offspring were produced, bred by three of the F_0 adults. Thus, 17% (3/18) of the founder quail that had received mixed gonadal cells from recessive white quail were identified as germline chimeras. Other progeny of these matings were from the non-donor derived endogenous germ cells were brown wild-type quail that displayed a large white breast patch and white primary wing feathers, termed a "tuxedo" (Wentworth *et al.*, 1989).

Other investigators fluorescently-labeled gonadal PGCs from the gonadal anlagen of five-, seven-, and ten-day-old White Leghorn embryos, and injected them into the dorsal aorta of 2-day-old recipient embryos. After 5 days incubation, fluorescently labeled PGCs collected from five- and seven-day-old embryos were found in the gonads of the recipient embryos, but PGCs from ten-day-old embryos were not present in recipient gonads (Yamamaka *et al.*, 1998). These experiments suggest that the migratory ability of chicken gonadal PGCs is lost between 7 and 10 days of embryonic age.

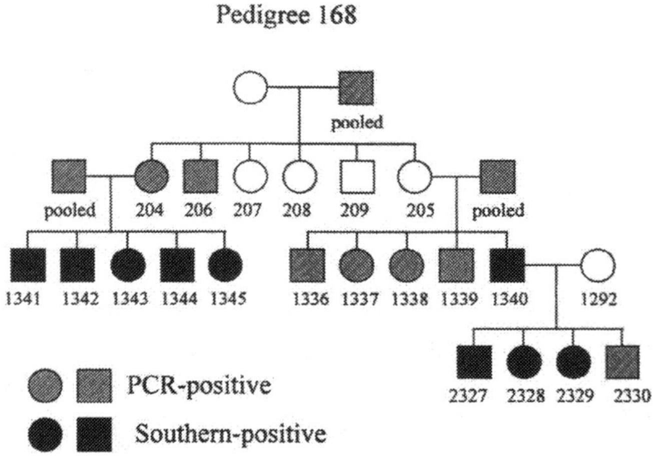

Figure 2. *Transmission of an antisense prolactin DNA construct in pedigree 168. Females are represented by circles and males are represented by squares. Hatched or filled symbols indicate the DNA samples were positive by PCR analysis only or by both PCR and genomic Southern blot analyses, respectively. Open symbols indicate that the DNA samples were negative by PCR analysis. The founder female was not tested. From Wong* et al., *1999.*

Studies have also demonstrated that chicken gonadal PGCs can be co-cultured *in vitro* with gonadal ridge stromal cells for extended periods and retain the ability to migrate to the gonads of recipient embryos and produce offspring. Cells from the germinal ridge of Stage 27 (5.5-day-old) Korean native ogol chicken (KNOC) embryos were cultured *in vitro* for 5 days and transferred into the vascular system of stage 14 White Leghorn (WL) recipients (Chang *et al.*, 1997). Six out of 60 recipients were identified by feather color as germline chimeras (KNOC/WL) after mating the progeny to the recessive KNOC black phenotype. The frequency of transmission was from 1.3 to 3% regardless of sex. Han *et al.* (1999) have demonstrated that gonadal PGCs may be cultured *in vitro* for two months and result in production of germline chimeric chicks following injection into Stage 17 embryos. Five KNOC/WL chimera were identified out of 19 hatched embryos by progeny testing, but the efficiency of germ line transmission by these F_0 animals was relatively low. Cultured gonadal PGCs have been shown to proliferate 3.8 fold during four days of culture, while PGCs isolated from the blood of two-day-old embryos will also proliferate *in vitro* if similarly co-cultured with gonadal stromal cells (Chang *et al.*, 1995), thus providing a substantial supply of PGCs for transfection. Hong *et al.* (1998) have transfected cultured gonadal PGCs by electroporation with high efficiency (80%) and detected the presence of the exogenous *lacZ* DNA in the gonads (but not the liver or heart) of 41% of hatched recipient chicks.

Finally, Wentworth *et al.* (1996) and Wong *et al.* (1999) have reported the germline transmission in turkeys of an antisense prolactin construct carrying the amp^r gene. Gonadal PGCs derived from Stage 26-28 turkey embryos, co-cultured with somatic gonadal cells, were transfected using Lipofectin®, a liposome-mediated transfection agent. These cells were injected into the vasculature of Stage 17 recipient embryos. After the hatched poults reached sexual maturity, they were mated and the foreign DNA was identified by PCR and by Southern analyses in blood and/or tissue samples from some of the F_1 through F_3 embryonic and hatched progeny. Figure 2 demonstrates pedigreed germline transmission of foreign DNA from the founder (F_0) through the F_3 generation. The pedigree was established by mating an untested F_0 female with pooled semen from 12 transgenic and non-transgenic F_0 males. Transmission of vector DNA did not follow normal Mendelian inheritance, suggesting that it was not integrated into the genome.

3.6 SPERM-MEDIATED GENE TRANSFER

Early reports that transgenic mice (Lavitrano *et al.*, 1989) and pigs (Gandolfi, 1989) had been produced by *in vitro* fertilization of oocytes with sperm that had been incubated with naked DNA stimulated interest in using sperm as a vehicle for DNA delivery. However, these and subsequent reports have been controversial, as results have varied among species and laboratories. Recently, Lavitrano *et al.* (1999a,b) have reported the production of transgenic pigs expressing human genes using sperm-mediated gene transfer. Several reviews of this field have recently been published (Smith, 1999; Wall, 1999; Gandolfi, 2000). The use of sperm-mediated gene transfer in avian species is likewise controversial (Nakanishi, 1993), and the generation of transgenic poultry by this technique has not been confirmed.

4. Chromosomal Integration and Expression of Transgenes in Poultry

The goal in producing transgenic poultry is for stable integration of the transgene into the genome of the bird and expression of the gene at times, and in amounts, that are appropriate for its intended function. This goal has yet to be achieved in poultry. Love *et al.* (1994); Thoraval *et al.* (1995) and Sherman *et al.* (1998) have successfully demonstrated germline transmission of foreign DNA in chickens for up to three generations, while Wong *et al.* (1999) have reported germline transmission of vector DNA to the F_3 generation in turkeys.

4.1 EPISOMAL AND CHROMOSOMAL INTEGRATION

The chicken genome consists of 78 chromosomes (39 pair), while the turkey genome has 80 chromosomes. Six of these are comparable in size to mammalian chromosomes, while the remaining chromosomes are much smaller *microchromosomes* (Bloom,

1993). The work of Wong *et al.* (1999) suggests that foreign DNA may be transmitted episomally for several generations in birds. Analysis of blood DNA of antisense-positive F_3 poults, described above, showed no evidence of junction fragments in Southern blots, but rather a single band was observed that co-migrated with the original plasmid. When this DNA was used to transform *E.coli*, the DNA conferred ampicillin resistance, while no bacteria transfected with control DNA grew on ampicillin-containing agar plates. Partial sequencing of the DNA extracted from colonies taken at random from positive plates provided convincing evidence that the plasmid DNA recovered had not changed from the original plasmid used for transfection of F_0 embryos. These results, coupled with the non-Mendelian pattern of inheritance shown in Figure 2, suggest that the DNA incorporated into the proliferating germ cells was passed through three generations episomally.

Episomal persistence of exogenous DNA has been demonstrated in several experiments that attempted to introduce exogenous DNA into avian species (Sang and Perry, 1989; Naito *et al.*, 1994b; Squires and Drake, 1994; Watanabe *et al.*, 1994; Li *et al.*, 1995). Notably, Squires and Drake (1994) also reported a non-Mendelian pattern of inheritance through three generations. Episomal transmission of transgenes has also been observed in a number of other 'transgenic' animals, including *Drosophila* (Steller and Pirrotta, 1985), *C.elegans* (Stinchcomb *et al.*, 1985), *Xenopus* (Etkin *et al.*, 1984), fish (Guyomard *et al.*, 1989) and the silky earthworm (Nilolaev *et al.*, 1993). The mechanism by which vector DNA replicates episomally in transgenic poultry is unknown, but Wong *et al.* (1999) speculated that the presence of numerous microchromosomes in avian species may promote episomal replication of foreign DNA.

In contrast, the results of Sherman *et al.* (1998) suggest that the *Drosophila* transposable element *mariner* can integrate foreign DNA into the chicken genome. The high frequency of transmission to F_2 birds suggests that this element may provide an efficient vector for stable transgene integration into avian chromosomes.

4.2 EXPRESSION OF TRANSGENES IN THE POULTRY GENOME

Production of pharmaceutical products, or genetic engineering of poultry to enhance productive traits, will require targeting transgene expression to specific tissues or organs. Use of the ovalbumin or vitellogenin gene promoters may target protein production to the oviduct and liver, respectively, where these proteins are normally produced in large quantities for packaging into the egg. Thus, a flock of transgenic Leghorn chickens, each laying an average of 280 eggs per year, could theoretically produce massive quantities of recombinant protein.

5. Conclusions

The production of transgenic poultry lags behind that of other livestock because of daunting technical and biological challenges presented by the avian species. More than a decade of dedicated effort has greatly advanced the ability of researchers to access the avian genome to produce chimeric or transgenic offspring. A totipotent cell type has not yet been identified in the bird, but must exist in some early stage of development. The primordial germ cell has been shown to provide a good vehicle for the germline transmission of foreign genes. This cell can now be isolated from several embryonic sources, cultured and propagated *in vitro*, efficiently transfected using retroviral or non-viral vector DNA, and injected into founder embryos that subsequently develop normally and hatch successfully. Primordial germ cells may also be cryopreserved for germline preservation. The direct injection of DNA into the germinal disc of the recently fertilized ova has also recently achieved excellent success in integrating DNA into the poultry genome. The many technical difficulties that have previously limited success at each of these steps have been overcome through meticulous basic research in numerous laboratories around the world. Thus, it seems likely that the quest for transgenic poultry must now be poised for greater success. The advantages of poultry for the application of transgenic technology, including a short generation time, small body size, a large number of progeny, and the potential application of production improvements to the billions of animals produced annually, should greatly enhance interest in transgenic poultry in the years ahead.

References

Al-Thani, R. and Simkiss, K. (1991) Effects of an acute in vivo application of concanavalin A on the migration of avian primordial germ cells, *Protoplasma 161*, 52-57.

Bakst, M. R. and Howarth, B., Jr. (1977) Hydrolysis of the hen's perivitelline layer by cock sperm *in vitro*, *Biol. Reprod. 17*, 370-379.

Bloom, S.E., Delany, M.E. and Muscarella, D.E. (1993) Constant and variable features of avian chromosomes, In Etches R.J. and Gibbins A.M.V. (ed.), *Manipulation of the Avian Genome*, CRC Press, Boca Raton, pp. 39-59.

Bosselman, R.A., Hsu, R.Y., Boggs, T., Hu, S., Bruszewski, J., Ou, S., Kozar, L., Martin, F., Nicolson, M., Schultz, J.A., Semon, K.M. and Stewart, R.G. (1989) Germline transmission of exogenous genes in the chicken, *Science 243*, 533-535.

Briskin, M.J., Hsu, R.Y., Boggs, T., Schultz, J.A., Rishell, W. and Bosselman, R.A. (1991) Heritable retroviral transgenes are highly expressed in chickens, *Proc. Natl. Acad. Sci. U S A 88*, 1736-1740.

Carsience, R.S., Clark, M.E., Verrinder Gibbins, A.M. and Etches, R.J. (1993) Germline chimeric chickens from dispersed donor blastodermal cells and compromised recipient embryos, *Development 117*, 669-675.

Chang, I.K., Jeong, D.K., Hong, Y.H., Park, T.S., Moon, Y.K., Ohno, T. and Han, J.Y. (1997) Production of germline chimeric chickens by transfer of cultured primordial germ cells, *Cell Biol. Int. 21*, 495-499.

Chang, I.K., Tajima, A., Chikamune, T. and Ohno, T. (1995) Proliferation of chick primordial germ cells cultured on stroma cells from the germinal ridge, *Cell Biol. Int. 19*, 143-149.

Charreau, B., Tesson, L., Soulillou, J.P., Pourcel, C. and Anegon, I. (1996) Transgenesis in rats: technical aspects and models. *Transgenic Res* 5, 223-234.

Colman, A. (1999) Dolly, Polly and other 'ollys': likely impact of cloning technology on biomedical uses of livestock, *Genet. Anal. 15*, 167-173.

Etches, R.J., Clark, M.E., Toner, A., Liu, G. and Gibbins, A.M. (1996) Contributions to somatic and germline lineages of chicken blastodermal cells maintained in culture, *Mol. Reprod. Dev. 45*, 291-298.

Etches, R.J., Clark, M.E., Zajchowski, L., Speksnijder, G., Verrinder Gibbins, A.M., Kino, K., Pain, B. and Samarut, J. (1997) Manipulation of blastodermal cells, *Poult. Sci. 76*, 1075-1083.

Etkin, L., Pearman, B., Roberts, M. and Bektesh, S.L. (1984) Replication, integration and expression of exogenous DNA injected into fertilized eggs of Xenopus laevis, *Differentiation* 26, 194-202.

Evans, M.J. and Kaufman, M.H. (1981) Introduction of a rabbit beta-globulin gene into the mouse germ line, *Nature 292*, 154-156.

Flytzanis, C.N., McMahon, A.P., Hough-Evans, B.R., Katula, K.S., Britten, R.J. and Davidson, E.H. (1985) Persistence and integration of cloned DNA in postembryonic sea urchins, *Dev. Biol. 108*, 431-442.

Fofanova, K.A. (1964) Morphologic data on polyspermy in chickens. *Zhurnal Obshchei Biologii 25*, 239-247.

Frisby, D.P., Weiss, R.A., Roussel, M. and Stehelin, D. (1979) The distribution of endogenous chicken retrovirus sequences in the DNA of galliform birds does not coincide with avian phylogenetic relationships, *Cell 17*, 623-634.

Fritts-Williams, M.L. and Meyer, D.B. (1972) Isolation and culture of homogenous populations of primordial germ cells in chick, *Exp Cell Res 75*, 512-514.

Gandolfi, F., Lavitrano M., Camaioni A., Spadafora C., Siracusa G. and Lauria, A. (1989) The use of sperm-mediated gene transfer for the generation of transgenic pigs, *J. Reprod. Fertil.*

Gandolfi, F. (2000) Sperm-mediated transgenesis, *Theriogenology 53*, 127-137.

Gilbert, S.F. (1991) *Developmental Biology*, 3rd ed. Sinauer Associates, Sunderland, U.K.

Ginsburg, M. (1994) Primordial germ cell formation in birds, *Ciba Found Symp 182*, 52-61.

Gonzales, D. (1989) Gonadal germ cell chimeras produced by microinjecting primordial germ cells into the vasculature of early chicken embryos, M.S. Thesis, University of Wisconsin.

Gordon, J.W. and Ruddle, F.H. (1981) Integration and stable germ line transmission of genes injected into mouse pronuclei, *Science 214*, 1244-1246.

Gordon, J.W., Scangos, G.A., Plotkin, D.J., Barbosa, J.A. and Ruddle, F.H. (1980) Genetic transformation of mouse embryos by microinjection of purified DNA, *Proc. Natl. Acad. Sci. U S A 77*, 7380-7384.

Guyomard, R., Chourrout, D., Leroux, C., Houdebine, L.M. and Pourrain, F. (1989) Integration and germ line transmission of foreign genes microinjected into fertilized trout eggs, *Biochimie 71*, 857-863.

Hamburger, V. and Hamilton, H.L. (1951) A series of normal stages in the development of the chick embryo, *J. Morphol. 88*, 49-92.

Han, J., Hong, H.Y., Jeong, D.K. and Park, T.S. (1999) Production of germ-line chimeric chickens using long-term cultured gonadal primordial germ cells, *Proc. Transgenic Anim. Res.*, Tahoe City, CA, pp. 40.

Heasman, J., Holwill, S. and Wylie, C.C. (1991) Fertilization of cultured Xenopus oocytes and use in studies of maternally inherited molecules, *Methods Cell Biol. 36*, 213-230.

Heath, J.K. (1978) Characterization of a xenogeneic antiserum raised against the fetal germ cells of the mouse: cross-reactivity with embryonal carcinoma cells, *Cell 15*, 299-306.

Hong, Y.H., Moon, Y.K., Jeong, D.K. and Han, J.Y. (1998) Improved transfection efficiency of chicken gonadal primordial germ cells for the production of transgenic poultry, *Transgenic Res. 7*, 247-252.

Inada, S., Hattori, M.A., Fujihara, N. and Morohashi, K. (1997) In vivo gene transfer into the blastoderm of early developmental stage of chicken, *Reprod. Nutr. Dev. 37*, 13-20.

Iyengar, A., Muller, F. and Maclean, N. (1996) Regulation and expression of transgenes in fish -- a review, *Transgenic Res. 5*, 147-166.

Johnson, A.L. (1986) Reproduction in the Female, in Sturkie P.D. (ed.), *Avian Physiology*, Springer-Verlag, New York, pp. 403-431.

Kagami, H., Tagami, T., Matsubara, Y., Harumi, T., Hanada, H., Maruyama, K., Sakurai, M., Kuwana, T. and Naito, M. (1997) The developmental origin of primordial germ cells and the transmission of the donor-derived gametes in mixed-sex germline chimeras to the offspring in the chicken, *Mol. Reprod. Dev. 48*, 501-510.

Lavitrano, M., Cappello, F., Lazzereschi, D., Stoppacciaro, A., DiStefano, C., Marfe, G., Turchi, V., Bacci, J.L., Forni, M., Wang, H.J., Renzi, L., Zannone, A., Stassi, G., Nuti, M., Seren, E. andFrati, L. (1999a) Efficiency of transgenesis using sperm mediated gene transfer: generation of hDAF transgenic pigs, *Proc. Transgenic Anim. Res.*, Tahoe City, CA, pp. 99.

Lavitrano, M., Stoppacciaro, A., Bacci, M.L., Forni, M., Fioretti, D., Pucci, L., Di Stefano, C., Lazzereschi, D., Rughetti, A., Ceretta, S., Zannoni, A., Rahimi, H., Moioli, B., Rossi, M., Nuti, M., Rossi, G., Seren, E., Alfani, D., Cortesini, R. and Frati, L. (1999b) Human decay accelerating factor transgenic pigs for xenotransplantation obtained by sperm-mediated gene transfer, *Transplant Proc. 31*, 972-974.

Lavitrano, M., Camaioni, A., Fazio, V.M., Dolci, S., Farace, M.G. and Spadafora, C. (1989) Sperm cells as vectors for introducing foreign DNA into eggs: genetic transformation of mice, *Cell 57*, 717-723.

Lee, M.R. and Shuman, R.M. (1990) Transgenic quail produced by retrovirus vector infection transmit and express a foreign marker gene, *4th World Cong. Genet. Appl. Livest. Prod. 16*, pp. 107-110.

Li, Y., Behnam, J. and Simkiss, K. (1995) Ballistic transfection of avian primordial germ cell in ovo, *Transgenic Res. 4*, 26-29.

Li, Z.D., Deng, H., Ma, Y.Z., Liu, C.H., Song, H.H., Wang, N., Sha, J. and Wei, H. (1999) Production of duck-chicken chimeras by transferring early blastodermal cells, *Proc. Transgenic Anim. Res.* Tahoe City, CA, pp. 99.

Love, J., Gribbin, C., Mather, C. and Sang, H. (1994) Transgenic birds by DNA microinjection, *Biotechnology (N Y) 12*, 60-63.

Martin, G.R. (1981) Isolation of a pluripotent cell line from early mouse embryos cultured in medium conditioned by teratocarcinoma stem cells, *Proc. Natl. Acad. Sci. U S A 78*, 7634-7638.

Marzullo, G. (1970) Production of chick chimaeras, *Nature 225*, 72-73.

Meyer, D.B. (1964) The migration of primordial germ cells in the chick embryo, *Develop. Biol. 10*, 154-190.

Muniesa, P. and Dominguez, L. (1990) A morphological study of primordial germ cells at pregastrular stages in the chick embryo, *Cell Differ. Dev. 31*, 105-117.

Naito, M., Agata, K., Otsuka, K., Kino, K., Ohta, M., Hirose, K., Perry, M.M. and Eguchi, G. (1991) Embryonic expression of beta-actin-lacZ hybrid gene injected into the fertilized ovum of the domestic fowl, *Int. J. Dev. Biol. 35*, 69-75.

Naito, M., Matsubara, Y., Harumi, T., Tagami, T., Kagami, H., Sakurai, M. and Kuwana, T. (1999) Differentiation of donor primordial germ cells into functional gametes in the gonads of mixed-sex germline chimaeric chickens produced by transfer of primordial germ cells isolated from embryonic blood, *J. Reprod. Fertil. 117*, 291-298.

Naito, M., Nirasawa, K. and Oishi, T. (1990) Development in culture of the chick embryo from fertilized ovum to hatching, *J. Exp. Zool. 254*, 322-326.

Naito, M., Nirasawa, K. and Oishi, T. (1992) Preservation of quail blastoderm cells in liquid nitrogen, *Br. Poult. Sci. 33*, 449-453.

Naito, M., Sasaki, E., Ohtaki, M. and Sakurai, M. (1994b) Introduction of exogenous DNA into somatic and germ cells of chickens by microinjection into the germinal disk of fertilized ova, *Molec. Reprod. Dev. 37*, 321-325.

Naito, M., Tajima, A., Yasuda, Y. and Kuwana, T. (1994a) Production of germline chimeric chickens, with high transmission rate of donor-derived gametes, produced by transfer of primordial germ cells, *Mol. Reprod. Dev. 39*, 153-161.

Nakamura, M., Maeda, H. and Fujimoto, T. (1991) Behavior of chick primordial germ cells injected into the blood stream of quail embryos, *Okajimas Folia Anat. Jpn. 67*, 473-477.

Nakamura, M., Yoshinaga, K. and Fujimoto, T. (1992) Histochemical identification and behavior of quail primordial germ cells injected into chick embryos by the intravascular route, *J. Exp. Zool. 261*, 479-483.

Nakanishi, A. and Iritani, A. (1993) Gene transfer in the chicken by sperm-mediated methods, *Mol. Reprod. Dev. 36*, 258-261.

Nilolaev, A.L., Tchkonia, T.T., Kafiani-Eristavi, C.A. and Tarantul, V.Z. (1993) Preferential extrachromosomal localization of exogenous DNA in transgenic silkworm *Bombyx mori L.*, *Molec. Gen.. Genet. 236*, 326-330.

Ono, T., Murakami, T., Mochii, M., Agata, K., Kino, K., Otsuka, K., Ohta, M., Mizutani, M., Yoshida, M. and Eguchi, G. (1994) A complete culture system for avian transgenesis, supporting quail embryos from the single-cell stage to hatching, *Dev. Biol. 161*, 126-130.

Pain, B., Clark, M.E., Shen, M., Nakazawa, H., Sakurai, M., Samarut, J. and Etches, R.J. (1996) Long-term in vitro culture and characterisation of avian embryonic stem cells with multiple morphogenetic potentialities, *Development 122*, 2339-2348.

Perry, M.M. (1987) Nuclear events from fertilisation to the early cleavage stages in the domestic fowl (*Gallus domesticus*), *J. Anat. 150*, 99-109.

Perry, M.M. (1988) A complete culture system for the chick embryo, *Nature 331*, 70-72.

Perry, M.M., Morrice, D., Hettle, S. and Sang, H. (1991) Expression of exogenous DNA during the early development of the chick embryo, *Roux's Arch. Develop. Biol. 200*, 312-319.

Petitte, J.N., Clark, M.E., Liu, G., Verrinder Gibbins, A.M. and Etches, R.J. (1990) Production of somatic and germline chimeras in the chicken by transfer of early blastodermal cells, *Development 108*, 185-189.

Pollock, P.P., Kutzko, J.P., Birck-Wilson, E., Williams, J.L., Echelard, Y. and Meade, H.M. (1999) Transgenic milk as a method for the production of recombinant antibodies, *J. Immunol. Methods 231*, 147-157.

Pursel, V.G., Pinkert, C.A., Miller, K.F., Bolt, D.J., Campbell, R.G., Palmiter, R.D., Brinster, R.L. and Hammer, R.E. (1989) Genetic engineering of livestock, *Science 244*, 1281-1288.

Pursel, V.G. and Rexroad, C.E. Jr. (1993) Recent progress in the transgenic modification of swine and sheep, *Mol. Reprod. Dev.* 36, 251-254.

Reynaud, G. (1976) Capacities reproductrices et descendance de poulets ayant subi un transfert de cellules germinales primordiales durant la vie embryonnaire, *Roux's Arch. Develop. Biol. 179*, 85-110.

Rosenblum, C.I. and Chen, H.Y. (1995) *In ovo* transfection of chicken embryos using cationic liposomes, *Transgenic Res. 4*, 192-198.

Rowlett, K. and Simkiss, K. (1987) Explanted embryo culture: *in vitro* and *in ovo* techniques for the domestic fowl, *British Poult. Sci. 28*, 91-101.

Rudolph, N.S. (1999) Biopharmaceutical production in transgenic livestock, *Trends Biotechnol. 17*, 367-374.

Salter, D.W., Smith, E.J., Hughes, S.H., Wright, S.E. and Crittenden, L.B. (1987) Transgenic chickens: insertion of retroviral genes into the chicken germ line, *Virology 157*, 236-240.

Salter, D.W., Smith, E.J., Hughes, S.H., Wright, S.E., Fadly, A.M., Witter, R.L. and Crittenden, L.B. (1986) Gene insertion into the chicken germ line by retroviruses, *Poult. Sci. 65*, 1445-1458.

Sang, H. and Perry, M.M. (1989) Episomal replication of cloned DNA injected into the fertilised ovum of the hen (*Gallus domesticus*), *Mol. Reprod. Dev. 1*, 98-106.

Sherman, A., Dawson, A., Mather, C., Gilhooley, H., Li, Y., Mitchell, R., Finnegan, D. and Sang, H. (1998) Transposition of the Drosophila element mariner into the chicken germ line [see comments] [published erratum appears in *Nat. Biotechnol* (1999) 17, 81], *Nat. Biotechnol. 16*, 1050-1053.

Shuman, R.M. and Shofffner, R.N. Potential genetic modification in the chicken (*Gallus domesticus*), *Proc. 2nd World Cong. Genet. Appl. Livest. Prod. 6*, 157-163.

Simkiss, K., Vick, L., Luck, G., Page, N. and Savva, D. (1990) Infection of primordial germ cells with defective retrovirus and their transfer to developing embryo, *Proc. 4th World Congress Genet. Appl. Livest. Prod. 13*, 111-114.

Simkiss, K., Rowlett, K., Bumstead, N. and Freeman, B.M. (1989) Transfer of primordial germ cell DNA between embryos, *Protoplasma 151*, 164-166.

Simon, D. (1960) Contribution a' l'etude de la circulation et du transport des gonocytes primaires dans des blastodisques d'Oiseau cultives in vitro, *Arch. Anat. Microsc. Morphol. Exp. 49*, 93-176.

Singh, R.P. and Meyer, D.B. (1967) Primordial germ cells in blood smears from chick embryos, *Science 156*, 1503-1504.

Smith, K.R. (1999) Sperm cell mediated transgenesis: a review, *Anim. Biotechnol. 10*, 1-13.

Souza, L.M., Boone, T.C., Murdock, D., Langley, K., Wypych, J., Fenton, D., Johnson, S., Lai, P.H., Everett, R., Hsu, R.Y. and Bosselman, R. (1984) Application of recombinant DNA technologies to studies on chicken growth hormone, *J. Exp. Zool. 232*, 465-473.

Squires, E.J. and Drake, D. (1994) Transgenic chickens by liposome-sperm-mediated gene transfer, *Proc. 5th World Congress Genet. Appl. Livest. Prod. 21*, pp. 350-353.

Steller, H. and Pirrotta, V. (1985) Fate of DNA injected into early Drosophila embryos, *Dev. Biol. 109*, 54-62.

Stinchcomb, D.T., Shaw, J.E., Carr, S.H. and Hirsh, D. (1985) Extrachromosomal DNA transformation of Caenorhabditis elegans, *Mol. Cell Biol. 5*, 3484-3496.

Su, H.Y., Jay, N.P., Gourley, T.S., Kay, G.W. and Damak, S. (1998) Wool production in transgenic sheep: results from first-generation adults and second-generation lambs, *Anim. Biotechnol. 9*, 135-147.

Swartz, W.J. and Domm, L.V. (1972) A study on division of primordial germ cells in the early chick embryo, *Am. J. Anat. 135*, 51-70.

Tajima, A., Naito, M., Yasuda, Y. and Kuwana, T. (1993) Production of germ line chimera by transfer of primordial germ cells in the domestic chicken (*Gallus domesticus*), *Theriogenology 40*, 509-519.

Thoraval, P., Afanassieff, M., Cosset, F.L., Lasserre, F., Verdier, G., Coudert, F. and Dambrine, G. (1995) Germline transmission of exogenous genes in chickens using helper-free ecotropic avian leukosis virus-based vectors, *Transgenic Res. 4*, 369-377.

Tsai, H. (1995) Blastodermal cells- the way to the avian germline, Ph.D. Dissertation, University of Wisconsin.

Tsai, H., Hunt, H.D., Young, E.E., Wentworth, B.C. and Bacon, L.D. (1999) Utilization of extensively cultured blastodermal stem cells for development of chimeric chickens, *Proc. Transgenic Anim. Res.*, Tahoe City, CA, pp. 57.

Vick, L., Li, Y. and Simkiss, K. (1993) Transgenic birds from transformed primordial germ cells, *Proc. Royal Soc. Lond. B Biol. Sci. 251*, 179-182.

Waddington, D., Gribbin, C., Sterling, R.J., Sang, H.M. and Perry, M.M. (1998) Chronology of events in the first cell cycle of the polyspermic egg of the domestic fowl (*Gallus domesticus*). *Int. J. Dev. Biol. 42*, 625-628.

Wall, R.J. (1999) Sperm-mediated gene transfer: advances in sperm cell research and applications, *Transgenic Res. 8*, 313-315.

Wall, R.J., Rexroad, C.E., Jr., Powell, A., Shamay, A., McKnight, R. and Hennighausen, L. (1996) Synthesis and secretion of the mouse whey acidic protein in transgenic sheep, *Transgenic Res. 5*, 67-72.

Ward, K.A. and Brown, B.W. (1998) The production of transgenic domestic livestock: successes, failures and the need for nuclear transfer, *Reprod. Fertil. Dev. 10*, 659-665.

Watanabe, M., Kinutani, M., Naito, M., Ochi, O. and Takashima, Y. (1992) Distribution analysis of transferred donor cells in avian blastodermal chimeras, *Development 114*, 331-338.

Watanabe, M., Naito, M., Sasaki, E., Sakurai, M., Kuwana, T. and Oishi, T. (1994) Liposome-mediated DNA transfer into chicken primordial germ cells in vivo, *Mol. Reprod. Dev. 38*, 268-274.

Wentworth, A.L., Frank, A. and Wentworth, B.C. (1992a) Transfection of Japanese quail primordial germ cells with DNA expressing lacZ gene product, *Poult. Sci. (Suppl. 1) 73*, 71.

Wentworth, B., Tsai, H., Wentworth, A., Proudman, J. and El Halawani, M. (1996) Primordial germ cells for genetic modulation in poultry, in Miller R.H., Pursel, V.G. and Norman, H.D. (ed.), *Beltsville Symposium in Agricultural Research, XX, Biotechnology's Role in the Genetic Improvement of Farm Animals.* American Society of Animal Science, Savoy, IL.

Wentworth, B.C., Tsai, H., Hallett, J.H., Gonzales, D.S. and Rajcic-Spasojevic, G. (1989) Manipulation of avian primordial germ cells and gonadal differentiation, *Poult. Sci. 68*, 999-1010.

Wentworth, B.C., Wentworth, A.L., Tsai, H., Lee, T., Wong, E.A. and Proudman, J.A. (1992b) Avian primordial germ cells as a vehicle for gene transfer, Proc. 5th *Inter. Symp. Avian Endocrinol.*, pp. 24.

Wong, E.A., Wentworth, A.L., Wentworth, B.C., Proudman, J.A. and El Halawani, M.E. (1999) Generation of transgenic poultry by transfection of primordial germ cells, in Murray, J.D., Anderson, G.B., Oberbauer, A.M. and McGloughlin. M.M. (ed.), *Transgenic Animals in Agriculture*, CABI Publishing, Boca Raton, Fl, pp. 117-129.

Yamamaka, K., Tajima, A.. Kuwama. T. and Kanai, Y. (1998) Migrating ability of the gonadal primordial germ (gPGCs) in domestic chicken, *Poult. Sci. (Suppl. 1) 77*, 93.

Zeng, Q., Foo, N.C., Funkhouser, J.M., Carter, D.A. and Murphy, D. (1994) Expression of a rat vasopressin transgene in rat testes, *J. Reprod. Fertil. 102*, 471-481.

Ziomek, C.A. (1998) Commercialization of proteins produced in the mammary gland, *Theriogenology 49*, 139-144.

ENZYMES AS DIRECT-FEED ADDITIVES FOR RUMINANTS

RODE L.M., MCALLISTER T.A., BEAUCHEMIN K.A., MORGAVI D.P., NSEREKO V.L., YANG W.Z., IWAASA A.D. AND WANG Y.
Agriculture and Agri-Food Canada, Lethbridge, AB, Canada, T1J 4B1

Abstract

Fibrolytic enzymes hold great potential to improve feed utilization and productivity in ruminants. In the past, it was believed that the endogenous activity against plant cell walls could not be augmented by supplementary exogenous enzymes. However, when diets of dairy and beef cattle are supplemented with commercial xylanases and cellulases, animal performance is significantly improved. The most likely site of action is the rumen rather than in the small intestine as is the case for poultry. Because of the complexity of the rumen environment, it has been difficult to identify the exact mode of action for this beneficial response. Since xylanases and cellulases are the main activities that occur in efficacious enzyme mixtures, it may be assumed that the enzymes are having a direct, additive effect on the hydrolysis of plant fiber in the rumen. However, evidence to date suggests that the benefits of exogenous enzymes is synergistic to ruminal endogenous enzymes. This synergy may explain why relatively small amounts of enzyme can have such large effects on animal productivity. Limitations to the exploitation of this technology are the development of an adequate screening system for new enzymes, and the identification of the specific enzyme activities that are critical for efficacy.

1. Introduction

Since the advent of intensive livestock production, there has been a continual search for new additives that will enhance feed utilization so that the greater nutrient demands of ever increasing animal productivity could be met. More recently, there has been rising public resistance to many of these additives such as antibiotics and anabolic implants. This has led to a drive to develop new replacement technologies. There is an interest in the potential use of enzymes in ruminant diets.

Exogenous enzymes have been used extensively to remove anti-nutritional factors from feeds and to increase the digestibility of existing nutrients for poultry (Classen *et al.*, 1991; Bedford, 1993). However, they have not been used routinely in adult

R. Renaville and A. Burny (eds.), Biotechnology in Animal Husbandry, 301-332.

ruminant diets. It has usually been assumed that exogenous enzymes cannot survive in the highly proteolytic environment of the rumen (Chesson, 1994). In addition, endogenous enzyme activities against plant fibre and phytate within the rumen environment are normally very high, and presumably not easily increased by a simple addition of exogenous enzyme products. The focus of most enzyme-related research for ruminants has been on plant fibre degrading enzymes, primarily cellulases and xylanases. Ruminant phytase research has been limited to the exploitation of the rumen as a novel source of phytase enzymes (Selinger *et al.*, 1999). The primary emphasis of this discourse will be the potential benefits of fibrolytic enzymes in ruminant diets, including production responses observed to date, as well as possible mechanisms by which these products may improve nutrient utilization by ruminants. However, the role that supplementary phytases may have for ruminants will also be addressed.

The use of exogenous enzymes in ruminant diets is not a new idea. In the 1960s, a number of studies were conducted to explore the potential of supplementing ruminant diets with enzyme preparations. Several studies showed the use of feed enzymes substantially improved feed digestibility and animal performance (Burroughs *et al.*, 1960; Clark *et al.*., 1961; Rovics and Ely, 1962; Van Walleghem *et al.*, 1964; Rust *et al.*., 1965; Perry *et al.*., 1960; Galiev *et al.*., 1982), although other studies reported no effects, and even negative responses (Burroughs *et al.*, 1960; Theurer *et al.*, 1963; Perry *et al.*, 1966). Little effort was made to describe the enzymes used in these early studies and no effort was made to determine their mode of action. Furthermore, production of exogenous enzymes was expensive at the time and it was not economically feasible to apply these preparations at the concentrations necessary to elicit a positive animal response. Recent reductions in fermentation costs, together with more active and better defined enzyme preparations, have prompted researchers to re-examine the role of exogenous enzymes in ruminant production (Chen *et al.*, 1995; Beauchemin *et al.*, 1995, 1997; McAllister *et al.*, 1999). Several studies have attempted to define possible modes of action of these additives (Judkins and Stobart, 1988; Feng *et al.*, 1996; Krause *et al.*, 1998; Morgavi *et al.*, 1999; Nsereko *et al.*, 1999; Yang *et al.*, 1999). Exogenous enzymes could exert a number of effects, both on the gastrointestinal microflora and on the ruminant animal itself. Therefore, it is highly probable, that physiological responses to exogenous enzymes are multi-factorial in origin.

2. Sources of Enzymes

Although enzyme products marketed for livestock (primarily poultry) number in the hundreds, they are virtually all derived from the bacterial (*Lactobacillus acidophilus*, *L. plantarum*, and *Streptococcus faecium*, spp.) and fungal (*Aspergillus spp.*, *Trichoderma reesei, Penicillium spp.*, and *Saccharomyces cerevisiae*) species (Pendleton, 1998). Moreover, it is unlikely that this list of source organisms will expand substantially, given the restrictive stance taken by the European Union and the U.S. Food and Drug Administration to add new organisms (Pendleton, 1998).

It is also important to recognize that many probiotic and direct-fed microbial products are marketed, at least partly or implicitly, upon their residual enzymatic content. These products contain relatively little actual fibrolytic enzyme activity (Kung, 1998) and fall outside of the scope of this discussion. The amount of enzyme added to the diet in the majority of the studies with exogenous enzymes (primarily derived from *Trichoderma* spp.) would appear to be considerably greater (>10 times) than what has been supplemented from crude fungal preparations such as *A. oryzae* (Newbold, 1995). However, when exogenous enzyme activities are reported (Beauchemin *et al.*, 1995, 1997, 1998; Rode *et al.*, 1999; Rode *et al.*, 1999), they are typically measured at the manufacturers recommended optima, which is a lower pH and higher temperature than that encountered in the rumen. Thus the activities quoted are considerably higher than those which would be measured at a pH and temperature similar to that in the rumen. Indeed, Newbold (1997) suggests that when measured under rumen like conditions (pH 6.5, 39°C) the dietary enzymes described by Beauchemin and Rode (1996) are likely to supply enzyme activities of a similar magnitude to those used in his experiments with *A. foetidus*. What specific enzyme activities are being added is the unanswered question.

Undoubtably, it is something more than just total cellulase or xylanase activity alone. However, as the mode of action for direct-fed ruminant enzymes becomes clearer, it is anticipated that a better understanding will be developed to explain why specific enzyme-containing probiotic and direct-fed microbial products are efficacious in ruminant diets.

Complete digestion of complex ruminant feedstuffs such as hay or grain requires literally hundreds of enzymes. Enzyme preparations for ruminants are evaluated primarily on the basis of their capacity to degrade plant cell walls. Typically, these enzymes fall into the general classification of cellulases or xylanases. However, most commercial preparations are not single gene products, containing a single enzyme activity. Secondary enzyme activities such as amylases, proteases, or pectinases are invariably present. Degradation of cellulose and hemicellulose alone requires a number of enzymes, all of which may loosely be termed cellulases or xylanases (Table 1). Differences in the relative proportions and activities of these individual enzymes will have an impact on the efficacy of cell wall degradation by the marketed products. Even within a single microbial species, the types and activity of enzymes produced can vary widely depending on the strain selected and the growth substrate and culture conditions employed for enzyme production (Considine and Coughlan, 1989; Gashe, 1992; Lee *et al.*, 1998).

The diversity of enzyme activities within commercially available enzyme preparations is probably advantageous, in that a single product can target a wide variety of substrates. However, this presents problems in terms of quality control and extrapolation of research findings among different preparations. For ruminants, blending crude enzyme extracts to obtain specified levels of one or two defined enzyme activities, such as xylanase and/or cellulase usually standardizes enzyme products. These products are not currently standardized for secondary activities. In fact, these activities, which may well be affecting the overall effectiveness of a given product, are seldom even measured. For example, Table 2 shows selected enzyme activities measured in some commercial enzyme products. There is no discernable relationship between xylanase or cellulase activity and other measured activities in these products.

between xylanase or cellulase activity and other measured activities in these products. To date, commercial enzyme products used for ruminants were originally developed for other industries (i.e. detergent, textile, pulp/paper, and monogastric feed industries). As single-gene enzyme products become more readily available, there is the real possibility of enzyme products targeted specifically for ruminants. However, this requires a clear understanding of the mode of action for these enzymes in ruminant diets. Mode of action for enzymes in ruminant diets is an extremely complex issue and it is unlikely, at least in the short term, that single-gene products will prove to be superior to cruder products containing a broad array of enzyme activities.

Table 1a. *Some plant cell wall degrading enzyme activities contained in commercial cellulases and xylanases.*

EC Number[1]	Official Name[2,3]	Alternative Name(s)[3]
Cellulases		
3.2.1.4	Cellulase	Endoglucanase; Endo-1,4-beta-glucanase; Carboxymethyl cellulase
3.2.1.6	Endo-1,3(4)-beta-glucanase	Endo-1,4-beta-glucanase; Endo-1,3-beta-glucanase; Laminarinase
3.2.1.21	Beta-glucosidase	Gentobiase; Cellobiase; Amygdalase
3.2.1.39	Glucan endo-1,3-beta-D-glucosidase	Endo-1,3-beta-glucanase; Laminarinase
3.2.1.58	Glucan 1,3-beta-glucosidase	Exo-1,3-beta-glucanase
3.2.1.73	Licheninase	Lichenase; Beta-glucanase; Endo-beta-1,3-1,4 glucanase
3.2.1.74	Glucan 1,4-beta-glucosidase	Cellobiase; Exo-1,4-beta-glucosidase
3.2.1.91	Cellulose 1,4-beta-cellobiosidase	Exoglucanase; Exocellobiohydrolase: 1,4-beta-cellobiohydrolase
Others		
3.1.1.1	Carboxylesterase	Cinnamoylesterase, feruloylesterase
3.1.1.6	Acetylesterase	Rhamnogalacturonan acetylesterase, pectin acetylesterase
3.1.1.11	Pectinesterase	Pectin methylesterases; Pectin demethoxylase; Pectin methoxylase
3.1.1.72	Acetylxylan esterase	
3.2.1.15	Polygalacturonase	Pectin depolymerase; Pectinase
3.2.1.126	Coniferin beta-glucosidase	aryl beta-glycosidase
4.2.2.10	Pectin lyase	

[1]Enzyme Commission Number
[2]Nomenclature Committee of the International Union of Biochemistry and Molecular Biology (IUBMB)
[3]from the ExPASy Molecular Biology Server, Swiss Institute of Bioinformatics (http://www.expasy.ch/).

3. Measurement of Enzyme Activity

Table 1b. *Some plant cell wall degrading enzyme activities contaned in commercial cellulases and xylanases.*

EC Number[1]	Official Name[2,3]	Alternative Name(s)[3]
Xylanases		
3.2.1.8	Endo-1,4-beta-xylanase	1,4-beta-D-xylan xylanohydrolase
3.2.1.32	Xylan endo-1,3-beta-xylosidase	Xylanase; Endo-1,3-beta-xylanase
3.2.1.37	Xylan 1,4-beta-xylosidase	Beta-xylosidase; 1,4-beta-D-xylan xylohydrolase; Xylobiase
3.2.1.55	Alpha-L-arabinofuranosidase	Arabinosidase
3.2.1.72	Xylan 1,3-beta-xylosidase	Exo-1,3-beta-xylosidase
3.2.1.89	Arabinogalactan endo-1,4-beta-galactosidase	Arabinogalactanase
3.2.1.90	Arabinogalactan endo-1,3-beta-galactosidase	Arabinogalactanase
3.2.1.99	Arabinan endo-1,5-alpha-L-arabinosidase	Endo-1,5-alpha-L-arabinanase
3.2.1.120	Oligoxyloglucan beta glycosidase	
3.2.1.136	Glucuronoarabinoxylan endo-1,4-beta-xylanase	Feraxan endoxylanase

[1]Enzyme Commission Number
[2]Nomenclature Committee of the International Union of Biochemistry and Molecular Biology (IUBMB)
[3]from the ExPASy Molecular Biology Server, Swiss Institute of Bioinformatics (http://www.expasy.ch/)

The first problem one confronts when attempting to measure the activity of plant fiber degrading enzymes, is the heterogenous nature of the substrate itself. This has resulted in plethora of assays used to measure fibrolytic enzyme activity. Some of these methods have been standardized by the International Union of Pure and Applied Chemists, Commission on Biotechnology (IUPAC, 1984). The most commonly used substrate for measuring cellulase activity is the soluble derivative carboxymethyl cellulose (CMC). This activity is often mistakenly assumed to be total or true cellulase. What is actually being measured in endo-1,4 α glucanase (EC 3.2.1.4) as exocellulases such as to measure a specific amount (2.0 mg) of glucose release rather than initial rate of reaction, one cannot extrapolate between FPU and assays that measure initial rate of digestion (Wood and Bhat, 1988). The occurrence of α glucosidase/cellobiase (EC 3.2.1.21) will have a profound effect on the measurement of cellulase, due to glucose and cellobiose acting as end product inhibitors of cellobiohydrolases. Wood and Bhat (1988) provide an excellent critique of the methods used for cellulase determination. Further problems are encountered in xylanase measurements where incomplete hydrolysis of xylans can occur unless acetylxylan esterase (EC 3.2.1.72) or other esterases and debranching enzymes are present (Johnson *et al.*., 1988; Williamson *et al.*., 1998).

Table 2. *Selected enzyme activities measured in some commercial plant fiber degrading enzyme products (Nsereko, Rode, Beauchemin, McAllister; unpublished results).*

Product	Protein Content[1]	Xylanase[2]	Cellulase[3]	b-D-glucosidase[4]	b-D-mannosidase[5]	a-L-arabinofuranosidase[6]
1	24.0	2.90	1.30	183.7	53.8	49.7
2	25.5	1.93	0.80	120.1	ND	ND
3	34.2	2.77	1.80	332.6	24.0	35.0
4	42.4	3.57	1.03	863.5	213.3	5.00
5	46.3	2.97	1.27	2189	4209	314
6	49.0	2.17	0.77	1352	404.9	501
7	50.0	4.90	2.13	4485	12.2	6.60
8	61.7	5.27	1.33	2445	2.18	0.06
9	69.7	5.63	2.77	7978	242.6	120
10	78.9	4.53	2.33	5268	214.0	ND
11	104	2.40	1.97	3946	ND	ND
12	109	5.53	2.93	6922	610.1	83.4
13	115	2.60	2.20	5900	5.20	0.08
14	138	5.33	2.33	19642	162.6	14.9
15	181	5.47	1.97	7424	220.7	169
16	199	7.83	1.40	15479	1861	185
17	210	3.67	0.40	5901	26.5	169
18	302	6.70	2.40	76320	672.6	78.5
19	305	5.73	2.10	41096	9027	79.9
20	522	9.03	3.53	156635	288.6	255

ND, not detected.

[1]Protein content expressed as mg protein/ml enzyme product.

[2]Xylanase activity measured as mg reducing sugars released from oat spelt xylan /ml enzyme/min; pH 6.5, 39oC.

[3]Cellulase activity measured as mg reducing sugars released from carboxymethyl cellulase /ml enzyme/min; pH 6.5, 39oC.

[4, 5, 6] _-D-glucosidase, _-D-mannosidase and _-L-arabinofuranosidase activities measured as _mol *p*-nitrophenyl (*p*NP) released from *p*NP-_-D-glucopyranoside, *p*NP-_-D-mannopyranoside and *p*NP-_-L-arabinofuranosidase, respectively; pH 6.5, 39oC

Enzyme activity can also be assessed using synthetic substrates, which usually consist of chromophores linked to molecules chemically similar to the natural substrate. Enzyme activity is measured as the release of the dye or chromophore (Biely *et al.*, 1985). These synthetic substrates offer uniformity among assays, but are subject to criticism in that they do not represent the substrate found in intact feeds such as cereal grains or forages. Therefore, these methods are suitable for measuring relative activity among enzymes but these measurements may have little meaning in terms of enzyme efficacy.

Activity measurements must be conducted under conditions closely defined with respect to temperature, pH, ionic strength, substrate concentration, and substrate type, as all of these factors can also affect the activity of an enzyme (Headon, 1993). Enzyme manufacturers will undoubtedly define the activity units of their products under conditions most favorable for their particular product. Assay conditions will have a significant impact upon enzyme selection. This makes it difficult to select among products that have different optimal conditions for activity. For example, the pH-temperature profiles for different commercial cellulases are shown in Table 3. Furthermore, the assay conditions used to assess enzyme activity are not representative of the conditions in the digestive tract where ultimately the level and persistence of enzyme activity may be most important. Whereas a temperature of approximately 60°C and pH between 4 and 5 are the optimal conditions for most commercial enzymes, normal ruminal conditions consist of a temperature of 39°C and pH closer to 6.7 (Russell and Baldwin, 1979). For these reasons, measurement of enzyme activity using traditional assay techniques may have little relevance to the potential efficacy of an enzyme as a feed additive for ruminants.

Table 3. *pH and temperature cellulase activity profiles[1] of three commercial plant degrading enzyme products (Nsereko, Rode, Beauchemin, McAllister; unpublished results).*

	Product 1			Product 2			Product 3		
T°	39	50	60	39	50	60	39	50	60
pH									
4.0	81.0	81.6	100	65.8	79.0	93.6	43.3	67.1	75.8
5.0	57.6	57.4	81.7	69.0	85.3	100	43.1	70.7	100
6.0	33.7	40.9	49.3	67.3	76.4	73.1	32.8	38.9	40.8
7.0	20.2	26.1	30.9	43.3	52.9	16.8	18.3	13.7	1.23

[1] Activity profiles determined using remazolbrilliant blue dyed carboxymethyl cellulose and expressed as a percentage of the maximum activity.

Researchers have attempted to develop biological assays that may be more indicative of the value of a given enzyme preparation for ruminants. These methods usually involve some form of *in vitro* incubation of enzyme and feed with ruminal contents, and measurement of the disappearance of substrates (e.g., cereal grain, straw, hay) representative of those consumed by the animal (Forwood *et al.*, 1990; Varel *et al.*, 1993; Hristov *et al.*, 1996). Alternatively, the amount of gas produced by the mixed rumen microbial culture can be used as an indication of digestion (Iwaasa *et al.*, 1998), which enables rapid screening of different enzyme products and application rates. These procedures are routinely used in ruminant nutrition. However, our experience to date indicates that these biological assays can only be related to efficacy when the amount of enzyme used *in vitro* is orders of magnitude greater than has been shown to be efficacious *in vivo*. Extrapolation of *in vitro* results to whole animal situations is limited by (i) variations in microbial composition among inocula from different donor animals; (ii) differences in growth of microbial populations in the *in vitro* system versus in the rumen and (iii) artifactual accumulation of end-products that alter enzyme activity. Additionally, these assays do not consider the possible impact of exogenous enzymes on biological parameters such as feed intake, rate of passage or post-ruminal digestion of nutrients. At present, the value of enzymes for ruminants can only be assessed through expensive, time-consuming production experiments with beef or dairy cattle. This makes screening large numbers of products impractical. The lack of an adequate bioassay for assessing the value of exogenous enzymes is perhaps the greatest impediment to the development of more efficacious enzyme products for ruminants.

4. Production Responses to Exogenous Enzymes

4.1. BEEF CATTLE

Evidence that exogenous enzymes could improve average daily gain and feed efficiency in beef cattle was first recorded in a series of feeding trials reported almost 40 years ago (Burroughs *et al.*, 1960). When given a range of diets treated with an enzyme cocktail containing amylolytic, proteolytic, and cellulolytic activities (Agrozyme®, Merck Sharp and Dohme Research Laboratories), cattle gained 7 to 24% more body weight and exhibited improved efficiency of converting feed into liveweight gain (F/G ratio) of 6 to 21%, relative to cattle fed untreated control diets. In the same year, four different enzyme preparations (Agrozyme®, Zymo-Pabst®, Rhozyme®, and Takamine®; Merck and Company, Rahway, NJ) were shown to increase liveweight gain in cattle fed a maize-lucerne hay diet by 14% (Nelson and Damon, 1960).

Further studies confirmed that enzyme supplements could improve average daily gain (ADG) and F/G ratio in cattle fed silage-based diets (Rovics and Ely, 1962), but not all responses to enzyme supplementation were positive. Leatherwood *et al.* (1960) added a fungal enzyme (Enzyme 19AP®, Rohm and Hass Co.) to a grain supplement for calves fed a lucerne hay-based diet and found no improvement in the ADG or F/G ratio of the calves. Two enzyme preparations containing primarily amylase and protease activities also failed to increase ADG by cattle given a diet comprising 80% maize concentrate and 20% chopped lucerne hay (Clark *et al.*, 1961). In a separate study, Agrozyme® even reduced the ADG of cattle by 20% when it was fed with a maize carrier to beef cattle given a maize silage diet (Perry *et al.*, 1960). Similarly, Kercher (1960) found that ADG was reduced when Zymo-Pabst® was fed with a maize carrier to cattle given a diet of steam-rolled barley, lucerne hay and maize silage.

Although these early studies provided valuable information on the potential benefits of enzymes for beef cattle, they did little to address the impact on animal responses of factors such as the composition of the diet, types and levels of enzyme activities present, or the method of enzyme application. More recent studies have been designed specifically to address these issues. Different feed types (Beauchemin *et al.*, 1995; Beauchemin *et al.*, 1997, Beauchemin *et al.*., 1999), application levels (Beauchemin and Rode, 1996; McAllister *et al.*, 1999; Michal *et al.*, 1996), enzyme products (Pritchard *et al.*, 1996) and enzyme application methods (Beauchemin *et al.*, 1998; Rode *et al.*, 1999; Hristov *et al.*, 1998b) have been compared under controlled conditions. Application of different levels (0.25 to 4.0 L tonne-1) of a mixture of xylanase and cellulase products (Xylanase B, Biovance Technologies Inc., Omaha, NE) and cellulase (Spezyme CP®, Genencor, Rochester, NY) increased ADG of steers fed lucerne hay or timothy hay cubes by 30 and 36%, respectively, but had no effect on ADG when applied to barley silage (Beauchemin *et al.*, 1995). When this same mixture was applied to a 95% grain diet, feed efficiency of cattle fed barley was improved by 11% but performance of cattle fed maize was unaffected (Beauchemin and Rode, 1996). Application of a mix of fungal enzyme preparations (Cellulase A, Xylanase B, Finnfeeds International Ltd. Marlborough, UK) at rates up to 5.0 L $tonne^{-1}$ increased the final weight and ADG of feedlot cattle given diets based on lucerne silage (Michal *et al.*, 1996; Pritchard *et al.*, 1996) or barley silage (McAllister *et al.*, 1999). Treatment of 83% maize diets with Amultiple stabilized enzymes@ increased ADG and F/G ratio by feedlot cattle by 10 and 8%, respectively (Weichenthal *et al.*, 1996), and similar improvements in feed efficiency have been reported for cattle fed sorghum-based diets treated with amylase (Krause *et al.*, 1989; Boyles *et al.*, 1992).

4.2. DAIRY CATTLE

The effect of exogenous enzymes on milk production in dairy cows was first examined in the mid 1990s (Chen *et al.*, 1995; Lewis *et al.*, 1995; Stokes and Zheng, 1995) and recently there has been a flurry of research activity in this area (Sanchez *et al.*, 1996; Luchini *et al.*, 1997; Nussio *et al.*, 1997; Kung *et al.*, 1998; Rode *et al.*, 1999; Yang *et al.*, 1999; Beauchemin *et al.*, 1998; Beauchemin *et al.*., 1999). As in studies using beef cattle, production responses by dairy cattle to exogenous enzymes have also been variable. While this variability may be viewed as an indication that fibrolytic enzymes is not a suitable technology for improving ruminant diets, we believe that the variability can be attributed to factors such as enzyme type, level of supplementation and particularly, method of enzyme application. Our group has conducted various feeding studies with a range of enzyme products using our knowledge of enzyme type, level and method of application to clearly demonstrate the potential of enzymes as feed additives for ruminants.

Table 4. Effects of supplementing a dairy cow diet with Low or Medium levels of a fibre degrading enzyme mixture applied to alfalfa hay cubes or cubes and concentrate (Yang et al., *1999).*

Item	Control cubes	Low enz. in cubes	Medium enz. in cubes	Medium enz. in cubes + conc.	SE
Dry matter intake, kg/d	20.4	20.7	20.7	20.8	0.7
Milk production, kg/d	23.7b	24.6ab	25.6a	25.3ab	0.6
4% fat corrected milk, kg/d	22.4b	22.9ab	24.6a	24.2a	0.7
Milk fat content, %	3.79	3.70	3.78	3.76	0.11
Milk protein content, %	3.36	3.41	3.48	3.49	0.04
Milk lactose content, %	4.56b	4.61ab	4.60ab	4.62a	0.02
Kg milk / kg dry matter intake	1.21	1.21	1.29	1.25	0.05
Ruminal organic matter (OM) corrected digestibility[1], %	54.1	54.3	58.4	57.2	3.0
Total tract OM digestibility, %	64.4b	65.9ab	67.0a	66.5a	0.7
Ruminal NDF digestibility, %	30.7	34.9	36.9	35.6	4.8
Total tract NDF digestibility, %	38.8b	41.2ab	43.6a	42.4ab	1.3

[a,b]Means in the same row with different superscripts differ (P < 0.05). [1]Organic matter truly fermented in the rumen.

In one such study, the xylanase/cellulase enzyme mixture, Pro-Mote®, (Biovance Technologies Inc., Omaha, NE), was applied to processed lucerne hay cubes with the cubes comprising 45% of the total diet (dry matter basis; DM) (Yang *et al.*., 1999). When the enzyme-enhanced cubes were fed to dairy cows with ruminal and duodenal cannulae, the digestibility was enhanced and milk production was increased (Table 4). Two levels of enzyme addition were used (Low = 0.5 g/kg; Medium = 1.0 g/kg dietary DM). The medium level was attained by adding the enzyme supplement to the cubes or to the cubes and concentrate. Milk yield was increased by about 1 kg/d for the Low enzyme level, and by 2 kg/d for the Medium enzyme level, with no effects of enzyme on feed intake. The increase in milk yield with Low and Medium levels of enzyme represented a 4 and 8% increase in milk yield, respectively. When compared at the same level of enzyme, the response was similar whether the enzyme was added to the cubes or to both the cubes and concentrate. Increased milk yield due to enzymes did not change the fat or protein content of milk, but lactose content was increased. Because the enzymes improved feed digestibility without affecting feed intake, feed conversion (kilograms of milk yield/kilogram dry matter intake) was 2 to 8% higher for cows fed enzyme enhanced cubes.

We speculated that increasing energy availability of feed using enzymes would lead to a substantial increase in milk yield of cows in early lactation due to their negative energy balance. Thus, we conducted a lactation study (Study 1) with 20 cows in early lactation fed diets treated with the same enzyme mixture (Biovance Technol. Inc., Omaha, NE) (Rode *et al.,* 1999). The enzyme was applied to the concentrate (1.3 g/kg dietary DM) and the diet was formulated to contain 24% maize silage, 15% lucerne chopped hay, and 61% concentrate mainly consisting of steam-rolled barley grain (DM basis). During the first 12 weeks of lactation, cows fed the enzyme-enhanced diets produced 4 kg/d more milk than cows fed the control diet, yet feed intake was unchanged (Table 5). Digestibility of nutrients in the total tract was dramatically increased by enzyme treatment. Lactose and protein yields were consistently higher throughout the experiment for cows fed the enzyme supplemented diet than for cows fed the control diet, but milk fat yield was only higher during the first 3 weeks of the study for cows fed the diet supplemented with enzymes. Afterwards, milk fat yield of those cows dropped below that of the cows fed the control diet.

The substantial reduction in milk fat yield and content due to enzyme supplementation was not expected considering the significant increase in feed digestion. There are a number of possible reasons for this decrease in milk fat. We have observed a change in ruminal protein metabolism with the use of fibrolytic enzymes (Yang *et al..*, 1999): ruminal degradablity of feed protein increases with a concomitant increase in microbial protein synthesis. If, in this study, there was an increase in protein degradability without a compensatory increase in microbial protein, it is possible that the substantial effect of enzymes resulted in inadequate levels of metabolizable protein. Alternatively, we observed a lower ratio of acetate to propionate in ruminal fluid from cattle fed diets supplemented with the same enzyme mixture (Krause *et al..*, 1998). We also reported increased post-ruminal digestion for dairy cows fed a diet supplemented with an enzyme mixture similar to that used in the current study (Beauchemin *et al..*, 1999). Increases in ruminal propionic acid and increased glucose from increased post-ruminal digestion may have stimulated insulin release. The net effect would be to depress milk fat synthesis by increasing adipose tissue lipogenesis. Another possibility is that increased fibre digestion due to enzymes reduced the effective ND content of the diet, indicating higher fibre levels may be needed to maintain a high milk fat content when enzyme supplements are used.

In the second lactation study (Study 2), cows in early lactation were fed diets treated with a modified version of Pro-Mote® (equivalent to1.5 g/kg DM with respect to xylanase and 0.4 g/kg DM with respect to cellulase; Biovance Technologies Inc.,

Omaha, NE) (Beauchemin, Rode and Yang; unpublished data). The enzyme was applied to either the concentrate or sprayed daily onto the TMR. The diet was similar to that used in Study 1, except that the diet was formulated to supply higher metabolizable protein levels.

During the first 15 weeks of lactation, cows fed the diet with enzyme applied to the concentrate produced 2 kg/d more milk than cows fed the control diet, without a change in feed intake (Table 5). In contrast, there was no effect on milk production when the enzyme was applied to the TMR. This study clearly illustrates that method of enzyme delivery is crucial in obtaining improvement in digestibility and milk production. Unlike in lactation Study 1, enzyme supplementation did not affect milk composition, indicating that dietary effective fibre levels were adequate even after enzyme supplementation.

Table 5. *Effects of supplementing diets fed to cows in early lactation with an enzyme mixture.*

Item	Study 1		Study 2		
	Control	Enzyme in conc.	Control	Enzyme in TMR	Enzyme in conc.
Dry matter intake, kg/d	18.7	19.0	19.1	19.8	19.4
Milk production, kg/d	35.9 a	39.5 b	35.3 b	35.2 b	37.4 a
Milk composition, %					
Fat	3.87 c	3.37 d	3.34	3.14	3.18
Protein	3.24	3.03	3.18	3.13	3.13
Lactose	4.73 e	4.62f	4.65	4.56	4.65
Milk component yield, kg/d					
Fat	1.35	1.32	1.14	1.13	1.15
Protein	1.13	1.19	1.09	1.15	1.14
Lactose	1.68	1.85	1.61	1.68	1.7
Body wt. change, kg/d	-0.63	-0.60	0.15	0.14	0.04
DM digestibility, %	61.7	69.1	63.9a	65.7ab	66.6a
NDF digestibility, %	42.5	51.0	N/A	N/A	N/A

[a,b]*Means within a study differ* ($P < 0.11$); [c,d]*Mean within a study differ* ($P = 0.02$); [e,f]*Mean within a study differ* ($P = 0.09$); *N/A Not available.*

Lack of effect of enzyme supplementation on intake seems to be consistent for dairy studies using the Biovance product, but this is not the case for all enzyme products. Sanchez *et al.*. (1996) reported that adding an enzyme mixture (Fanfolds International Ltd., Marlborough, UK) to forage (approximately 0.6, 1.25, and 2.5 L/t dietary DM), increased DMUS of dairy cows in early lactation by 7, 7, or 9%, respectively, compared with cows fed a control diet. However, the milk production response was quadratic: milk production only increased at the 1.25 L/t level. Using the same enzyme mixture applied to forage at 0.39, 0.67, or 0.95 L/t dietary DM, Lachine *et al.*., (1997) reported increased intake for cows from 6 to 17 wk post calving fed the intermediate and high enzyme levels. Also using the same enzyme mixture applied to forage at 0, 0.32, 0.54, and 0.77 L/t dietary DM and fed to cows in early and mid stages of lactation, Nussio *et al.*., (1997) reported a 7, 9, or 11% increase in DMUS, for the three levels, respectively. In that study, milk yield was not affected by the enzyme treatments, but milk yield of the few cows in early lactation fed the highest level of enzyme averaged 3.6 kg/d higher than the milk yield of cows in early lactation fed the control diet. More recently, Kung *et al.*., (1998) used what appears to be a different enzyme mixture from Fanfolds International than used in previously reported studies. Increasing the level of enzyme applied resulted in a quadratic response in milk yield, without a change in dry matter intake.

We conducted a study to determine the effects of an enzyme product containing xylanase and beta-glucanase activities designed for poultry diets (Natugrain®, BASF, Ludwigshafen, Germany), on intake and digestion in dairy cattle (Beauchemin and Rode; unpublished data). Cows received either: a) control (no enzyme); b) low level of enzyme (1.22 L/t of dietary DM); or c) high level of enzyme (3.67 L/t of dietary DM). The enzyme mixture was added to the concentrates at the time of manufacturing and the diet consisted of barley-based concentrate (55%), barley silage (22.5%), and alfalfa haylage (22.5%; DM basis). Low enzyme increased intake of dry matter by 7.5% and high enzyme increased intake by 5.2%. A quadratic effect of enzyme addition on digestion was observed: total tract digestion of dry matter increased by 2.6 percentage units using the low level of enzyme, but there was no effect on digestion for the high level. These studies suggest that some enzyme mixtures may increase feed intake of dairy cows, but increased intake may only lead to higher milk yield when cows in early lactation that are in negative energy balance are used. The varying effects of enzyme products on feed intake indicate that the mode of action likely differs for the different supplements used. Feng *et al.*., (1996) observed that ruminal particulate passage rate was 31% faster for enzyme-treated hay than for control hay (Cornzyme, Fanfolds International Ltd., Marlborough, U.K.), while this effect was not observed in studies using the Biovance product (Rode *et al.*., 1999; Beauchemin *et al.*., 1999).

The effect of enzymes on increasing intake suggests that the mode of action may relate to the collapse and breakdown of the cellular structure of fibre. The physical volume of plant material provides a structure which houses the cellular contents. Van Soest (1992) proposed an "hotel theory" of fibre digestion. The theory uses the analogy of a hotel made up of many rooms: even if the rooms are vacant, the hotel still provides structure. An enzyme mixture that hydrolyzes cellular contents and soluble matter removes the inner "hotel rooms", but leaves the hollow cell wall structure, or "hotel". This type of enzyme will increase digestibility but not necessarily intake. However, an enzyme mixture that increases intake may help collapse the 'hotel"structure. This type of enzyme will lead to an increase in intake but not necessarily an increase in digestibility if the rate of passage of the "hotel walls" is faster than the rate of digestion.

4.3. LEARNING FROM ANIMAL EXPERIMENTS

The positive effects of exogenous enzymes on growth production both by beef and by dairy cattle have been demonstrated definitively, but the information required to improve the consistency and increase the magnitude of these responses needs refinement. Comparisons among experiments are exceedingly difficult, because many enzyme products are poorly defined. Further, several studies have shown that over-application of enzyme is possible, such that increased application costs are not recovered by corresponding improvements in animal performance (Beauchemin *et al.*, 1996; Sanchez *et al.*., 1996; McAllister *et al.*, 1998). Thus, application of one enzyme preparation at a given concentration provides little information with regard to the potential effect on animal performance of a different application level, l*et al.*one a different product. Method of application has a major impact on production responses; they have been shown to differ between dry forage, fresh forage, and silage (Feng *et al.*, 1996; Beauchemin *et al.*, 1995), and if the enzyme is infused directly into the rumen, applied to the complete diet or to the concentrate component only (Lewis *et al.*, 1996; McAllister *et al.*, 1998; Beauchemin *et al.*, 1999). It is obvious that many factors may influence enzyme efficacy in ruminants. Therefore, an understanding of the modes of action by which enzymes improve nutrient utilization in ruminants is key to obtaining consistent positive responses to enzyme additives over a broad range of diets and animal types.

5. Modes of Action

Upon initial consideration, exogenous enzymes might be expected to alter feed utilization in ruminants either through their effects on the feed prior to consumption, or through their enhancement of digestion in the rumen and/or in the post-ruminal digestive tract (Figure 1). In actuality, however, all of these possible modes of action are intertwined and enzyme-mediated alteration of the feed prior to consumption likely has ramifications on ruminal and post-ruminal digestion of nutrients. Preconsumptive effects of exogenous enzymes may be as simple as the release of soluble carbohydrate or as complex as the removal of structural barriers that limits the microbial digestion of feed in the rumen. Within the rumen, exogenous enzymes could act directly on the feed or could indirectly stimulate digestive activity through synergistic effects on ruminal microorganisms.

Exogenous enzymes may remain active in the lower digestive tract, contributing to the post-ruminal digestion of fiber or could indirectly improve nutrient absorption in the lower tract by reducing viscosity of intestinal digesta. Ultimately, the goal of enzyme supplementation is to improve the efficiency of feed utilization in ruminants and reduce waste production. Undoubtedly, the mode of action of exogenous enzymes in ruminants is exceedingly complex and continues to be a major focus of the research presently being conducted with these additives.

5.1 Pre-ingestive Effects

From the previous discussion, it is clear that when fibrolytic enzymes, in a liquid form, are applied to feeds prior to consumption, they can have a positive effect on animal performance. Therefore, it is not unreasonable to assume that the mode of action is through some form of pre-ingestive attack of the enzymes upon the plant fibre. This would suggest that exogenous enzymes should be more efficacious when applied to high moisture feeds such as silages compared to dry feeds. However, there is evidence that exogenous enzymes are more effective when applied to dry forage as opposed to wet forage (Feng *et al.*, 1996; Beauchemin *et al.*, 1995, Beauchemin *et al.*, 1998). At first this seems improbable, given that the role of water in the hydrolysis of soluble sugars from complex polymers is a fundamental biochemical principle (Lehninger, 1982).

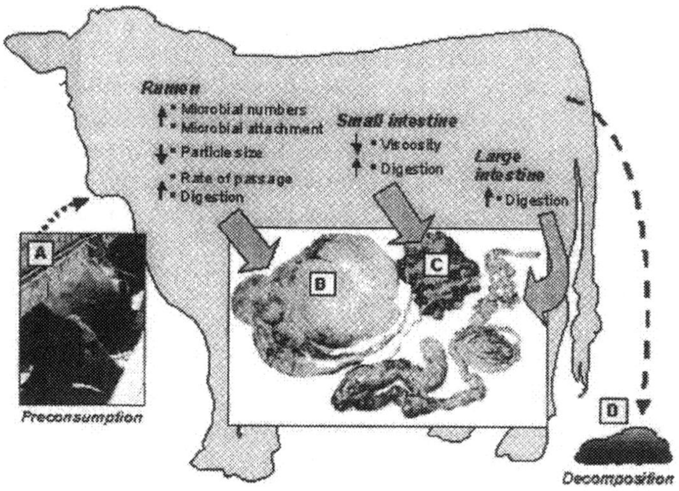

Figure 1. *Possible modes of action of exogenous enzymes in ruminants. (A) Pre-ingestion: partial digestion of feed or weakening of structural barriers that impede rumen microbial digestion. (B) Ruminal: hydrolysis of feed directly or in synergy with ruminal microorganisms to enhance feed digestion. (C) Post-ruminal: improvement of nutrient absorption by reducing intestinal viscosity, or by hydrolyzing substrates that escape ruminal digestion. (D) In the feces: may increase the rate of decomposition (McAllister* et al., *1999).*

However, feed offered to ruminants is seldom absolutely dry; even feeds that nutritionists would describe as "dry" (e.g., grain, hay) contain 6 to 10% moisture. Release of soluble sugars from these "dry" feeds suggests that their free water content is sufficient to enable hydrolysis. However, it is virtually impossible to measure solubilisation of dry matter or fibre without some incubation period. Therefore, it cannot be ruled out that the observed solubilisation of dry feed with enzymes is simply an artefact of the extraction method. Release of sugars from feeds arises at least partially from the solubilisation of NDF and ADF (Hristov *et al.*, 1996; Gwayumba and Christensen, 1997). This is consistent with observed increases in the soluble fraction and rate of *in situ* digestion (Feng *et al.*, 1996; Hristov *et al.*, 1996; Yang *et al.,* 1999). However, most studies have not found exogenous enzymes to improve the extent of *in situ* or *in vitro* DM digestion (Feng *et al.*, 1996; Hristov *et al.*, 1996). These results suggest that enzyme additives only degrade substrates that would be naturally digested by the endogenous enzymes of the rumen microflora.

Although exogenous enzymes do affect release of soluble carbohydrates, the amount liberated represents only a minute portion of the total carbohydrate present in the diet. It is difficult to attribute observed enzyme-associated production responses solely to the generation of soluble carbohydrates prior to consumption, given that comparable increases in yield were not seen when up to 9% of total dietary DM was supplied as molasses (Wing *et al.*, 1988). Additionally, there is ample evidence that through associative effects soluble carbohydrates can actually depress fiber digestion in ruminants (Huhtanen, 1991). A pre-ingestive effect also does not account for why exogenous enzymes can be effective in improving overall fibre digestion when they are only applied to the concentrate (low fibre) portion of the diet (Beauchemin *et al.*, 1997; Beauchemin *et al.*, 1999; Yang *et al.*, 1999). Thus, it is most likely that the major portion of the positive production responses observed to accompany enzyme supplements is due to post-ingestive effects.

5.2 RUMINAL EFFECTS

5.2.1 Direct Hydrolysis

Until recently, it was assumed that the proteolytic activity in the rumen ecosystem would rapidly inactivate unprotected enzyme feed additives (Chesson, 1994; Kung, 1998). This was substantiated by an earlier work from Kopecny *et al.*. (1987) reporting the rapid inactivation of a *Trichoderma reesei* cellulase preparation by rumen bacterial proteases. More recently, the fate of specific feed enzyme additives has been tested for their resistance to rumen degradation *in vitro* and *in vivo*. Exogenous enzymes in the rumen are generally more stable than previously thought. However, there are significant differences depending on the type of enzyme preparation and activity measured (Hristov *et al.*, 1998a,b). Xylanases have been reported to resist proteolysis (Fontes *et al.*, 1995), possibly due to their high degree of glycosylation (Gorbacheva and Rodionova, 1977).

We have observed that cellulase and xylanase activities of some enzyme preparations to be remarkably stable with little or no decline in enzyme activity when incubated in rumen fluid from four cows for up to six hours. (Morgavi, Rode, Beauchemin, McAllister, unpublished results). However, the hydrolytic activity within the rumen is higher in some cows than in others. Stability of the various exogenous enzymic activities measured also varied among cows. This effect was more evident for glycosidases, particularly ß-xylosidase (Figure 2). In activation of ß-xylosidase for most enzyme preparations was rapid in two of the animals (Figure 2.b1.) but slowin the other two (Figure 2.b2.) suggesting that survival of the enzymic activities provided with the feed will vary depending on ruminal conditions in the host animal.

Enhanced stability of feed enzymes after feeding may be the result of several factors including reduced number of microorganisms and presence of readily available soluble nutrients (Dehority and Orpin, 1997). After feeding, the rumen pH in dairy cows is below optimal for most rumen microbial proteases (Wallace *et al.*, 1997; Krause *et al.*, 1998) and soluble proteins liberated from feed also contribute to enzyme stability (Morgavi, Newbold, Beever and Wallace, unpublished results). The effect of protein concentration on stability may explain reduced proteolysis when extremely high exogenous enzyme concentrations are used (Hristov *et al.*, 1998a,b). Differences in stability between enzyme preparations can be explained by their origin, as secondary and tertiary protein conformation can influence their susceptibility to proteases (Fontes *et al.*, 1995). However, our observed differences in the stability of different *T. longibrachiatum* preparations can only be attributed to factors such as carriers and stabilizers contained in the commercial products. Inborr and Grönlund (1993) also reported differences in survivability for two preparations from the same fungus indicating that manufacturing practices may enhance stability. In practice, feed enzyme additives are administered with the feed and the presence of the feed substrate is a known factor increasing enzyme resistance to proteolytic inactivation (Fontes *et al.*, 1995).

The fact that exogenous enzymes remain active in the rumen raises the possibility that they may improve digestion through the direct hydrolysis of ingested feed within the rumen environment. Given that exogenous enzymes are stable in the rumen and the nature of ruminant diets, this is the most obvious mode of action for enzyme efficacy in ruminant diets. Several researchers have shown that exogenous enzymes can enhance fiber degradation by ruminal microorganisms *in vitro* (Forwood *et al.*, 1990; Varel *et al.*, 1993; Hristov *et al.*, 1996; Feng *et al.*, 1996) and *in situ* (Lewis *et al.*, 1996). This effect has been confirmed in some (Beauchemin *et al.*, 1998; Rode *et al..*, 1999) but not in all (Firkins *et al.*, 1990; Varel and Kreikemeier, 1994) studies conducted using cattle with ruminal and duodenal cannulae. The contribution of added exogenous enzymes to total ruminal activity is small (less than 15%; Beauchemin *et al.*, 1998) and even this may be an inflated amount depending upon how enzyme activity is measured. Total endogenous cellulase or xylanase activity is a highly variable measurement from animal to animal (Morgavi, unpublished data). These variations may be attributed to differences in rumen fluid pH, microbial population, and the animal eating behavior. Adding exogenous enzymes may increase the activity of xylanases and cellulases in ruminal fluid, enzyme activity in the fluid usually represents less than 30% of the total enzyme activity in the rumen, the remainder being associated with the feed particles (Minato *et al.*, 1966; Brock *et al.*, 1982). Applying fibrolytic enzymes to a grass hay diet for sheep prior to consumption increased endoglucanse activity and xylanase activity in ruminal fluid, but this activity accounted for only 0.5% of the total endoglucanase activity in the rumen (Table 6; Dong, 1998). However, active enzymes are difficult to extract from insoluble substrates. Therefore, both endogenous and exogenous enzyme activities may be seriously underestimated.

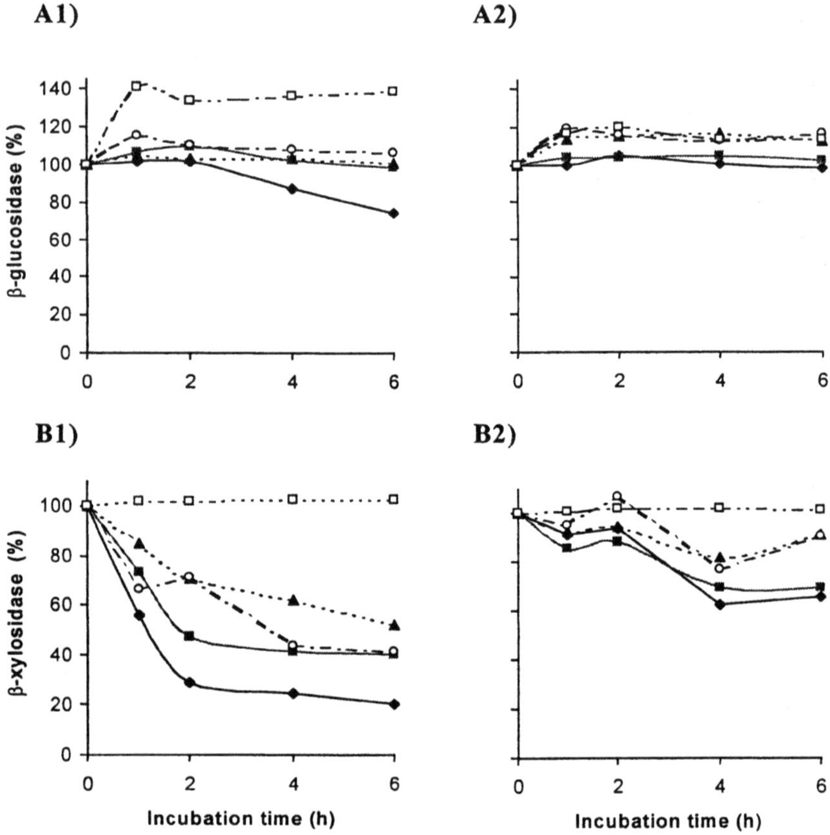

Figure 2. Stability of b-glucosidase (A) and b-xylosidase (B) activities from different enzyme preparations when incubated with rumen fluid taken from dairy cows after feeding. Mean of two cows displaying high (1) or low (2) proteolytic activity. Activity at zero time is set at 100%. T. longibrachiatum *preparations A (♦) and B (□), preparation 1 (□), preparation 2 (O), and* A. niger *preparation (□).*

Given that exogenous enzymes represent only a small fraction of the ruminal enzyme activity, and that the ruminal microbiota is inherently capable of readily digesting fibre (McAllister *et al.*, 1994), it is difficult to envision how exogenous enzymes would enhance ruminal fibre digestion through direct hydrolysis. This, of course, is assuming that measured activities such as xylanase or cellulase are causing the observed improvement in animal performance. However, if the causative activity is one of the minor activities contained in the enzyme products, it is then likely that enzyme supplementation can significantly increase the overall level in the rumen.

5.2.2. Synergism with Ruminal Microorganisms

Enhancement of fibre digestion in the rumen would seem more feasible if these products are working synergistically with ruminal microbes. Logically, this concept implies that exogenous enzyme preparations contain enzymatic activities that would normally be limiting to digestion of plant cell walls by ruminal microorganisms. Limitations to plant cell wall digestion in the rumen could result from insufficient quantities or types of enzyme production by the ruminal microbes, from an inability of degradative enzyme(s) to interact with the target substrates, or from conditions in the rumen not being optimal for enzyme activity (e.g., low ruminal pH). At least 21 different enzymatic activities have been identified as being involved in the hydrolysis of the structural polysaccharides of the plant cell wall, all of which are produced by a normally functioning ruminal microflora (White *et al.*, 1993). If exogenous enzymes are responsible for improvements in feed digestion in the rumen, presumably the enzymes must function within a few hours of feeding before being degraded by the proteolytic activity of the rumen microbes (Kopecny *et al.*., 1987) or leaving the rumen as part of the normal digesta flow to the small intestine (Beauchemin *et al.*, 1999). Also, the quantities of added enzyme activity is small compared to endogenous microbial activities in the rumen (Newbold, 1992b). The enzymes present in fungi must therefore act synergistically with rumen microbial activities rather than additively (Newbold, 1997).

5.2.2.1. Stimulation of Rumen Microbial Populations. Extractions from *Aspergillus oryzae* (AO) cultures can increase the number of ruminal bacteria (Newbold *et al.*. 1992a,b) and can work synergistically with extracts from ruminal microorganisms to enhance release of soluble sugars from hay (Newbold, 1995). Newbold (1997) observed that a high molecular weight (>10,000 MW) component from *A. oryzae* and *Aspergillus foetidus* stimulated bacterial numbers and fiber digestion. The same fraction had the highest carboxymethylcellulase and amylase activity. When the high molecular weight fraction was incubated with a wide spectrum protease prior to feeding, its ability to stimulate fibre digestion and bacterial numbers was destroyed. Thus it appears that proteins are important factors in the ability of *A. foetidus* to stimulate rumen digestion. In view of the enzyme activity in the high molecular weight fraction, the involvement of polysaccharidase enzymes in the efficacy of *A. foetidus* seems likely. Nsereko *et al.* (1999) observed similar effects on rumen microbial populations when exogenous enzymes were fed to lactating dairy cows. In this study, relatively low levels of enzyme supplementation increased non- fibrolytic as well as fibrolytic bacteria. The increase in non-fibrolytic bacterial numbers may have been due to an enzyme-induced release of small molecular weight polysaccharides that can be utilised by these bacteria. Interestingly, this response disappeared when higher levels of enzyme supplementation were used. This response may explain the non linear dose-responses observed in vivo (Beauchemin *et al.*, 1995; Sanchez *et al.*, 1996; Kung *et al.*, 1996). Arambel *et al.*. (1987) speculated that endogenous polysaccharidase enzymes might play an important role in the stimulation of ruminal digestion when ruminants are fed Aspergillus spp. If supplemental enzymes stimulate rumen microbial numbers as shown here, then the resultant greater microbial biomass would provide more total polysaccharidase activity to digest feedstuffs

Table 6. *Effect of exogenous enzymes on endoglucanase and xylanase activitiesz in the rumen of sheep fed grass hay*[z]*.*

Activity	ENZ	CON	% of Total Ruminal Activity ENZ	CON
Activity in liquid phase (units ml-1h-1)				
Endoglucanase	0.026	0.029b	-	-
Xylanase	0.315a	0.344b	-	-
Activity in particulate phase (units g-1 DM h-1)				
Endoglucanase	3.14	3.12	-	-
Xylanase	2.37	28.27	-	-
Total activity in liquid phase (units x 103)				
Endoglucanase	0.14	0.165	3.6	4.6
Xylanase	1.76	1.95	5.4	6.4
Total activity in particulate phase (units x 103)				
Endoglucanase	3.79	3.45	96.4	95.4
Xylanase	30.5	28.54	94.6	93.6

[a,b]*Within a row, means bearing unlike superscripts differ* ($P < 0.05$)

Endoglucanase activity was standardized against a commercial enzyme preparation from Penicillum funiculosm *(EC 3.2.1.4, Sigma Chemical Co., St. Louis, MO) and xylanase activity was standardized against a commercial xylanase activity from* Aspergillus niger *(EC 3.2.18, Sigma Chemical). Incubations were conducted in sodium phosphate buffer (pH 6.5) at 39oC for 30 min. Adapted from Dong, 1998.*

5.2.2.2. Bacterial Attachment. Hydrolysis of cellulose and hemicellulose is accomplished either by free enzymes or by cellusomal structures comprising multiple enzymes bound non-covalently to form an organized complex (Teeri, 1997). Aerobic fungi, the principal commercial source of exogenous enzymes, hydrolyze plant cell walls by means of free enzymes, whereas hydrolysis of plant cell walls by *Clostridium spp.* and the anaerobic ruminalfungi involves cellulosomal structures (Beguin *et al.*, 1998; Ljungdahl *et al.*, 1998). There is also evidence that the ruminococci may rely on a cellulosome-like multi-enzyme complex for fibre degradation (Flint *et al.*, 1998; Ohara *et al.*, 1998). Destruction of these multi-enzyme complexes during the extraction process may explain why enzymes from mixed ruminal microorganisms failed to release as much soluble sugar from hay and straw as extracts from *A. oryzae* (Newbold, 1995).

Close association between microorganisms and fibre is essential for the digestion of feedstuffs in the rumen (McAllister *et al.*, 1994; 1996) and the rate of degradation of fibre in vivo depends on the rate at which the adherent cellulolytic microbial population develops (Silva *et al.*, 1987). If exogenous enzymes can stimulate the attachment of rumen microbes to plant fiber, it may explain how small quantities of enzymes can have such a significant effect on fiber degradation *in vivo.*

Cellulosomes and cellulose binding domains may play important roles in adhesion of microbial cells to their substrates (Pell and Schofield, 1993; Beguin *et al.*, 1998). Newbold (1997) observed that initial attachment of total and cellulolytic rumen bacteria were both enhanced by in vitro inclusion of *A. foetidus*. These results were confirmed in vivo where the inclusion of *A. foetidus*, in the diet of sheep, stimulated the attachment of cellulolytic bacteria to straw incubated in the rumen. Treating lucerne hay with exogenous enzymes, prior to ingestion, increased bacterial colonization and in situ DM disappearance of forage between 3 h and 24 h of ruminal incubation (Yang *et al.*, 1998). These responses were supported by concurrent increases in digestibility of fibre in the rumen and total gastrointestinal tract (Rode *et al.*, 1999). The mechanism by which fibrolytic enzymes stimulate the attachment of rumen bacteria to plant fibers remains unknown. However, chemotactic responses to soluble sugars released from plant fibers have been shown to initiate the attachment of fungi and protozoa to plant cell walls in the rumen (Orpin ,1979). It is possible that these enzymes release soluble sugars from the fiber thus increasing the chemotactic attraction and eventual attachment of fibrolytic rumen bacteria to the plant surface. Enzymic attack may also alter the surface structure of the plants making them more suitable for microbial colonization (Newbold, 1997).

5.2.2.3. Supplying Critical Enzyme Activities. The extent of cross-linking by p-coumaryl and feruloyl groups to arabinoxylans has been identified as one factor that limits the digestion of plant cell walls (Hatfield, 1993). *Aspergillus oryzae* has been shown to produce an esterase capable of breaking the ester bridges forming between ferulic and p-coumaric acids and arbinoxylan (Tenkanen *et al.*, 1991); these are the activities that have been proposed to function synergistically with ruminal microorganisms (Varel *et al.*, 1993). However, many of the ruminal fungi (e.g., *Neocallimastix* spp., Borneman *et al.*, 1990) and some species of ruminal bacteria (e.g., *Fibrobacter succinogenes*, McDermid *et al.*, 1990; *Butyrivibrio fibrisolvens*, Dalrymple *et al.*, 1996) produce esterases capable of hydrolyzing linkages with phenolic acids. The fact that exogenous enzymes usually only increase the rate and not the extent of digestion (Varel *et al.*, 1993; Feng *et al.*, 1996; Hristov *et al.*, 1996) suggests that the activities contributed by these additives are not novel to the ruminal environment. Recent work in which oligonucleotide 16S rRNA genes were cloned and sequenced from the rumen fluid of dairy cattle (Forster and Whitford, 1998). Only 8% of the cloned sequences could be attributable to known genus species combinations and 31% were not close to any known bacteria. Clearly, there is a large unidentified genetic pool in the rumen that is contributing to the diverse array of enzyme activities required for efficient digestion of plant cell walls.

Administering an aqueous solution of mixed fibrolytic enzymes directly into the rumen lowered DM digestion in sheep (McAllister *et al.*, 1998) and beef steers (Lewis *et al.*, 1996). It is unclear if the negative effects on digestion, observed with ruminally infused enzymes, are associated with the in vivo observations that indicate that exogenous enzymes are not as efficacious when applied to moist compared to dry feed (Beauchemin *et al.*, 1995; Beauchemin, Yang, and Rode, unpublished results; Table 6). Alternatively, these negative effects may be due to the relatively high levels of exogenous enzymes that have been used in ruminal infusion experiments. In vivo responses to enzyme supplementation are typically nonlinear (Beauchemin *et al.*, 1995; Sanchez *et al.*, 1996; Kung *et al.*, 1996) suggesting that extremely high enzyme supplementation levels may induce negative in vivo responses.

5.2.2.4. Ruminal pH. Considering the low fiber content of high concentrate diets, it is surprising that fibrolytic enzymes have improved feed digestion (Krause *et al.*, 1998) and performance of cattle fed high cereal grain diets (Beauchemin *et al.*, 1997; Iwassa *et al.*, 1997). An explanation of this phenomenon may come from comparing the pH optima of the fibrolytic enzymes produced by ruminal microorganisms with the pH optima of commercial fibrolytic enzymes (Table 3). It is well documented that growth of fibrolytic bacteria is inhibited (Russell and Dombrowski, 1980), and that fibre digestion is severely compromised when pH falls below 6.2 (Hoover *et al.*, 1984). Isolated and purified fibrolytic, enzymes from pure cultures of rumen bacteria have pH optima above 6.2 (Greve *et al.*, 1984; Matte and Forsberg, 1992) in contrast to the pH optima of less than 6 for many commercial fibrolytic enzymes (Gashe, 1992). For example, the extent to which *T. longibrachiatum* enzymes enhances gas production is shown to increase as the pH declined from 6.5 to 5.5 (Table 7; Morgavi, Rode, Beauchemin, McAllister, unpublished data). Further, although a decline in pH from 6.5 to 5.5 reduced dry matter disappearance from maize silage in mixed ruminal cultures supplemented with *T. longibrachiatum* enzymes, the negative effect of low pH on dry matter disappearance was more pronounced in the absence of added enzyme (Table 7). Ruminal pH can be below 6.0 for a significant portion of the day in dairy cattle Beauchemin *et al.*, 1999; Rode *et al.*, 1999) and in feedlot cattle (Krause *et al.*, 1998). Under these conditions, exogenous enzymes could make a significant contribution to ruminal fibre digestion.

5.2.2.5. Non-enzymic Factors. Some evidence also suggests that non-enzymatic factors in crude enzyme extracts may work synergistically with ruminal microorganisms. Varel *et al.* (1993) showed that autoclaved *A. oryzae* extract enhanced cell wall degradation in vitro. However, the researchers attributed the effect to the presence of soluble substrate in the extract, but concluded that it would not be relevant at the concentration of extract expected at recommended *in vivo* dosages. These results are contrary to observations of Newbold (1997) and reported here (Table 7; Morgavi *et al.*, 1999). Enzyme extracts often contain preservatives to prolong their shelf life, as well as emulsifying agents (e.g., surfactants) that maintain the enzymes in suspension and facilitate application of the product to the feed. Unfortunately, few enzyme manufacturers list the non-enzymatic components of their products and the consequent difficulty in distinguishing between non-enzymic and enzymic effects of enzyme products continues to hamper progress toward defining the mode of action of these products.

5.2.3. Post-ruminal Effects

Alteration of intestinal viscosity is the primary mode by which feed enzymes function to improve animal performance in monogastric animals (Bedford, 1993). Because viscosity of duodenal digesta increases with increasing levels of grain in the diet (Mir *et al.*, 1998), enzyme-mediated reductions in viscosity could improve nutrient absorption in the small intestine of cattle fed grain diets. Reduced intestinal viscosity was associated with 1.2% and 1.5% increases in total tract digestibility of DM when enzymes were applied to the feed or infused into the abomasum, respectively (Hristov *et al.*, 1998a). However, intestinal viscosity in cattle is only between 1 and 2 cPoise (Mir *et al.*, 1998) whereas intestinal viscosity in poultry may exceed 400 cPoise (Bedford, 1993). Improved growth performance in poultry supplemented with enzymes is often associated with 10-fold reductions in intestinal viscosity (Bedford, 1993; Graham, 1996). Consequently, it is difficult to comprehend how the relatively modest declines in intestinal viscosity observed in ruminants supplemented with high levels of enzymes results in a substantial improvement in nutrient absorption in the small intestine.

Table 7. *Effect of pH and* Trichoderma longibrachiatum *enzyme preparations on* in vitro *gas production and dry matter disappearance from corn silage during 48 h of incubation with mixed ruminal cultures*[Z].

Item	pH	No enzyme	Enzyme	
			Autoclaved	Unautoclaved
Gas production (ml)	6.5	9.4	10.4	11.9
	6	7.3[a]	8.0[a]	9.8[b]
	5.5	6.4[a]	7.1[a]	9.0[b]
DM disappearance (%)	6.5	32.1[a]	31.5[a]	36.2[b]
	6	23.6[a]	23.5[a]	31.8[b]
	5.5	23.2[a]	22.8[a]	32.7[b]

[a,b]Within a row, means bearing unlike superscripts differ (P<0.05). [Z]Consecutive batch culture techniques were used to adapt mixed ruminal microorganisms to each respective pH prior to incubation. Morgavi, Rode, Beauchemin, McAllister, unpublished data.

In studies with dairy cows fed barley grain diets, improvements in total tract digestion were attributed, at least partly, to an improvement in the digestibility of fiber and starch in the lower tract (Beauchemin *et al.*, 1998a). Hydrolysis of complex carbohydrates by exogenous enzymes in the small intestine and subsequent absorption of released sugars would offer energetic and nitrogen balance benefits to the animal that would not be accessible if these substrates remained undigested or were fermented by microbial populations residing in the large intestine. It is possible that exogenous enzymes work synergistically with the microbes even in the large intestine. This raises the possibility that exogenous enzymes may even still be active in feces and could aid in reducing overall manure output from cattle. However, we have been unable to see any increases in the fibrolytic enzyme activity, when feeding exogenous enzymes, even at 10 times the recommended level of supplementation.

6. Phytases for Ruminants

The majority of phosphorous (P) in seeds is contained in phytate (Nelson *et al.*, 1968). The inositol ring of faded must be hydrolyzed by fetuses before the P in faded can be available for intestinal absorption (Wise, 1983). The use of fades as dietary adjunct is a relatively new development in animal nutrition compared to other enzyme supplements because it has generally been cheaper to supplement with extra inorganic P than to add fetuses to feedstuffs. However, environmental concern over high P levels in manure has led to the use of fades in monogastric diets in order to increase P availability, thereby reducing P excretion in manure. To our knowledge there have been no studies with the use of fetuses in ruminant diets. This is primarily because it has always been assumed that the rumen is a major source of natural fades activity and that faded-P is completely available (Morse *et al.*, 1992). However, faded availability in cattle may be as low as 50% in some situations (Shinoda *et al.*, 1996).

Satter and Wu (1999) have estimated that over-supplementation of P costs the U.S. dairy industry US$100 million per year. This over-supplementation is largely due to the fact that inorganic P is relatively inexpensive and that the consequences of under-supplementation are related to reduced reproductive performance (Satter and Wu, 1999). The primary source of this perception is nearly 60 years old (Hignett and Hignett, 1951). Estimates of phosphorous availability in ruminant diets ranges from 50 to 70% (Tamminga, 1992). However, if bioavailability is low and fecal phytate is less than complete as suggested by Shinoda *et al.* (1996), there may be potential for the use of commercial fetuses in ruminant diets, particularly in regions where there are legislative restrictions on disposal of manure P.

7. Toward Improving Exogenous Enzymes for Ruminants

7.1 MATCH THE ENZYME TO THE FEEDING SITUATION

Not all exogenous enzymes are equally effective at digesting complex substrates such as lucerne, maize and barley. Feedstuffs are exceedingly complex structurally, and our lack of knowledge of the factors that limit the rate and extent of feed digestion impedes our engineering of enzyme preparations designed to overcome constraints to feed digestion. Additionally, our lack of understanding of the precise mode of action for ruminant feed enzymes limits our ability to target specific feeding scenarios with specific enzymes.

If exogenous enzymes improve ruminant performance through a direct hydrolytic effect on the feed, we could identify specific targets for specific feeds. In maize, for example, the protein matrix surrounding the starch granules, as opposed to the properties of the starch itself, dictates the extent and rate of starch digestion in that grain (McAllister *et al.*, 1993). Thus, exogenous enzymes designed to improve the utilization of maize should contain proteases capable of digesting the protein matrix and exposing starch granules to digestion by endogenous ruminal or host enzymes. An exogenous preparation containing amylase but not protease activity would not be expected to substantially improve utilization of maize by ruminants. In straw, the major barriers to microbial digestion are apparently silica, wax and cutin (Bae *et al.*, 1997).

In poultry nutrition, exogenous enzymes are included to add enzyme activities lacking in the avian digestive tract. In contrast, it would appear that in ruminant systems, the opportunity to improve animal performance lies in complementing a biological system that already has an active fibrolytic system. In modern feeding scenarios, the environmental conditions (e.g. pH) within the rumen often fall outside the range under which this particular biological evolved. This means that new feed enzymes will need to be synergistic with the endogenous rumen system. This provides the greatest opportunity for improvement. However, it is also the most difficult to manipulate as it requires an understanding of both the structural components of the substrate and the underlying endogenous enzyme system itself.

To date most enzyme preparations are currently being used with no attempt to define the types or activities of the enzymes they contain. Such random employment of enzymes on feeds, without consideration for specific substrate targets, will only discourage or delay adoption of exogenous enzymes for more standard use in the animal production industry. Ultimately, enzyme cocktails should be designed specifically to overcome the constraints limiting digestion of a particular type of feed. Component enzymes in such cocktails might vary for given feedstuffs and class of animal being fed. Recent developments in biotechnology make it feasible to engineer such enzyme cocktails containing xylanase and □-glucanase activities, but we presently lack the technology for specific production of many other potentially important enzymes (e.g., cutinase, ferulic acid esterase, acetylxylan esterase, arabinofuranosidase). However, until the specific critical enzyme activities are identified, enzyme cocktails containing an array of activities offer the best opportunity to take advantage of this technology.

7.2. LOWER ENZYME COST

Once the mechanisms of action and specific targets for degradative exogenous enzymes have been identified, steps can be taken to optimize the application of these preparations. Application concentrations can be minimized by ensuring that the preparations contain the enzyme activities most likely to elicit an improvement in feed utilization. In some instances, the activity of crude enzyme preparations may be increased by including specific enzymes most likely to overcome the constraints to feed digestion. Recently, application of recombinant DNA technology has enabled manufacturers to increase the volume and efficiency of enzyme production, and to create new products (Ward and Conneely, 1993; Hodgson, 1994). Genes encoding superior enzymes can be transferred from organisms such as anaerobic bacteria and fungi, typically impractical for commercial production, into well-characterized industrial microbial production hosts (e.g., *Aspergillus* and *Bacillus* spp.). Expression of genes encoding novel enzymes in plants such as canola and potato may be a particularly effective means of lowering the cost of enzyme production (Pen *et al.*, 1993; van Rooijen and Moloney, 1994).

8. Conclusion

The use of feed enzymes in the monogastric animal production industry has increased dramatically in recent years and numerous commercial products are presently being marketed. In many instances, the mechanisms and modes of action of these preparations have been defined. In contrast, few commercial preparations of exogenous enzymes exist for ruminants and many of these are just now entering the marketplace. Although positive responses in animal performance have been observed, results have been inconsistent. Characteristics of the ruminant digestive tract (e.g., complex microbial populations producing numerous endogenous enzymes) complicates elucidation of the mechanisms of exogenous enzyme action in ruminants. There is evidence that exogenous enzymes initiate digestion of feeds prior to consumption and that they can improve feed digestion in the rumen and lower digestive tract. The challenge for researchers is to determine the modes of action, singly or in combination, that enable exogenous enzymes to improve feed efficiency and increase growth and milk production.

References

Arambel, M.J., Wiedmeier, R.D. and Walters, J.L. (1987) Influence of donor animal adaptation to added yeast culture and/or Aspergillus oryzae fermentation extract on in vitro rumen fermentation, *Nutr. Reports Inter. 35*, 433-436.

Bae, H.D., McAllister, T.A., Kokko, E.G., Leggett, F.L, Yanke, L.J., Jakober, K.D., Ha, J.K., Shin, H.T., and Cheng, K.-J. (1997) Effect of silica on the colonization off rice straw by ruminal bacteria, *Anim. Feed Sci. Technol. 65*, 165-181.

Beauchemin, K.A., Rode, L.M. and Sewalt, V.J.H. (1995) Fibrolytic enzymes increase fiber digestibility and growth rate of steers fed dry forages, *Can. J. Anim. Sci. 75*, 641-644.

Beauchemin, K.A. and Rode, L.M. (1996) Use of feed enzymes in ruminant nutrition, in Rode, L.M. (ed.), *Animal Science Research and Development--Meeting Future Challenges,* Minister of Supply and Services Canada, Ottawa, ON, pp. 103-131.

Beauchemin, K.A., Jones, S.D.M., Rode, L.M. and Sewalt, V.J.H. (1997) Effects of fibrolytic enzyme in corn or barley diets on performance and carcass characteristics of feedlot cattle, *Can. J. Anim. Sci. 77*, 645-653.

Beauchemin, K.A., Rode, L.M., Yang, W.Z. and McAllister, T.A. (1998) Use of feed enzymes in ruminant nutrition, *Proc. 33rd Pacific Northwest Nutr. Conf.*, Vancouver, British Columbia, pp. 121-135.

Beauchemin, K.A., Yang, W.Z. and Rode, L.M. (1999) Effects of grain source and enzyme additive on site and extent of nutrient digestion in dairy cows, *J. Dairy Sci. 82*, 378-390.

Bedford, M.R. (1993) Mode of action of feed enzymes, *J. Appl. Poult. Res.* 2, 85-92.

Beguin, P., Chauvaux, S., Guglielmi, G., Matuschek, M., Leibovitz, E., Chaveroche, M.-K., Miras, I., Alzari, P., and Gounon, P. (1998) The Clostridium thermocellum cellulosome: organization and mode of attachment to the cell, *MIE Bioforum, Genet. Biochem. Ecol. Cellulose Degrad.*, Suzuka, Japan, p. 107.

Biely, P., Mislovicova, D., and Toman, R. (1985) Soluble chromogenic substrates for the assay of endo-1,4-ß-xylanases and endo-1,4-ß-glucanases, *Analyt. Biochem. 144*, 142-146.

Borneman, W.S., Hartley, R.D., Morrison, W.H., Akin, D.E. and Ljundahl, L.G. (1990) Feruloyl and p-coumaroyl esterase from anaerobic fungi in relation to plant cell wall degradation, *Appl. Microbiol. Biotech. 33*, 345-351.

Boyles, D.W., Richardson, C.R., Robinson, K.D. and Cobb, C.W. (1992) Feedlot performance of steers fed steam-flaked grain sorghum with added enzymes, *Proc. Western Section Amer. Soc. Anim. Sci. 43*, 502-505.

Brock, F.M., Forsberg, C.W. and Buchanan-Smith, J.G. (1982) Proteolytic activity of rumen microorganisms and effects of proteinase inhibitors, *Appl. Environ. Microbiol. 44*, 561-569.

Burroughs, W., Woods, W., Ewing, S.A., Greig, J. and Theurer, B. (1960) Enzyme additions to fattening cattle rations, *J. Anim. Sci. 19*, 458-464.

Chen, K.H., Huber, J.T., Simas, J., Theurer, C.B., Yu, P., Chan, S.C., Santos, F., Wu, Z. and Swingle, R.S. (1995) Effect of enzyme treatment or steam-flaking of sorghum grain on lactation and digestion in dairy cows, *J. Dairy Sci. 78*, 1721-1727.

Chesson, A. (1994) Manipulation of fibre degradation: an old theme revisited, in Lyons, T.P. and Jacques, K.A. (eds.), *Biotechnology in the feed industry*, Proc. Alltechs 10th Ann. Symp., Nottingham University Press, Loughborough, UK, pp. 83-98.

Clark, J.D., Dyer, I.A. and Templeton, J.A. (1961) Some nutritional and physiological effects of enzymes for fattening cattle, *J. Anim. Sci. 20*, 928.

Classen, H. L., Graham, H., Inborr, J. and Bedford, M. R. (1991) Growing interest in feed enzymes to lead to new products, *Feedstuffs 63*, 22-24.

Considine, P.J. and Coughlan, M.P. (1989) Production of carbohydrate-hydrolyzing enzyme blends by solid-state fermentation, in Coughlan, M.P. (ed.), *Enzyme Systems for Lignocellulose Degradation*, Elsevier Applied Science, New York, pp. 273-281.

Dalrymple, B.P., Swadling, Y., Cybinski, D.H. and Xue, G.P. (1996) Cloning of a gene encoding cinnamoyl ester hydrolase from the ruminal bacterium Butyrivibrio fibrisolvens E14 by a novel method, *FEMS Microbiol. Lett. 143*, 115-120.

Dehority, B.A. and C.G. Orpin (1997) Development of, and natural fluctuations in, rumen microbial populations, in Hobson, P.N. and Stewart, C.S. (eds.), *The Rumen Microbial Ecosystem*, Second Edition, Chapman & Hall, London, pp. 196-245.

Dong, Y. (1998) Reducing methane emissions from ruminant animals, Ph. D. Dissertation, University of Alberta, Edmonton, Alberta.

Feng, P., Hunt, C.W., Pritchard, G.T. and Julien, W. E. (1996) Effect of enzyme preparations on in situ and in vitro degradation and in vivo digestive characteristics of mature cool-season grass forage in beef steers, *J. Anim. Sci. 74*, 1349-1357.

Firkins, J.L., Weiss, W.P., Eastridge, M.L. and Hull, B.L. (1990) Effects of feeding fungal culture extract and animal-vegetable fat on degradation of hemicellulose and on ruminal bacterial growth in heifers, *J. Dairy Sci. 73*, 1812-1822.

Flint, H.J., Aurillia, V., Kirby, J., Miyazaki, K., McCrae, S.I., Rincon, M.T. and Martin, J.C. (1998) Organization of plant cell wall degrading enzymes in the ruminal anaerobic bacteria Ruminococcus flavefaciens and Prevotella bryantii, *MIE Bioforum, Genetics, Biochemistry and Ecology of Cellulose Degradation*, Suzuka, Japan, p. 115.

Fontes, C.M.G.A., Hall, J., Hirst, B.H., Hazlewood, G.P. and Gilbert, H.J. (1995) The resistance of cellulases and xylanases to proteolytic inactivation, *Appl. Microbial Biotech. 43*, 52-57.

Forster, R.J. and Whitford, M.F. (1998) Investigations into rumen microbial diversity using molecular cloning and probing techniques, *MIE Bioforum, Genetics, Biochemistry and Ecology of Cellulose Degradation*, Suzuka, Japan, p. 93.

Forwood, J.R., Sleper, D.A. and Henning, J.A. (1990) Topical cellulase application effects on tall fescue digestibility, *Agron. J. 82*, 900-913.

Galiev, B., Kusova, T., Pribylov, V. and Sechin, V. (1982) Feermenty v ratsionakh molodnyka myasnykh porod, [Enzymes in diets for young beef cattle], *Molochnoe I Myasnoe Skotovodstvo No. 2:10-11* in *Nutr. Abstr. Rev. - Series B 52*, 711.

Gashe, B.A. (1992) Cellulase production and activity by Trichoderma sp. A-001, *J. Appl. Bact. 73*, 79-82.

Gorbacheva, I. and Rodionova, N.A. (1977) Studies on xylan degrading enzymes. I. Purification and characterization of endo-1,4-ß-xylanase from Aspergillus niger str. 14, *Biochem. Bioph. Acta 484*, 79-93.

Greve, L.C., Labavitch, J.M. and Hungate, R.E. (1984) ß-l-Arabinofuranosidase from Ruminococcus albus 8: Purification and possible role in the hydrolysis of alfalfa cell wall, *Appl. Environ. Microbiol. 47*, 1135-1140.

Gwayumba, W. and Christensen, D.A. (1997) The effect of fibrolytic enzymes on protein and carbohydrate degradation fractions in forages, *Can. J. Anim. Sci. 77*, 541-542.

Hatfield, R.D. (1993) Cell wall polysaccharide interactions and degradability. Phenolic constituents of plant cell walls and wall biodegradability, in Jung, H.G., Buxton, D.R., Hatfield, R.D. and Ralph J. (eds.), *Forage Cell Wall Structure and Digestibility*, American Society of Agronomy, Crop Science Society of America, Soil Science Society of America, Madison, Wisconsin, pp. 285-313.

Headon, D.R. (1993) Activity analysis of enzyme under field conditions, in Wenk, C. and Boessinger, M. (eds.), *Enzymes in Animal Nutrition, Proc. 1st Symp.*, Kartause, Switzerland, pp. 233-240.

Hignett, S.L. and Hignett, P.G. (1951) The influence of nutrition on reproductive efficiency in cattle. I. The effect of calcium and phosphorus intake on fertility of cows, *Vet. Rec. 63*, 603-609.

Hodgson, J. (1994) The changing bulk biocatalyst market, *Bio/Technology 12*, 789-790.

Hoover, W.H., Kincaid, C.R., Varga, G.A., Thayne, W.V. and Junkins, L.L.Jr. (1984) Effects of solids and liquid flows of fermentation in continuous cultures. IV. pH and dilution rates, *J. Anim. Sci. 58*, 692-699.

Hristov, A.N., Rode, L.M., Beauchemin, K.A. and Wuerfel, R.L. (1996) Effect of a commercial enzyme preparation on barley silage in vitro and in sacco dry matter degradability, *Proc. Western Sect. Amer. Soc. Anim. Sci. 47*, 282-284.

Hristov, A.N., McAllister, T.A. and Cheng, K.-J. (1998a) Stability of exogenous polysaccharide-degrading enzyme in the rumen, *Anim. Feed Sci. Technol. 76*, 161-168.

Hristov, A.N., McAllister, T.A., Treacher, R.J. and Cheng, K.-J. (1998b) Effect of dietary or abomasal supplementation of exogenous polysaccharide-degrading enzyme supplementation on rumen fermentation and nutrient digestibility, *J. Anim. Sci. 76*, 3146-3156.

Huhtanen, P. (1991) Associative effect of feeds in ruminants, *Norwegian J. Agricul. Sci. Suppl. 5*, 37-57.

Inborr, J. and Grönlund, A. (1993) Stability of feed enzymes in physiological conditions assayed by in vitro methods, *Agricul. Sci. Finlandia 2*, 125-131.

International Union of Pure and Applied Chemistry, Commission on Biotechnology (1984) Measurement of cellulase activity (Ghose, T.K, ed.), Biochemical Engineering Research Centre, Indian Institute of Technology, New Delhi, India.

Iwassa, A.D., Rode, L.M., Beauchemin, K.A. and Eivemark, S. (1997) Effect of fibrolytic enzymes in barley-based diets on performance of feedlot cattle and in vitro gas production, in *Evolution of the Rumen Microbial Ecosystem, Joint RRI-INRA Rumen Microbiology Symposium*, Aberdeen, Scotland, Poster 39.

Iwaasa, A.D., Rode, L.M. and Beauchemin, K.A. (1998) Cumulative gas production of alfalfa forage treated with different cell wall-degrading enzymes, *J. Dairy Sci. 81 Suppl. 1*, 291.

Johnson, K.G., Fontana, J.D. and Mackenzie, C.R. (1988) Measurement of acetylxylan esterase in Streptomyces, in Wood, W.A. and Kellogg, S.T. (eds.), *Methods in Enzymology Volume 160, Biomass Part A Cellulose and Hemicellulose*, Academic Press Inc., New York, NY, pp. 551-560.

Judkins, M.B. and Stobart, R.H. (1988) Influence of two levels of enzyme preparation on ruminal fermentation, particulate and fluid passage rate and cell wall digestion in wether lambs consuming either a 10% or 25% grain diet, *J. Anim. Sci. 66*, 1010-1015.

Kercher, J. (1960) Value of diallylstilbesterol and Zymo-pabst enzyme preparation for fattening yearling steers, *J. Anim. Sci. 19*, 966.

Kopency, J., Marounek, M. and Holub, K. (1987) Testing the suitability of the addition of Trichoderma viride cellulases to feed rations for ruminants, *Zivocisna vyroba 32*, 587-592.

Krause, M., Beauchemin, K.A., Rode, L.M., Farr, B.I. and Norgaard, P. (1998) Fibrolytic enzyme treatment of barley grain and source of forage in high grain diets fed to growing cattle, *J. Anim. Sci. 96,* 1010-1015.

Krause, O.G., Richardson, R. and Cobb, C.O. (1989) Biological responses of steers and rats fed grains with an added microbial enzyme mixture, *J. Anim. Sci. 67 Suppl. 1,* 280.

Kung, L.Jr. (1998) Direct-fed microbial and enzyme feed additives, in Muirhead, S. (ed.), *Direct-Fed Microbial, Enzyme and Forage Additive Compendium,* The Miller Publishing Company, Minnetonka, Minnesota, pp. 15-20.

Kung, L.Jr., Treacher, R.J. and Cohen, M.A. (1998) Enzyme-treated forages for lactating cows, *J. Anim. Sci. 76 Suppl. 1,* 196.

Leatherwood, J.M., Mochrie, R.D. and Thomas, W.E. (1960) Some effects of a supplementary cellulase preparation on feed utilization by ruminants, *J. Dairy Sci. 43,* 1460-1464.

Lee, B., Pometto, A.L., Demici, A. and Hinz, P.N. (1998) Media evaluation for the production of microbial enzymes, *J. Agricult. Food Chem. 46,* 4775-4778.

Lehninger, A. L. (1982) *Principles of Biochemistry,* Worth Publishers, Inc., New York, NY.

Lewis, G.E., Sanchez, W.K., Treacher, R., Hunt, C.W. and Pritchard, G.T. (1995) Effect of direct-fed fibrolytic enzymes on lactational performance of mid-lactation Holstein cows, *Proc. Western Section Am. Soc. Anim. Sci. 46,* 310-313.

Lewis, G.E., Hunt, C.W., Sanchez, W.K., Treacher, R., Pritchard, G.T. and Feng, P. (1996) Effect of direct-fed fibrolytic enzymes on the digestive characteristics of a forage-based diet fed to beef steers, *J. Anim. Sci. 74,* 3020-3028.

Ljungdahl, L.G., Blum, D.L., Chen, H., He, Y., Kataeva, I., Li, X. and Ximenes, E. A. (1998) The cellulase/hemicellulase system of the anaerobic fungus *Orpinomyces* and aspects of further cellulase research, *MIE Bioforum,Genetics, Biochemistry and Ecology of Cellulose Degradation,* Suzuka, Japan, p. 19.

Luchini, N.D., Broderick, G.A., Hefner, D.L., Derosa, R., Reynal, S. and Treacher, R. (1997) Production response to treating forage with fibrolytic enzymes prior to feeding to lactating cows, *J. Dairy Sci.* 80 *Suppl. 1,* 262.

Matte, A. and Forsberg, C.W. (1992) Purification, characterization, and mode of action of endoxylanases 1 and 2 from *Fibrobacter succinogenes* S85, *Appl. Environ. Microbiol. 58,* 157-168.

McAllister, T. A., Phillippe, R. C., Rode, L. M., and Cheng, K.-J. (1993) Effect of the protein matrix on the digestion of cereal grains by ruminal microorganisms, *J. Anim. Sci. 71,* 205-212.

McAllister, T. A., Bae, H. D., Jones, G. A., and Cheng, K.-J. (1994) Microbial attachment and feed digestion in the rumen, *J. Anim. Sci. 72,* 3004-3018.

McAllister, T. A. and Cheng, K.-J.. (1996) Microbial strategies in the ruminal digestion of cereal grains, *Anim. Feed Sci. Technol. 62,* 29-36.

McAllister, T. A., Oosting, S. J., Popp, J. D., Mir, Z., Yanke, L. J., Hristov, A. N., Treacher, R. J., and Cheng, K.-J. (1999) Effect of exogenous enzymes on digestibility of barley silage and growth performance of feedlot cattle, *Canadian Journal of Animal Science* (in press).

McAllister, T.A., Hristov, A.N., Beauchemin, K.A., and Rode, L.M. (1999) Recent advances in enzymes for ruminants, in M. Bedford and G. Partridge (eds.), *Enzymes in Farm Animal Nutrition,* CABI Publishing, Wallingford, UK, (in press).

McDermid, K. P., McKenzie, C. R., and Forsberg, C. W. (1990) Esterase activities of *Fibrobacter succinogenes* subsp. *succinogenes* S85, *Journal of Bacteriology* 170, 2914-2922.

Michal, J. J., Johnson, K. A., Treacher, R. J., Gaskins, C. T., and Sears, O. (1996) The impact of direct fed fibrolytic enzymes on the growth rate and feed efficiency of growing beef steers and heifers, *Journal of Animal Science* 74, Suppl. 1, 296.

Minato, H., Endo, A., Ootomo, Y., and Uemura, T. (1966) Ecological treatise on the rumen fermentation. II. The amylolytic and cellulolytic activities of fractionated bacterial portions attached to the rumen solids, *Journal of General Applied Microbiology* 12, 53-69.

Mir, P. S., Mears, G. J., Mir, Z., and Morgan Jones, S. D. (1998) Effects of increasing dietary grain on viscosity of duodenal digesta and plasma hormone and glucose concentrations in steers, *Journal of Animal Science* 76, Suppl. 1, 247.

Morgavi D. P., Rode, L. M., Beauchemin, K. A., McAllister, T. A., Nsereko, V., Wang, Y., Iwaasa, A. D., and Yang, W. Z. (1999) Interaction between rumen hydrolytic enzymes and those of *Trichoderma longibrachiatum* in degrading fiber substrates, *Journal of Dairy Science* 82, Suppl. 1 (in press).

Morse, D., Head, H.H., and Wilcox, C.J. (1992) Disappearance of phosphorus in phytate from concentrates in vitro and from rations fed to lactating dairy cows, *Journal of Dairy Science* 75, 1979-1986.

Nelson, F. and Damon, V. (1960) Comparison of different supplemental enzymes with and without diethylstilbestrol for fattening steers, *Journal of Animal Science* 19, 1279.

Nelson, T.S., Ferrara, L.W., and Storer, N.L. (1968) Phytate phosphorus content of feed ingredients derived from plants, *Poultry Science* 47, 1372-1380.

Newbold, C. J., Brock, R., and Wallace, R. J. (1992) The effect of Aspergillus oryzae fermentation extract on the growth of fungi and ciliate protozoa in the rumen, *Letters in Applied Microbiology* 15, 109-112.

Newbold, C. J., Frumholtz, P. P., and Wallace, R. J. (1992b) Influence of *Aspergillus oryzae* fermentation extract on rumen fermentation and blood constituents in sheep given diets of grass hay and barley, *Journal of Agricultural Science Cambridge* 119, 423-427.

Newbold, C. J. (1995) Microbial feed additives for ruminants, in R. J. Wallace and H. C. Chesson (eds.), *Biotechnology in Animal Feeds and Animal Feeding*, VCH Publishers Inc., New York, pp. 259-278.

Newbold, C.J. (1997) Proposed mechanisms for enzymes as modifiers of ruminal fermentation, *Proceedings of the 8th Annual Florida Ruminant Nutrition Symposium* (Ed Staple, C.R.), University of Florida, USA, pp. 146-159.

Nsereko, V.L., Rode, L.M., Beauchemin, K.A., McAllister, T.A., Morgavi, D.P., Furtado, A., Wang, Y., Iwaasa, A.D., and Yang, W.Z. (1999) Effects of feeding a fungal enzyme preparation from *Trichoderma longibrachiatum* on the ruminal microbial population of dairy cows, *IX International Symposium on Rumen Physiology* (ISRP), Pretoria, October 17-22,(in press).

Nussio, L. G., Huber, J. T., Theurer, C. B., Nussio, C. B., Santos, J., Tarazon, M., Lima-Filho, R. O., Riggs, B., Lamoreaux, M., and Treacher, R. J. (1997) Influence of a cellulase/xylanase complex (C/X) of lactational performance of dairy cows fed alfalfa hay based diets, *Journal of Dairy Science* 80, Suppl. 1, 220.

Ohara, H., Karita, S., Kimura, T., Sakka, K., and Ohmiya, K. (1998) Multicellulase complex of *Ruminococcus albus* F40, *MIE Bioforum, Genetics, Biochemistry and Ecology of Cellulose Degradation*, Suzuka, Japan, p. 140.

Orpin, C.G. (1979) Chemotaxis in rumen ciliate protozoa. *Society of General Microbiology Quarterly.* 7, 32.

Pell, A. N. and Schofield, P. (1993) Microbial adhesion and degradtion of plant cell walls, in H. G. Jung, D. R. Buxton, R. D. Hatfield and J. Ralph (eds.), *Forage Cell Wall Structure and Digestibility,* American Society of Agronomy, Crop Science Society of America, Soil Science Society of America, Madison, Wisconsin, pp. 397-423.

Pen, J., Verwoerd, T. C., van Paridon, P. A., Beudeker, R. F., van den Elzen, P. J. M., Geerse, K., van der Klis, J. D., Versteegh, H. A. J., van Ooyen, A. J. J., and Hoekema, A. (1993) Phytase-containing transgenic seeds as a novel feed additive for improved phosphorus utilisation, *Bio/Technology* 11, 811-814.

Pendleton, B. (1998) The regulatory environment, in S. Muirhead (ed.), *Direct-Fed Microbial, Enzyme and Forage Additive Compendium,* The Miller Publishing Company, Minnetonka, Minnesota, pp. 47-52.

Perry, T. W., Cope, D. D., and Beeson, W. M. 1960) Low vs high moisture shelled corn with and without enzymes and stilbestrol for fattening steers, *Journal of Animal Science* 19, 1284.

Perry, T. W., Purkhiser, E. D., and Beeson, W. M. (1966) Effects of supplemental enzyme on nitrogen balance, digestibility of energy and nutrients and on growth and feed efficiency of cattle, *Journal of Animal Science* 25, 760-764.

Pritchard, G., Hunt, C., Allen, A., and Treacher, R. (1996) Effect of direct-fed fibrolytic enzymes on digestion and growth performance in beef cattle, *Journal of Animal Science* 74, Suppl. 1, 296.

Rode, L.M., Yang, W.Z., and Beauchemin, K.A. (1999) Fibrolytic enzyme supplements for dairy cows in early lactation, *Journal of Dairy Science* (in press).

Rovics, J. J. and Ely, C. M. (1962) Response of beef cattle to enzyme supplement, *Journal of Animal Science* 21, 1012.

Russell, J. B. and Dombrowski, D. B. (1980) Effect of pH on the efficiency of growth by pure cultures of rumen bacteria in continuous culture, *Applied and Environmental Microbiology* 39, 606-610.

Russell, J.B. and Baldwin, R.L. (1979) Comparison of maintenance energy expenditures and growth yields among several rumen bacteria grown on continuous culture, *Applied and Environmental Microbiology* 37, 537-545.

Rust, J. W., Jacobsen, N. L., McGilliard, A. D., and Hotchkiss, D. K. (1965) Supplementation of dairy calf diets with enzymes. II. Effect on nutrient utilization and on composition of rumen fluid, *Journal of Animal Science* 24, 156-160.

Sanchez, W. K., Hunt, C. W., Guy, M. A., Pritchard, G. T., Swanson, B. I., Warner, T. B., and Higgins, J. M. (1996) Effect of fibrolytic enzymes on lactational performance of dairy cows, *Journal of Dairy Science* 79, Suppl. 1, 183.

Satter, L.D. and Wu, Z. (1999) New strategies in ruminant nurtiion: getting ready for the next millenium,Proceedings of the 14th Annual Southwest Nutrition and Management Conference, University of Arizona, Tucson, USA, pp. 1-25.

Selinger, B. L., Liu, J. H., Yanke, L. J., Bae, H. D., McAllister, T. A., Hsu, Y. C., and Cheng, K.-J. (1999) Novel phytases from ruminal microorganisms and their use in industry, *Journal of American Oil Chemists Society* (in press).

Shinoda, M., Iwasaka, K., Abe, A., and Okamoto, M. (1996) Apparent digestibilities in ruminants of phosphorus, calcium, and phytate phosphorus in rations composed of hay, grass silage, and corn grain, *Animal Production* 89, AB-17.

Silva, A.T., Wallace, R.J. and Orskov, E.R. (1987). Use of particle bound microbial enzyme activity to predict the rate and extent of fibre degredation in the rumen. *British Journal of Nutrition.*, 47, 407-415.

Stokes, M. R. and Zheng, S. (1995) The use of carbohydrase enzymes as feed additives for early lactation cows, *Proceedings of the 23rd Conference on Rumen Function*, Chicago, Illinois, p. 34.

Sturkie, P. D. (1970) Avian digestion, in M. J. Swenson (ed.), *Duke=s Physiology of Domestic Animals*, Cornell University Press, New York, pp. 530-533.

Tamminga, S. (1992) Nutrition management of dairy cows as a contribution to pollution control, *Journal of Dairy Science* 75, 345-357.

Teeri, T. T. (1997) Crystalline cellulose degradation: new insight into the function of cellobiohydrolase, *Trends in Biotechnology* 15, 160-167.

Tenkanen, M., Schuseil, J., Puls, J., and Poutanen, K. (1991) Production, purification, and characterization of an esterase liberating phenolic acids from lignocellulosics, *Journal of Biotechnology* 18, 69-84.

Theurer, B., Woods, W., and Burroughs, W. (1963) Influence of enzyme supplements in lamb fattening rations, *Journal of Animal Science* 22, 150-154.

van Rooijen, G. J. H. and Moloney, M. M. (1994) Plant seed oil-bodies as carriers for foreign proteins, *Bio/Technology* 13, 72-77.

Van Soest, P.J. (1982) Nutritional ecology of the ruminant. O & B Books, Inc., Corvallis, OR.

Van Walleghem, P. A., Ammerman, C. B., Chicco, C. F., Moore, J. E., and Arrington, L. R. (1964) Enzyme supplements and digestibility of protein and energy in rations high in dried citrus pulp, *Journal of Animal Science* 23, 960-962.

Varel, V. H., Kreikemeier, K. K., Jung, H. G., and Hatfield, R. D. (1993) In vitro stimulation of forage fiber degradation by ruminal microorganisms with *Aspergillus oryzae* fermentation extract, *Applied and Environmental Microbiology* 59, 3171-3176.

Varel, V. H. and Kreikemeier, K. (1994) Influence of feeding *Aspergillus oryzae* fermentation extract (Amarferm) on the in situ fiber degradation, ruminal fermentation, and microbial protein synthesis in nonlactating cows fed alfalfa of bromegrass hay, *Journal of Animal Science* 72, 1814-1822.

Wallace, R.J., Onodera, R., and Cotta, M.A. (1997) Metabolism of nitrogen-containing compounds, in P.N. Hobson and C.S. Stewart (eds.), *The Rumen Microbial Ecosystem*, Second Edition, Chapman & Hall, London, pp. 283-328.

Ward, P. P. and Conneely, O. M. (1993) Using biotechnology to improve enzyme yields: From DNA to the marketplace, in C. Wenk and M. Boessinger (eds.), *Enzymes in Animal Nutrition, Proceedings of the 1st Symposium*. Schriftenreihe aus dem Institut für Nutztierwissenschaften Gruppe Ernährung, Kartause Ittingen, Switzerland, pp. 16-21.

White, B. A., Mackie, R. I., and Doerner, K. C. (1993) Enzymatic hydrolysis of forage cell walls, in H. G. Jung, D. R. Buxton, R. D. Hatfield, and J. Ralph (eds.), *Forage Cell Wall Structure and Digestibility*, American Society of Agronomy, Crop Science Society of America, Soil Science Society of America, Madison, Wisconsin, pp. 455-484.

Williams, G., Kroon, P.A., and Faulds, C.B. (1998) Hairy plant polysaccharides: a close shave with microbial esterases, *Microbiology* 144, 2011-2023.

Wing, J. M., Van Horn, H. H., Sklare, S. D., and Harris, B. Jr. (1988) Effects of citrus molasses distillers solubles and molasses on rumen parameters and lactation, *Journal of Dairy Science* 71, 414-420.

Wise, A. (1983) Dietary factors determining the biological activities of phytate, *Nutrition Abstracts and Reviews: Series A, Clinical Nutrition* 53, 791-806.

Weichenthal, B., Rush, I., and Van Pelt, B. (1996) An enzyme-microbial feed product for finishing steer, in *Nebraska Beef Cattle Report*, University of Nebraska, Lincoln, Nebraska, pp. 68-69.

Wood, T.M. and Bhat, K.M. (1988) Methods for measuring cellulase activity, in W.A. Wood and S.T. Kellogg (eds.), *Methods in Enzymology Volume 160, Biomass Part A Cellulose and Hemicellulose*, Academic Press Inc., New York, NY, pp. 87-112.

Yang, W. Z., Beauchemin, K. A., and Rode, L. M. (1999) Effects of enzyme feed additives on extent of digestion and milk production of lactating dairy cows, *Journal of Dairy Science* 82: 391-403.

Yang, W. Z., Rode, L. M., and Beauchemin, K. A. (1998) Effects of fibrolytic enzyme additives on milk production of dairy cows, *Journal of Animal Science* 76, Suppl. 1, 320.Suppl. 1, 320.

Yu, B. and Tsen, H.Y. (1993) An *in vitro* assessment of several enzymes for the supplementation of rabbit diets, *Animal Feed Science and Technology* 40, 309-320.

MOLECULAR TRACEABILITY OF ANIMALS AND THEIR PRODUCTS

HAEZEBROECK V., RENAVILLE R., BERTOZZI C., PARMENTIER I., PIRARD M. AND PORTETELLE D.
Animal and Microbial Biology Unit, Gembloux Agricultural University, 5030 Gembloux, Belgium

Abstract

Adulteration of meat products by addition of substances of lower value is prohibited for fair trading, consumer protection, religious reasons or public health. Several methods are used for the identification of meat species and for the adulteration in meat products. Two important groups of methods exist: immunological methods and DNA-based analyses. Immunological methods are currently used but failed in analysis of treated meat samples or food products with complex matrices.

The present review report methods based on DNA analysis. Principal methods used to reveal DNA polymorphism are described as their applicability in species identification and meat traceability.

1. Introduction

For the last few years, people are more and more concerned about their feeding. The banned hormones in Europe, BSE and dioxine crises shattered consumer confidence in meat and precipitated acute public concern for the safety of the human food chain. Species identification of animal tissues and meat products are important for economical, religious or reasons concerning public health. Today, breeders, retail chain, meat processor and consumer are in need of a reliable system that assure animals and species identification and meat products authenticity.

Scientists have developed methods based on biochemical polymorphism like typing of blood groups. Today, these methods are always applied, but fail in precision. Recent developments in DNA technologies provide solutions. DNA fingerprinting enables the identification and characterisation of any animal and its derived products.

R. Renaville and A. Burny (eds.), Biotechnology in Animal Husbandry, 333–344.

2. Genetic Traceability

A variety of methods are available for an individual or a species identification. These methods reveal polymorphisms or variations that can be used in this way. Some of these markers are used from a long time but are limited. The simplest and oldest are anatomical and physical polymorphisms like coat colour, horns or tattoo. Identification methods such as typing of blood groups and biochemical polymorphisms have proved their usefulness (Grobet and Hanset, 1993). Other studies looked for phenotypic polymorphisms by different techniques such as isoelectric focusing (Skarpeid *et al.*, 1998), agar gel immunodiffusion, counter immunoelectrophoresis (Sherikar *et al.*, 1988) and enzyme-linked immunosorbent assay (Patterson *et al.*, 1983).

These protein markers are more discriminating than morphological markers but have great disadvantages. These methods are all based on the protein fraction which may vary in composition between tissues. Furthermore, heat treatment causes denaturation or methods are of limited use for heat treated meat products analyses. Another great disadvantage of immunological methods is that antisera show cross-reactions (Sherikar *et al.*, 1988). The deficiencies of the methods described above warrant the development of new methods based on other compounds than proteins.

Since DNA is the only basis of genetic differences between distinct organisms, DNA fingerprinting seems to be the ultimate method of biological individualization (Krawczak and Schmidtke, 1998). The advantages of DNA - based analysis are manifold. First, the ubiquity of DNA: all cell types of an individual contain identical genetic information rendering analysis independent of the origin of the sample (blood, muscle etc...). Second, the information content of DNA is higher than that of proteins because of the degeneracy of the genetic code. Third, DNA is a rather stable molecule, allowing its extraction from many different samples even after technological processes including thermal treatment and marinating (Wolf *et al.*, 1999).

3. Studying DNA Polymorphism

Regions of the genome are known to differ frequently between individuals. They are termed polymorphic sites (Krawczak and Schmidtke, 1998). In the past decade, various methods have been developed for the identification and typing of procaryotic and eucaryotic organisms at the DNA level. These methods differ in their taxonomic range, discriminatory power, reproducibility, ease of interpretation and standardisation. The ideal genotyping method produces results that are invariable from laboratory to laboratory and allows unambiguous comparative analyses and the establishment of reliable databases (Savelkoul *et al.*, 1999). DNA-based markers could be grouped into two types (Dodgson *et al.*, 1997): the clone/sequence-based markers and the fingerprint markers.

The first category requires the isolation of a cloned DNA fragment and often determination of some, if not all, of its DNA sequence. The clone/sequence-based markers include microsatellites, RFLP (Restriction Fragment Length Polymorphism), STS (Sequence Tagged Site) and EST (Expressed Sequence Tags). The fingerprinting markers require no knowledge of the sequence of the polymorphic region or isolation of a cloned DNA fragment. They include RAPD (Random Amplified Polymorphic DNA), arbitrarily primed polymerase chain reaction, VNRT (Variable Number of Tandem Repeats) and AFLP (Amplified Fragment Length Polymorphism).

DeVienne *et al.* (1998) realised also a molecular classification of the polymorphism with on the one hand sequence polymorphism and on the other hand repeated DNA polymorphism. The next of this chapter will give more details about the major techniques used to reveal DNA polymorphism.

3.1. DNA SEQUENCING

Individual identification can be done by sequencing portion of the individual genomes where a difference can be expected to show up. However, DNA sequencing methods require time and appear to be not routinely feasible.

3.2. RESTRICTION FRAGMENT LENGTH POLYMORPHISM (RFLP)

The RFLP is the first DNA polymorphism to be exploited in genetic studies.
DNA restriction enzymes recognize specific sequences in DNA and catalyse endonucleolytic cleavages, yielding fragments of defined lengths. Fragments are separated by electrophoresis according to their molecular size. Differences of fragments length among individuals result both from single bp polymorphims and insertions or deletions of blocks of DNA within a given fragment resulting in loss of cleavage site or formation of a new one (Botstein *et al.*, 1980).

Although RFLPs are widely distributed throughout the genome, their utility is limited by low informativeness. Since most RFLPs are only dimorphic (either the enzyme cuts or it doesn't), the maximum heterozygosity of 50% can only occur when both alleles are equally represented in a population. More RFLPs have lower heterozygoties (Payne, 1997). RFLP have also the disadvantage of being labor-intensive in both the development stage and in the typing stage (Dodgson *et al.*, 1997).

3.3. REPEATED DNA SEQUENCE POLYMORPHISM

This polymorphism is also called Simple Sequence Length Polymorphisms (SSLPs). SSLPs are arrays of repeat sequences that display length variations, different alleles containing different numbers of repeat units. Unlike RFLPs, SSLPs can be multiallelic as each can have a number of different length variants (Brown, 1999). The functioning of all this DNA is obscure and it is named selfish DNA or parasite DNA (Ramel, 1997). We can recognize two major types of repeated DNA sequences - minisatellites have repeat units up to 100 bp, but mostly about 9 to 30 bp and microsatellites have repeat units from 1 to 6 bp (Ramel, 1997).

Minisatellites also called variable number of tandem repeats (VNRT) have been used to carry out the first human DNA fingerprinting (Jeffreys *et al.*, 1985). It are the most commonly used fingerprint markers.

Jeffreys *et al.* (1985) detected and developed DNA probes that are able to simultaneously detect large numbers of hypervariable minisatellite loci. Hybridization to digested and electrophoresed DNA with these core sequences at low stringency detects a pattern of fragments that is unique for unrelated individuals. These multilocus probes (MLPs) hybridize with a family of minisatellites sharing the same core sequence. This produces the multi-band fingerprint pattern, often likened in appearance to a bar-code (Rysiecki *et al.*, 1997). The majority of minisatellites are clustered at some subtelomeric ends of chromosomes (Debrauwere *et al.*, 1997)

Microsatellites also called simple sequence repeats (SSR) or short tandem repeats (STR) are tandem repeats of one to six bp, which are interspersed throughout the DNA of animal genome (Tautz, 1989). Microsatellites have the advantages of being abundant, multi-allelic, codominant and uniformly distributed throughout the genome of numerous species (Alexander *et al.*, 1996). In the pig genome the number of CA repeats was estimated between 65 000 and 100 000 copies highly dispersed in a major part of the genome (Wintero *et al.*, 1992). The most common microsatellite repeat motifs are A, AC, AG, AT although the best characterised are dinucleotide (dC-dA/dG-dT) repeats (Payne, 1997).

Given their high degree of polymorphism, it can be argued that mutation events leading to new allelic variants would be relatively common at these repeated sequences (Ellegren, 1995). Microsatellites loci are small enough to be analyzed using the polymerase chain reaction (PCR) and the ability to resolve PCR products differing in size by just one base on polyacrylamide gels allows precise allele designation (Kimpton *et al.*, 1993). In addition to their suitability for mapping and linkage analysis STRs provide a source of highly informative loci for use in the identification of individuals. DNA profiling based on PCR amplification of STRs has the advantage of being more sensitive than conventional techniques. Furthermore, because of their small allele sizes, STR systems are more likely to be successful on samples that contain degraded DNA (Kimpton *et al*, 1993)

Microsatellite-based markers overcome many of the difficulties associated with the other marker types. They generally have higher heterozygoties than RFLPs and, in arising from defined loci, results are more easily interpreted than patterns generated by minisatellites (Usha *et al.*, 1995).

3.4. RAPD

The random amplified polymorphic DNA (RAPD) or arbitrarily primed PCR (AP-PCR) is a recent technique based on the polymerase chain reaction. This method reveals sequence polymorphisms between different template DNAs based on the selective amplification of DNA sequences that are flanked, by chance, by sequences matching an arbitrarily chosen primer.

This method has been proven to be a sensitive and efficient method to generate a large number of molecular markers in a short time (Mathieu - Dauché *et al.*, 1997)

Because a large number of RAPD primers can be purchased at reasonable cost from commercial suppliers, the upfront investment for RAPD mapping is low (Dodgson *et al.*, 1997). However, a major problem with RAPD patterns is their dependence on the exact PCR conditions employed which can lead to reproducibility problems. This is obviously due to the less stringent annealing temperature and shorter primers used. RAPD is more sensitive to temperature differences than conventional PCR (He *et al.*, 1994).

3.5. AMPLIFIED FRAGMENT LENGTH POLYMORPHISM ANALYSIS (AFLP)

AFLP is a newer fingerprinting technique which has become the fingerprint marker of choice for plant genome and begin to be more and more used in animal genome analysis (Ajmone-Marsan *et al.*, 1997). The AFLP technique is based on the amplification of subsets of genomic restriction fragments using PCR. DNA is cut by two restriction enzymes and double-stranded adapters are ligated to the ends of the fragments to generate template DNA for amplification. With selective primers, the PCR products result in predominant amplification of those restriction fragments (Vos *et al.*, 1995).

AFLP simultaneously screens high numbers of loci for polymorphism and detects a greater number of polymorphic DNA markers than any other PCR-based detection system (Vos *et al.*, 1995). Compared to RFLP, the AFLP technique is expected to permit a more accurate estimate of the genetic similarity between individuals and populations (Ajmone-Marsan *et al.*, 1997). But animals microsatellites-based genotyping is probably more useful for linkage analysis, parentage testing and individual identification (Peelman *et al.*, 1998; Savelkoul *et al.*, 1999).

4. DNA Fingerprinting: Applications in Meat Traçeability

4.1. LEGISLATION IN MEAT TRACEABILITY

After the BSE crisis, the European community wants to improve the transparency of the conditions for the production and marketing of the meat products. They decided to establish a traceability system for the identification and registration of bovine animals at the production stage resulting in a Community labelling system in the beef sector based on objective criteria at the marketing stage.

The Council Regulation (EC) N°820/97 includes the general rules for the compulsory beef labelling system. This system shall be obligatory in all Member States, from 1 January 2001. All animals shall be identified by an eartag with a code number applied to each ear. However, Member states impose a labelling system for beef meat which shall contain, among other things, the identification number of the animal.

Since 1994, belgian bovine production is secured by a system in agreement with the principle of the council Regulation. In the SANITEL system, the identity of each animal is recorded in a computer data base. Each animal is identified by an eartag applied on each ear and accompanied by a passport throughout any movement. All those systems that impose the labelling of animals and meat can be very effective but it remains a falsification risk. The permanent nature of the eartag, the passport and the label is not garanteed. DNA fingerprinting can eliminate this risk by creating a biological link between animals and recorded data.

4.2. IDENTIFICATION OF THE SPECIES ORIGIN OF MEAT AND PROCESSED MEAT

The possiblity of detecting animal diversity at various levels is one of the many applications of DNA technology. DNA fingerprinting can be used to distinguish between individuals within families and to distinguish samples of different species. Several techniques have been developed using DNA markers to identify the origin of an unknown sample.

Genomic DNA probes have been applied for distinction between DNA samples of most species. These probes are generally found in highly repetitive DNA sequences. The specificity of DNA hybridization has been investigated by Chikuni *et al.* (1990). Although they succeeded in differentiating meat from distantly related species such as pig and cattle, they reported difficulties in differentiating between beef, meat from sheep and meat from goat because of cross hybridization. Ebbehoj and Thomsen (1991) demonstrated that cross hybridization between probe and DNA sequences from closely related species is reduced by addition of unlabelled DNA from the cross hybridizing species. Other studies of the sensitivity of species differentiation by DNA hybridization have been conducted on mixtures of raw pork and beef. As little as 0,5% raw pork could be detected using total genomic DNA as well as a cloned pig-specific DNA fragment as DNA probe (Wintero and Thomsen, 1990).

These methods based on the presence of species-specific DNA sequences in meat detected by techniques such as DNA hybridization can be very interesting but have the disadvantages of being relatively laborious, costly, time consuming and radioactivity use.

A fast procedure for species identification in heated meat using mini-satellite DNA probes is described by Buntjer *et al.* (1995). The probes investigated in this study are highly specific for species and the complete test is performed within four hours without special equipment.

Highly repeated satellite sequences are easily observed following restriction digestion because of their very high copy number and their tandem arrangement. Guglich *et al.* (1994) described a straightforward method of species identification that is less expensive and faster than DNA hybridization or sequencing. Highly repetitive DNA markers have been used for determining the species origin of animal tissues in cases of illegal commercialization and poaching of game animals. DNA from white-tailed deer, moose, other game species and commercial species was examined following digestion with 15 restriction enzymes. Agarose electrophoresis revealed unique banding patterns for each species.

According to Blackett *et al.* (1992), the DNA restriction size class bands observed with ethidium bromide staining discriminate between some species but not between related species. Another disadvantage with the analysis of visually assessing repetitive DNA bands is that analysis is not possible for low yields of DNA (Guglich *et al.*, 1994).

The polymerase chain reaction discovery opened the way to many applications particulary in food analysis. Rapid amplification of specific DNA sequences by PCR is a method of considerable interest for species differentiation considering its simplicity, specificity, but also its low cost and high speed. The analysis of restriction fragment length polymorphism of PCR fragments was successfully applied for species differentiation by Meyer *et al.* (1994, 1995).

First, PCR has been considered for species-specific amplification of GH genes in pig, cattle, sheep and goat. The porcine PCR assay was demonstrated to be highly specific for porcine DNA. This test detected pork in fresh or heated meat mixtures of pork in beef at levels below 2% (Meyer *et al.*, 1994). Meyer *et al.* (1995) have used an universal pair of primers amplifying a fragment of the vertebrate mitochondrial cyt b gene. They detected restriction fragment length polymorphisms specific for pig, cattle, wild boar, buffalo, sheep, goat, horse, chicken and turkey. PCR-RFLP appeared to be a simple method that can also quickly analyse exotic animals because it does not require preliminary sequencing of the investigated fragment. Wolf *et al.* (1999) used PCR-RFLP for identification of the same species. They investigated intraspecies polymorphism which occurs in some species at a frequency of roughly one base-pair substitution per 100 bp and can affect RFLP. However, it does not allow individual identification.

The use of the PCR followed by RFLP studies has also proven to be an efficient tool for the identification of fish species. Céspedes *et al.* (1998) applied the PCR-RFLP of a conserved region of the mitochondrial cytochrome b for identification of flatfish species. Because it is a simple and cheap method, PCR-RFLP appeared for these authors to be very useful for routine analysis. Unequivocal RFLP results can sometimes occur with highly degraded DNA (Wagner *et al.*, 1994). Chun and Chang (1994) demonstrated that RAPD-PCR fingerprints can be used to identify animals in forensic biological samples. According to these authors, RAPD-PCR fingerprints can be species-specific, subspecies-specific or individual-specific.

Species identification and particularly identification of bovine materials in animal feedstuffs is also essential to control a potential source of bovine spongiform encephalopathy. Tartaglia *et al.* (1998) report a PCR method that allows rapid detection and identification of a bovine-specific mitochondrial DNA sequence. This method detects the presence of bovine material at less than 0,125 % in feedstuffs.

In the surveillance of food quality, DNA analytical methods based on the polymerase chain reaction can also be used for the detection of genetically modified organisms. Methods have been developed for transgenic plants particularly for genetically modified soy (Meyer *et al.*, 1996; Van Hoef *et al.*, 1998). They could probably be adapted to the detection of transgenic animals like salmon.

4.3. ANIMALS AND MEAT TRACEABILITY USING MICROSATELLITE MARKERS

Peelman *et al.* (1998) evaluated 23 microsatellites in terms of their ability to identify individuals in the four main cattle breeds bred in Belgium and to construct a panel containing the nine to 12 most useful microsatellites for identification purposes in all four breeds. The probability of finding two genotypically identical individuals from two different breeds, taking into account all 23 microsatellites rise between 2.892 x10^{-23} and 2.930 x 10^{-24}. The idea is these microsatellites to develop a fingerprint method to use certify animals, carcasses or foodstuffs in the meat channel.

Mortiaux *et al.* (1997) studied 11 of these 23 loci (figure 1). These 11 microsatellites are issued from the Stockmarks Kit (Perkin Elmer) and they are located on distinct chromosomes. Easiness of interpretation, allelic frequencies and informativeness have been evaluated and calculated for each marker in the four main cattle breeds bred in Belgium. The probabilities for two random genome to be identical are better than 10^{-9} for the 11 markers simultaneously in the 4 breeds.

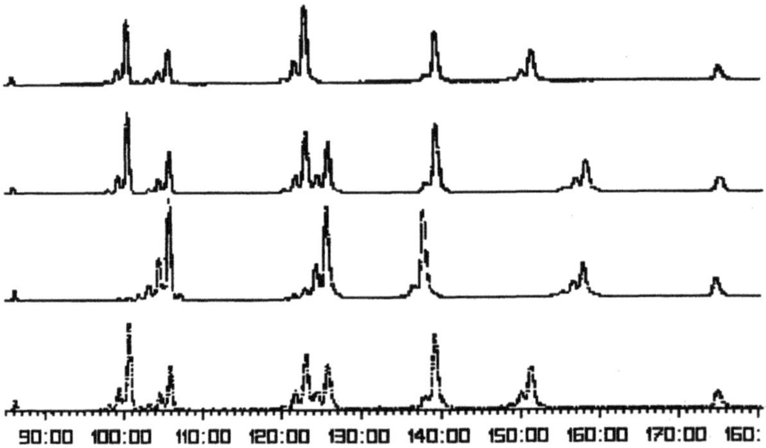

***Figure 1.** Amplification of bovine microsatellite loci TGLA122, TGLA126 and TGLA227 in 4 related bovine animals (Mortiaux* et al.*, 1998).*

The same "Hair Library" can be applied to other species for which sequences of microsatellites are widely reported in the literature like for swine (Ellegren *et al.*, 1994; Rohrer *et al.*, 1994; Alexander *et al.*, 1996), chicken (Crooijmans *et al.*, 1997), sheep (Cushwa *et al.*, 1996), ostrich (Kimwele *et al.*, 1998), horses (Blasi *et al.*, 1999), … .
Finally, microsatellites markers find solid utilisation not only at the meat production level to control the conformity of the passport and the eartag in the meat chanel but also at the breeding level to control family parentage (Figure 3).

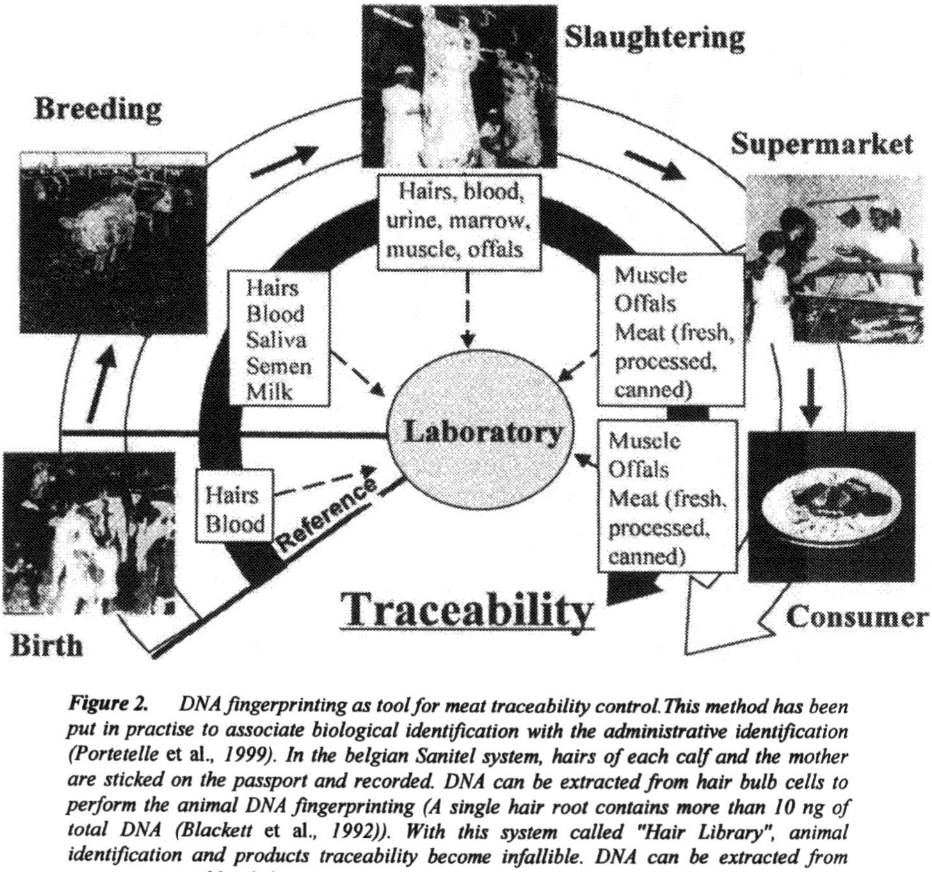

Figure 2. *DNA fingerprinting as tool for meat traceability control. This method has been put in practise to associate biological identification with the administrative identification (Portetelle* et al., *1999). In the belgian Sanitel system, hairs of each calf and the mother are sticked on the passport and recorded. DNA can be extracted from hair bulb cells to perform the animal DNA fingerprinting (A single hair root contains more than 10 ng of total DNA (Blackett* et al., *1992)). With this system called "Hair Library", animal identification and products traceability become infallible. DNA can be extracted from many sources: blood, hair roots, muscle, milk, semen,.... and used in the meat chanel certification as illustrated on figure 2.*

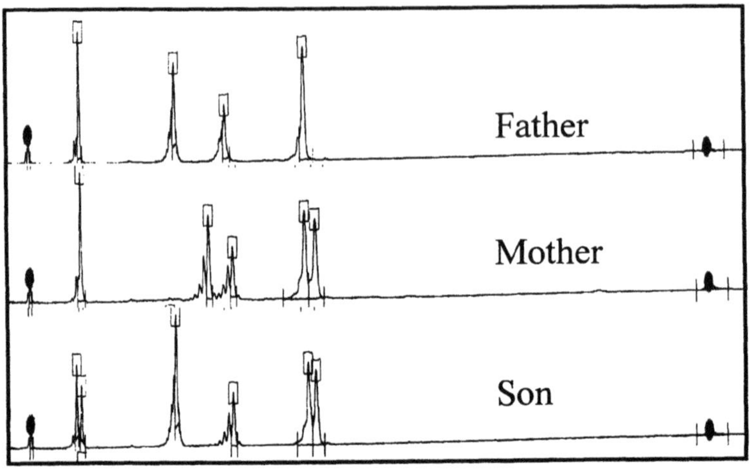

Figure 3. DNA fingerprinting as tool for familly parentage control.

5. Conclusion

A lot of DNA-based markers exist and are used to reach different goals like linkage map, parentage testing, species and individual identification or identification of the geographical origin. These techniques give new and alternative approaches of animal and species characterisation in meat products. Various events like the BSE and dioxine crises assure that such procedures will be more applied in the future to search for more accurate food traceability methods and detection of falsifications. These methods will be, as much as possible, specific, discriminating, reproducible, but also easy to use and economically interesting

Acknowledgements

This research was supported by the Federal Belgian Ministry of Small Enterprises, Traders and Agriculture- DGVI (Brussels, Belgium),

References

Aimé - Sempé, C. (2000) Petit poisson deviendra grand, trop grand?, *Biofutur 199*, 34-37.

Ajmong-Marsan, P., Valentini, A., Cassandro, M., Vecchiotti-Antaldi, G., Bertoni, G. and Kuiper, M. (1997) AFLP marker for DNA fingerprinting in cattle, *Anim. Genet. 28*, 418-426.

Alexander, L.J., Rohrer, G.A. and Beattie, C.W. (1996) Cloning and characterization of 414 polymorphic porcine microsatellites, *Anim. Genet. 27*, 137-148.

Blackett, R.S. and Keim, P. (1992) Big game species identification by deoxyribonucleic acid (DNA) probes, *J. Forenic. Sc. 37*, 590-596.

Botstein, D., White, R. L., Skolnick, M. and Davis, R.W. (1980) Construction of a genetic linkage map using restriction fragment.length polymorphisms, *Am. J. Hum. Genet. 32*, 314-331.

Buntjer, J.B., Lenstra, J.A. and Haagsma, N. (1995), Rapid species identification in meat by using satellite DNA probes, *Z. Lebensm. Unters Forsch. 201*, 577-582.

Brown, T.A. (1999) Chapter 2: Maing genomes by genetic techniques, in *Genomes,* Bios Scientific Publishers Ltd, 472 pages.

Céspedes, A., Garcia, T., Carrera, A.E., Gonzales, J., Sanz, B., Hernandez, P.E. and Martin, R. (1998), Polymearse chain reaction - restriction fragment length polymorphism analysis of a short fragment of the cytochrome b gene for identification of flatfish species, *J. Food Prot. 61*, 1684-1685.

Chun, I.L. and Chang, J.G. (1994) Random amplified polymorphic DNA polymerase chain reaction (RAPD-PCR) fingerprints in forensic species identifcation, *Forensic Sci. Int. 67*, 103-107.

Crooijmans, R.P., Dijkhof, R.J., Vander Poel, J.J. and Groenen, M.A. (1997) New microsatellites markers in chicken optimized for automated fluorescent genotyping, *Anim. Genet. 28*, 427-437.

Debrauwere, M., Gendrel, C.G., Lecath, S. and Dutreix, M. (1997) Differences and similarities between various tandem repeat sequences: minisatellites and microsatellites, *Biochimie 79*, 577-586.

DeVienne, D., Santoni, S. (1998), Les principales sources de marqueurs moléculaires, in DeVienne D. (ed) *Les marqueurs moléculaires en génétique et biotechnologies végétales*, INRA Edit., 15-44.

Dodgson, J.B., Cheng, H.H. and Okimoto, R. (1997) DNA marker technology: a revolution in animal genetic, *Poultry Sci. 76*, 1108-1114.

Ebbehoj, K. and Thomsen, B. (1991) Differentiation of closely related species by DNA hybridization, *Meat Sci. 30*, 359-366.

Ellegren, H. (1995) Mutation rates at porcine microsatellite loci, *Mamm. Genome 6*, 376-377.

Ellegren, H., Chowdhary, B.P. and Johansso, M. (1994). A primary linkage map of the porcine genome reveals a low rate of genetic recombination, *Genetics 137*, 1089-1100.

Grobet, L. and Hanset R. (1993) Identification et diagnostics de filiation par empreintes génétiques chez les animaux domestiques, *Ann. Méd. Vét. 137*, 123-132.

Guglich, E.A., Wilson, P.J. and White, B.N. (1994) Forensic application of repetitive DNA markers to the species identification of animal tissues, *J. For. Sci. 39*, 353-361.

Haezebroeck, V., Renaville, R., Parmentier, I., Bertozzi, C., Sindic, M. and Portetelle, D. (2000) Identification par empreintes génétiques des espèces animales entrant dans une chaîne alimentaire. Cinquième carrefour des productions animales:Quels systèmes de productions animales pour le 21e siècle? Gembloux, le 26 janvier 2000.

He, Q., Viljanen, M.K. and Mertsola, J. (1994) Effects of thermocyclers and primers on the reproducibility of banding patterns in randomly amplified polymorphic DNA analysis, *Mol. Cell. Probes 8*, 155-160.

Jeffreys, A.J., Wilson, V. and Thein, S.L. (1985) Hypervariable minisatellite region in human DNA, *Nature 314*, 67-73.

Kimpton, C.P., Gill, P., Walton, A., Urquhart, A., Millican, E.S. and Adams, M. (1993) Automated DNA profiling employing multiplex amplification of short tandem repeat loci, *PCR Met. Appl. 3*, 13-22.

Krawczak, M. and Schmidtke, J. (1998) DNA fingerprinting; second edition. BIOS Scientific Publishers, 101 pages.

Mathieu-Daudé, F., Ralph, D. and McMelland, M. (1997) Arbitrarily primed PCR fingerprints, in Taylor G.R. (ed.) *Laboratory methods for the detection of mutations and polymorphimss in DNA* CRC Press, pp5-15.

Meyer, R., Candrian, U. and Lüthy, J. (1994) Detection of pork in heated meat products by the polymerase chain reaction, *J. AOAC Int.. 77*, 617-622.

Meyer, R., Höfelein, C., Lüthy, J. and Candrian, U. (1995) Polymerase chain reaction-restriction fragment length polymorphism: a simple method for species identification in food, *J. AOAC Int.. 78*, 1542-1551.

Meyer, R., Chardonnens, F., Hübner, P. and Lüthy, J. (1996) Polymerase chain reaction (PCR) in the quality and safety assurance of food detection of soya in processed meat products, *Z. Lebensm. Unters Forsch. 203*, 309-344

Mortiaux, F., Renaville, R., Gengler, N., Parmentier, I., Bertozzi, C., Burny, A. and Portetelle, D. (1997) Certification in the meat channel, *J. Anim. Sci. 75 Suppl 1*, 136.

Mortiaux, F., Renaville, R, Peelman, L., VanZeveren, A., Bouquet, Y., Massart, S., Parmentier, I., Bertozzi, C., Burny, A. and Portetelle, D. (1998) Empreintes génétiques et certification en production bovine. Troisième carrefour des productions animales: les biotechnologies: Enjeu de société? *Proc. 3° Carrefour Prod. Anim. Gembloux*, pp 15-23.

Patterson, M.R. and Spencer, T.L. (1983) A rapid on-site test for speciation of meat, *Austr. Vet. J. 60*, 381-382.

Payne, S.J. (1997) Microsatellite analysis, in Taylor G.R. (ed.) *Laboratory methods for the detection of mutations and polymorphimss in DNA* CRC Press, pp 31-44.

Peelman, L.J., Mortiaux, F., Van Zeveren, A., Dansercoer A., Mommens, G., Coopman, F., Bouquet, Y., Burny, A., Renaville, R. and Portetelle, D. (1998) Evaluation of the genetic variability of 23 bovine microsatellite markers in four Belgian cattle breeds, *Anim. Genet. 29*, 161-167.

Portetelle, D., Haezebroeck, V., Mortiaux, F. and Renaville, R. (1999). Traçabilité moléculaire des viandes. Belgian Association for meat science and technology (BAMST). Journée d'étude "traçabilité": un moyen de garantir la qualité des viandes et produits de viande, Gembloux, Belgium, 11-31.

Ramel, C. (1997) Mini- and microsatellites, *Environ. Health Perspect. 105*, 781-789.

Rorher, G.A., Alexander, L.J., Keele, J.W., Smith, T.P. and Beattie, C.W. (1994) A microsatellite linkage map of the porcine genome, *Genetics 136*, 231-245.

Rysiecki, G., Childs, W.P. and Duxbury, N. (1997) Minisatellite analysis, in Taylor G.R. (ed.) *Laboratory methods for the detection of mutations and polymorphimss in DNA* CRC Press, 333 pages.

Savelkoul, P.H.M., Aarts, H.J.M., DeHaas, J., Dijkshoorn, L., DUIM, B., Otsen, M., Rademaker, J.L.W., Schouls, L. and Lenstra, J.A. (1999) Amplified-Fragment Length Polymorphism analysis: the state of an art, *J. Clin. Microbiol. 33*, 1716-1719.

Sherikar, A.T., Khot, J.B., Jayrao, B.M. and Pillari, S.R. (1988) Use of species-specific antisera to adrenal heat-stable antigens for the identification of raw and cooked meats by agar gel diffusion and counter immunoelectric techniques, *J. Sci. Food Agric. 44*, 63-73.

Skarpeid, H.J., Kvaal, K. and Hildrum, K.I. (1998) Identification of animal species in ground meat mixtures by multivariate analysis of isoelectric focusing protein profiles *Electrophoresis 19*, 3103-3109.

Tartaglia M., Saulle E., Pestalozza S., Morelli L., Antonucci G. and Battaglia P.A. (1998) Detection of bovine mitochondrial DNA in ruminant feeds: a molecular approach to test for the presence of bovine-derived materials, *J. Food Prot. 61*, 513-518.

Tautz, D. (1989) Hypervariability of simple sequences as a general source for polymorphic DNA markers, *Nuc. Acids Res. 17*, 6563-6571.

Usha, A.P., Simpson, S.P. and Williams J.L. (1995) Probability of random sire exclusion using microsatellite markers for parentage verification, *Anim. Genet. 26*, 155-161.

Van Hoef, A. M.A., Kok, E.J., Bouw, E., Kuiper, H.A. and Keijer, J. (1998) Development and application of a selective detection method for genetically modified soy and soy-derived products, *Food Add. Contam. 15*, 767-774.

Vos, P., Hogers, R., Bleeker, M., Reijans, M., Vandelee, T., Hornes, M., Frijters, A., Pot, J., Peleman, J., Kuiper, M. and Zabeau, M.. (1995) AFLP: a new technique for DNA fingerprinting, *Nuc. Acid Res. 23*, 4407-4414.

Wagner, V., Schild, T.A. and Geldermann, H. (1994) Application of the polymorphic DNA sequences to differentiate the origin of decomposed bovine meat, *J. For. Sci. Int. 64*, 89-95.

Wintero, A.K., Fredhom, M. and Thomsen, P.D. (1992) Variable (dG-dT)n (dC-dA)n sequences in the porcine genome, *Genomics 12*, 281-288.

Wolf, C., Rentsch, J.and Hubner, P. (1999) PCR-RFLP analysis of mitochondrial DNA: a reliable method for species identification, *J. Agric. Food Chem. 47*, 1350-1355.

CONSUMER ATTITUDES TO ANIMAL BIOTECHNOLOGY:

BROADER ETHICAL CONSIDERATIONS

SPENCER S
Center for Agricultural Bioethics, Catholic University of Leuven, Kardinaal Mercierlaan 92, B-3001 Leuven, Belgium.

Abstract

Despite the potential benefits there is a sizeable consumer opposition to genetically modified foods and other biotechnologies. This article looks at some of the reasons for this antipathy.

In recent years, people have lost their faith in the impartiality of scientists, consequently their reluctance to accept biotechnologies is based on gut feelings rather than scientific rationale. These gut feelings represent consumers' ethics and include, in any combination: religious, safety, moral, social, political, economic, welfare and ecological concerns – as well as straightforward prejudices. Various ethics philosophies relevant to animal biotechnologies are outlined. The (un)acceptability of biotechnologies to various religions and philosophical opinions is discussed, as is the background to consumer concerns about food safety. Animal welfare issues concern some people, while others are opposed to biotechnology because of worries about the ecology, particularly if genetically modified species escape into the wild. Although advocates of biotechnologies point out the enormous potential for improving food supplies and nutrition in underdeveloped countries, the evidence of the 'Green Revolution' suggests that, even if the new techniques were affordable and practical in these countries, it would probably serve to widen the gulf between rich and poor in the Third World. Similarly, in more developed economies the introduction of biotechnologies may lead to big changes in rural societies, including unemployment. These social implications are related to economic and political affects of the biotechnologies at local, national and global levels. Everyone is affected in some way or another by one or more of these factors, so influencing the acceptability of biotechnologies to consumers.

R. Renaville and A. Burny (eds.), Biotechnology in Animal Husbandry, 345–357.

1. Introduction

The industrial revolution in the 19th century is a spark compared with the potential explosion that biotechnology will bring to the 21st century. Modern day consumers, without necessarily being neo-Luddite, are concerned that the implications of the new biotechnologies should be seriously considered and debated.

Various aspects of biotechnology have generated public concern: genetically modified organisms (particularly crops and animals, but also microbes), reproductive technologies such as cloning, the use of recombinantly derived hormones such as bovine somatotropin, and the use of animals as factories for pharmaceuticals, or as sources of organs for transplantation. Today's consumers (at least in affluent countries) are concerned about the intrinsic morality of biotechnologies and, even more, are worried about health, the ecological, social, and economic consequences of embracing the technology of genetic manipulation.

The issues and ethical viewpoints that have shaped consumer attitudes to biotechnology cover a wide spectrum, each of which in itself may be a subject for a complete book. Inevitably, therefore, this chapter will be superficial. There is no intention to be prescriptive, but rather to stimulate and encourage thinking around the subject. It is important to look beyond the science to the various downstream consequences.

Consumer attitudes to transgenesis have been generally formed on the basis of information available from genetically modified plants – largely because the technologies that have been applied to plants for a quarter of a century have only relatively recently being applied to animals on farms. Consequently, much of the debate on biotechnology has been one-dimensional – relating to a single, particular application – rather than to concerns that arise horizontally and can be applied across the whole spectrum of animals, plants, microorganisms and, indeed, humans. This article will attempt a broader view by drawing examples from different areas of biotechnology to demonstrate how the implications of all these technologies may be evaluated. Issues that have come to prominence with commercial application of transgenic plants can serve to demonstrate how we might approach the evaluation of the broader social, economic and ethical implications of animal biotechnology.

2. Consumer Acceptability

Public attitudes to the biotechnologies are related to risk. It would be comforting to think that views were made on rational evaluation of the science, but usually they are made on values and emotion – a gut feeling. One UK survey of attitudes to genetic modification of foods found 70% of those questioned thought it was morally wrong. The figure was somewhat less in a US survey (45%), although 82% of US respondents expressed concerns about eating genetically modified foods. A similar equal split on acceptability was recorded in a New Zealand survey in 1990. However, with greater exposure of the topic in recent years, public concern seems to have increased more recently.

Commonly in the debates on this subject, objectivity is largely lacking. Having cited the figure of 82% of US respondents being concerned about genetically modified foods, it should also be noted that another US study indicated that 82% thought genetic studies should be continued; this was cited as public support for biotechnology. However, support for continued research is not quite the same as not being concerned about genetically modified foods. Such conflicting reports and biases only serve to make consumers suspicious. It is not practical to list the many surveys of public attitudes to genetically modified foods, but whatever the true percentage of people worried about biotechnology but, as reported by Bredahl, it is certainly a sizeable proportion of the population in many countries. There are, of course, numerous websites dealing with consumer concerns over biotechnology. Among the more rational are the various Consumer Association sites (*e.g.* www.which.net), and Greenpeace International (www.greenpeace.org).

3. Disenchantment with Science

The public's concerns with biotechnology may have started with visions of mutant animals, and uneasiness over cloning of humans, but once the debate started many ethical issues were raised including: human health, ecology, multinational corporations, unemployment and Third World agriculture.

Particularly for issues of health and ecological safety the public have traditionally looked to scientists to provide objective reassurance. However, the debate about genetic modification has come at a time when the public is disenchanted with scientists. In Europe, in particular, science has fallen from its pedestal, helped by issues like BSE. The image of the absent-minded professor working altruistically to cure the world's ills have been supplanted by idea of an egotist who capitalises on the (short-term) advantages of bending to political, or corporate, will. The public feels that they have lost their unbiased source of information and are left having to figure out important matters for themselves.

Probably one of the biggest disservices to science was the attempt to silence (and when that failed, to discredit and remove) Arpad Pusztai for disseminating his work that showed large changes in the gastrointestinal tract of rats fed genetically modified potatoes. The ferocity of the attacks appeared to be born of prejudice and self-interest, and was out of all proportion to the deficiencies in the study. This attempt to suppress controversial evidence sent shivers through the scientific community and caused enormous damage to the public's view of science. The campaign also caused more suspicion over transgenic foods than could possibly have resulted from allowing publication of the data without comment.

4. Ethics

It is probably useful to consider what is meant by 'ethical' and 'moral' (at least in the context of this article). Morals are intrinsically held beliefs and may be on any subject.

One may have a moral objection to bull-fighting, or to sexist jokes. Such objections are not always based on careful reflection and analysis, but are often gut feelings that something is not right. Ethics is a narrower concept and suggests standards by which a society regulates its behaviour. Ethics suggests a critical analysis of the basis of morality and includes consideration of all the consequences of an action. Broader appreciation of outcomes has, to date, been largely absent in scientists' evaluation of the acceptability of biotechnology.

Some of the moral debate has involved discussion over whether the biotechnologies are unnatural. Ultimately, all farming (and indeed our own lifestyles) can be considered to be unnatural; accepting this reality makes the whole argument over unnaturalness baseless. New technologies, even if they are unnatural, are not necessarily ethically wrong.

There may be risks involved, but risks are not invariably of ethical concern. There may be a risk associated with stepping out into traffic without looking, but is it ethically wrong? The ethics come with the weighing of the consequences of the action: stepping into traffic may result in death or injury, thus engendering family grievances, health service costs, psychological trauma on others involved in the accident *etc*.

What is the basis for consideration of the acceptability of biotechnology? There are theological, moral, ethical, intrinsic and extrinsic aspects to the issue. The largely intrinsic, theological and moral aspects (should we be interfering in God's work?) will not be covered in depth. This chapter will concentrate on the ethical implications of the consequences of introducing genetically modified organisms into general farming practice. However, it is worth recognising that different religious beliefs may have problems with different aspects of biotechnology. Islam holds that God created life in its best form – therefore we should not alter it. Hindu and Buddhism also revere Nature. Judeo-Christian beliefs hold that it is acceptable for humans to use nature if they protect and preserve it, but Jews might have a problem if pig genes were used in the wheat used to make their bagels. Moslems may also have concerns about pig genes, Hindus about bovine genes, and Jains might object to eating any vegetable in which an animal gene has been incorporated (*e.g.* toad genes in potatoes).

Three main philosophical arguments will be applied to the ethics of biotechnology: utilitarianism, Kantian and Aristotlean. Utilitarianism counts the costs and the benefits to decide acceptability on the balance of the consequences (not the intrinsic moral factors) of the issue. In utilitarianism it is important that the various sides are dealt with impartially – something which proves difficult in this field. Kant suggested that people should not be used solely as a means without consideration of their requirements. Kant specifically excluded animals, but other, such as Regan, have extended the Kantian philosophy to animals and produced an abolitionist philosophy. Different again, but in some ways between these two views, is the philosophy that we have a duty to not cause pain or suffering (a 'do no harm' philosophy).

Coming from another position is Aristotle who advocated that an animal's essential being (its *telos*) must be considered – that a pig must always have the essence of being a pig. Dressing a pig in clothes, sitting it in front of the fire, or tucking it into a warm comfortable bed would not be respecting the *telos* of a pig, although anthropomorphic sentiment may make it appear 'kinder' to the animal than letting it suffer the possible dangers and discomforts of being a wild pig. All of these philosophies may be applied to biotechnological advances.

5. Human Food Safety

Self-interest is, of course, a powerful factor in determining attitudes. The issue of human safety from eating genetically modified foods is the current focus of the news media in the debate on the acceptability of biotechnologies. Whether the press have responded to this concern or generated it, is unclear, but it is a central aspect in determining consumer acceptability.

There are both short-term and long-term implications of genetically modified foods on human safety. The previously mentioned study by Ewen and Pusztai, was purely experimental and the result was probably predictable given what we already knew of the effects of the lectin introduced into the potatoes. However, another example, which was thought to be safe, is the use of a brazil nut gene in modified soya plants. The nut gene was able to produce an allergic reaction in people sensitive to nuts. The biotechnologists' response to such concerns is that the testing of the new crops is intensive, and they rightly point out that this effect was recognised and corrected before the product was marketed. But it highlights the difficulty that we do not always think of all the implications of a modification. Genetically modified products are analysed to ensure that the composition is similar to unmodified crops, but the public is concerned that we are only testing what we think might be important, and not what *could* be important. Many argue that we simply don't know about all potential allergens and adverse effects.

There are also consumer concerns about potentially longer term hazards to health from antibiotic resistance. The majority of genetically modified organisms are selected using a marker gene for antibiotic resistance. All these modified organisms exhibit antibiotic resistance and the concern is that this resistance may spread. Fortunately, nearly all of these modifications use the same antibiotic resistant marker gene, but it does mean that the usefulness of this antibiotic (at least) in our armoury against disease, is now severely compromised. The bigger concern is the potential for cross-resistance to other antibiotics, as is seen in clinical practice, is also more likely.

The use of recombinantly produced somatotropin to enhance lactation has also met with consumer resistance on the basis that the hormone could be present in the milk. The amounts of bST in milk are low, and it has been argued that the bovine hormone does not bind to the human growth hormone receptor even if it can get across the gut undigested. The downstream effects are the ones on which the opponents to bST are now focusing. Levels of IGF-1 are higher in the milk of bST-treated cows, and bovine IGF-1 *does* cross-react with the human receptor. This hormone also directly affects the gut of young animals and so the spectre of unwanted effects in babies remains a concern of consumers, although there appears to be a sizable safety margin. Similar concerns about 'residues' of inappropriately high hormone levels are also voiced for meats. Animal scientists have long argued that natural oestrogen levels in female animals can be many times higher than any possible level of administered oestrogen, but such logic falls upon deaf ears at the butcher's counter. Many consumers oppose the use of hormones.

6. Animal Welfare

Transgenic animals are not as close to full-scale farm use as are plants. It is more difficult to make transgenic animals, it costs more, takes longer, and has come under more stringent ethical scrutiny than transgenesis in plants. Even so, many hundreds of different genetic animal mutations have been prepared in the laboratory (usually in mice). These laboratory experiments have been valuable tools in elucidating many physiological processes, but have often resulted in weak, deformed, sick, infertile, or non-viable animals. Some of the mutations have been directed towards producing antibodies or drugs in animals (particularly in the milk), but others have been more directly aimed at increasing farm productivity – usually by increasing growth hormone(s). A classic example of an attempt to improve animal productivity, was the introduction of the human growth hormone gene into pigs at the USDA station at Beltsville. Growth hormone stimulates both growth and catabolism, and the effect in the Beltsville pigs was to produce very sick animals with a high incidence of mortality, arthritis, infertility, lethargy, gastric ulcers, degenerative joint disease, and susceptibility to infection. Such outcomes have been eagerly grasped by animal rights groups to attack the development of animal biotechnologies. The Beltsville pigs did not provide a practical benefit to farmers, and so the adverse effects on these animals have passed with the death of the pigs. However, other aspects of biotechnology still provide ammunition for animal rights groups.

As mentioned above, growth hormone (bST) stimulates milk production in cows. However, it has been suggested that the use of exogenous bST may adversely affect the health and welfare of the animals. Soon after its introduction dairy farmers started reporting increased incidence of mastitis, decreased conception rates, inflammation due to repeat injections, heat sensitivity, arthritis and lameness in bST treated animals. Welfare activists argue that this is unethical use of biotechnology as it has a detrimental effect on the cows. They argue that this applies also to genetic modification to increase endogenous growth hormone and milk production through introduction of extra copies of the growth hormone gene, as it will have the same adverse influence. Indeed, sheep with extra growth hormone genes were found to be severely diabetic.

The challenge of ignoring animal welfare is often levelled at modern animal husbandry in general. In today's herds, cows selected for increased milk yield are metabolically challenged even when not treated with bST, and the increased stimulus from either exogenous or endogenous bST leaves the cows in poor condition and susceptible to disease. The modern hen (whether broiler or layer), may also be close to its practical maximal efficiency, and it seems likely that attempts to radically improve performance further through biotechnology may have severe consequences (*e.g.* leg weakness).

Reproductive biotechnologies are also advancing in both human and animal science. The moral concerns expressed by some religions over cloning and *in vitro* fertilisation (IVF) in humans can be extended to animals. It is recognised that (as presently practised) IVF increases the incidence of multiple births, and results in increased fetal and neonatal losses with financial, physiological and emotional costs in humans. Similar losses occur in animals with implications for animal welfare. It may be difficult to quantify any psychological distress in mother animals due to neonatal losses, but reducing the interpartum interval by treatment with exogenous hormones increases metabolic stresses and may shorten the life (certainly the agriculturally 'useful' life) of animals. When applying biotechnologies to pregnant animals without due regard for all the effects on the mother, it can be argued that we are using the animal "solely as a means without regard for their requirements", thus failing Regan's criteria for ethical treatment of animals.

7. Ecology

After human safety, the biggest public concern with biotechnology is the potential environmental consequences. In the 20th century the splitting of the atom brought great potential for good – and for harm. It delivered an enormous capability for producing electricity and power, but also created problems with safety, disposal of waste and the horrors of nuclear warfare. Transgenics (creating new 'designer' organisms by altering their genetic make up) is just as powerful – and potentially just as dangerous. The public have been made well aware of this, and to ignore or deny this fact weakens the case for application of biotechnology.

Particularly for the non-scientist, one of the frightening aspects of radioactivity is that it cannot be seen. We could not see the radioactive fallout from atmospheric nuclear tests (or more recently from the Chernobyl accident). Like radioactivity, biotechnology is largely unseen, and this is one of the things that make the public uneasy. However, radioactivity and biotechnology do differ considerably in other respects. The pollution from nuclear testing and Chernobyl was finite and limited to the radioactivity released at that time. Pollution with transgenics may not be finite. While transgenics are restricted to non-pathogenic microbes under rigorous laboratory protocols, they are a powerful tool to study biology, but when they are let out of the bottle (or the laboratory) and dispersed, it may be very difficult to control the spread. Unlike escaped radioactivity, the organisms will be able to reproduce, and perpetuate both the good and the bad.

Clearing up oil spills (*eg* Exxon Valdez) and other chemical disasters (*eg* Seveso and Bhopal) is difficult enough, but when the pollutant reproduces itself the task is almost impossible. Any gardener will know how difficult it is to prevent weeds coming back time and time again, and the public is anxious that this might be the case if genetically modified organisms were to "escape". There are plenty enough examples of ecological problems caused inadvertently through biological intervention: the difficulty of containing the possum population in New Zealand, and the expanding invasion of Thai waterways by the water hyacinth, are both examples of the difficulty in controlling biological agents introduced into an inappropriate environment. There are also examples of human tragedies – devastation of indigenous native populations through transmission of syphilis, tuberculosis and smallpox to isolated peoples who were immunologically naïve to these diseases.

Transgenesis in plants has been practised for almost two decades and, until recently, had not produced much public concern. The more recent transgenesis of animals has stirred some anthropomorphic sentiment, and ignited the debate on the ethics of scientists creating genetically new organisms. Paradoxically, it may now be the transgenic plants that are of more worry to the public – because of the growing awareness of the importance and fragility of the ecology. From a field containing many hundred of thousands of genetically manipulated plants the probability of constraining all the genetic material is vanishingly small. Despite the protests that the chances of cross-fertilisation outside of closely related species is small, the possibility of herbicide resistance in weeds remains a major concern of ecology groups and needs to be taken seriously if consumers are to be reassured about the introduction of genetically modified organisms in agriculture.

On the other side, the supporters of biotechnology point out that such changes are unlikely and, in any event, would probably be less damaging to the environment than the continuing destruction of forests with the subsequent effects on the atmosphere, rainfall, temperature, soil erosion and societies. However, the debate over the morality of destruction should be about stopping all deleterious actions, *not* whether the existence of one form justifies the introduction of another.

The arguments against transgenic plants have formed the basis of much of the anti-biotechnology campaign, but animal scientists must learn from these criticisms. Transgenic animals are easier to control, and it is relatively easy to constrain them to a pen or a barn. Furthermore, animals can be physically sterilised to preclude the transmission of the altered genes, but sterilisation of all the plants in the field can only be reliably achieved through further genetic manipulation. However, there is concern over the possible ecological effects of transgenic animals that might escape. Escaped farmed salmon now outnumber wild salmon in Norway by 5 to 1. Ecologists are concerned that if genetically modified, farmed salmon escape to this degree, the potential dominance of the superior transgenic variety could lead to a loss of genetic diversity, and also a disturbance in the ecology through disturbing food supplies for other aquatic species.

8. Social

Another pressure group is concerned with the ethics of the social consequences of biotechnologies. The widening gap between rich and poor farmers is not confined to the Third World. Less advanced farms, in Eastern Europe, and small family farms in Western Europe may also fail to benefit from the biotechnological revolution. The competitive efficiency of the large farms, together with the profit advantages of using biotechnology, will cause the demise of smaller farms. This may be economically efficient from the agri-business viewpoint, but not necessarily so for regional

economics. Furthermore, small farms have an important place in maintaining our psychological association with the land, and with our heritage.
Although amalgamation of small farms into a larger corporate business structure improves efficiency, there are unwelcome social, economic and environmental implications. Increased unemployment of agricultural workers is one undesirable outcome. The absolute reduction in rural workers will be small, particularly compared with the massive efflux from the countryside over recent years. However, because of the past changes rural economies are extremely fragile, and even small changes may have large social effects.

Another likely result of commercial transgenics and increased corporatisation of agriculture is a decrease in diversity and choice. The variety of animal breeds and plant cultivars available at present is largely due to the prejudices and fancies of small farmers in choosing to produce particular breeds of sheep or pigs, or particular varieties of tomatoes or potatoes to grow. In large corporate farms the decision on what to produce is made by an accountant. The decision is not made on behavioural characteristics, or on the intrinsic organoleptic properties of the variety but solely on the best profit obtainable. This will inevitably be the genetically improved variety. The consumers' opportunity for choice between Cox's, Golden Delicious or Braeburn apples will be lost as the more profitable "corporate" apple becomes the accountants' choice.

The debate over 'efficiency' of agriculture affects animal husbandry as much as it does agronomy. It is conceivable – even likely – that the thousands of pig producers in the UK will soon be reduced to single figures as they try to compete with huge US pig factories the size of cities. Application of biotechnology will favour the large, more cost-effective enterprises and cause the loss of smaller farms. Small family farms are central to the maintenance of animal gene stocks through organisations such as the Rare Breeds Survival Trust and similar preservation groups around the world.

9. Global Issues

At the Seattle meeting of the World Trade Organisation in 1999, sizeable and effective demonstrations against biotechnology were organised. Two of the major arguments underpinning the demonstrations were: political control of food supplies, and the morality of gene ownership. The third was exploitation of Third World resources.

9.1. CONTROL OF FOOD SUPPLY

Clearly a large number of people were concerned about control of food production. If it comes to pass that food production is concentrated on a few big farms that owe their profitability to the biotechnology companies (and may indeed be owned by the biotechnology companies), then food supply will be controlled by these companies. When the companies are large multinationals making very significant contributions to either local or national economies and employment, their influence over Governments can be powerful. This potential threat is well recognised in the regulations controlling monopolies and by the anti-Trust legislation in the USA. Such laws are intended to prevent control of business sectors ending up in the hands of a few companies, but the accumulation of land and food production seem to have escaped regulation. A strong Government might be prepared to address some of these problems but this would be difficult and ineffective in the absence of a universal agreement. This is one of the major areas of disagreement that have dogged the WTO talks. The US government – heavily influenced by the large US-based biotechnology companies – insist that there be no restrictions on importation of genetically modified crops. On the other side, are nations that want to be more protective – not of their markets *per se*, but of their social and ultimately political independence.

9.2. MORALITY OF GENE OWNERSHIP AND EXPLOITATION OF THIRD WORLD RESOURCES

One of the ethical considerations is the morality of owning genes. Initially there was little problem. A breeder could, for example, make a highly productive cross between two (or more) breeds of sheep. The breeder might refuse to sell any live breeding stock and keep the secret of the components of the cross. However, this would not prevent his neighbour from undertaking a similar cross. The patenting of manipulated DNA sequences by themselves was not considered to be any more than chemistry. Similarly, incorporation of the DNA sequences into micro-organisms was merely providing the 'mixing vessel' for the chemistry. At this stage there was insufficient thought given to who owns a manipulated human cell – the donor or the owner of the tiny sequence that has been introduced. The problem has now extended further and companies are scouring Third World countries for plant and animal products that may be valuable, and are collecting, modifying and patenting the genes. The multinational biotechnology companies have benefited from an unwillingness of authorities to regulate against the granting of patents over these biological processes. The patent laws were not designed with biotechnology in mind and some major biotechnology companies have used their political muscle to put pressure on governments over granting of patents.

There is a growing public concern that there must be urgent international agreement over patenting of biotechnology. This may be possible within the EU, but could such arguments be brought to bear in the USA where the power of dollars has already spoken so loudly to get things this far? Many, on religious grounds, have advocated a universal agreement to ban patenting of genetic material, but on July 6, 1998 the EU accepted the patenting of plant and animal biotechnological inventions.

10. Who Gains?

There is potential benefit from using biotechnology in agriculture, but who gains? The biotechnology companies will want to maximise their profits: this is not unreasonable. Quite logically, they will cost their products at a highest figure that the farmer will pay. This is the level where the farmer gets just a bit more profit than he would from using unaltered crops. The consumer sees little difference between normal and genetically modified soya flour, so will not pay extra for the genetically modified product. The consequences for the consumers are that they pay the same amount for food that they feel may be less safe, and may have worrying implications for society and for the environment. The gains from biotechnology go principally to the biotechnology companies, with some to the farmer, but there seem to be no benefits to the general population – only possible losses.

11. Funding

In many of the newspaper, magazine, radio and television pieces on biotechnology, scientists can be heard to complain that the antagonism to genetic modification is slowing progress. However, ultimately it is the public who pays the scientists – whether directly through government or university research, or indirectly through pension funds who hold shares in biotechnology companies. It seems reasonable that it is the paymaster who should dictate the conditions (rate of progress), not the ambition of the scientist. In the present circumstances where commercial exploitation of research is increasingly important, we look no further than how we can stress the financial benefits of the application of our research in a quest to generate future funding. Belatedly, we are now being forced to recognise the implications of our findings in areas far from the comfort of the laboratory. Philosophically science should serve, rather than undermine or control, the public interest.

It is appropriate to reflect upon how well the new biotechnologies do serve the public interest. Although biotechnology companies have invested in the commercial development of these technologies, the ground-breaking discoveries were from publicly funded research. It is not unreasonable, therefore, for the public to hope for some benefit from their investment, or at least to prescribe the conditions for acceptance of subsequent developments, including socio-economic and political acceptability, as well as health and environmental considerations. It is now well recognised that we have an absolute obligation to keep the world safe for future generations. In terms of the persistent risks, such as those posed by inappropriate or incompletely assessed commercial transgenics, the burden of proof must be well beyond the "reasonable doubt" that usually applies.

12. Role of Governments

The public's unease over genetic modification is heightened by inconsistencies in government policies. The selling of pharmaceuticals is simply not allowed until they have been thoroughly checked and been shown to be safe. This "precautionary principle" was largely supplanted by the "innocent until proven guilty" principle in the case of genetically modified crops. The European Union ban on the importation of beef treated with genetically modified substances was ruled illegal by the World Trade Organisation because 'harm' had not been conclusively demonstrated. Contrast this stance with the rigorous procedures for the introduction of a new drug for human (or animal) use. Apparent blind-spots are also evident in other situations. Even though the ecological impact of gei .ically modified crops has been recorded (decrease of butterfly numbers in USA, transference of DNA to wild plants in Canada), other governments are still arguing that open field trials must be undertaken to confirm "a *lack* of adverse effects".

Consumers are also suspicious of the attitude of biotechnology companies to labelling. The very fact that the biotechnology companies are fighting against labelling their products as having an association with genetically modified products, raises doubts in the minds of consumers. If the products (from genetically modified tomatoes to milk produced by recombinant bovine somatotropin treated cows) are perfectly safe and ethically acceptable, why not label them?

In summary, the advantages of biotechnology are attractive. In theory we can expect big increases in yields and increased resistance to diseases. Using biotechnology we can enhance normal processes (such as growth) by putting more copies of its own growth-promoting genes into a transgenic organism. This effectively short-cuts the natural selection procedures of conventional breeding strategies. The moral distinction between slow selection through breeding programmes and more rapid manipulation is arguably small. This may be a more generally acceptable aspect of biotechnology. However, are we, or should we, be so ready to accept xenogenetics - introducing genes from other species? How can we possibly know the long term effects of transferring toad genes into potatoes? Such manipulation is far removed from shortening conventional selection procedures and introduces the concept of scientists "playing God". There are probably very good reasons why in Nature it is extremely rare for even closely related species to interbreed. Xenogenetics is a real Pandora's box.

The public are in a dilemma. They see the enormous potential benefits from some aspects of biotechnology. There are advantages to be gained from application of biotechnology to health care – "pharming". The production of large quantities of rare biologicals (such as blood clotting factors and cytokines); the development of endless supplies of stem cells for use in leukaemia and in generating organs and tissues for transplantation; and the development of oral vaccines in appropriate practical vehicles such as bananas.

Genetic engineering has the potential to increase food production in both the developed and developing world through increased resistance to pests and decreased dependence upon fertilisers. However, these benefits can not be achieved without political will, and always require appropriate supply of water (which is more of a problem than anything else). The enrichment of Third World staple food by increasing vitamin and protein contents (for example Golden rice) could have enormous benefits on health, but these poor countries cannot afford to purchase the technological advances, and will the biotechnology companies provide all the modified crops free?

Despite these potential benefits, genetically modified organisms are not ethically acceptable to most consumers. Not all consumers have the same reasons for rejecting these technologies, but at heart each is a judgment about ethics. Whether it is about theology, ecology, human safety, animal welfare, unemployment, diversity of products, rural social disintegration or any combination of these, a majority of consumers feel that the harm potentially caused by recent biotechnological advances outweighs the benefits.

As mentioned in the introduction, this chapter is not intended to prescribe the acceptability or otherwise of biotechnologies. There is no doubt that research in the area of biotechnology is stimulating, clever and intellectually satisfying: some of it may even be beneficial. There is also no doubt that researchers have an ethical duty to consider *all* the consequences of their work on world communities and the environment. It is unethical to simply take the pay cheque and leave the hard decisions to others.

References and Suggestions for Further Reading

Eurobarometer survey 46.1 'Europeans and modern biotechnology'. European Commission DGVII, Brussels, 1997

Spencer, S. (1999) The animal scientist in a changing society, *Dom. Anim. Endocrinol. 17*, 95-100.

Ewen, S.W.B. and Pusztai, A. (1999) Effects of diets containing genetically modified potatoes expressing *Galanthus nivalis* lectin on rat small intestine, *Lancet 354*, 1353-1354.

Couchman, P.K. and Fink-Jensen, K. (1990) Public attitudes to genetic engineering in New Zealand. DSIR Crop Research Report No 138. Christchurch, New Zealand, DSIR Crop Research.

Nordlee, J.A., Taylor, S.L., Townsend, J.A., Thomas, L.A. and Bush, R.K. (1996) Identification of a Brazil nut allergen in transgenic soy beans, *New Engl J. Med 334*, 688-692.

Pursel, V.G., Pinkert, C.A., Miller, K.F., Bolt, D.J., Campbell, R., Palmiter, R.D., Brinster, R.L. and Hammer, R.E. (1989) Genetic engineering of livestock. *Science 244*, 1281-88.

O'Brien, T. (1995) Gene transfer and the welfare of animals. *A Compassion in World Farming Report.*

Steinbrecherm, R.A. (1996) From green to gene revolution – the environmental risk of genetically engineered crops, *The Ecologist 26*, 235-242.

Mackenzie, D. (1996) Can we make super salmon safe? *New Scientist.* 27-33

Our Genetic Future. BMA. OUP. Oxford 1992.

Orlans, F.B., Beauchamp, T.L., Dresser, R. (1998) *The Human Use of Animals.* OUP, Oxford, UK.

Reiss, M.J. and Straughan, R. (1998) Improving Nature?: The science and ethics of genetic engineering. CUP. Cambridge.

Ekesbo, I. (1991). Animal safety, health and welfare. In: P. van der Wal , G.M. Weber and F. van der Wilt (Eds). *Biotechnology for control of growth and product quality in meat production: implications and acceptability.* Pudoc. Wageningen pp 109-120.

Hoban, T.J. and Burkhardt, J. (1991) Determinants of public acceptance in meat and milk production: North America. . In: P. van der Wal , G.M. Weber and F. van der Wilt (Eds). *Biotechnology for control of growth and product quality in meat production: implications and acceptability.* Pudoc. Wageningen pp 229-244.

Kronfeld, D.S. (1988) Biologic and economic risks associated with use of bovine somatotropins. *J. Am. Vet. Med. Assoc. 192*, 1693-1696.

Bredahl, L (1999) Consumers' recognition with regard to genetically modified foods. Results of a qualitative study in four countries. *Appetite 33*, 343-360.

INDEX